民國建築工程期刊匯編

MINGUO JIANZHU GONGCHENG QIKAN HUIBIAN

8

《民國建築工程期刊匯編》編寫組 編

广西师范大学出版社

GUANGXI NORMAL UNIVERSITY PRESS

·桂林·

第八册目録

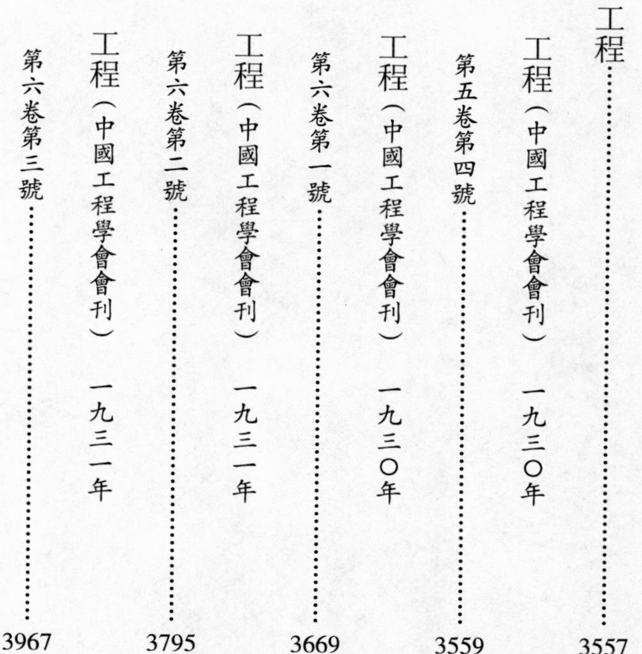

工程

民國十九年八月

工程

第五卷 第四號

中國工程學會會刊

THE JOURNAL OF THE CHINESE ENGINEERING SOCIETY

VOL. V, NO. 4 AUGUST 1930

中國工程學會發行 總會會所： 上海寧波路七號 電話 一九八二四
每册三角預定全年四册定價一元郵費本埠二分外埠五分國外一角八分

工程

中國工程學會會刊

季刊第五卷第四號目錄 ☆ 民國十九年八月發行

總編輯 周厚坤 總務 楊錫鏐

本刊文字由著者各自負責

中國工程學會發行

中國工程學會會章摘要

第二章　宗旨　本會以聯絡工程界同志研究應用學術協力發展國內工程事業爲宗旨

第三章　會員

(一)會員　凡具下列資格之一由會員二人以上之介紹再由董事部審查合格者得爲本會會員

(甲)經部認可之國內外大學及相當程度學校之工程科畢業生并確有二年以上之工程研究或經驗者

(乙)曾受中等工程教育并有六年以上之工程經驗者

(二)仲會員　凡具下列資格之一由會員或仲會員二人之介紹並經董事部審查合格者得爲本會仲會員

(甲)經部認可之國內外大學及相當程度學校之工程畢業生

(乙)曾受中等工程教育并有四年以上之工程經驗者

(三)學生會員　經部認可之國內外大學及相當程度學校之工程科學生在二年級以上者由會員或仲會員二人之介紹經董事部審查合格者得爲本會學生會員

(四)永久會員　凡會員一次繳足會費一百元或先繳五十元餘數於五年內分期繳清者爲本會永久會員

(五)機關會員　凡具下列資格之一由會員或其他機關會員二會員之介紹並經董事部審查合格者得爲本會機關會員

(甲)經部認可之國內工科大學或工業專門學校或設有工科之大學

(乙)國內實業機關或團體對於工程事業確有貢獻者

(八)仲會員及學生會員之升格　凡仲會員或學生會員具有會員或仲會員資格時可加繳入會費正式請求升格由董事部審查核准之

第四章　組織　本會組織分爲三部(甲)執行部(乙)董事部(丙)分會(本總會事務所設於上海)

(一)執行部　由會長一人副會長一人書記一人會計一人及總務一人組織之

(三)董事部　由會長及全體會員選出之董事六人組織之

(七)基金監　基金監二人任期二年每年改選一人

(八)委員會　由會長指派之人數無定額

(九)分　會　凡會員十人以上同處一地者得呈請董事部認可組織分會其章程得另訂之但以不與本會章程衝突者爲限

第六章　會費

(一)會員會費每年國幣五元入會費二十元　(二)仲會員會費每年國幣三元入會費六元

(三)學生會員會費每幣國年一元　(四)機關會員會費每年國幣十元入會費二十元

中國工程學會職員錄

（會址上海寗波路七號）

歷任會長

陳體誠（1918—20）　吳承洛（1920—23）　周明衡（1923—24）　徐佩璜（1924—26）
李垕身（1926—27）　徐佩璜（1927—29）

民國十八年至十九年職員錄

董事部

凌鴻勛　南京鐵道部　　陳立夫　南京中央執行委員會祕書處
李垕身　南京鐵道部　　吳承洛　南京工商部
徐佩璜　上海新西區楓林路市政府參事室　　薛次莘　上海南市毛家弄工務局

執行部

（會長）胡庶華　吳淞同濟大學　　（副會長）徐恩曾　南京建設委員會祕書處
（書記）朱有騫　上海新西區楓林路公用局　　（會計）李俶　上海徐家滙交通大學
（總務）楊錫鏐　上海寗波路七號楊錫鏐建築事務所

基金監

惲震　南京建設委員會　　裘燮鈞　杭州大方伯浙江省公路局

各地分會

分會	職務	姓名	地址	姓名	地址
上海分會	（會長）	黃伯樵	上海新西區楓林路公用局		
	（副會長）	薛次莘	上海南市毛家衖工務局		
	（書記）	王魯新	上海九江路22號新通公司		
	（會計）	朱樹怡	上海四川路215號亞洲機器公司		
南京分會	（委員）	吳承洛	南京工商部	薛紹清	南京中央大學工學院
		胡博淵	南京農礦部		
蘇州分會	（委員）	魏師達	蘇州吳縣建設局		
北平分會	（幹事）	王季緒	北平西四北溝沿189號王寓		
天津分會	（會長）	李書田	唐山交通大學土木工程學院		
	（副會長）	稽銓	天津良王莊津浦路工務處		
	（書記）	顧毅成	天津西沽津浦機廠		
	（會計）	邱凌雲	天津法界拔柏葛鍋爐公司		
武漢分會	（會務委員）	石瑛	武昌武漢大學	陳彰琯	漢口工務局
	（書記委員）	孔祥鵝	武昌建設廳	朱樹馨	武昌建設廳
	（會計委員）	繆恩釗	武昌武漢大學建築工程處		
青島分會	（會長）	林鳳岐	青島膠濟路四方機廠		

(書　記)　嚴宏溎　青島公用局
(會　計)　孫寶墀　青島膠濟鐵路工務處

杭州分會　(會　長)　張可治　杭州浙江大學工學院
(副會長)　陳體誠　杭州浙江省公路局
(書　記)　茅以新　杭州裏西湖三號杭江鐵路局
(會　計)　楊耀德　杭州浙江大學工學院
(幹　事)　吳琢之　杭州浙江省公路局

太原分會　(會　長)　唐之肅　山西太原育才鍊鋼廠
(副會長)　董登山　山西軍人工藝實習廠計核處
(文　牘)　曹煥文　山西太原山西火藥廠
(庶　務)　郗三善　山西大學校

梧州分會　暫告停頓

濟南分會　(會　長)　張含英　濟南山東建設廳
(副會長)　于皞民　濟南山東建設廳
(書　記)　宋文田　濟南山東建設廳
(會　計)　王洵才　濟南膠濟路工務總段

瀋陽分會　(會　長)　張潤田　瀋陽北甯鐵路港務處
(副會長)　王孝華　瀋陽兵工廠
(書　記)　余稚松　瀋陽東北大學
(會　計)　胡光燾　瀋陽東北大學

美洲分會　(會　長)　張乙銘　526 W. 123rd St., N. Y. C.
(副會長)
(書　記)　陶葆楷　67 Hammond St., Cambridge Mass. U.S.A.
(會　計)　李嗣綿　Room 905, 105 Broadway, N. Y. C.

委員會

建築工程材料試驗所委員會

委員長　沈　怡　上海南市毛家衖工務局

委　員　徐佩璜　上海新西區市政府參事室　　薛次莘　上海南市毛家衖工務局
李垕身　上海福煦路成和邨605號　　徐恩曾　南京建設委員會祕書處
支秉淵　上海甯波路七號新中公司　　顧道生　上海福州路九號公利營業公司
裘燮鈞　杭州大方伯浙江省公路局　　黃伯樵　上海新西區楓林路公用局

工程教育研究委員會

委員長　金問洙　江灣復旦大學

委　員　楊孝述　上海亞爾培路309號中國科學社　　戴　濟　上海法界遇爾西愛路永慶坊一號
茅以昇　鎮江江蘇省水利局　　陳茂康　唐山交通大學
張含英　濟南山東建設廳　　梅貽琦　清華大學駐美監督處

周子競　上海亞爾培路205號中央研究院　陳廣沅　天津西沽津浦機廠
李熙謀　杭州浙江大學工學院　許應期　上海徐家匯交通大學
程干雲　江灣勞動大學　孫昌克　南京建設委員會
阮介藩　上海法租界環龍路志豐里四號　俞同奎　北平北平大學第一工學院
譚伯羽　吳淞同濟大學　鄒恩泳　上海新西區楓林路公用局
鄭肇經　青島港務局　李昌祚　上海西愛咸斯路 55 號
陳懋解　南京中央大學　唐藝青　長沙湖南大學
笪遠倫　北平清華大學　徐名材　上海徐家匯交通大學
徐佩璜　上海新西區楓林路市政府參事室

會員委員會

委員長　黃炳奎　上海高廊橋申新第五廠

委　員　上海　徐紀澤　南市十六浦荳市街恆吉巷五號　上海　黃元吉　愛多亞路38號凱泰建築公司
蘇州
南京　徐百揆　蚌埠市政籌備處
杭州　朱耀庭　工務局　北平
天津　邱凌雲　法界拔柏葛鍋爐公司　濟南　張含英　山東建設廳
青島　王節堯　膠濟路工務處　武漢　孔祥鵝　湖北建設廳
廣州　桂銘敬　粵漢鐵路株韶段工程局　山西　唐之肅　太原育才鍊鋼廠
瀋陽　張潤田　北寧鐵路港務處　美國　薛楚書　500 Riverside Drive, New York City.

經濟設計委員會

委　員　徐佩璜　上海新西區楓林路市政府　朱樹怡　上海四川路215號亞洲機器公司
胡庶華　吳淞同濟大學　張延祥　安慶安徽全省公路管理局
李　俶　上海徐家匯交通大學

編譯工程名詞委員會

委員長　程瀛章　上海梅白克路三德里639號

委　員　張濟翔　廣州光椏中國電氣公司　尤佳章　上海寶山路商務印書館編譯所
馮　雄　上海寶山路商務印書館編輯所　徐名材　上海徐家匯交通大學
張輔良　上海福開森路378號中央研究院社會科學研究所　孫洪芬　北平南長街 22 號中華教育文化基金董事會
藍春池　上海膠州路大夏大學　錢昌祚　南京中央陸軍軍官學校航空隊
林繼庸　江灣俞涇廟大南製革廠　鄒恩泳　上海新西區市政府公用局
葛敬新　李伯芹
胡衡臣　南京建設委員會　錢福謙
吳欽烈　南京軍政部兵工署

工程研究委員會

委員長　徐恩曾　南京中央黨部　主任委員　胡博淵　南京農礦部
化工組主任委員　徐鳳石　南京工商部　土木組主任委員　許心武　南京復成橋導淮委員會

電機組主任委員　惲蔭棠　南京建設委員會
委員　鍾兆琳　上海徐家匯交通大學

機械組主任委員　鈕甸受　南京交通部技術委員會
委員　周坤厚　上海福州路一號德士古火油公司

礦冶組主任委員　楊公兆　南京農礦部
委員　吳稚田　上海九江路六號沙利貿易司公

建築組任委員　齊兆昌　南京金陵大學建築部

建築條例委員會

委員長　薛次莘　上海南市毛家衖工務局

委　員　朱耀廷　杭州工務局
徐百揆　蚌埠市政籌備處
許守忠　青島工務局
薛卓斌　上海江海關五樓濬浦總局
李　鏗　上海圓明園路愼昌洋行

本會辦事細則起草委員會

委員長　薛次莘　上海南市毛家衖工務局

委　員　張延祥　安慶安徽全省公路管理局
徐佩璜　上海新西區楓林路市政府
徐恩曾　南京建設委員會祕書處
惲　震　南京建設委員會

職業介紹委員會

委員長　朱有騫　上海新西區楓林路公用局

委　員　馮寶齡　上海圓明園路愼昌洋行
徐恩曾　南京建設委員會祕書處

職業介紹審查委員會

委　員
化學工程　徐佩璜　上海新西區楓林路市政府
徐名材　上海徐家滙交通大學
建築工程　薛次莘　上海南市毛家衖工務局
裘燮鈞　杭州大方伯浙江省公路局
橋梁工程　馮寶齡　上海圓明園路愼昌洋行
許貫三　上海南市毛家衖工務局
電氣工程　鄭葆成　上海新西區楓林路公用局
機械工程　支秉淵　上海甯波路七號新中公司
水利工程　朱有騫　上海新西區楓林路公用局
無線電工程　王崇植　青島工務局
土木工程　朱有騫　上海新西區楓林路公用局
道路工程　鄭權伯　青島港務局
鐵路工程　洪嘉貽　杭州平海路 37 號

材料試驗委員會

委員長　王繩善　上海徐家滙交通大學

委　員　康時淸　上海徐家滙交通大學
盛祖鈞　上海徐家滙交通大學

科學咨詢處

常務委員長　徐佩璜　上海新西區市政府參事室

常務委員　胡庶華　吳淞同濟大學
裘維裕　上海交通大學
黃炳奎　上海高郎橋申新第五紡織廠
周厚坤　上海外灘德士古火油公司
支秉淵　上海寧波路七號新中工程公司

孫多𤩊先生遺像

孫君景蘇傳略

柴福沅

君諱多熲號景蘇.其先爲安徽壽縣望族.祖秋潭先生,父次青先生,俱任清爲縣令,所至有聲.次青先生故名孝廉,嘗主各校講席,門下稱盛,君生於清光緒中,思想新穎爲文踔厲風發,往往數千言,不能自休.年十六,入揚州民立學堂此爲君受新教育之始.君聰明好學,成績冠儕輩,諸教習交相稱許.其後肄業蘇省鐵路學堂,南京上江公學,文辭,科學,益見進境.繼入交通部工業專門學校,即前南洋公學.治土木工程科,畢業後,交通部派滬杭甬鐵路實習.旣調周襄及株欽鐵路從事測量,自西安,道漢中,入川,遍歷險阻,見聞日廣,術亦大進.繼供職於交通部,技術委員會.旋上海濬浦總局任以工程書記,謹愼從公,勞績優異.先後十一年,漸擢升至濬工科副工程師之職.民國十九年五月卅日晨間,君偕濬工科職員五人,攜現欵二萬餘元,至第一工作處海蝎挖泥船旁.發放工資.正在解囊之始.突有盜輪踪至,躍上劇盜九人,持盒子砲,手鎗,開火轟擊,受傷者共五人,而君獨以彈中咽喉登時殞命,哀哉!君天性厚重,處世接物,藹然有仁者之風,親友得其資助者甚衆.聞君死,識與不識,均同聲嗟悼,以爲天之報善人者,何其酷也!君時年始四十.妻姜氏.有子三人女二人,俱幼.

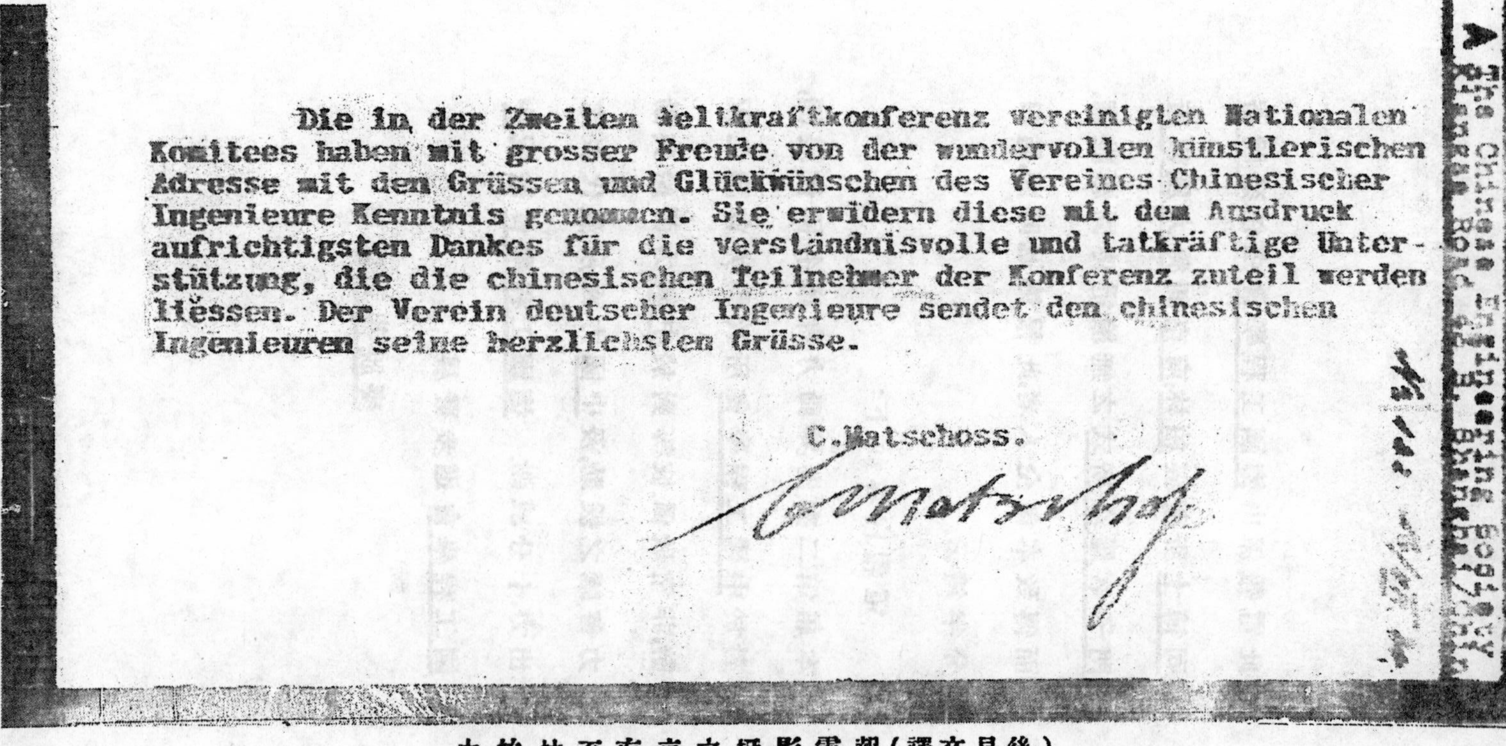

The Chinese Engineering Society
Kiangse Road 42 B, Shanghai/China

Die in der Zweiten Weltkraftkonferenz vereinigten Nationalen Komitees haben mit grosser Freude von der wundervollen künstlerischen Adresse mit den Grüssen und Glückwünschen des Vereines Chinesischer Ingenieure Kenntnis genommen. Sie erwidern diese mit dem Ausdruck aufrichtigsten Dankes für die verständnisvolle und tatkräftige Unterstützung, die die chinesischen Teilnehmer der Konferenz zuteil werden liessen. Der Verein deutscher Ingenieure sendet den chinesischen Ingenieuren seine herzlichsten Grüsse.

C.Matschoss.

由柏林至南京之攝影電報（譯文見後）

此電報係由德國工程學會主任麥脫蓄司博士所發,經柏林,南京間得力風根西門子客羅洛斯式之無線電傳圖影設備,而傳致本會仝人者,此報傳遞時間,僅四分半鐘云.

電報內容如下:

本屆第二次世界動力會議各國代表團,對於中國工程學會所致動力會議之祝辭,非常欣佩.彼等對於中國方面給與動力會議之援助,及中國方面所派代表之出力,十分感謝. 德國工程學會敬祝中國工程學會進步發展.

麥脫蓄司謹啓

中國工程師之使命

著者：程文勳

本屆日本開萬國工業會議,各國著名工程師,技術家,參加者頗多.雖討論的問題甚少,會議不免屬於形式,但這會的論文中,有價值的很多.有好多論文,是關於工程師本身『說是社會上最重要的物質文明進步,完全靠着工程師發明,歐戰時發明更多,所以各國近來對工程師的地位,日益尊崇』.此次會議中,有人提議,聯合萬國工程界,組織一大規模團體,猶之政治問題,有國際聯盟會議.此項提案,經議決由各參加人員,先同本國組織,再於下屆會議討論萬國結合辦法.瑞典工程師恩斯德姆 Euston 有一文論『工程師之職務』,立意尤爲深切.兄弟今天選擇這個題目,大都取意於恩君之文,再參酌吾國情形,加以發揮.

工程師又稱爲技術家,在西文原意,係多才多藝之人,其地位在社會之重要,既如上述.恩君云『經此次會議之後,希望各國工程師更加增自重之觀念』.此語是要他增加責任心,並不是要傲慢世俗,妄自尊大,因明白責任,便是謙遜.吾國以前以『士』列四民之首,而范文正公爲秀才時,便以天下爲已任,亦即此意

工程師既負增進人生之幸福,其第一重要任務,須開闢交通:交通爲文化所繫,從古民族,均沿江河而蕃衍,次則沿海岸,均得交通捷便之故.(即如吾國之文化,發源自黃河流域,漸移於長江流域以及沿海).科崙布尋獲美洲,貫通重洋.自蒸汽動力發明,輪舟之於海洋,鐵路之於大陸,效用益偉.近則汽車,飛機,更足以飛渡重山,以補鐵路之所不及.此外電報,電話,無線電,尤足發揚文化,貫通各種民族固有之發明,闢未來之新智,闢交通上之新紀元.

第二重要任務,在保持及發展固有之生產:工業上的生產力量,終要保持,或設法增進,方足以改良人類之生活程度.人以舒適爲幸福,過度之舒適,便是奢華.所以奢華與幸福,原無絕對之區別,要視個人之經濟狀況爲斷.(昔時帝皇寶座.雖加盛飾,實一硬板凳,遠不如今日彈簧椅之舒適;此雖例外,但足證使人舒適,必藉科學之進步).大致奢侈品是價貴的,所以富人用以享受之品,貧人祇能目爲奢侈品,而不敢問津.儉固美德,然必儉到若何程度,亦難得一適當之標準.物質文明發達之旨,在人人能多得享受,然此非生產增多,價值低廉不爲功.工程師之任務,在能使各種工業出品,成本低廉,應用者可以普及.

自發明機器以代人工,出品多而價值漸低.機器廠中各雇用工人所做之工作早定有分工之制,近年美國退拉氏,又研究煉鋼廠運鋼工人動作之多寡,並泥水匠砌牆動作之多寡,設法減少,以增進其成績,此即所謂科學管理法.今日福特廠製造汽車,所用工人,每日工作之時間少,工價貴,而成本反輕實藉特種管理之法.其製造汽車用 Conveyer 循環旋轉機,懸掛工作之品,工人都站立一定地點,不必移動,用機械規定工作之品,(車件)移至某處,卽應由某種工人於若干時間以內,做某種工作,設或某處工人,不按時勤作,則時間一滿,其品卽移至第二步工作之處,不能强其等候也.科學管理法之精意,不外(一)濟經時間,(二)經濟材料,(三)免去枉費之工作.苟能隨時隨事,於此三點注意,則出品之成本可減.成本輕,售價自廉,售價廉,銷路方足普通,使奢侈之品,改爲實用品,人人得以享受.觀乎美國凡有汽車之人極多,當工人者亦得自備汽車以代步,而吾國則中人之家,尙屬難能,可想見矣.

工程師重要之工作,便是研究.研究天地間之眞理,並支配方法,以求進步,務使生產之品,日益增多,價值日益低廉,並求種種新發明,以造福社會,此種務神,堪稱爲文化之進步.

工程師於工業界中,具有領導工人之責任,應時之求,增進所屬人員之幸福,同事相處,尤當具公開誠實之情感,時時交換經驗及智識,以求共同之利益,至個人之利益,有時雖犧牲而不辭,處事忠實,不貪一時利益之工程師方謂之智,反是謂之愚蠢,愚蠢者於經濟上必難有所成就.總之,治工業之最要者,在具有高尚之志願及辦事之精神.

中國現謀建設,發展工業,兄弟以爲多非難事,而最難者爲『人』之組織.人謂中國人好作弊,好作首領,鄙意此係政無軌道,故有此種現象,尚非國民之惡劣根性.佛法無我,而今人則惟我主義充溢社會,此實爲國民之惡劣根性.惟其有我,則名也,利也,權也,勢也,均我所欲也,倖進之心,貪得之念,種種慾望,均由此而起.惟我主義盛,便難成大組織,無大組織,便難成大事業.普通之人,但咎政局不定,百業不振,然執政之人,來自民間,以惟我主義之國民,即時局平靖,亦難成大業.試觀機關也,工廠也,商店也,其內部無不有爭權奪利之事.由小合併而結合成大業者,在所罕有.大事業中,因傾軋而分裂者,却所恒見,又何咎乎在上者?

青年學子,以惟我主義之陶冶,出校之後,不能俯就社會致難立足.資本之家,亦以惟我主義之濡染,不能羅致人才,以改進其事業,更見萬難之人,任務過繁,不能專一,亦惟我主義有以致之也.兄弟此次在日本參加萬國工業會議,見各國之工程師,均白髮蕭蕭,終其生以研究學問.回顧吾國代表,類皆風塵中人,奔走衣食者,故對於研究學問,言之足滋慚汗.諸君皆工程專家,在今日物質進化之世界,當爲社會上第一流人物.鄙意欲求中國工業之發展,必先由各工程界挺身而出,打倒惟我主義,陶冶國民根性,固當經一番嚴格訓練,應自小學起,但受過高等教育之人,自己刻苦修養,亦足移風易俗,不識諸君,以爲何如?

國產製皂油脂之標準及其實用試驗法

著者：張雪楊

溯自近世工業革命以還,凡百製造,皆由小規模之手藝,奪趨於大規模之機械化.質與量既並重,而效率尤特別關要.故於原料,半製品,出產及副產物等局部之科學管理,莫不首貴試驗,而須有一定之標準,藉資比較.所謂失之毫釐,謬以千里,此歐美各國所以有標準專局之設立以釐訂之者也.我國自遜清末葉起,即漸有建樹近世工業基礎之傾向.惜頻年變亂,內憂外患,在在皆足阻其滋長.迨革命政府奠都南京,工商部乃首先奉行　總理實業救國政策,而有發展全國各種基本工業之大計劃,同時並有工業法規之編纂,與夫劃一度量衡制度法令之頒布,誠屬當務之亟.不過建設伊始,萬端待舉.全恃中央獨擎,焉能一蹴而幾,奏功於瞬息,幸地方行政機關及學術團體亦有鑒於此,故年來成立各種檢驗研究機關多所.惟彼此分工,各事專業,缺少聯絡而乏統計上之價值.以故工程學會本研究之精神,作實地之調查.曾以屬於化工範圍之編訂國產材料各種常數表,及確定國產材料各種標準試驗法,與一般的常數之比較上優劣問題,見委於予.顧自維淺薄,安敢有所貢獻.不過以製皂油脂一項,類皆國產.且有柏油一種,則係中華特產,而為他國所無.近年出口之鉅,駸駸日上,幾有與桐油並駕齊驅之勢.所試既多,乃彙列其結果而平均之.雖不能即據此為標準,但亦足為釐訂標準常數之一助.尚望諸同志更進而匡正之,則幸甚矣.

油脂之種類固多,但除桐油及亞麻子油等極少數之乾性油外,餘皆適用於製皂.茲擇十八年度一年中在五洲固本廠關於各種製皂油脂之試驗結果,分類彙載於次.其試品悉取原樣,不加任何處置,藉存其真,而便於實地工作,故與一般書籍上所載自不能不有所出入焉.

(一) 常數彙錄

(甲) 桕油

桕油或稱皮油,係由烏桕種子外皮製得.烏桕屬大戟科 Euphorbiaceae 植物學名 Stillingia Sebifera Russ.為落葉喬木,葉卵形而尖.夏日開黃花.結實至八九月而熟.實外部蔽以樓褐色殼.熟後此殼破裂現扁圓形種子三粒.種子外面裹以白色蠟狀脂肪層.其內又有硬殼,殼內有淡黃色核仁.清油或梓油 Stillingia Oil即由此仁壓榨所得者也.清油為一種極佳之乾性油,設壓榨得法,不難與桐油爭一日之短長.烏桕為我國特產,山陽平澤,沿江溫煖之區,無不產之.就中以四川東部,湖北西部及江西,湖南,安徽,浙江為尤盛.入秋霜葉楓紅,極寒豔之致.落木後壘壘滿樹者皆桕子也.待桕子採下,則枯幹枯枝,連岸遍野,又別饒逸趣.吾國山水畫中,固數見不鮮.而曾有溯江而上之舟行者亦類能道之也.大約烏桕生長四年後始結實.平均每樹結實四五十斤,可產油二十餘斤,依現在市價每擔二十圓計算,則每樹每年可獲四圓.如此大利所在,焉得不竭力提倡,庶國計民生兩有裨益,又豈特防止水患點綴風景而已哉?

牌號或來源	水分	雜質	鹼化數	融點	游離脂肪酸	酸數	碘數	試驗日期
萬順豐	0.43%	0.25%	217.7	41°C	2.26%	4.49	——	1-12-29
荆州蘇城	0.19%	0.20%	214.8	41°C	5.36%	10.66	——	1-13-29
向春和	0.30%	0.23%	211.1	40°C	6.36%	12.62	——	1-14-29
益豐	0.32%	0.10%	211.1	40°C	7.19%	14.31	——	1-14-29
恆生銓	0.28%	0.20%	216.0	39.5°C	6.20%	12.34	——	1-14-29
蘇城	0.04%	0.26%	211.1	39°C	2.96%	5.89	——	1-14-29
林瑞豐	0.68%	0.25%	200.6	39°C	7.61%	15.15	——	1-14-29
義成春	0.34%	0.21%	212.6	41°C	3.38%	6.73	——	1-14-29
志成	0.74%	0.25%	211.1	40.5°C	3.38%	6.73	——	1-14-29
同興茂	0.34%	0.22%	218.0	41°C	5.08%	10.05	——	1-14-29
永茂協	0.35%	0.12%	216.6	39°C	4.79%	1.54	——	1-14-29
蘇城	0.40%	0.28%	212.6	38.5°C	7.70%	3.37	——	1-14-29

牌號或來源	分水	雜質	鹼化數	融點	游離脂肪酸	酸數	碘數	試驗日期
洽成永	0.12%	0.10%	218.1	39°C	3.38%	6.73	——	1-17-29
洽成永	0.03%	0.13%	218.0	41°C	2.26%	4.49	——	1-30-29
萬順豐	0.15%	0.10%	212.7	40°C	2.82%	5.61	——	2-21-29
方協和	1.10%	0.05%	207.2	43°C	3.81%	7.57	——	2-25-29
孫洪曙	0.13%	0.02%	206.6	41°C	2.54%	5.05	——	2-23-29
洽成永	0.24%	0.11%	207.2	39°C	1.69%	3.37	——	3- 3-29
同　源	0.36%	0.06%	208.0	42°C	2.82%	5.61	——	3- 6-29
裕美成	0.22%	0.15%	208.9	38°C	7.47%	14.87	——	3- 9-29
裕昌元	0.24%	0.07%	211.2	40°C	4.79%	9.54	——	3- 9-29
和　合	0.31%	0.07%	206.7	40°C	3.38%	6.73	——	3- 9-29
恆生銓	0.27%	0.30%	213.2	41°C	3.53%	7.01	——	3- 9-29
恆興長	0.24%	0.35%	212.8	41°C	4.23%	8.41	——	3- 9-29
蔴　城	0.27%	0.15%	212.8	41°C	1.69%	3.37	——	3- 9-29
熊正興	0.04%	0.41%	213.8	39°C	1.27%	2.53	——	3-14-29
丁和記	0.05%	0.15%	212.4	39°C	1.41%	2.81	——	3-14-29
洽成永	1.18%	0.30%	215.2	38°C	2.82%	5.61	——	3-15-29
朱大順	0.52%	0.13%	208.5	38.5°C	10.01%	19.92	——	3-15-29
同發源	0.36%	0.16%	208.0	41°C	2.82%	5.61	——	3-15-29
洽成永	1.80%	0.61%	206.7	40°C	4.51%	8.98	——	3-18-29
金　華	1.08%	0.13%	212.4	41°C	6.63%	13.18	——	3-20-29
大　有	1.54%	0.14%	202.4	39°C	4.09%	8.13	——	3-19-29
永　順	0.20%	0.27%	204.4	40°C	1.69%	3.37	——	3-19-29
高壽記	0.59%	0.33%	203.3	40°C	1.97%	3.93	——	3-19-29
德　順	0.08%	0.20%	210.1	40°C	3.53%	7.01	——	3-23-29
朱大順	0.05%	0.19%	208.9	41°C	2.82%	5.61	——	3-23-29
穗　源	1.40%	0.36%	206.3	41°C	4.51%	8.98	——	3-28-29
穗　源	1.48%	0.27%	214.2	41°C	2.82%	5.61	——	3-28-29
王隆昌	0.27%	0.19%	203.5	42°C	6.63%	13.18	——	3-30-29
怡　和	0.12%	0.09%	205.8	41°C	2.54%	5.05	——	3-30-29
義　記	1.78%	0.31%	215.6	40°C	4.23%	8.42	——	3-30-29
王恆豐	2.56%	0.32%	205.8	42°C	2.40%	4.77	——	3-30-29
合　興	2.02%	0.27%	207.2	42°C	3.53%	7.01	——	3-30-29

牌號或來源	水分	雜質	鹼化數	融點	游離脂肪酸	酸數	碘數	試驗日期
吉昌仁	0.28%	0.37%	207.2	40°C	6.06%	12.01	——	3-30-29
茂林春	0.09%	0.19%	205.8	42°C	9.45%	18.79	——	3-30-29
大道生	0.19%	0.21%	210.0	42°C	3.05%	6.17	——	3-31-29
同興盛	1.07%	0.09%	207.2	38°C	15.23%	30.29	——	3-31-29
和順鮮	1.37%	0.04%	200.9	41°C	9.17%	18.23	——	3-31-29
永順福	0.12%	0.18%	210.0	40°C	2.82%	5.61	——	3-31-29
永順興	0.03%	0.12%	203.5	42°C	2.68%	5.33	——	3-31-29
宏榮發	1.77%	0.26%	202.3	42°C	4.51%	7.95	——	3-31-29
李長興	0.18%	0.15%	209.3	40°C	8.58%	17.67	——	4- 4-29
長順和	0.20%	0.12%	209.3	40°C	3.95%	7.85	——	4- 4-29
德順正	0.09%	0.26%	207.9	40°C	5.08%	9.60	——	4- 4-29
洽成永	3.17%	0.32%	204.4	43°C	4.89%	9.32	——	4- 6-29
洽成永	2.37%	0.25%	206.8	41°C	8.74%	17.34	——	4- 6-29
洽成永	0.19%	0.20%	211.8	40°C	2.68%	5.33	——	4- 6-29
天成亨	0.84%	0.32%	210.4	39°C	9.31%	18.51	——	4- 7-29
保合	0.25%	0.11%	205.1	40°C	7.47%	14.87	——	4- 7-29
李恆裕	2.31%	0.40%	206.9	41°C	6.35%	12.62	——	4- 9-29
裕美成	0.12%	0.26%	203.0	39°C	4.79%	9.54	——	4- 9-29
裕興	0.12%	0.22%	207.2	40°C	2.96%	5.89	——	4- 9-29
大有	0.62%	0.32%	210.4	39°C	10.15%	20.19	——	4- 9-29
洽成永	0.35%	0.12%	211.4	40°C	3.67%	7.29	——	4-13-29
洽成永	0.11%	0.14%	210.0	40°C	3.53%	7.01	——	4-13-29
徵昌	2.24%	0.42%	198.8	41°C	6.35%	12.62	——	4-15-29
徵昌	1.68%	0.21%	203.0	42°C	5.22%	10.38	——	4-15-29
童純甫	2.35%	0.15%	200.2	41°C	2.82%	5.61	——	4-15-29
全丈	0.12%	0.18%	210.0	42°C	1.83%	3.65	——	4-18-29
同興久	0.21%	0.18%	210.7	42°C	2.82%	5.61	——	4-18-29
童純甫	10.44%	0.17%	199.5	43°C	3.38%	6.73	——	4-18-29
徵昌	19.77%	0.22%	192.5	41.5°C	5.01%	9.96	31.72	4-18-29
徵昌	16.32%	0.61%	189.7	42°C	4.94%	9.82	29.93	4-18-99
胡發鮮	0.07%	0.29%	201.7	39.5°C	1.83%	3.65	——	4-26-29
徵昌	15.89%	0.41%	180.3	41°C	4.37%	8.70	31.81	5- 5-29

牌號或來源	水分	雜質	鹼化數	融點	游離脂肪酸	酸數	碘數	試驗日期
徽 昌	16.98%	0.36%	178.9	42°C	4.44%	8.84	29.63	5- 5-29
三 星	0.02%	0.22%	204.0	40°C	2.12%	4.21	——	5-10-29
荆 州	0.71%	痕 跡	207.2	40°C	4.23%	8.42	33.75	5-12-29
徽 昌	12.89%	0.43%	——	——	——	——	——	5-12-29
徽 昌	10.58%	0.40%	——	——	——	——	——	5-12-29
荆 州	0.15%	0.28%	204.3	40°C	2.12%	4.21	——	5-26-29
穗 源	2.31%	0.14%	201.7	41°C	5.59%	11.50	——	6-14-29
沈鳳章	0.14%	0.26%	211.4	40°C	3.03%	7.01	——	7-31-29
恒春元	0.17%	0.24%	211.3	40°C	3.17%	6.03	——	8-13-29
恆春元	0.20%	0.27%	211.4	40°C	1.97%	3.92	——	8-13-29
同 源	1.31%	1.45%	204.1	42C°	5.36%	10.66	——	8-15-29
同 源	2.57%	0.33%	208.0	39°C	4.79%	9.54	——	8-15-29
同 源	4.11%	0.24%	207.5	40°C	6.63%	13.18	——	8-18-29
同 源	2.79%	0.33%	207.4	40°C	5.08%	10.10	——	8-18-29
同 源	0.63%	0.34%	207.4	40°C	12.69%	25.25	——	8-19-29
同 源	1.18%	0.26%	206.9	41°C	2.40%	4.77	——	8-19-29
同 源	0.85%	0.26%	213.1	42°C	6.91%	13.75	——	8-21-29
同 源	0.05%	0.25%	211.4	41°C	3.95%	7.85	——	8-21-29
同 源	0.19%	2.24%	212.7	41°C	3.81%	7.57	——	8-21-29
萬順豐	1.24%	0.34%	207.0	41°C	3.67%	7.29	——	9- 8-29
萬順豐	0.11%	0.23%	205.4	41°C	2.68%	5.33	——	9-13-29
萬順豐	0.02%	0.18%	201.3	41°C	2.54%	5.05	——	9-13-29
萬順豐	1.54%	0.29%	195.9	40°C	4.05%	9.26	——	9-13-29
萬順豐	0.66%	0.23%	204.6	40°C	5.50%	10.94	——	9-13-29
天泰昌	0.16%	0.18%	206.2	40°C	5.08%	10.10	——	9-17-29
和生祥	0.18%	0.29%	205.6	39..5°C	3.38%	6.73	——	9-17-29
和生祥	0.28%	0.27%	210.2	40°C	3.67%	7.29	——	9-21-29
天泰昌	0.03%	0.24%	211.1	39°C	3.10%	6.17	——	9-21-29
同泰祥	1.48%	0.54%	201.8	40.5°C	11.99%	23.84	——	9-25-29
謙泰益	1.09%	0.38%	201.4	39°C	4.51%	8.98	——	9-25-29
潤記泳	0.11%	0.26%	205.1	39°C	4.23%	8.42	——	9-27-29
胡愼遠	0.08%	0.35%	213.4	39°C	4.23%	8.42	——	9-29.29

牌號或來源	水分	雜質	鹼化數	融點	游離脂肪酸	酸數	碘數	試驗日期
謙泰益	0.13%	0.28%	210.5	38°C	3.24%	6.45	——	9-29-29
同　興	0.15%	0.35%	212.2	39°C	4.09%	8.13	——	9-29-29
鼎　春	0.45%	0.32%	210.0	38°C	5.50%	10.94	——	9-29-29
同發源	0.64%	0.26%	219.8	40°C	4.09%	8.13	——	10- 2-29
恒　昇	0.44%	0.28%	217.8	39°C	6.99%	12.90	——	10- 4-29
陳復興	0.46%	0.41%	219.8	38°C	4.51%	8.98	——	10- 4-29
同發源	1.14%	0.28%	214.2	39°C	3.37%	6.73	——	10- 4-29
金　華	0.04%	0.35%	218.4	39°C	5.36%	10.66	——	10- 4.29
吉和祥	1.58%	0.46%	215.6	39°C	8.18%	16.27	——	10- 4-29
元利昌	0.37%	0.36%	212.8	40°C	3.38%	6.73	——	10- 4-29
萬順恒	0.92%	0.35%	203.0	39°C	3.53%	7.01	——	10- 4-29
福　興	0.33%	2.26%	214.2	39°C	8.74%	17.39	——	10- 4-29
金　華	0.51%	0.28%	215.6	39°C	8.18%	16.27	——	10- 6-29
洽成永	1.75%	0.35%	210.0	40°C	6.63%	13.18	——	10- 8-29
同　源	0.12%	0.31%	196.0	39°C	4.23%	8.41	——	10-15-29
同發祥	0.32%	0.39%	218.4	39°C	5.64%	11.22	——	10-18-29
慶祥興	0.32%	0.46%	197.4	39°C	5.22%	10.38	——	10-18.29
蘇　城	0.12%	0.32%	191.8	38°C	3.17%	6.31	——	10-19-29
溫　州	0.10%	0.31%	207.2	39°C	5.36%	10.66	——	10-20-29
荆　州	0.09%	0.28%	210.0	38°C	2.54%	5.05	——	10-20-29
金　華	2.66%	0.27%	207.6	39°C	5.36%	10.66	——	10-20-29
同　源	0.82%	0.36%	207.6	38°C	5.78%	11.50	——	10-20-29
同　源	0.32%	0.29%	217.0	39°C	3.81%	7.57	——	10-27-29
洽成永	0.60%	0.36%	214.2	39°C	5.92%	11.78	——	10-27-29
洽成永	0.63%	0.43%	212.8	39°C	8.32%	16.55	——	10.27-29
洽成永	0.47%	0.38%	214.2	39°C	2.54%	5.05	——	10-27-29
洽成永	0.10%	0.28%	210.0	40°C	3.95%	7.85	——	10-27-29
萬順豐	0.07%	0.25%	214.2	40°C	3.53%	7.01	——	10-27-29
萬順豐	0.50%	0.37%	215.6	39°C	6.63%	13.18	——	10.27-29
萬順豐	0.27%	0.36%	203.0	39°C	3.10%	6.17	——	10-27-29
穗　源	2.57%	0.45%	210.0	39°C	5.22%	10.38	——	10-30-29
萬順豐	1.25%	0.33%	208.2	38°C	2.96%	5.89	——	10-30-29

牌號或來源	水分	雜質	鹼化數	融點	游離脂肪酸	酸數	碘數	試驗日期
萬順興	0.01%	0.24%	212.8	39°C	2.12%	4.21	——	11- 3-29
萬順興	0.03%	0.37%	207.2	38°C	1.97%	3.93	——	11- 3-29
李恒裕	1.67%	0.48%	214.2	38°C	10.72%	21.32	——	11- 3-29
萬順豐	0.66%	0.27%	208.4	39°C	5.81%	11.55	——	11-30-29
萬順豐	0.02%	0.50%	204.3	39°C	7.15%	14.21	——	11-30-29
萬順豐	0.07%	0.18%	206.0	40°C	3.68%	7.15	——	11-30-29
程達記	3.53%	0.25%	207.2	40°C	4.23%	8.42	——	12-13-29
平均值	1.37%	0.27%	208.2	40.1°C	4.66%	9.28	31.37	
最低值	0.02%	0.00%	178.9	38°C	1.27%	2.53	27.63	
最高值	19.77%	1.45%	219.8	43°C	15.23%	30.29	33.75	

綜計以上所試桕油,共一百四十七樣.牌號紛歧,來源不一,故其平均值當必近似.各常數中在製皂上最關重要者自屬鹼化數一項.其平均值爲208.2.惟此數係含有平均水分1.37%及雜質0.27%(合計共有外物1.64%)之試品所有之平均值.設二者均爲零,卽純粹桕油,則其鹼化數由推算而知當爲211.67也.碘數爲乾性油試驗中最重要之一項.此處以時間問題,故未全行試驗,僅於特種貨樣及有待考核時時間或行之而已.

上列結果與Allen氏所測得之融點36°—46°C,鹼化數179—203,碘數23—38,及Lewkowitsch氏之融點43°—46°C鹼化數200.3,碘數32.1—32.3相較,亦差不甚鉅耳.

(乙) 牛油

牛油Beef Tallow各地皆有,而品質隨處不同.南方多水牛,北地多黃牛,但農家恃以作耕種之具者則南北皆一也.法律旣有禁宰耕牛之明文,而禮俗又以牛爲大牲,非逢祭祀不殺.因此牛油之產量遂亦不多.惟漠北游牧之邦,與夫信奉回教者廣播之區反是.牛羊則爲重要肉食,而油脂之用於製蠟燭也由來已久,一如長江以南各省之利用桕油者然.迨海禁大開,歐風東漸.我國人民始稍稍知牛之爲用不僅限於耕種祭祀而已.牛肉既清潔而易於消化,牛乳復滋補而有益衛生.牛皮因製革術之進步而爲用愈廣,而牛油更緣苛性鈉電化製造方法之成功,其需要之驟增,尤有一日千里之勢.他如牛痘漿,又爲防止天花之唯一血清,推而至於一切牛乳棚中之製作物,如牛乳油,牛乳酪及冰激冷等,亦莫不飛越重洋而來,飽我素嗜猪味之華人口腹,且漸有奪席之概.於是通商大埠,乃始有乳牛食牛之蓄養,深望各地農民而能羣起遍營之,非但財用可足,而間接嘉惠於民族體格,公衆衛生,豈淺鮮哉?牛油與飼牛之食料,及所在地之氣候,亦有極大之關係.大抵東北多飼以荳餅,故所產牛油,特别豐腴.北地氣候乾燥,所產亦佳,輙堅潔而呈魚卵狀結晶,此爲上品.至於南方所產,或取自水牛,油量既少,兼以氣候溫濕之多雨,故所含水分多而腐敗易.加以商人不顧道德,時以有機雜質混入.且有以低價之礦物油摻入,或調和漆油及棉花子油,而施其魚目混珠之伎倆者焉.雖然,天產有限,工業之需求無窮,不得已自不得不設法補救.如人造脂肪,即藉氫化鎳之接觸作用,使鯨油或其他流動性之荳油棉花子油等,經氫化而成,惟此硬化油不得即謂之曰牛油無疑.其理如 Margarin 之不得稱爲牛乳油,及糖精之不得稱爲甘蔗糖,正復相同.可恨我國一般無知商人,祗圖近利而無遠大計劃.故此種重要之工業,僅於十年前曇花一現於南通州,迨商公物化竟成絕響,可不惜哉?

牌號或來源	水分	雜質	鹼化數	融點	游離脂肪酸	酸數	碘數	試驗日期
德　茂	0.85%	0.35%	204.4	44°C	5.54%	11.22	——	1- 2-29
廣　和	2.49%	0.30%	207.9	44°C	10.15%	20.20	——	1- 3-29
庚　興	7.89%	0.55%	204.3	45°C	10.15%	20.20	33.64	1- 4-29
庚　興	7.96%	0.23%	201.9	44.5°C	14.06%	27.77	——	1- 4-29
同　源	0.13%	0.12%	209.0	38.5°C	3.67%	7.29	——	1- 5-29
同　源	5.11%	0.73%	207.8	47°C	14.57%	28.26	——	1- 6-29
同　源	5.34%	0.67%	200.5	47°C	15.44%	30.72	——	1- 6-29
庚　興	4.94%	0.20%	207.2	47°C	5.08%	10.05	——	1- 7-29

牌號或來源	水分	雜質	鹼化數	融點	游離脂肪酸	酸數	碘數	試驗日期
庚 興	5.32%	0.30%	200.8	46°C	7.46%	14.87	——	1- 7-29
元 亨	1.83%	0.02%	194.6	47°C	1.13%	2.24	——	1-13-29
李裕順	0.95%	0.20%	207.0	45.5°C	2.68%	5.33	——	1-13-29
庚 興	2.17%	0.20%	204.5	45°C	1.97%	3.93	——	1-13-29
李裕順	0.98%	0.25%	200.1	46°C	4.09%	8.14	——	1-13-29
德 茂	1.20%	——	——	——	——	——	——	1-14-29
德 茂	5.35%	0.48%	202.8	44°C	6.49%	12.90	——	1-14-29
德 茂	0.50%	0.53%	208.7	41°C	2.82%	5.61	——	1-14-29
德 茂	9.14%	——	——	——	——	——	——	1-17-29
無 錫	7.81%	0.81%	200.0	46°C	17.79%	31.42	——	1-17-29
庚 興	9.83%	0.30%	197.0	46°C	8.46%	16.83	——	1-30-29
庚 興	9.55%	0.78%	208.5	45°C	8.04%	15.99	——	1-30-29
同 源	3.48%	0.75%	201.7	46°C	11.99%	23.84	——	2-18-29
同 源	9.09%	0.72%	200.0	45°C	2.82%	5.61	——	2-18-29
同 源	6.80%	0.67%	200.1	45°C	6.06%	12.06	51.30	3- 3-20
同 源	0.56%	0.06%	207.5	43°C	3.53%	7.01	61.17	3- 3-29
同 源	1.33%	0.35%	205.3	47°C	1.55%	3.04	——	3- 6-29
裕 豐	1.63%	0.21%	200.2	44°C	9.73%	19.35	39.56	3-12-29
同 源	3.53%	0.57%	198.5	44°C	8.74%	17.39	56.90	3-15-29
同 源	4.08%	0.42%	201.0	45°C	5.64%	11.22	——	3-20-29
大 茂	0.99%	0.20%	204.6	43°C	5.92%	11.78	——	3-22-29
大 茂	1.42%	0.10%	203.9	43°C	7.05%	14.03	——	3-22-29
李裕順	1.55%	0.21%	200.5	45°C	7.61%	15.15	——	3-24-29
大 茂	0.50%	0.13%	205.8	43°C	6.35%	12.62	——	3-24-29
同 源	4.69%	0.45%	196.0	45.5°C	6.35%	12.62	——	3-27-29
同 源	5.57%	0.32%	194.6	45°C	5.64%	11.22	——	3-27-29
大 茂	1.41%	0.12%	201.6	44°C	1.97%	3.93	——	3-30-29
大 茂	0.83%	0.10%	195.3	44°C	2.12%	4.21	——	3-31-29
大 茂	1.63%	0.33%	202.7	43°C	3.53%	7.01	——	4- 7-29
萬順豐	6.90%	0.94%	194.6	45°C	6.35%	12.63	51.80	4-10-29
大 茂	2.11%	0.84%	196.0	43°C	7.20%	14.31	——	4-10-29
大 茂	1.19%	0.07%	205.1	46°C	7.98%	15.85	——	4-18-29

牌號或來源	水分	雜質	鹼化數	融點	游離脂肪酸	酸數	碘數	試驗日期
大　茂	1.63%	0.61%	194.9	47°C	4.51%	8.98	——	4-26-29
大　茂	0.70%	0.23%	195.2	44°C	3.10%	6.17	——	5- 4-29
大　茂	0.29%	0.22%	197.5	43°C	6.35%	12.62	——	5-16-29
大　茂	0.52%	0.31%	194.7	43°C	7.05%	14.03	——	5-19-29
大　茂	0.50%	0.32%	198.2	43°C	5.92%	11.78	——	5-26-29
同　源	4.41%	0.82%	188.7	44°C	9.02%	17.95	——	6- 6-29
同　源	3.82%	0.57%	190.8	44°C	10.15%	20.20	——	6- 6-29
同　源	4.14%	0.51%	190.8	44°C	15.23%	30.29	——	6-26-29
同　源	6.76%	0.26%	173.4	44°C	11.00%	21.88	——	6-26-29
蚌　埠	3.80%	0.32%	196.0	47°C	8.88%	17.67	——	7-27-29
蚌　埠	3.81%	0.46%	201.6	47°C	12.97%	25.31	——	7-27-29
李炳發	0.35%	0.35%	207.2	46°C	1.41%	2.81	——	7-31-29
李炳發	0.50%	0.43%	207.0	46°C	5.96%	11.78	——	7-31-29
蚌　埠	6.24%	0.41%	185.0	47°C	11.70%	23.28	——	8- 2-28
蚌　埠	2.80%	0.40%	193.5	47°C	8.60%	17.11	39.87	8- 6-29
蚌　埠	2.60%	0.36%	195.7	47°C	9.82%8	19.64	——	8- 6-29
蚌　埠	5.40%	0.41%	192.2	47°C	13.25%	26.37	——	8- 7-29
蕭綬記	1.67%	0.43%	203.1	47°C	9.56%	19.01	——	8-14-29
蕭綬記	3.53%	0.36%	202.3	47°C	8.74%	17.39	——	8-15-29
同　源	2.32%	0.47%	207.6	47°C	9.02%	17.95	——	8-21-29
蚌　埠	2.83%	0.67%	195.9	46°C	13.11%	26.09	——	8-31-29
織昌祥	3.89%	0.42%	191.1	46°C	11.28%	22.44	——	9-17-29
織昌祥	4.15%	0.46%	195.8	46°C	13.40%	26.65	——	9-24-29
織昌祥	3.53%	0.41%	200.4	46°C	10.01%	19.92	——	9-24-29
織昌祥	4.77%	0.43%	188.7	46°C	14.81%	29.45	——	9-24-29
萬順豐	2.38%	0.36%	194.8	46°C	8.60%	17.11	——	9-29-29
萬順豐	2.19%	0.43%	204.4	47°C	8.60%	17.11	——	70- 6-29
蕭綬記	4.28%	0.46%	201.6	46°C	10.72%	21.32	——	10-13-29
同　源	5.06%	0.76%	190.4	46°C	18.19%	36.18	53.94	10-13-29
同　源	1.93%	0.29%	200.0	47°C	6.77%	13.46	——	10-13-29
織昌祥	2.37%	0.45%	200.2	46°C	8.46%	16.83	——	10-15-29
織昌祥	2.09%	0.57%	205.8	46°C	7.47%	1487	——	10-15-29

牌號或來源	水分	雜質	鹼化數	融點	游離脂肪酸	酸數	碘數	試驗日期
張竹記	4.33%	0.36%	204.4	47°C	13.68%	27.27	——	10-17-29
張竹記	1.45%	0.36%	204.6	44°C	1.41%	2.80	——	10-19-29
蕭綬記	2.83%	0.37%	205.8	45°C	9.59%	19.07	——	10-20-29
孫德源	1.13%	0.46%	190.4	44°C	1.41%	2.81	——	10-20-29
同源	1.38%	0.35%	200.2	45°C	4.23%	8.42	——	10-20-29
同源	3.13%	0.46%	207.2	46°C	12.97%	25.81	51.51	10-25-29
德茂	1.18%	0.45%	205.6	44°C	3.67%	7.29	——	10-27-29
無錫	6.27%	0.77%	198.8	44°C	20.45%	40.67	——	10-29-29
沈增記	1.10%	0.36%	200.2	45°C	4.09%	8.13	——	10-29-29
德茂	3.00%	0.43%	200.2	44°C	9.87%	19.64	——	10-29-29
顧福記	0.15%	0.48%	202.8	46°C	1.83%	3.65	——	10-29-29
李裕順	0.44%	0.35%	207.2	46°C	2.40%	4.77	——	10-30-29
李裕順	7.43%	0.45%	200.2	45°C	20.59%	40.95	——	10-30-29
泰森永	1.11%	0.25%	198.8	45°C	4.51%	8.48	——	11- 2-29
沈增記	1.40%	0.36%	202.8	46°C	3.81%	7.57	——	11- 2-29
沈增記	1.28%	0.44%	197.4	45°C	6.06%	12.06	——	11- 5-29
沈增記	1.40%	0.37%	184.8	46°C	4.23%	8.42	——	11- 5-29
沈增記	6.22%	0.36%	187.6	46°C	33.56%	66.76	——	11- 5-29
費興發	0.30%	0.64%	205.5	46°C	1.83%	3.65	——	11- 6-29
李裕順	1.01%	0.45%	201.4	45°C	1.55%	3.09	——	11- 6-29
費興發	0.48%	0.45%	193.2	45C°	2.68%	5.28	——	11- 6-29
童泰昌	0.82%	0.29%	207.2	44°C	3.67%	7.29	39.07	11- 7-29
德茂	11.16%	0.38%	204.4	44°C	21.43%	42.64	50.88	11- 7-29
德茂	11.71%	——	——	——	——	——	——	11- 7-29
德茂	15.45%	——	——	——	——	——	——	11- 9-29
德茂	3.79%	0.49%	193.2	44°C	31.73%	63.11	——	11- 9-29
李裕順	0.55%	0.47%	204.2	46°C	1.13%	2.24	——	11- 9-29
沈增記	0.93%	0.23%	201.4	44°C	3.38%	6.73	——	11-10-29
沈增記	1.57%	0.30%	205.6	44°C	7.05%	14.03	39.56	11-10-29
同源	1.94%	0.38%	200.0	45°C	14.38%	28.61	——	11-10-29
德茂	9.50%	0.55%	184.8	43.5°C	28.48%	56.66	——	11-17-29
德茂	4.61%	0.48%	204.4	44°C	31.16%	61.99	——	11-17-29

牌號或來源	水分	雜質	鹼化數	融點	游離脂肪酸	酸數	碘數	試驗日期
大　茂	0.64%	0.26%	204.2	43°C	12.13%	24.12	——	11-17-29
同　源	1.60%	0.24%	203.0	44°C	18.05%	35.90	——	11-17-29
李裕順	8.46%	0.53%	200.2	43°C	26.37%	52.45	35.43	11-17-29
繼昌祥	1.31%	0.46%	187.6	44°C	7.61%	15.15	——	11-17-29
李裕順	0.91%	0.35%	201.6	43°C	6.91%	13.74	——	11-17-29
童泰昌	0.30%	0.45%	204.2	44°C	3.24%	6.45	——	11-21-29
德　茂	2.37%	0.43%	205.6	43°C	11.84%	23.56	——	11-21-29
大　茂	1.24%	0.38%	205.8	42°C	11.99%	23.84	——	11-21-29
李裕順	0.22%	0.22%	200.0	43°C	2.26%	4.49	——	11-21-29
同　源	2.45%	——	——	——	——	——	——	11-21-29
申昌裕	5.89%	0.45%	204.4	45°C	25.10%	49.93	59.82	11-22-29
童泰昌	0.69%	0.46%	201.4	43°C	4.65%	9.26	——	11-24-29
童泰昌	5.61%	0.00%	184.2	43°C	5.37%	10.68	——	11-24-29
童泰昌	0.35%	0.55%	200.0	43°C	4.37%	8.70	——	11-27-29
張竹記	5.19%	0.45%	208.6	45°C	24.32%	49.37	——	11-27-29
泰森永	2.00%	0.36%	202.8	43°C	2.82%	5.61	——	11-27-29
大　茂	1.01%	0.35%	208.0	42°C	8.74%	17.39	51.41	11-28-29
李裕順	1.68%	0.46%	203.2	44°C	4.94%	9.82	44.81	11-28-29
沈增記	1.95%	0.24%	196.4	43°C	3.05%	6.07	——	11-30-29
張竹記	4.77%	0.71%	151.3	45°C	18.01%	35.82	——	12- 3-29
李裕順	0.83%	0.63%	204.2	43°C	6.35%	12.62	——	12- 6-29
童泰昌	0.28%	0.35%	205.6	46°C	0.28%	0.56	——	12- 6-29
陸紹泉	2.40%	0.64%	205.6	43°C	7.33%	14.59	——	12- 7-29
松大仁	0.48%	0.34%	202.8	42°C	0.14%	0.28	——	12- 7-29
童泰昌	0.51%	0.54%	208.5	43°C	7.05%	14.03	——	12-10-29
大　茂	0.64%	0.15%	208.6	42°C	10.09%	19.07	——	12-10-29
同　源	2.33%	0.35%	205.8	46°C	6.06%	12.06	——	12-13-29
童泰昌	2.34%	0.57%	208.6	42°C	2.68%	5.33	——	12-13-29
同　源	1.00%	0.45%	204.4	46°C	3.24%	6.45	——	12-15-29
同　源	2.42%	0.58%	208.8	47°C	5.22%	10.38	48.36	12-15-29
泰森永	2.51%	0.34%	201.4	44°C	3.10%	6.17	——	12-15-29
同　源	3.94%	0.30%	203.0	45°C	2.11%	4.27	——	12-19-29

牌號或來源	水分	雜質	鹼化數	融點	游離脂肪酸	酸數	碘數	試驗日期
德 茂	7.91%	0.36%	197.4	46°C	14.66%	29.17	45.82	12-19-29
董泰昌	0.21%	0.33%	207.2	43°C	1.97%	3.93	——	12-20-29
泰森永	1.83%	0.24%	207.0	44°C	1.27%	2.52	——	12-20-29
繼昌祥	2.50%	0.46%	201.4	46°C	7.30%	14.59	——	12-21-29
李裕順	2.59%	0.78%	207.0	43°C	3.24%	6.45	——	12-21-29
李裕順	1.42%	0.23%	200.0	42°C	3.59%	7.13	——	12-21-29
同 源	2.32%	0.31%	207.2	46°C	2.40%	4.77	——	12-21-29
同 源	1.49%	0.27%	205.6	45°C	3.24%	6.45	——	12-21-29
大 通	3.71%	0.35%	207.2	45°C	2.26%	4.49	——	12-22-29
張竹記	2.30%	0.26%	209.8	44°C	10.86%	21.60	——	12-22-29
大 茂	2.25%	0.31%	200.2	42°C	19.60%	38.99	——	12-23-29
同 源	3.93%	0.50%	205.6	44°C	9.45%	18.49	70.61	12-23-29
同 源	1.23%	0.34%	200.0	46°C	1.27%	2.54	——	12-23-29
德 茂	17.45%	11.41%	144.2	47°C	18.05%	35.90	——	12-25-29
大 茂	0.65%	0.35%	200.2	42°C	6.49%	12.90	——	12-24-29
同 源	2.53%	0.38%	207.0	45°C	3.10%	6.17	——	12-27-29
大 茂	1.47%	0.25%	203.0	42°C	20.59%	40.95	——	12-27-29
同 源	2.60%	0.26%	204.2	43°C	10.86%	21.60	——	12-29-29
董泰昌	1.50%	0.34%	207.0	43°C	3.67%	7.29	——	12-29-29
同 源	4.32%	0.25%	203.0	43°C	5.92%	11.78	——	12-29-29
同 源	0.50%	0.36%	209.8	43°C	3.81%	7.57	——	12-29-29
同 源	4.92%	0.28%	205.6	49°C	74.73%	148.73	——	12-29-29
同 源	1.98%	0.43%	200.0	43°C	10.29%	20.48	——	12-29-29
沈增記	1.05%	0.44%	201.2	45°C	3.24%	6.45	——	12-30-29
董泰昌	0.14%	0.48%	200.0	43°C	2.96%	5.89	39.27	12-30-29
朱銀綬	1.41%	0.46%	208.1	43°C	5.50%	10.94	38.01	12-30-29
董泰昌	1.79%	0.33%	208.4	46°C	8.52%	17.95	——	12-30-29
廣 成	3.25%	0.27%	208.6	49°C	85.31%	169.70	——	12-30-29
平均值	3.05%	0.47%	200.7	44.7°C	9.23%	18.33	48.18	
最低值	0.13%	0.00%	144.2	38.5°C	0.14%	0.28	33.64	
最高值	17.45%	11.41%	209.8	49°C	85.31%	169.70	70.61	

以上試驗爲數一百六十四.牌號雖較桕油略少,但品質異常複雜.色澤白黃棲黑無不具備.按牛油原屬一種混合脂肪,故其中難免有膺品摻入.著者曾以棉花子油與漆油對和而一如常法試驗之其結果爲

品名	水分	雜質	鹼化數	融點	游離脂肪酸	酸數	碘數	試驗日期
混合脂肪	0.49%	0.04%	195.7	45°C	2.73%	5.42	72.10	12-20-29

細考各數大致與牛油之平均值相近.惟碘數獨大耳.由是逆推,凡碘數過大之各樣,即不能不有懷疑之處焉.次核其平均鹼化數爲200.7,苟計算至純粹牛油,則應爲208.09.與桕油之平均鹼化數相差,極有限耳. Allen氏關於牛油之測定爲融點38°—50°C 鹼化數 193—198 碘數 33—48.證以上述結果,頗稱脗合.

（待續）

本刊啟事

徵稿　本刊爲吾國工程界之惟一刋物,同人等鑒於需要之急,故力求精進,凡會員諸君及海內外工程人士,如有鴻篇鉅著,闡明精深學理,發表良善計劃,以及各地公用事業,如電氣,自來水,電話,電報,煤氣,市政等項之調查,國內外工業發展之成績,個人工程上之經營,務望隨時隨地,不拘篇幅,源源賜寄,本刋當擇要刋登,使諸君個人之珍藏,成爲全國工程界上之南針,本刋除分酬本刊自五本至十本外,并每期擇重要著作數篇,印成單行本若干,酬贈著者,以答雅誼.

推銷　凡海內外各機關,各學校,各書局欲代銷本刊者,請函致本會事務所上海寗波路七號接洽是荷.

總務楊錫鏐啓

論實業及擬發展我國電氣事業之辦法

著者: 朱瑞節

著者曾在國內工廠實習,後進英國電機製造廠做工,兩相比較,深感欲以實業救國,應羣起臥薪嘗膽,不恥下層工作,痛除一切舞弊.爰作此篇,願吾同志指敎之,共勉之.

(一) 人才;制度;賞罰.

(二) 實業分析:物質與人工.

(三) 我國實業不振之三大原因:

(甲) 學問與經驗之不調和,而洋行得以操縱一切.

(乙) 資本;智識,勞工,三界無合作之精神.

(丙) 國風日奢,生產不增,上下相賊;到處舞弊.

(四) 發展電氣事業之三步手續:

(甲) 調查時期.

(乙) 與外商合作製造時期,

(丙) 獨立製造,研究改良時期,

(五) 各時期需要之人才與其訓練方法:

(一)人才,制度,賞罰

制度由人定,賞罰由人行,是以制度之不良,賞罰之不明,皆歸罪于人.天下更不多不良之制度,而多執行之不妥:不多不明賞罰之人,而或被威武所屈,或為私利所誘,賞罰因此亂行.玆願舉例以證之,譬如鐵路運輸,機車之能力有限,橋樑之載重有度,負行車之職者,有限掛車輛之權,制度無不當.乃行于我國,則其弊無窮.他若戮盜于市,衆皆鼓掌而呼曰該死,明盜之結果如是,而踣盜之禍國害民,爲惡萬倍,何反不罰耶?于是消極者信宗教善惡報于後世之理,積極者圖革命改造之工.奈消極不問,而督促無人,國風愈下.除惡不盡,則以暴易暴,使後人而復恨後人,談救國而禍人民者,欲求雪恥救國,其可能乎?嗟夫!人無不知廉恥也,人無不欲方正也,然人亦莫不趨名利而遠禍患.是以爲惡者,亦見其利而未見其禍耳.今之社會,剝奪者可享物質之幸福,正直者反難以生存,無怪人人不欲放棄剝奪之機會,而背正直之天良.假使善惡因果,可以立刻證明,則天下之民,皆爲善矣.蓋爲善而有利,則人人爲善,爲惡而有利,則人人爲惡,宗教談死後之報,不足以奪今生之利,良知良心,終不能敵環境之迫,* 是以惟有執權管理者,審制度,明賞罰,以導民入規也.吾工程人才之訓練,豈僅技術上之學問與經驗哉?竊察我國人民,重私情,忽公事,而畏結怨于小人,小人乃得以用計,有以嚴厲手段,整頓圖治者,人或袖手坐視,或且非其不近人情,國家之危,至此莫甚.再觀外國,人心未必良善也,惟社會風俗,重規則,守秩序,一介不取,一介不與,此即足以防舞弊,而小人不能得志矣.願我國負治人之責者,無論其所轄範圍之小大,賞罰以公,行事以規,毋畏結怨于小人,毋待人行而後行,要見義勇爲,以救國之將亡,則人之問題,可先解決,然後從物質上工作,無往而不順矣.

*編者按 至誠爲善者,終其身必得善報,但同時須諸惡莫作,以塞其漏,否則一方作惡,一方行善,不過扯直,其效力自等於零耳.

(二)實業分析:物質與人工

造物不言,而至公正.人類之所以異于禽獸者,由觀察而起奇心,從奇怪而動思想,思想之精華,合而成理論,化而爲學術,是以發明偉人,可謂造物之信徒,而代其言所不言也.吾人從事工程事業,終日與造物爲伍,無時無地,必須服從其公正自然之理,苟有疏忽,則立受損失;若順其道,則無往不利;不然,則理論不必講,經驗不足貴,而物質文明之進步,何從而來耶?物質以其大者而觀,則地球不過空中一星耳,以其小者而觀之,則錙銖之微,不知含多少原子.化而爲原子,合而成萬物,變化天然之物,使人類得用增進生活幸福者,謂之實業.是以實業者,物質與人工之結合耳.物質既不增不減,不生不滅,則實業之進步,全視吾人之工作.不事經營,而欲提高生活;不勞心力,而求生存;有此理乎?所謂人工者,其先必由困難好奇或需要而有發明,因困難而發明者,黃帝被蚩尤迷途,而創指南車其例也.因好奇而發明者,牛頓見物下降,而定動體之理屬之.因需要而發明者,在我國史上,伏犧,神農,燧人,有巢,皆是也.而近世人類之需要日夥,此種發明,亦隨之增加.既有發明,次言計劃.蓋發明之範圍,在定原理,立制度,成機械;其如何應用,如何佈置,則因地異,因時而變.所謂凡事預則立,不預則廢,計劃者,即此意也.計劃既定,然後談製造;計劃無定,則製造期間,中途變更,費工傷材;計劃不妥,則實用期內,或不附所需,或發生危險,是以計劃工作,不可不慎也.再進則爲製造,試觀第一表.製造更分勞工,器械,材料三項.勞工指直接用人力或器械將材料工作者,其手藝精良者,亦須計劃而勞心,其未經訓練者,多勞力之責任.俗云勞心者使人,勞力者使于人.物質文明進步以來,勞力之責,漸屬於機械.而人不勞心,將無以立足,平民教育,可不努力哉?至于器械材料,爲已成之物品,如第一表中,將虛線連至購者,上推其原,仍不出物質與發明.如此循環,實業爲物質與人工之結合,可以證明.而物質爲造物所賜,本無價值,經人之工作,而各有代價,亦可想見矣.造成之物,按正當之手續,應受商場之支配,于是商人究供求之理,通運輸之法,爲造者購者之媒介.是以經濟原理,亦由實業發端,工商之關係如是,從

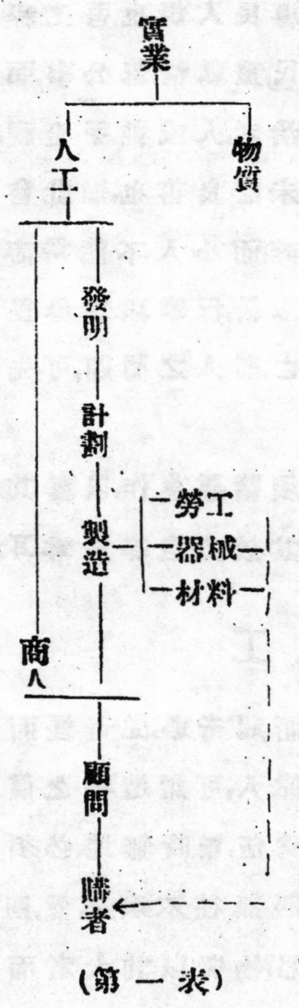

(第一表)

事于實業者,對于經濟學識,豈可忽哉?爾來實業之發達,製造者之衆多,購者不得不慎其選擇.蓋商人為營業競爭,利益所關,無不忠其所代理之廠家,對于購者,鮮有適當之勸告.於是于購者與商人之間,又有顧問任也.俗云,購物商諸內行,此內行者,即顧問也.為顧問者明應用,善計劃,知製造之難易,熟悉各廠之出品,各地之情形,而代購者計劃選擇,可得完滿之結果,其工作亦至要也.物品至購者,或消費,或復為器械,或材料,而終歸于消費,消費與生產,其平均出于自然,其數量為人工之代表,其結果為社會之幸福,實業之目的在提高吾人之生活,其成功全由于工作也明矣.

玆願再將英國某廠製造渦輪之預算例第二表,而論之,以切事實:

名　目	售價之百分數
直接工資	22½
材　料	22½
工廠雜費	30
營業費	15
淨　利	10

(第二表)

(1) 直接工資:　凡勞工之可歸于製造一指定之物品者皆屬之.由以前同樣或相似出品之工作單,推算而得.

(2) 材料:　購入材料,無論一釘之微,苟非自造者皆屬之.由計劃之需要,定其量;市價之高低,决其數.

(3) 工廠雜費:　包括製造權費,資本利息,廠房設備,機器之消耗折舊,保險等,及一切不能劃歸于製造一指定物品之薪金工資,從預算一年中之產額而分派之.

(4) 營業費:　廣告,出版,交際,各處樣子間,事務所之薪工雜費皆屬之.

(5) 淨利:　按表若預算與決算同,則得一十之利.查直接工資,與材料兩項,雖預算較易,然工作不慎而傷材料者有之;重要材料,于工作中途,發現劣點,供者祗能賠償原料,已費之工,無從收回.且工廠雜費,占最大部份,全由預算一年中之產額而定.是以營業發達,則每件之工廠雜費減輕,其結果為增加淨利,減低售價.反之則損失虧本,從事實業,可不慎歟?

購者除此而外,尙須付運輸佣金等費;且不之論,再察第二表,工廠雜費,與營業費兩項,當成本之半,較之我國從前之家庭工業,祇有人工與材料,則對工廠雜費,與營業費之能否減輕,可令人注意矣.吾人既知製造廠之一切開支,全由購者分任其責,則對于規模宏大之廠家,可否信任;及擴充工廠,有無限止;皆重要之問題也.蓋自機器代人勞力,而有工廠;其所製造之物品,愈巨大複雜,需人材愈多,器械愈精,工廠雜費乃愈大.是以此種費用之增加,使造昔日之所不能造者,並減省直接人工,不足以起驚.營業費為鼓勵購者,以增加產額,卽可減各件之工廠雜費,亦不可以為大.是以規模宏大之廠,常占優勝者,所費大而分任者.亦多.若以昔日之家庭工業為之,或不能製造,卽有可造者,其人工之增加,必數倍于用機械而有組織之工廠雜費及營業費也.欲減輕此兩項費用.惟有聯合同業,採標準出品,與分工之制.發端于美國,盛行一時,然至今有供過于求之現像,是以擴充當有限止,論者有謂其制度之不良,吾不信也.並以為發展我國實業,亦當採取其法,惟擴充之時,應謹愼考慮耳.

(三)我國實業不振之三大原因

(甲) 學問經驗之不調和,而洋行得以操縱一切: 用者不必造,造者不自用;用者專其用,造者專其造;此之謂分工專門,近世實業之發達,人類之進化,強國強種之由來也.雖然造吾所能造,而思用者之便利;購吾所欲用,而不知造者之難易;則造者必失其業,用者必傷其金,此供求相應之理,而負計劃建設事業者,更介乎造者用者之間;有精通學理,而不能勝任者,經驗之不足也;有積數十年之親自操作,雖天資過人,勤勞獨出,而見題束手者,學問之缺乏也.此果我國目前之情形,而為實業不振之一大原因矣.竊觀國中購置機器,常為洋行所操縱,佣金得隨購者之所欲,而貨物則聽洋行所支配.當事者以為外人道德之高,决不我欺,既省聘請顧問之費,又可得從心所欲之佣;而彼外人則正笑我私利之重,學識之淺,可得而任意調度;於是存貨多而式舊者,飾新而來,退貨劣而無主者,混雜以進.勢所不免.欲其開誠佈公,忠于購者,按余在國外觀察所及,不能信也.願我國人毋過自卑而尊人.須知燧人取火,而熱力生.黃帝作舟車,而機械創始.造指南針,而電學發端.土木遠始有巢.水利興自大禹,而外書不傳,豈不因其後裔不肖,智者雖有發明,祕而不傳,國政輕視物質,民間守舊不進,由來久矣.時至今日,雖未爲亡國奴,然普通人之生活狀況,較之歐美各國,不可同日而語矣.竊常以工程師比醫士,則洋行爲藥房.以工程問題商諸洋行,猶疾病之不請醫士開方,而向藥房購藥,其果可以想見.雖然,吾知洋行本不欲操縱其事,奈購者缺乏智識,又少學問經驗並長之才,致以全權委之耳.是以凡抱實業救國之志者,學問經驗尤當並重也.

(乙)資本,智識,勞工,三界無合作之精神: 欲早耕而暮獲,見利淺而不善用人者,我國之資本家也.好逸惡勞,趨勢利而不務實事者,我國之智識界也.欺詐弄巧,以維持地位者,我勞工界之匠目也.惟普通工人,終日勞動,而生活最苦,可謂仰不愧,俯不怍之神聖矣.是以國家至此,受痛苦者無罪,而在上者作惡.願投資于實業者,當知才學有專門,經營須時日,所謂良梓人者,不受主人之限止,合則就,不合則去;自古帝皇尙尊賢而重士,欲求資本之生利,應聽專家之計劃.以匠目代工程師者,所省之薪金,萬不足以抵損失.以工程問題,委于洋行者,私佣雖豐,後患無窮.欲利厚而虐待勞工者,勞工亦有以罰之.智識界爲資本生利,助勞工謀生,不可順資本家之所欲而奪勞工之利益,亦不應聽勞工之放肆,而使資本受損失;不勤勞不知工作之艱難,何以能盡管理之責,勝計劃之任.不得資本家之信任,爲匠目所愚弄者,禍亦自取,可不醒歟?至于勞工之匠目,須知機械複雜,誰不能効其一二,令起變化,使人難以接手.然此種行爲,在具有學問經驗者,何足爲懼.此後應謀工程教育之改良,(參閱各時期需要人才與其訓練方法一段)若有志於上進,當求學問以補其不足,豈可弄巧欺人哉?凡吾勞工,其知民生之艱,由于列強經濟侵略,抵制外貨,提倡國貨,全賴國中有較廉之人工.將來實業發達,勞工生活,自必提高,欲速則勞資分裂,危險實甚.是以資本之不易召集,人才之用非所學,勞工之生活艱難,皆出此資本,智識,勞工三界之無合作精神,而爲我國實業不振之又一大原因也.

(丙) 國風日奢,生產不增,上下相賊,創處舞弊: 夫人心不足,天演之理.其出邪徑而險速者,爲賊,爲盜,爲內亂,此軍閥政客以人與人爭,禍國害民,所以滿其不足也.至于學者窮究深思,發明原理,以供人之利用,或利用得法,而造成機械.以滿人類之所欲.此人與物爭,物質文明之所以進步,亦不知足之現像也.是以天下之變,皆由于不知足者,其爲軍閥爲政客,則民化爲兵丁盜賊,走險而冒死.其事發明,與實業,則民化爲工匠藝員,安居樂業.然則禍民者,不知足,興國者,亦不知足;國家對于不知足者,應如何注意,勉其興國,止其禍民,則普及教育,改良社會,不可緩也.奈察國風日奢,而生產不增,以致人心不足之氣,日益澎漲,走險無勇,建設無能,于是化爲舞弊,造成不公開之社會.此又爲異邦所少見,而亦中國實業不振之一大原因也.

他若關稅之保障,以及別種政治問題,論者既多,而本篇目的,在工程界之自身覺悟,其希望于他人者,不暇及之矣.

(四) 發展電氣事業之三步手續

(甲) 調查時期: 我國實業不振之原因飢詳,請進而言發展我國電氣事業之辦法.擬分三種時期,依次進行,可否實行,尙希執政當局,與負工程之責者,予以教正.其初曰調查時期.吾國交通不便,而各地匠目,皆有派別,守舊排外,調查困難,可以想見.然無切實之調查,萬難言改進,此則工程當從實事求是,非空談所及者也.方今民智未開,國家應負調查之責,則權力所及,事半功倍.對于各地電廠員司工匠,妥爲開導,解釋利害,聲明目的,毋使疑及奪權剝利,選派工程人才,親往各處實地調查.茲將所擬組織與經費等,詳解于下.

按表組織,定名爲全國電氣事業調查隊,直轄于建設委員會,其工作可先

員司薪水表

員司	人數	月薪人數	附註
土木工程師	1	$ 300	有工廠築建之經驗與學問
機械工程師	2	600	有各種原動機之經驗與學問
電機工程師	2	600	有發電廠輸電線路之經驗與學問
測繪員	2	300	助各工程師
渦輪司機	2	210	經驗宏富能修能用
蒸汽機司機	1	100	"
內燃機司機	1	100	"
鍋爐匠	1	100	"
攝影員	1	100	
冲洗員	1	100	
小工	4	160	兼測地夫
總數		$ 2700	

每年經費表

名目	每年經費	附註
薪水	$ 32,400	詳員司薪水表
雜費	12,000	膳宿照影材料等在內
儀器工具之消費折舊與利息	5,600	測量儀器電錶汽錶等
總數	$ 50,000	

在江,浙兩省交通便利各處進行.出發以前,令各地建設局或地方政府,先向電廠接洽,毋使誤會,而調查隊職員,以勤勞,耐苦,廉潔,爲最要.立誓以救國爲目的,所到各處,務于最短期間,完其工作,不帶遊歷性質,不受電廠任何供給.倘有舞弊,建設委員會當加以嚴罰,使不能立足于中國工程界上.每到一廠,務詳細調查,試驗各機效率,考察線路設備,究其利害,告以應與應革事宜.按調查隊之人數既多試驗是易,然儀器工具等,多則費大難運.少則試驗不準,

是以先在交通便利之處舉行,俾得調動靈快,而對于各種電廠性質,先有查考,然後推及全國各地,則赴交通不便之處,所需儀器工具,或覺缺乏材料,可先妥為預備.約計須時三年,共經費十五萬元.所得結果,分述于下:

(1) 由調查所得各地人口.出產,交通,負荷增加綫,機器之新舊等,可推算將來之需要.

(2) 由各廠之設備,各地之氣候水性,經過觸電等情形,可定電廠規則,適合我國情形之材料規定,選擇機器之要點,為將來擴充及開辦之參考.

(3) 由試驗結果,可增高電廠効率,改正各種錯誤.

(4) 採取各處所用中文專門名詞,著成完全中文之調查報告,推銷全國,名詞得以審定而歸統一.

(5) 可得人才缺乏之量,與支配失當之處,而定訓練人才之方法.

方今我國百廢待興,善為政者,應視事之緩急,支配經費.想此實地調查之工作,誠為當務之急,則其費用豈成問題哉.按上數端,最足注意者,為由確實之調查,而得將來之需要.依此即可進而為第二時期工作,曰

(乙) 與外商合作製造時期: 近來各國電氣事業之發達,可謂一日千里,然其在本國,必有極點;過此則供過于求,製造範圍,萬難擴充,而其對于國外貿易,非努力不足以維持其出產速度.是以我國將為各國競賣之商場.若能外以關稅為保障,內有較廉之人工,而以我國地大物博,將來電業之盛,可以想見,外商必願投資于我國,我政府亦應鼓勵外商與我合作,蓋工程人才之訓練,非由學識經驗並重不可.我國人在外留學者,不過略窺學理.實習者,終日與勞工為伍,重要之處,黃色者何能問徑,且材料等種種問題,至為複雜,欲其速成易舉,當出此計,而專門人才,乃可漸漸訓練成功.預計十年以後,不難收回,而入

(丙) 獨立製造,研究改良時期: 我國電氣事業始脫依賴然後努力前進,豈可限量.是則吾工程界所望者也.

(五) 各時期需要之人才與其訓練方法

人生之變化無窮,有初好學而後怠惰者;有本荒唐而後奮發者;有先爲惡而後覺悟者;有始光明而入昏亂者;其情境之複雜,誠難描寫.下例兩圖,略表

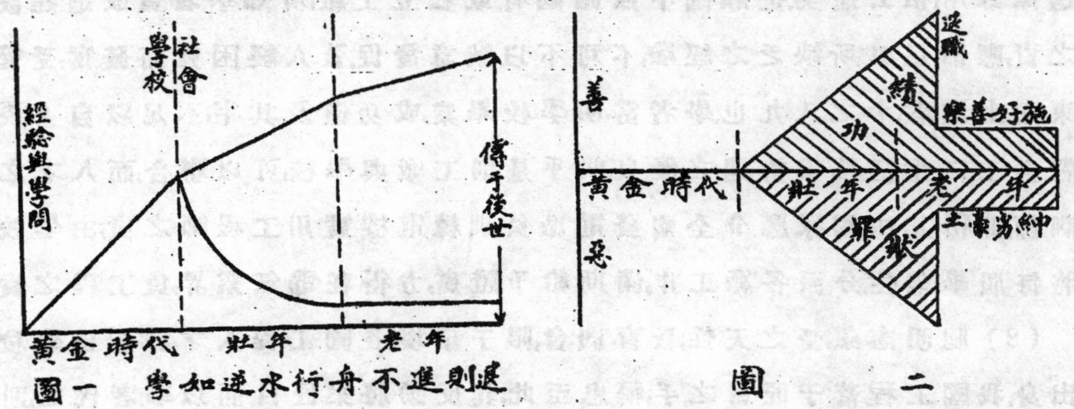

圖一　學如逆水行舟不進則退　　　圖　二

人生所經之途,各舉二種情形,其實曲折上下,決非簡單若是要不出此理耳.試觀圖一,在黃金時代,學問與時並進.及入社會,向上者得經驗,增學問;向下者茸廢之速,實可警人.學問與經驗增加之率,常隨年齡而減低,是以壯年及老年之線平,表明黃金時代之可貴也.次及圖二,假定學校生活,無能爲惡,無人爲善.及與社會接觸,小人得志,則罪惡日深;君子當道,則立功漸高;其成績各以面積計.而第一圖中學術之傳,則當量其高度也.

英雄爲時勢所造成,人才因需要而產生,原近來國中無線電之發達,可以知之矣.是以國家教育人才之方針,當視需要而變更.方今學術之繁,書籍之多,有志之士,欲用其才,而所學不附所需者,誠可嘆也.我寶藏富足而實業落後之中國,政局既定,建設事業,莫不待舉,所需人才之範圍,豈有限止.然察歐美各國,發明製造,一日千里,回顧我國,則購者不知其貨,用者不明其法.以致三相電流,用其二相者有之.(見工程四卷四號南寧電燈整理之成功及方法)塞汽錶之管,毀試驗之器者,在國中工廠,不知其數.是當調查期內,最要人才,在如何維持,如何購置.請將其訓練方法,分述于下:

(1)工程教育,與其他文學等之性質不同,應分兩種時期,在學校之時,授以基本學理,實習時期,補以各種經驗,有學問與經驗,方可負責任.是以外人注重實習,待遇全同工人;與我國之一紙空文,從此不加管束,亦不辨別勤惰,選擇錄用,相去遠矣.是願國中無論國有或私立工廠,須知學者負改造建設之責,應補足其所缺乏之經驗,不可不以誠意督促.蓋人經困難而發奮,受管束而上進者,十之八九也.學者當明學校畢業,成功僅及其半,不足以自大.要習勞工之生活,莫爭待遇之優良.明乎是則工廠與學校,可以聯合,而人才之訓練,庶得其法.國家應令全國發電廠負訓練電機實用工程師之責,由學校將每屆畢業生,分派各廠工作,滿期給予證書,方得在電氣業界,負工程之責.

(2)聰明志氣,受之天性;教育機會,限于環境.各國工程人才,非皆自學校出身.我國工程落于匠目之手,禍患至此,其反動將棄匠目而以學者代之,則匠目之欺詐弄巧,保障更嚴,學者與勞工之合作難行.施學者以實習,而不補勞工以教育,使聰明有志者,無以發展,而成貴族式之工程教育,非所宜也.竊察外國設立夜校,獎勵學金,使人人皆有求學之機,甚至善也.願我人急起效之,從事編訂中文工程課本,以普及勞工智識.其富有經驗之匠目,一經智識貫輸,眼界自開,而樂知經驗之所以然,未使不能勝重任,豈可因其不爲學校出身,而限止其地位哉?

(3)外人以機械輸入,雖不願告以製造計劃,維持保護之法可得問也.機器之改良進步,多由用者經驗而來.是以創辦購置,當由專家審定,經建設委員會批准而行.其價值較大者,應由公司選派人員,赴外實習,探其製造方法,査明維持保護之策,以歸國在該公司服務數年爲合同.

調査期畢,實用工程師而外,當注意計劃製造工程師之訓練.與外商合辦電機製造廠成立之初.上級職員,是屬外人.吾人應握管理之權,擇才赴外與在廠訓練,同時並進,漸將外排拆,至于完全收回之地步.他若國中材料之研究試驗,至此可求實用,而當盡力擴充矣.閱者或謂外人資料及此,豈肯與我我合作.茲特以英國茂偉電機製造廠爲證.方二十年前,創辦之初,人才資本,皆是美國西屋公司,今則全歸英人.現我國資本人材,皆難自辦,借助他山,外人得一時之利,我人樹永久之基.吾不恐外人之不願與我合作,而慮吾人之不能克苦勤勞,以達收回之期,其共勉之矣.

在完全自造,研究改良之時.學者始可各就性之所近,專一門以求高深.國家對于是種人材.乃得利用,而應竭力鼓勵.否則我國學術,將永遠步人之後,安能與世抗衡,爲民族爭光,爲國家爭榮,高深人才,用武之時也.今之用人者,常嘆人才缺乏,而負材者又恨用非所學.亦末勢非供求不應之理耳.

國外工程新聞二則

(一)土西鐵道通車:—　聯貫土耳其斯坦與西比利亞鐵路,已於本年四月二十三日全部竣工,二十八日上午九時,在伊亞爪布拉克舉行盛大通車典禮,參與者有勞農政府代表,各國大使館代表,及新聞記者.此鐵路之建築,乃俄國三大事業之一.完成後,不特交通便利,運輸迅速,卽產業之開發,商業之興盛,均與俄土兩國,有莫大之關係云.

(二)北海運河工竣:—　接通阿姆斯特丹與北海之運河,費荷幣一千八百萬盾,於十年內造成,最大商船,可由北海經此運河直達阿姆斯特丹,乃世界極大之工程.本年四月二十九日,荷女皇舉行伊穆登新閘門開幕禮,機紐一轉,閘門遂開,水道大臣演說,謂此工程之完成,係荷蘭技師,與荷蘭進取家共同之努力,可爲荷蘭國民魄力之紀念物云.

油頁岩工業及撫順頁油岩

著者:胡博淵

(一) 緒言

世界石油之需要,日甚一日,而世界石油儲量,則固有限,以美國爲世界最著名之產油國,按現在之消耗率,據專家之報告,亦僅足十七年之開採,即告罄盡,故專仰來源於天然產出之礦油,終有供不應求之一日,而探索油田以外之新油源,遂成近代工業界一最緊要問題矣.惟環顧已發明之代用品中,其儲量豐富,並適於工業的提油之用者,莫如油頁岩.英國蘇格蘭,於七十年前,即已盛用此物,以煉巨額之石油.最近美國坎拿大及瑞典等國,亦正在積極進行之中.我國油田雖不甚多,但油頁岩之儲量固甚豐富.已知者如遼甯之撫順,及熱河之凌源,皆產此物.前者之儲量,達五十五億噸,所含原油,約當美國石油總儲量四分之一,堪稱東亞希有富源,所惜撫順煤礦,久歸日人經營,而此項鉅大利源,遂亦連帶被佔矣.油頁岩(Oil Shale)含有一種稱爲Kerogen之瀝靑質,與普通油田地方之含油頁岩不同,乃完全不含有液狀石油之粘土頁岩也.其色黑褐有光澤,比重普通在一·七五左右,若將此物加以乾溜,可得頁岩油 Shale Oil,此油雖可即作燃料重油之用,但一般均依處理礦油之方法,以製各種石油製品,其處理方式之一,略如次表所示:

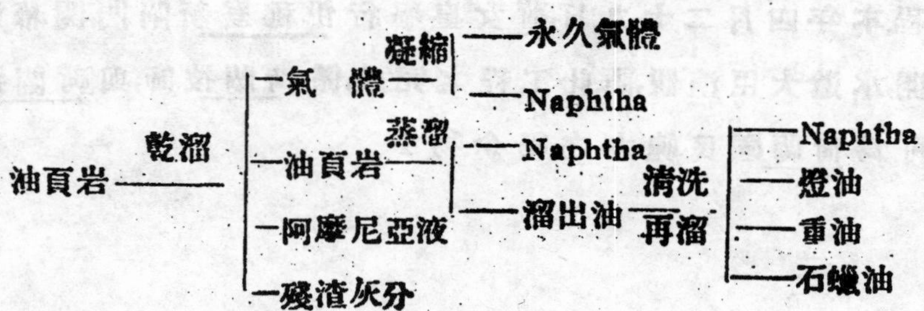

(二)蘇格蘭式提油裝置

蘇格蘭之油頁岩,作層狀產出,層厚一·五呎,其採掘法與採煤相同,鑛夫二人每日可採八噸,其平均成分如次:

水分	○·四六%	固定碳	五·四五%
揮發分	二三·二五%	灰分	七○·八五%

提油工場設置乾溜煉爐,(Retort)凝縮機,(Condenser)撒水摩洗機,(Water Scrubber)撒油摩洗機,(Oil Scrubber)汽油精煉裝置,(Gasoline plant)蒸溜釜及硫酸錏工場等.所採取之油頁岩.最初用碎機擊成二吋左右之大小,然後裝入煉爐中,煉爐形狀略如下圖所示:

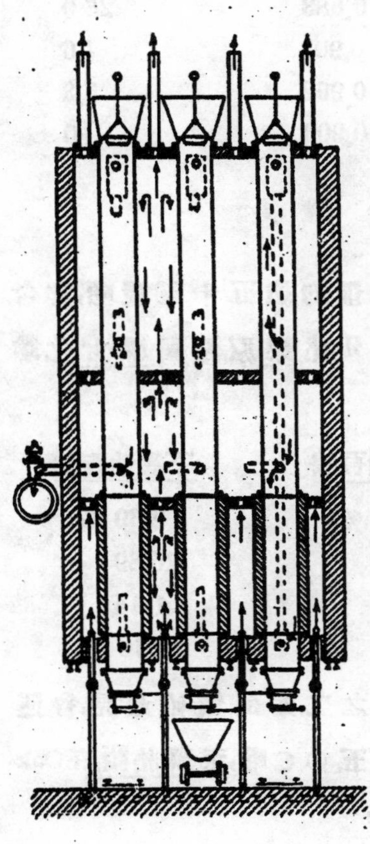

煉爐作直立圓筒形,上半部爲鑄鉄製,下半部由火磚砌成,其外週爲烟道包圍,乾溜時可將乾溜油頁岩所得之氣體,導入烟道下部,使作燃燒,而漸上升,以自外部加熱煉爐.

煉爐上部,備有原料裝入口,油頁岩由此口漸次降下,最初在上部被加熱,至攝氏三七○——四八○度,然後來至下部,而被加熱至攝氏七○○度,又在下部吹入水蒸汽,使固定碳生成一養化碳,或二養化碳,並令其與氮氣化合,以便形成阿摩尼亞,NH_3 油頁岩每隔六小時裝入一次,在爐內經過十八小時,然後由下部之排出口取出.

乾溜所得之氣體,由裝入口稍下之鉄管導入凝縮蛇管(Condensing Coil)內,使受冷却,藉使氣體中所含之石油,及阿摩尼亞液,凝縮分離.然後用排送機(Exhaus'er,吸取此氣,先用撒水摩洗機,加以洗滌,使氣中殘存之阿摩尼亞,形成液體,而被析出.次導此氣入於撒油摩洗機,將殘存氣中之最輕汽油,用重油洗去,然後更將此氣導入烟道內,作煉爐燃料之用.

由凝縮機及摩洗機中所分離之頁岩油,及阿摩尼亞液,可更在分離器內使兩者分離,將頁岩油蒸溜,以製各種製品,其阿摩尼亞液亦須行蒸溜,使成氣體,並用硫酸吸收此氣,製成硫酸錏,以供作肥料之用.

由一噸油頁岩之溜出油,(Once run Oil)及硫酸錏收得量,約如次表所示:

產 地	溜出油收得量(加倫)	溜出油比重	硫酸錏收得量(磅)
蘇格蘭 Baoxburn Curly Seam	26.7	0.861	39.4
蘇格蘭 Baoxburn Grey Seam	23.7	0.877	39.0
蘇格蘭 Dunnet Shale	30.0	0.883	25.0
美 國 Colorado	16.2	0.903	7.6
美 國 Utah	25.2	0.902	5.3
美 國 Wyoming	16.1	0.900	60

(三)撫順油頁岩

撫順油頁岩,位於主要煤層之上,厚約四十呎,儲量約達五十五億噸,其含油量平均為百分之六,所含原油,約達三億噸.其成分,據栗原鑑司博士化驗所得者,略示如次:

	上等油頁岩	中等油頁岩	下等油頁岩
水 分	2.95 %	2.59 %	2.69 %
揮發分	24.46	17.64	15.89
固定碳	4.72	3.59	1.94
灰 分	67.64	76.68	80.10

民國十三年,南滿鐵路公司,為決定撫順油頁岩之工業的價值起見,曾運上等油頁岩二五〇噸,中等油頁岩二五〇噸,共計五〇〇噸,至蘇格蘭,在Oakbank工廠,作大規模之營業試驗.其試驗成績如次:

由一噸原料收得之量

原　料	頁岩油(加倫)	粗汽油(加倫)	硫酸錏(磅)
上等油頁岩	17.02	2.30	30.03
中等油頁岩	11.59	1.20	29.71

又用此時所得之頁岩油,及粗汽油,在試驗室,行製品實驗之結果,則如次表所示:

(A) 頁岩油精製得

精製品	由上等原料頁岩油之收得量(%)	由中等原料頁岩油之收得量(%)
燈　油	40.14	38.56
○・八三五油	2.19	0.75
中間油	13.25	15.69
輕機油	4.23	4.48
機　油	7.86	9.66
石　蠟	8.32	5.78
殘　油	0.83	0.57
精製殘油	0.95	0.93
合　計	77.70	76.69

(B) 粗汽油精製得

精製品	收得量
Motor Naphtha	81.51
Naphtha	9.83
殘　油	2.57
合　計	93.91

南滿鐵路公司,當最初施行油頁岩乾溜之際,曾擬用蘇格蘭式提油裝置提油.其後經種種研究之結果,因用該式裝置,乾溜剩餘之殘渣中,尚殘存多量之油質,故已決定採用內燃式乾溜裝置.曾築四十噸乾溜爐爐一座,於十五年一年間,實行試驗,其結果甚爲圓滿.故由十六年四月起,已以八百五十

萬日金之預算,實行第一期工程計劃,預定二年內建築四十噸乾溜煉爐一百座,於竣工後,每日得處理四千噸之油頁岩,每年可獲原油五萬四千噸,硫酸錏一萬八千噸,粗石臘九千四百噸,及 Pitchcoker 四千九百噸.現在業已築成煉爐八十座,聞於年內,即可開爐提油云.

又內燃式乾溜法,係於乾溜爐之下部,另設瓦斯發生爐,Gasproducer 使乾溜殘渣落入此爐內,生成發生爐瓦斯,更將其導入乾溜爐內,利用其顯熱,以乾溜油頁岩,然後再與乾溜所得之含油氣體,共同流出爐外,以備提油.其特點在能用劣質之撫順油頁岩,乾溜提油,而不需使用任何補助燃料也.

最新會員錄已出版

本會對於會員通訊地址,力求準確,以期消息靈通,接洽便利,會員錄自去年第七次重印以來,不過一年,其中變更,已不勝枚舉.故特詳細校訂,修正重印,現已出版,分寄各會員.然人數逾千,散處各地,本會辦事人員,雖竭力設法校正,終不免有遺漏錯訛之處.會員諸君,如有未經收到,或發現其中錯處者,請即函知上海甯波路七號本會辦事處可也.

FUNDAMENTALS OF STROWGER AUTOMATIC TELEPHONE SYSTEM

史端喬式自働電話之述略

BY B. J. YOH (郁秉堅)

General. An automatic telephone system is one in which the calling party is enabled, without the aid of an operator, to complete a call through remotely controlled switches. The Strowger's Step by Step system is the one most widely used in the world. It accomplishes selection of a called line by successive stages—or steps—which correspond to operations performed by the calling subscriber, known as dialling. These dialling operations consist of the manipulation of a dial which is part of the subscriber's apparatus in addition to the transmitter, receiver and ringer.

Before studying the description of the apparatus employed, the requirements to be satisfied by the automatic system should be considered. First of all, when the subscriber lifts his receiver he must be provided at the exchange with a circuit to the automatic apparatus which will later respond to the impulses he dials. Secondly, when the subscriber has sent impulses he must be connected to the line whose number he has dialled. Thirdly, When the called line is found and is disengaged the called subscriber's bell must be rung. Fourthly, when the called subscriber answers the telephone, the ringing current must be disconnected and talking current supplied. Fifthly, while the conversation is in progress the lines connected together must be rendered engaged to other calls and safe-guarded against interference. Sixthly, while the calling subscriber replaces his receiver both subscribers must be left free to make or receive calls and the apparatus restored to normal so as to be available for use on further calls. Seventhly and lastly, when one subscriber calls another whose line as engaged the calling subscriber must receive a distinctive engaged signal to acquaint him with the fact.

The Dial. The dial and its mechanism may be seen from Figs. 1 and 2. To make a call, the calling party first lifts the receiver and listens for the "dial tone" which is a distinctive buzzing sound. This indicates that the equipment at the central office is ready to receive the call. He then inserts his forefinger into one of the holes in the dial through which the first figure of the required number appeares, and turns the front disc of the dial in a clockwise

Fig. 1. The Dial with Number Card.

Fig. 2. The Mechanism of the Dial.

direction until the finger strikes the finger stop. The disc is then released and, whilest being driven back to its original position by a main spring located within the dial, disconnects the subscriber's line a certain number of times corresponding to the digit shown on the number plate near the hole into which the forefinger was inserted. This operation is repeated for each digit of the number called. The dial is arranged so that the disconnections occur on the return journey of the dial, not on the forward journey, in order that the system may be rendered reasonably independent of any peculiarity in dialling on the part of the calling subscriber, such as hesitation in the middle of a digit. In order to avoid trouble which would accur if the dialling circuit included the variable resistance of the transmitter, and also to avoid annoyance to the subscriber from clicks in the receiver, it is usual to switch the speaking apparatus out of circuit during dialling.

Principal Switches. Three distinguished types of Strowger switches are generally in operation one after another, when the subscriber lifts his receiver and sends impulses by dialling. They are:

1. Pre-selector or Line Switch.
2. Group Selector (For Exchange over 100 Lines).
3. Final Selector or Connector.

A typical trunking diagram of a four figure system, having line switch, first group selector, second group selector and final selector or connector, is shown in Fig. 3.

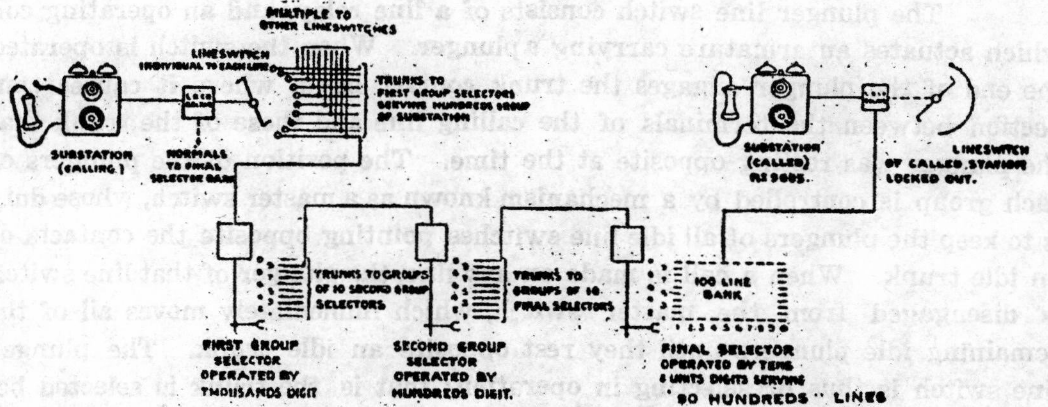

Fig. 3. Typical Trunking Diagram of Four Figure System.

Line Switch. It is an non-numerical switch, attached to the subscriber line, as shown in Fig. 3, which connects the line to a trunk leading to an idle selector. This is to avoid the necessity for having an expensive selector associated with each line.

There are two types of line switches in use in Strowger system; the plunger type or Keith line switch (Fig. 4), which requires a common trunk selecting mechanism for each group, and the rotary type line switch (Fig. 5), having individual control.

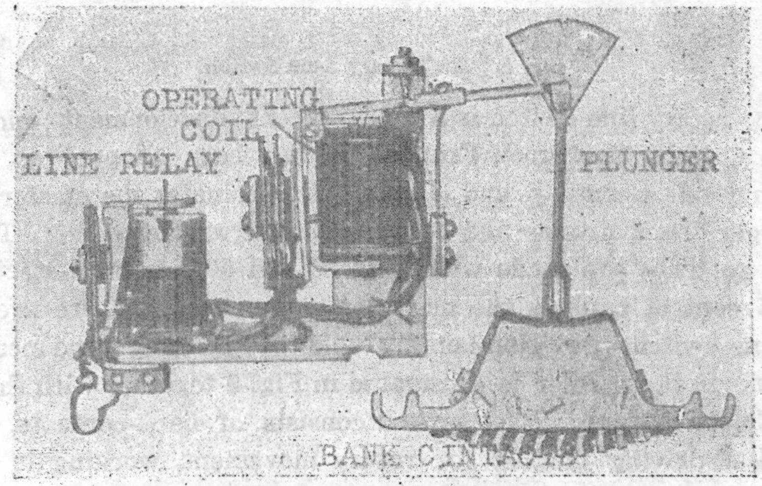

Fig. 4. The Plunger or Keith Line Switch.

The plunger line switch consists of a line relay, and an operating coil which actuates an armature carrying a plunger. When the switch is operated the end of the plunger engages the trunk contact bank, where it causes connection between the terminals of the calling line and those of the trunk that the plunger was resting opposite at the time. The position of the plungers of each group is controlled by a mechanism known as a master switch, whose duty is to keep the plungers of all idle line switches pointing opposite the contacts of an idle trunk. When a call is made on any line, the plunger of that line switch is disengaged from the master switch, which immediately moves all of the remaining idle plungers until they rest opposite an idle trunk. The plunger line switch is thus preselecting in operation; that is, the trunk is selected before the receiver is lifted from the hook.

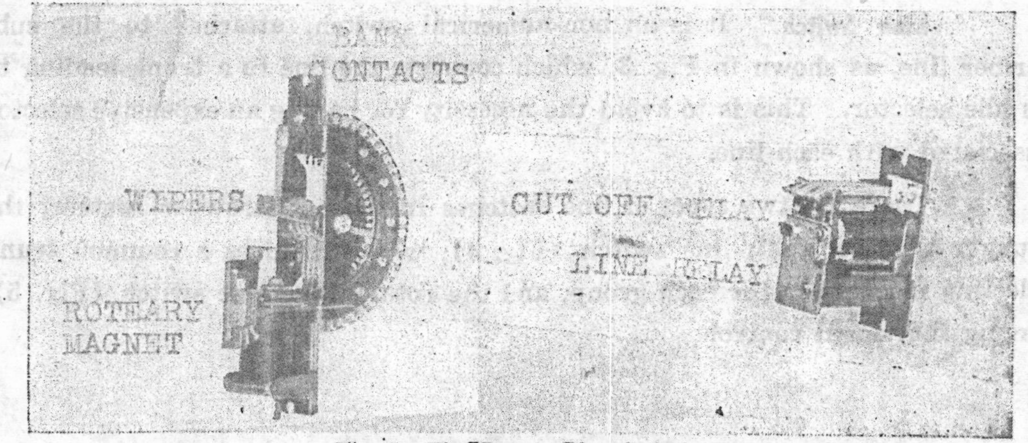

Fig. 5. The Rotary Line Switch.

The rotary line switch is a more recent development and was first used in the Orleans Exchange, France. It has been adopted as a standard type in England, Germany and Japan. It eliminates the master switch and permits larger trunk groups and a more flexible arrangement. The bank of rotary line switches are made with both 25 and 50 sets of contacts. At present the 25 contact bank is the most commonly used. There are two kinds of rotary line switch, the "Non-homing" and the "Homing", the mechanism and contact bank of the former is illustrated in Fig. 5 together with the associated Line and Cut-off Relays. Each switch consists of bank-contacts, double end wipers and a driving magnet for rotary movement, working in conjunction with its line relay and cut-off relay. The number of wipers may be four—

negative, positive, meter and private—or five—negative, positive, meter, private and normal private. The positive and negative wipers are in the talking circuit during the conversation. The private wiper is used for searching an un-engaged trunk and safe-guarding it against interference. The normal private wiper gives guarded homing feature in homing type rotary line switch.

Group Selector. Group selectors are in existance when the exchange has more than 100 Subscribers. Its vertical motion is controlled by dialling while the rotary motion is automatic. It is used for selecting trunk and extending line circuit through to next switch. It switches wipers on first idle trunk, and overflows busy when all trunks are engaged. It holds preceding switch or switches until the trunk is extended. The dial tone is provided from this switch.

Fig. 6. The Group Selector.

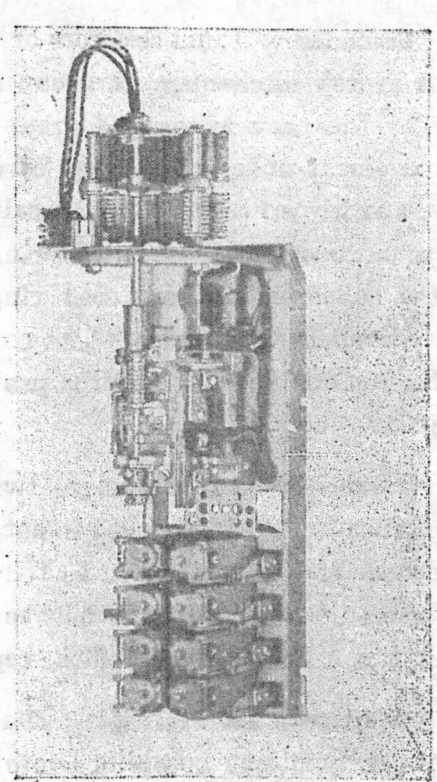

Fig. 7. The Connector.

Each selector consists electrically of a set of relays (Line or Impulse Accepting Relay, Release relay, Vertical movement relay, Switching relay or relays and Auto rotary impulse relay), a set of magnets (Vertical, Rotary and Release), contact-springs, line and private banks and wipers. The 10-level 10-trunk group selector has one line bank while the 10-level 20-trunk type has two line banks, making its private bank with the contacts in pairs, and arranging the circuit so that as the wipers rotate over both the banks at the same they hunt idle trunks and seize the first one found, whether it be in the upper bank or the lower one. When a small exchange or a small office in a multi-office exchange needs for more than one level or trunks to a certain destination, the multi-level group selector can be used. It requires horizontal-chain relays, one for each trunk in a level, vertical relays, one for each level and a vertical wiper and bank in addition to the ordinary type.

Connector. Final selectors or connectors (Fig. 7) have practically the same switch mechanism as group selectors. As soon as this switch is engaged by the pre-selector and group selectors, it executes both vertical and rotary movement under the control of calling device. It holds all preceding switches, and gives busy tone if the called subscriber is engaged and prevents intrusion on busy line. It gives ring-back tone to the calling subscriber whilst the called subscriber's bell rings. It feeds talking current to both calling and called subscribers. The connector is released when the last party hangs the receiver up, while all the preceding switches are controlled by the calling party only.

Besides those fundamental switches as mentioned above, Automatic Impulse Repeaters are introduced in out-going trunks from one office to another. The purposes of a repeater are as follows: Ground the release trunk to permit the use of two-wire trunks between offices; supply talking current to the calling station from the home office; repeat impulses to the distant office.

Adjustment and Testing. The adjustment of an automatic switch of the Strowger type may be divided into two parts, the relay adjustment and the adjustment of the motor magnets. Each relay is adjusted to a definite armature stroke with a fixed residual air gap. The amount of contact of each spring with its mate is likewise fixed by the position of the armature when the contact takes place. These distances are measured with thickness gages placed between the armature and the pole of the relay. The spring pressure of the contact springs of a relay is measured by the operating and non-operating currents. The final performance of a switch is tested by being operated under line conditions which are worse than any imposed by commercial use.

Standard impulse machines used in adjusting the central office switches may be in portable form with carrying case, in which are mounted all necessary resistances and capacities and means for making all changes in conditions required to "vary" any switch so that it will be tested under variations of resistance, capacity and speed greater than those ever experienced in receiving impulses from a subscriber's station.

As a result of the constant aim on the part of Strowger engineers to eliminate as far as possible, the human element, both as regards the establishing of connections as well as maintaining the apparatus in a perfect condition, the Strowger Automatic Routiners have been designed. These routiners are arranged to perform "routine tests and inspections" on the particular unit of Strowger equipment for which they are designed.

Trunking and Grading. The methods of trunking and grading are among the chief factors in the success of the automatic system. They covers the means of determining the amount of plant required in the exchange to carry the expected traffic, the best means of arranging that plant, of rearranging it when the traffic changes in total value or in distribution, and also, the measurement of traffic overflows. There is ample scope for investigating the subject from both the theoretical and practical standpoints.

In the investigation of traffic problems for proper trunking and grading, the bases are the mathematical theory of probabilities and matters of personal experience and opinion. The busy hour calling rate, the average time taken for a call, the probability of lost call and the load distribution on different stages of switches are chiefly considered.

Power Plant ant Supervisory Scheme. Like manual exchanges, the power plant of an automatic switchboard central office generally consists of one storage battery, two ringing equipments, two battery charging equipments, one power switchboard and one supervisory cabinet.

The proper scheme for supervission in an automatic exchange enables the attendants to inspect and clear the fault, if there is any, within the shortest time as possible. It consists of supervisory relays, coloured lamps and alarming bells for operation and indication.

Telephonic Transmission. So far as telephonic transmission is concerned, there is practically no difference between manual and automatic systems. The problem of increasing the range and reducing the cost of the telephone has been studied for over 40 years. The introduction of microphonic transmitters, the application of hard drawn copper wires and the improvement in receivers may be considered as the first stage of progress. In order to eliminate transmission losses and to make their lines efficient for long distances, the introduction of Coil Loading, Cable Balancing, Thermoinic Valve Repeater and Phantom Circuits makes the whole situation possible and practical. It will be quite possible to telephone from any one point to any other in the whole world.

Conclusion. In this paper, only a very general idea about the Straight Strowger Automatic System is given. For detailed informations, references should be made in a series of standard books and publications concerning this subject.

京滬漢三市改裝自動電話之經過

著者：莊智煥

(一)改裝之原由

科學進步,日異月新.以電話機件而論,由磁石式一進而爲共電式,再進而爲自動式.我國電話事業,視歐美各國,已爲落後;城市電話全國祇六九,〇八五號,長途電話祇一四,一六〇里(見附表),急起直追,猶恐望塵莫及,倘仍苟且因循,必致電話事業,永無發展可言.值此訓政建設時期,普設全國市內及長途電話以便利交通,尤爲當務之急.然一方固當力求擴充,一方則各大都市原有電話裝置之改良,亦屬要圖.

溯自國都旣定,人口驟增,政商事務,日益紛繁,原有電話設備,旣覺不敷,機件又已陳舊,與其再添舊機,孰若改裝自動.滬,漢兩地,華洋雜處,方言不一,接用電話早感困難,且上海租界電話,已有一部份改用自動新機,華界仍用舊機,未免相形見絀.武漢電話,除武昌局被焚,迄未恢復原狀外,漢口局則機額久滿,亟待擴充.此京,滬,漢急需改裝自動電話之情形也.

附表一

(1)部辦市內電話

地點	程式	容量	地點	程式	容量
北平	共電式 磁石式	一九,四〇〇 五四七	天津	自動式 共電式	九,〇〇〇 三,六〇〇
武漢	共電式 磁石式	四,五八〇 五二〇	首都	共電式 磁石式	二,八〇〇 三〇〇
上海	共電式 磁石式	二,〇〇〇 九八〇	青島	自動式	三,〇〇〇
蘇州	共電式	二,〇〇〇	鎮江	磁石式	五〇〇
揚州	磁石式	四〇〇	蕪湖	又	六〇〇

烟台	磁石式	五八〇	太原	磁石式	五〇〇
保定	又	五〇〇	吉林	又	六〇〇
長春	又	四〇〇	蚌埠	又	三〇〇
九江	又	一五〇	沙市	又	一〇〇
鄭州	又	三〇〇	洮南	又	二〇〇

自動式	一二,〇〇〇號
共電式	三四,三八〇號
磁石式	六,二九七號
共計	五二,六七七號

(2) 部辦長途電話

地點	距離里數	話線種類	回線數目
平津	二四〇里	四 二百磅銅線 二 半銅半鐵	一 六回線 四
津奉	一,三一〇里	四 二百磅銅線 二	一 二回線 一
吉長哈	六九〇里	二百磅銅線	四回線
濟青	八二〇里	又	二回線
京滬	六七〇里	二 甯錫三百磅銅線 滬錫二	二 三回線 二
京蕪	一八〇里	八號鉄線	一回線

話線共計長 一四,一六〇里

(3) 民營電話公司

公司名稱	程式	容量	公司名稱	程式	容量
常熟電話公司	自動式	四五〇	武進電話公司	共電式	八〇〇
無錫電話公司	共電式	一,〇〇〇	廈門電話公司	共電式 磁石式	六〇〇 三〇〇
太倉電話公司	磁石式	四〇〇	海門海胣電話公司	磁石式	二一〇
泰興電話公司	又	五〇	崑山電話公司	又	二〇〇
高郵電話公司	又	一五〇	溧陽電話公司	又	一〇〇
江陰電話公司	磁石式	四五〇	徐州電話公司	磁石式	一〇〇
盛澤電話公司	又	二〇〇	淞陽電話公司	又	二〇〇

南潯電話公司	磁石式	一五〇	南通大聰電話公司	磁式石	五七〇
嘉興中興電話公司	又	三五〇	崇明通利鄉村電話公司	又	一〇〇
松江電話公司	又	三〇〇	杭州電話公司	又	二,五四〇
泰縣電話公司	又	二〇〇	吳興電話公司	又	五〇〇
四明電話公司	又	一,三〇〇	吳興雙林鎮電話公司	又	五〇
硤石捷利電話公司	又	二四〇	定海電話公司	又	二〇〇
溫州東甌電話公司	又	三〇〇	嘉善電話公司	又	一〇〇
紹興電話公司	又	五〇〇	蕭山電話公司	又	
海甯斜川電話公司	又	一八	海寧電話公司	又	五〇
平湖永通電話公司	又	三〇	袁花四定電話公司	又	五〇
常德常徽電話公司	又	二〇〇	廣莲電話公司	又	六〇
漳州通敏電話公司	又	三〇〇	宜昌清新電話公司	又	二〇〇
張家口電話公司	又	八〇〇	豐鎮電話公司	又	一〇〇
歸綏電話公司	又	四〇〇	包頭電話公司	又	二〇〇
滄縣電話公司	又	一〇〇			
	自動式	四五〇			
	共電式	二,四〇〇			
	磁石式	一三,五五八			
	共　計	一六,四〇八號			

(二)機式之決定

自動電話之發明最早,應用較廣者,當推步進式,(即史瑞喬式)美,英,德,日諸國多採用之.歐戰前後,電話學者,孜相研究,於是機式倍增,其中以於轉式(即機動式之一種)爲最著.法,比,荷,瑞諸國多採用之.茲將兩式主要同異各點,略分述之.

(甲)大略相同諸點:

(一)用戶所裝機件,各傳音器,收音機,號數轉盤器等及其運用法大致相同.

(二)從用戶至話局,線路材料及設置無大殊異.

(乙) 主要不同諸點:

(一) 步進式機每部係壹百號,旋轉式機每部則爲二百號.

(二) 步進式局內機件端賴用戶,由轉盤所轉號數,依照十進數位次第選接,即可通話.旋轉式機件轉動之速度,與用戶處之轉盤速度不同,所需號數,須先至登記器,再傳達至各項選接機件,方可接通.

(三) 用戶所需號數,在步進式,則與機上所轉到之號數相同.在旋轉式,則先經一度更改,故與機上所轉到之號數不同.

(四) 步進式機件之動作,端賴用戶之轉數.旋轉式機件之動作,則由另置之小電動機行使之.

是故步進式與旋轉式,從技術方面觀察,固多所不同;從利弊方面研究,則各有長短,即歐,美先進,亦各執一詞.交通部爲此曾多方審查,再三研討,乃根據下列四項理由,爲京,滬,漢各地電話,採用步進式之決定.

(一)各步進式機件較爲簡單,故障礙不易發生,而維持亦較便利.現在部轄各局工人,對於機械智識,尙屬粗淺,均以採用機件簡單之程式爲佳.

(二) 步進式無論用戶多至百萬,少至數百,均能經濟適用.旋轉式則僅適宜於萬號以上之大城,否則即不經濟.京,滬,漢各話局,將來雖均可超過萬號,惟目前尙非其時,均以採用多寡咸宜之程式爲佳.

(三) 步進式祗在使用時費電,通話少時費電亦少,故用電較省.旋轉式則無論通話之多寡,全數機器,終日旋轉不息,費電自然較多,機件亦易損壞.

(四) 步進式電話用法較便,故用戶使用時不易錯誤,且其接通較便,故每次通話所需時間較短,局內機器常可敷用,而不能接通之電話亦可較少.

(三) 各局原有之設備及通話記載

京,滬,漢三局之原有設備及通話記載如下:

附表二

局名		機式	容量	平均每日每線用話次數	平均每線忙線小時內用話次數	平均每次通話時間
首都電話局	城內	共電	二,〇〇〇	二六·六二四	二·〇九八	一六〇秒
	下關	共電	八〇〇	三二·五九七	二·九七五	一六〇秒
	浦口	磁石	二〇〇			
	浦鎮	磁石	一〇〇			
上海電話局	南市	共電	二,〇〇〇	二〇·三一〇	一·五〇〇	一二〇秒
	閘北(一)	磁石	四八〇	一五·二七〇	一·二〇〇	一三一秒
	閘北(二)	磁石	二〇〇			
	江灣	磁石	一〇〇			
	吳淞	磁石	一〇〇			
	南翔	磁石	一〇〇			
武漢電話局	漢口特別區	共電	四,三〇〇	一九·三〇〇	一·九三〇	
	漢口華界(一)	磁石	四〇			
	漢口華界(二)	磁石	八〇			
	漢口華界(三)	磁石	一〇〇			
	武昌	磁石	四〇〇	二三·五〇〇	二·五〇〇	
	漢陽	共電	二八〇	九·〇〇〇	一·四〇〇	

(四)各局改裝之規定及機料價格

本部根據上述之記載,得各局改裝之規定如下:

附表三

局名		預定最大容量	現裝容量
首都	城內總局	五,〇〇〇	三,〇〇〇
	下關分局	二,〇〇〇	一,〇〇〇
	北分局	二,〇〇〇	一,〇〇〇
上海	南市總局	六,〇〇〇	三,〇〇〇
	閘北分局	五,〇〇〇	一,五〇〇
	浦東分局	一,〇〇〇	三〇〇
武漢	漢口總局	八,〇〇〇	四,〇〇〇
	漢口分局	八,〇〇〇	三,〇〇〇
	武昌分局	四,〇〇〇	一,五〇〇

上表所列南京之北分局,上海之浦東分局,均須添設.其南京原有之浦口,浦鎮,兩分局,及上海原有之江灣,吳淞,南翔三分局,則暫仍舊狀,均未列入.至武漢原有之漢陽分局,則撤銷之,該處電話由水線直達漢口.

京,滬,漢三處改裝自動電話所需機件及線料,均係美國自動電器公司承售.茲將各該處與天津,青島改裝之價格,以美金計算,比較如下:

地點	號 數	機件價目	線料價目	工程人員監工費	總 價	每線機件價目
首都	5,000	450,000	250,000	30,000	730,000	90.0
上海	4,800	241,000	303,000	26,000	570,000	*58.6
武漢	8,500	562,000	335,000	29,000	930,000	66.1
青島	3,000	146,000	——	10,000	156,000	48.7
天津	9,000	956,000	——	37,000	993,000	106.2

*上海機件價目,係以用戶話機作三,〇〇〇具計算.若以四,八〇〇號計算,則機件價目,應爲二八二,〇〇〇美金.

貨價之高低,原不僅根據材料,工價,及利益三項,運輸之距離,商家之競爭,亦復所關非淺.且自動電話機件之多少,隨當地通話次數與通話時間而定,故每線價目亦因之而不等.大概吾國人接聽電話,多假手僕役,各機關團體尤甚.其結果,則平均通話時間加增,即機件不能不加多.首都之機關,較其他各市爲多,故平均通話時間亦較多.每機價目之加多,此亦一因也.滬,漢二市電話訂購在後,各商家競爭較劇,遂使我政府得有廉價之機件,實堪欣慰.現在各市拆卸之共電式機件,擬加以修理,運裝較小都市.依照交通部既定計劃,逐步推進,則五年之內,全國應有五萬號自動電話之增添,各大都市長途電話之聯接,電訊交通,當有可觀之發展與便利也!

A DISCUSSION OF THE FUTURE ELECTRIC INDUSTRY IN CHINA

BY M. S. KWEI (桂銘新)

It is a well extablished fact that in countries where the fruits of materialistic civilization are being enjoyed, the extent of electrification of industries requiring electro-mechanical or electro-chemical applications serves as dependable barometer of the nation's economic progress. When viewed in this light, the United States, endowed by nature in her vast potential wealth, her people excelling in organization of intensive productive enterprises, hold an undisputed lead among civilized nations of the globe. The advanced nations of Europe, such as England, Germany, France and Switzerland, though having accomplished much in pioneering work in the practical application of scientific methods, have yet to attain a plane of economic enlightenment on a par with that of the United States.

Before entering into a discussion of the importance of an electric industry in China during the next few years, it would be well to examine the conditions in the United States. Inasmuch as China and United States possess many similarities as regards potential wealth, area of the land, climate and to a less extent the total population. The marked difference between the two countries stands out clearly when a study is made of the economic welfare of two peoples. In the United States mechanization of productive industries has almost completely replaced ancient manual methods. Both physical effort and skill have been transferred to machines under perfect human control. In the operation and regulation of these mechanical contrivances, electricity plays no small part.

That the extent of electrical application is indicative of a community's economic welfare may be gathered from a study of the following figures. The domestic consumption of electric energy has increased steadily from year to year in the United States, hand in hand with the increase in population and as progress is being made in selling more and more electric service to territories heretofore unexploited. The average cost of energy for domestic purposes is extremely moderate. This is accomplished through the intelligent employment of gigantic units of high thermal efficiencies and further economy in capital investment resulting from inter-connecting the various main sources of supply. The following tabulation shows the actual kilowatt-hour consumption per year per capita and the average unit cost of energy as supplied to domestic users in the principal cities:—

City	KWH. Consumption per year per capita 1925	1926	1927	Ave. Cost in Cents per KWH.
Boston	388	417	426	—
Buffalo	672	734	743	3.38
Philadelphia			408	6.58
Washington, D.C.			503	6.08
Chicago	499	516	537	5.12
Cleveland	500	525	547	—
Detroit	470	519	544	4.72
Milwaukee	408	456	469	5.23
Seattle	400	690	1286	—
Spokane	547	612	676	4.25

On examination of the above figures, one will be struck with the cheapness of the energy made available to the general public, though it entails great expense to pay for metering, billing and service to customers etc. The cost of energy for industrial loads is even lower; this may vary from a quarter to one half of the rate charged for domestic use. It will be noticed that the rates at Buffalo and other industrial centers such as Detroit, Chicago etc, are lowest, whereas in territories lacking in industrial development, notably Washington, D. C. and to some extent Philadelphia, the rate charged for domestic service is highest. The proximity of a large industrial load greatly influences the cost of energy to domestic users. However, Buffalo owes its extremely low rate to the extensive Niagara Power Development in its vicinity.

The electric industry in China, if it may be called such, cannot be said to have as yet emerged from its embryo stage. Excluding several haphazard electric lighting installations at Peking, Tientsin, Hankow, canton and other smaller cities on the East coast and in the south, which are but of a few hundred kilowatts capacity, China can hardly boast of a modern electric power station of large size. The Municipal Electric Power and Light Plant in Shanghai, though being the largest in the Orient, is unfortunately of foreign ownership. The total kilowatt-hour consumption for domestic use in the United States in the year 1928 was some 8,489,000* corresponding to a population of about 100,000,000 people. This gives an average domestic consumption of about 84.89 K.W.H. per year per capita. If in future, China should have her electric industries as fully developed as the United States, with her population of over 400,000,000 souls, the annual consumption will amount to the neighborhood of four times this quantity, or about 33,956,000,000 K.W.H. At an average cost of five cents per K.W.H. charged for electric service, this will be an annual electrical business of $1,697,800,000, which is for energy only exclusive of amount involved in appliancees and motor sales, and possible maintenance costs.

In spite of tremendous handicaps, the future of China's electric industry appears comparative bright. Her wealth in mineral resources, particularly in the form of coal, is not inferior to that of the United States. Having been supplied with an abundance of cheap labor, she will soon be able to produce electric energy at a cost equal to, if not less than in the latter country. When future hydro-electric power is consistered, the potential wealth of the upper reachest of the Yangtzee and other rivers cannot he neglected. Men who are responsible for the nation's economic reconstruction program should not overlook the fact that electric industrial development is as vital to China's progress as the building of railways, highways, radio and telegraph stations, air ports, water supply systems and flood control undertakings. The electric industry is the nucleus around which all other productive enterprises are centered. This cannot be more evident than to observe the intimate relationship which exists between electric power and the various large industries in the United States. A specific instance may be taken for comparison by noting that in the Detroit industrial load area, the predominating products being motor cars, 68.5 per cent of a total electric output is supplied to industrial uses, and only 31.5 per cent of this is employed for lighting and domestic appliances. The daily load on the Detroit Edison Company's electric system is in the neighborhood of 531,000 kilowatts during peaks. Some idea of the vast number of related industries which owe their existence to the availability of cheap electric power may be obtained from the following tabulation of important products:—

*"Electrical World," Dec. 21, 1929.

Motor vehicles, commercial and passenger
Motor vehicle bodies, and parts
Chemicals, chemical preparations
Manufacture of machinery
Printing and publishing
Meats and meat products
Engines, internal combustion
Foundry products
Drugs and pharmaceutical products
Brass goods
Bread and bakery products
Iron and steel
Machine tools
Stoves and heating appliances
Electric appliances and supplies
Paints and varnishes
Refrigeration Units, electric
Cigars, tobacco
Paper and paper products
Lumber, millwork and wood products
Beverages
Metal specialties
Dairy products
Leather Goods etc.

When an analysis of the load of the Niagara Falls Power district is made, and even greater portion of the total output is furnished to the electro-chemical and other industrial plants o` size.

In the heart of the Yangtzee valley, particularly the region near Hankow, Wuchang and Hanyang or the Nankng dstrict, electric power stations of larger dimensions should now be contemplated. Since larger installations are conducive to lower unit cost and increased reliability of service, greater effort should be made to combine all smaller independent installations into a large group. The inefficient and minor power stations now existing, which supply the textile industries, local lighting plants, government telegraph and radio stations and self-contained electic service to hotels and department stores etc. should be all pooled together to form one efficient and modern power supply system. As most electric lighting plants in China usually remain inoperative until the time for turning on lights, pooling of energy sources will result inevitably in advantages inherent to increased diversity of connected loads and a possible betterment of load factor. In many localities, it will enable the power plant to operate on a twenty-four hour basis, which will encourage use of industrial power as is the case in the United States.

The comparative backwardness of China's industrial development may be traced to many causes which are somewhat deeply felt. These may be named as follows: 1. natural conservatism of the people, 2. lack of capital, 3. scarcity of scientifically-minded or technically trained personnel in the higher governmental positions and lastly, 4. frequent disruption of peace from selfish political party leaders, with its attending evils that invariably reflect unfavorably upon the nation's economic welfare and trade. Allowing that a stable government can be maintained, the investment of people with capital may be easily induced to assist in a program of intensive reconstruction. It is the hope of the writer that the government should take concerted action in the actual encouragement of home industries and the utilization of electric energy wherever possible. As long as the leaders have steadfastly resolved to take this important step, there are numerous native industries in which electricity can be advantageously utilized. Its application will either bring about an increase in production or an improvement in the quality of the product. The existing native industries in which electricity is destined to play an important part are as follows:—

Paper mills
Textiles, cotton and silk products
Lumber mills
Tanning industry
Manufacture of leather goods, shoes, bags, etc.
Manufacture of vegetable products, bean cake etc.
Wood oils, Tung oil, ground nut oil, crushing and refining
Minerals, extraction and manufactures thereof.
Antimony; tin ingots
Medical products
Domestic manufacture of sundries for native consumption, etc.

Contrary to the opinion that is often entertained by the uninformed, the judicious application of electricity does not produce unemployment of China's laboring masses, but it will rather increase the lattitude of unskilled workers who havenot been capable of contributing their share to the community heretofore, except as human "beasts of burden." In short, the employment of electricity in industry general results in increase of productivity of the community and improvement of its economic welfare. It is now the time for the authorities who are in charge of the welfare of each community to follow a definite policy. Encouragement in the use of electricity and installation of medium sized but efficient power plants should be started. Improvement or enlargement of the existing stations in the principal cities will be another important step.

For a city the size of Nanking, with a future population of at least about 1,000,000 souls by the end of the next few years, a steamelectric power station of approximately 50,000 kilowatts may be contemplated granting that all important industries in the surrounding districts are thoroughly electrified. At a unit cost of $125 per kilowatt of intalled capacity, a plant of this size will cost no more than 6,250,000. A portion of this sum will be sufficient to commence the project. The fixed charges on such a plant at 15% will not be more than 936,000 per annum. Of course, the transmission and distribution equipments will cost a similar amount. The government may see fit to finance such a project and it can also take it under control. An alternative will be to make this a private enterprise, to be sponsored by leading citizens of the community, with the rates for energy under the supervision of the government. Other electric generating stations of similar dimensions may be established in a dozen or more other localities sufficiently close to independent fuel sources. The modernizing effect of these enterprises will soon react favorably either directly or indirectly upon the livelihood of the adjacent populace.

Before concluding, the writer wishes to point out that the application of electric machinery by China to her native industries will ultimately come despite all prevalent handicaps. In order to reach the same plane in industrial development as other nations have already attained, and to accomplish this in the minimum of time, the employment of electricity offers the only solution towards bridging the gap that has existed between her economic prosperity and that of the other industrial nations.

SPECIFICATIONS FOR ASPHALT CEMENT

BY H. K. CHOW (周 厚 坤)

Asphalt as a road building material is finding greater favor every day. In America, for the twelve large cities, up to Jan. 1, 1926, the proportions of different road surfaces are as follows:

Asphaltic Types	54.6%
Stone Block	14.9%
Macadam, Gravel & others	13.3%
Brick	12.8%
Portland Cement Concrete	2.7%
Wood Block	1.7%
	100.0%

Since then the proportion for Asphalt has steadily increased.

The advantages of an asphalt road surface are:

(1) Low abraision factor, and therefore dustless.

(2) Water-proofness, on account of both monolithic and water shedding properties.

(3) Ease of opening up, and repair, a factor of considerable importance in city pavements where public utilities make use of the streets.

(4) Its over-all economy, i.e., lowest combined first cost and maintenance charges.

In China, this material is also coming into general use in large cities like Shanghai, Hankow, Canton, Tsingtao, Tientsin, Peiping, Mukden and Dairen. Minor cities such as Hanchow, Chinkiang are also turning to it partly for appearance. Increasing interest is therefore manifested by all city engineers in this road material. Unfortunately, Asphalt is a comparatively new material, and its properties are not so well understood as other road materials. Books on the subject are both expensive and scarce. Much of the latest technical information are in the hands of few consulting road engineers, the refiners and the marketers of the product, and are not available to the public. The result is that most purchases are made on the basis of brand names, ill-advised recommendations, or in other haphazard unscientific way. Add to it the inexperience in its application and you get a real unsatisfactory job at the end of its use.

The writer now proposes to present a series of articles on asphalt and its application. The following is the general specification for Asphalt Cement for road work. It is sufficiently broad to cover most of the products on the market but narrow enough to exclude certain brands which are either especially susceptible to temperature changes (i.e. unstable under summer heat), or lack in ductility in freezing weather.

A. Specifications for Asphalt Cement

Impurities. — The asphaltic cement shall contain no water, decomposition products, granular particles or other impurities, and it shall be homogeneous.

Ash passing the 200-mesh screen shall not be considered an impurity but if greater than 1%, correction in gross weights shall be made to allow for the proper percentage of bitumen.

Specific Gravity. — The specifi gravity of the asphaltic cement shall not be less than 1.000 at 77°F.

Fixed Carbon. — The fixed carbon shall not be greater than 18%.

Solubility in Carbon Bisulphide. — The asphaltic cement shall be soluble to the extent of at least 99% in chemically pure cabon bisulphide at air temperature and based upon ash Free material.

Solubility in Carbon Tetrachloride. — The asphaltic cement shall be soluble to the extent of at least 98.5% in chemically pure carbon tetrachloride at air temperature and based upon ash free material.

Melting Point. — The melting point shall be greater than 128°F. and less than 160°F. (General Electric method).

Flash Point. — The flash point shall be not less than 400°F. by a closed test.

Penetration. — The asphaltic cement shall be of such consistency that at a temperature of 77°F a No. needle weighted with 100 grams acting 5 seconds shall not penetrate more than 9.0 nor less than 5.0 millimeters. Sellers should also be required to submit penetration figures at 32°F, 200 grams,

1 minute; at 115°F, 50 grams, 5 seconds. For asphaltic cement containing ash, 0.2 millimeter may be added for each 1.0% of ash to give the true penetration.

Loss by Volatilization. — The loss by volatilization shall not exceed 2%, and the penetration after such loss shall be more than 50% of the original penetration. The ductility after heating as above shall have been reduced not more than 20%, the value of the ductility in each case being the number of centimeters of elongation at the temperature at which the asphaltic cement has a penetration of 5.0 millimeters. The volatilization test shall be carried out essentially as follows:

Fifty grams of the asphaltic cement in a cylindrical vessel 55 millimeters in diameter and 35 millimaters high shall be placed in an electrically heated oven at a temperature of 325°F. and so maintained for period of 5 hours. The oven shall have one vent in the top I centimeter in diameter, and the bulb of the thermometer shall be placed adjacent the vessel containing the asphaltic cement.

Ductility. — When pulled vertically or horizontally by a motor at a uniform rate of 5 centimeters per minute in a bath of water, a cylinder of asphaltic cement 1 centimeter in diameter at a temperature at which its penetration is 5 millimeters shall be elongated to the extent of not less than 10 centimeters before breaking. The lowest temperature at which ductility becomes zero should also be submitted by the seller.

B. Epitome of the Purposes ot Certain Specifications for Asphaltic Cement

Impurities are a measure of the with which the asphaltic cement has been refined and handled. Usually the presence of impurities in large quantities indicates a poor grade of asphalt. Water as an impurity would act as diluent and would cause foaming in the kettle. Ash or mineral matter is not considered an impurity if it is a natual constituent of the asphaltic cement, but the mix and cementing value must be figured on the bitumen alone.

Specific Gravity of the asphaltic cement should be over 1.000. The advantage of a specific gravity more than 1.000 is that there will be less tendency for water to float out the asphaltic cement. The specific gravity is raised by the presence of mineral matter. Asphaltic oils of a penetration satisfactory for paving purposes always have a specific gravity greater than 1.000. Paraffin base oil and air-blown products usually have a specific gravity less than 1.000.

Fixed Carbon is a measure of the chemical constitution of an asphalt to some extent. Certain types of asphalt such as Mexican have naturally a constitution that yields a large amount of fixed carbon. Fixed carbon is largely used for determining the source and uniformity of an asphalt. Fixed carbon is not free carbon, but includes free carbon, which is ordinarily absent in asphaltic cements.

Solubility in Carbon Bisulphide is a measure of the purity of an asphaltic cement. The cementing value, other things being equal, is proportional to the carbon bisulphide solubility. Any carbonaceous material such as coal tar or pitch is detected by the carbon bisulphide solubility test.

Solubility in Carbon Tetrachloride is very nearly the same as the solubility in carbon bisulphide. It is claimed that an asphalt having more than 1½% difference in the solubility in carbon bisulphide and carbon tetrachloride has been subjected to excessive heat in refining.

Melting Point is the temperature at which the asphaltic cement will flow readily. The melting point desired is dependent upon the mixture. If the amount of fine dust in the mineral aggregate is low, the asphalt should have a melting point higher than the highest temperature to which the pavement is subjected.

Flash Point is a measure of the amount of volatile hydrocarbons that are present in the asphalt and its readiness to decompose by heat.

Penetration is a measure of the consistency of the asphaltic cement. It is merely a quick, convenient test for checking up numerous individual samples. The penetration is expressed in degrees and in accordance with the method of the American Society for Testing Materials, each degree representing 1-10 of a millimeter or 1-250 of an inch. The penetration, then, is the number of degrees that a No. 2 A.S.T.M. needle when weighted with 100 grams will pass vertically into the asphaltic concrete at a temperature of 77 F. (25 C.) in 5 seconds. The penetration to be desired will depend upon the climate, the nature of the traffic, the grading of the mineral particles, the amount of voids, the amount of compression attainable, the ductility and cementing strength of the asphaltic concrete and the amount of dust filler.

Loss by Volatilization is a measure of the amount of light hydrocarbons that are present in asphalt and is also a measure of the tendency of an asphalt to oxidize and to lose its ductility and penetration. Asphaltic cement which has no ductility after this volatilization test will not be satisfactory for paving purposes.

Ductility is the measure of the ability of an asphaltic cement to expand and contract without breaking or cracking. The same asphalt at a higher penetration should have a higher ductility, so all ductility tests should be based on a certain definite penetration regardless of the temperature, or should be based upon a temperature of 32 F. Ductility is also a measure of the cementing strengh.

Viscosity is a measure of ability of the asphaltic cement to impart plasticity and malleability.

Engineering News

World's Largest Arch Bridge is Nearly Finished — The largest arch bridge in the world, which British engineers are building over the harbor at Sydney, Australia, is nearing completion. This impressive piece of engineering spans a channel 1,670 feet wide and rises above water level to seventy-five feet higher than the summit of Saint Paul's Cathedral. The two ends are about to be joined and are now only inches apart. Each weighs 14,000 tons net. Despite this huge weight, they are within an inch of alignment and the width of the gap is exactly as calculated by the designer over two years ago.

After the two halves have been joined next week the work will begin of placing a floor on the bridge to carry four lines of railway, a fifty-seven feet roadway and two footways. The work is being done by Dorman Long and Company, engineers, of Middlesborough, and the cost will be nearly £6,000,000. Nearly 60,000 tons of steel, eighty per cent of which was made at Middlesborough is being used.

工業館改建工業博物院之芻議

著者：張可治

浙省近數年來，對于建設事業，異常努力，去夏又舉辦西湖博覽會，搜羅宏富，世界聞名，尤足使全浙人民，一新耳目。治以猥才，蒙李振吾館長任以工業館幹事之職，乃亦得躬與斯會之盛，誠幸事也。會期五月，各地來會者踵相接，觀膽參攷，其獲益何可計哉？顧盛會不常，瞬已閉幕，匪特人民，將失此良好提倡實業，啓發智識之機關，即會中宏壯之建築，與完美之設備，廢置亦殊可惜。故治不揣冒昧，建議當局，俟閉幕後，將工業館改建爲工業博物院，幸邀允可。惟工業博物院，與工程學實有莫大之關係，茲特摭抄原呈，投登本刊，以請敎于國內工程界諸位先進也。

世界愈文明，則事物愈複雜，而其歷史上所遺留之發明與發現亦愈多。但人類因受生活之逼迫，與環境之支配，往往不得不專精一藝，以圖存于現代分工合作之社會。故吾人之智力，每致受一種畸形之發展，而學校教育，則又限于地位，厄于經濟，敎授則祇憑書本，實驗則缺乏機會，故青年學子，每患以耳代目，尙空談而忽實際，是誠現代教育家所引爲深慮者也。惟其如是，故各國之政府，及文化機關，每亟亟于博物院，展覽會，博覽會及通俗演講會之設立，以圖灌輸充量之常識，並維持文化于不墜，制至善，意至良也。

顧博覽會之規模，每苦過于浩大，其所耗費又必過鉅，荷國家非遇有特殊之慶典不能辦也。展覽會之範圍較小，其費用亦輕，且無論工業，農業，敎育，藝術皆可隨時隨地，分別舉辦之。惜其會期短者只有數日，長者亦祇有一二月，其籌備期間，每苦短促，布置每欠周密，誠難收普遍之效也。通俗演講，正犯空談之弊，亦無足取。惟博物院，係帶永久之性質，其于物品之搜羅也，不必亟亟，而可以隨其經於能力之需充,與夫徵集機緣之良慳,以爲緩急.其于院內之陳列與布置也,則分門別類,旁徵博引,解剖之,烘托之,連貫之,尤以爲未足,則加以說明,補以圖案,助以模型,務期雅俗共賞,少長咸宜,而使其增長知識,于不知不覺之間.故在歐美各國城市,無論大小,莫不各有博物院也.

博物院有以自然科學名者,有以藝術或考古名者,有以軍事或工業名者,其名稱不同,其性質各異,惟其以啓發民智爲目的則一也.自然科學,上極洪荒,遠窺星辰日月,近察鳥獸蟲魚,草木花卉,地脈河流,使民衆認識人類在宇宙間所處之地位,並了解天人合作之奧旨,故各國以設立自然科學博物院者爲尤多.第二十世紀,乃應用科學昌明之時代,歐美各國之興盛與幸福,幾莫不以工業爲基礎.其機器之靈巧,交通之迅速,工程之偉大.與夫社會組織之複雜,實爲亘古所未有.故吾人處今之世,必須明瞭乎聲光電化之原理,與其在實用上無窮之變化,庶行事不致迂緩,思想不致落後.况夫科學之爲物,更非一成不變,則尤必賴乎人類之努力,以繼續發揚而光大之.然則提倡應用科學,灌輸工業常識,實爲今日之要圖,而不可忽視者也,故工業博物院尙焉.

工業博物院之歷史甚短,且各國之已設立者亦甚少,惟各國現已感覺人民對于工業常識需要之急切,而努力以從事之矣.工業博物院之範圍,極爲廣大,舉凡衣食住行所用之工具事物,與其由最初發明,以迄臻于實用之過程,及其所依爲根據之理論,莫不蒐集採納,以充陳列.至其所分門類,則係依其應用之程序,而爲先後,茲請將工業博物院應具之內容,約略表列于下:

(一) 採礦工程　內分礦層,礦井,礦室,礦用機械,保安設備,及礦石整理方法等類.

(二) 冶金工程　內分冶爐構造,冶廠設備,普通冶金,及冶鐵錬鋼等類.

(三) 金屬工作　內分金屬之熔,鑄,煅,拉,壓,軋,剪,鋸,鑽,車,銼,鉋,磨,拋,點,銲等各項工作,及金屬性質之測驗,與改良等類.

(四) 原動機 內分人力,風力,水力,潮力,浪力,蒸汽力,與氣力等原動機;蒸汽機,及分汽鍋,往復機關,汽渦輪,龍頭,機關等目;氣力機,又分爲熱空氣機關,瓦斯機關,油機關,提士機關,瓦斯渦輪等目.

(五) 代步工具 內分人力與馬力之車轎,自行車,汽車,街車,火車等類.

(六) 道路工程 內分上古道路,中古道路,近世道路等類,鑽山工程,地道工程,亦屬之.

(七) 軌道建築 內分平地軌道,爬山軌道,鐵索軌道等類,鐵路所用之信號,與保安設備,亦屬之.

(八) 橋梁 內分天然橋梁,木橋,石橋,索橋,鋼筋橋,鋼橋等類.

(九) 河海工程 內分測量器具,建築機械,水流控製方法,塘壩,閘提,運河等項工程,海岸信號等類.

(十) 造船 內分原人用船,帆船,舵船,蒸汽機船,汽油船,仗艦,潛艇,渡船,郵船,貨船,航海儀器,船塢構造及設備,救生設備,及船用機械等類.

(十一) 航空 內分天然飛行,人工飛行,飛機,飛船,氣體流動學,航空用具等類.

(十二) 物理 內分空間重量與時間之測量,固體力學,流體力學,波浪與震學動之原理,熱,光學,聲學,音樂等類.

(十三) 電學 內分靜電動電,電磁感應,電流,有線電,無線電等類.

(十四) 化學 內分古代化學,中古化學,近代化學,物質構造化學,工業製藥學,滋養料學,釀酒等類.

(十五) 工程材料 內分木,石,泥土,磚,瓦,水泥,石灰,紙張等材料之製造與應用,及最近關于材料研究之工作.

(十六)市政　內分房屋與街市,自原人時代以迄現代逐漸進化之狀況,工廠房屋,辦公房屋,私人住宅,室內光線,用水,取暖,避暑,溝渠排洩,公園設備等類.

(十七)陶瓷　內分磁器,陶器,玻璃,搪磁之應用製造及試驗等類.

(十八)電機工程　內分發電機,馬達,變壓器,蓄電器,測電表,電力輸送,中央發電廠之一切設備等類.

(十九)纖維工業　內分棉,麻,絲,毛,苧,草等原料品之應用,人造纖維品之製用紡紗,梭織,針織,氈毯,刺綉縫紉,印花染色等工作,與其所用之機械等類.

(二十)仿印工業　內分抄寫,印刷,電刻,雕鐫,照相,曬圖等項工作之歷史與現狀,及其所用之機械等類.

難者曰,工業博物院之需要固急矣,奈其規模過大,需費過鉅,何毋乃心有餘而力不足乎?答者曰,唯唯否否,語云,爲山者必因邱陵;又云,雖有滋基,不如待時,是蓋言成事者,必須有天時與地利也.今者,西湖博覽會籌備經年,會期數月,集天下之精華,攬湖山之全勝,非惟中國所罕有,實亦萬邦所希逢.若于閉會之後,選擇各館之出品,聚于一處,加以整理,添具說明,妥爲陳列,工業館之機械,特種陳列所之模型,農業館之農具,絲綢館之圖案,率皆出類拔萃,是則工業博物院內容備矣.工業館新建館屋,倚山臨湖,地位優越,建築宏固,面積廣大,且馬達線,蓄水池,自流井,水塔,自來水等裝置,甚爲完備,是則工業博物院之院址有矣.陳列品與院址既具,則開辦費可以不用矣.院務管理,只需一二幹事而已.是其經常費亦極省矣.至于陳列品之補充,固需經費,然當省欵支絀之時,盡可緩辦,是則臨時費亦不成問題矣.故工業館之改建工業博物院,實具莫大之便利,蓋其所費雖微,而其收效則大也.邦人士其不以斯言爲河漢耶?盍興乎來!

維司丁好司推进工廠實習概况

(Westinghouse Steam Turbine)

著者：陳宗漢

維司丁好司推进，以前在該公司之東壁工廠製造.歐戰期內，始於費城南十英哩，建立專廠；一九一八年，各種汽力機之製造，悉遷於此，今稱南督工廠.該廠前濱德拉瓦河，後臨數家鐵路幹線及費城與潵斯特間之電車線，交通甚便.廠址面積頗大，現有斜E字形廠，小推进廠，火車用黑油機廠，推进葉片廠，翻砂廠，發電廠，公事房各一棟，隙地尙多.全廠職工共約二千五百人，每週工作五日，每日九時三十六分.

斜E字形廠係四廠相連成左斜E字形.其三橫廠各稱一號二號三號廠，一號廠製造推进汽缸及凝汽器，二號廠製造大推进輪軸，三號廠銲接鋼鐵，惟後二者餘地甚多.其相當於E字直畫之廠，則爲全廠最忙冗最擁擠最重要之部分，推进汽缸及輪軸之嵌鑲葉片，離心力抽水機之裝配，各種汽箱汽門之配合，汽嘴之製造，輪軸之試平衡，大牙輪之裝配與修磨，以及一千馬工率左右以上乃至十萬馬工率之推进之全部裝配與荷重試驗，咸萃於此.

小推进廠係製造，裝配幷試驗大約一千馬工率以下之推进，大推进之零件亦多在此製造.火車用黑油機，原亦在此廠佔用一小部分，最近則另遷於就已廢之打鐵廠現改造之新廠.推进葉片廠有割片機各種衝壓機銑葉機；又葉機之製造，葉胚之鍛冶，亦在此.翻砂廠規模頗大，大推进之汽缸，與凝汽器之外殼，均在此鑄造.發電廠僅有小推进發電機三部，均一千五百啓羅瓦德上下，幷不常用，因廠中動力用電由費城電力公司購買；鍋爐則常供蒸汽，備試驗推进用.

3634

此廠實習學生有數種;一爲南費工廠自設之大學畢業生科,實習期限一年,分派在工廠或公事房各若干時;一爲東壁工廠派來之大學畢業生,爲期僅月半至二三月不等,大都均在工廠;一爲寒暑假短期實習之大學畢業生或肄業生.平均同時約共有學生二三十人,工作及待遇,大都相同.畢業生起初六月工資每小時五角,六月後增爲五角半,肄業生工資略少.廠中設有學務處專員一人,管理學生工作及課務.

學生工作部分,普通爲凝汽器,抽水機及大小推进之裝配部,推进輪軸平衡部,推平試驗部,黑油機部等.各種製造部與葉片廠,鮮有學生實習.惟學生如欲考察各種製造手續及方法,可向工頭取得允許,隨意往各部分觀察.至於工作,裝配部較忙,學生須實地操作,所得亦較多.平衡及試驗諸部則較閒.

學生入廠時,即由學務處發給各種推进,凝汽器與加煤機之樣本與說明書多種,并問題一本,內有關於大推进,小推平,凝汽器及加煤機之問題四種,每種有問題五六十至百餘不等.學生須在家參閱各種說明書,預備問題答案.但遇工作不忙時,可向工頭取得允許,持書往學務處閱看.如有疑難,可隨時請教學務專員或往廠中詢問工頭或監工員.每星期二及四兩日上午十時半至十二時,在學務處上課兩次,有工程部之工程師主講,學務專員亦在旁襄助.將每種問題依次令諸生解答,有欲討論或問難者,可儘量提出.每種問題須數次始能講完.每種畢後,舉行試驗一次,收卷記分,均甚認眞.又平時問題中有關於計算者,亦須算出交卷.每換一種問題,主講員亦改由該種機器之工程師擔任.四種問題共須二月半之時間,始能完畢.因學生時常調動,新陳代謝,故諸課周而復始.(加煤機之設計與製圖均在此廠,製造則另在他處.又牙輪舊爲諸課之一,今已廢置.聞不久將增黑油機課一種).

此廠平時無中國學生,今夏則同時有顧毓琇,馬德建及作者三人.顧君係暑假實習,馬君及作者則由東壁工廠派來.

此廠所製各種機器之種類及構造,非本文範圍所及,惟維司丁好司之推平與凝汽器較他家特異之點,請舉其一二.就推迸言,(一)調速器舊有遠心調速器及遠心與油力并用之調速器二種.近年來則後者已廢用而代以專用油力之調速器,現在每分鐘千八百轉之推迸,倘仍用遠心調速器,而每分鐘三千六百轉之推迸,則均用油力調速器.(二)輪軸與汽缸合口間之密封,他家大部用炭質塡料圈,維司丁好司則用輪擊水作墊.就凝汽器言,(一)銅管四周,均留隙地,蒸汽得以輻湊流入,可使凝水溫度幾與蒸汽相同.(二)凝水箱內有密封裝置,可除去凝水內所含空氣.此外特點及足供研究者尙多,如有專習此科,作者甚願與之通信討論.

『推迸』二字,係作者試譯,肯賜指敎或欲詢問取其『推』字有衝動之義,『迸』字有噴射之義,兩者均可附會於汽輪發動之原理.一如前人之譯蒸汽機爲『引擎』.究竟是否妥當,作者不敢武斷,甚望高明指敎.

中國國內蒸汽旋輪(透平)發電機之調查(補遺)

去年本刋四卷三號第509頁登有張延祥袁丕烈二君所著『中國國內蒸汽旋輪(透平)發電機之調査』一篇,該篇彙集各公司各洋行之記錄,排成表格,旣可利企業者參考觀察之用,又足供工程家研究比較之需,其有益焉,殊非淺鮮.故自發行後,不逾數月,該號工程,卽告售罄,可知社會人士,對於此種有價值之調査,均極注意.現由萬泰洋行(B. T. H.)交到補充表一紙,亟爲登入,以補闕焉.

THE BRITISH THOMSON-HOUSTON CO., LTD. ENGLAND

Name of Plant	Location	No. of units	Capacity each Unit	R.P.M.	Volts	Cycles	Year Installed	Notes
The Hongkong Electric Co.	Hongkong	2	1,500 K.W.	3000	6600	50	1919	High Pressure
,, ,, ,, ,,	,,	1	5,000 ,,	3000	6600	50	1924	,,
,, ,, ,, ,,	,,	1	10,000 ,,	3000	6600	50	1929	,,
The Taikoo Sugar Refining Co.	,,	3	750 ,,	3000	440	50	1920 to	Extraction
,, ,, ,, ,, ,,	,,	2	750 ,,	3000	440	50	1925	Back Pressure
The China Light & Power Co. (1918) Ltd. Kowloon	,,	2	750 ,,	3600	2200	60	1922	High Pressure
,, ,, ,,	,,	1	3,000 ,,	3600	2200	60	1924	,,
,, ,, ,,	,,	2	5,000 ,,	3000	6600	50	1930	,,
Nanchang Electric Light Co.	Nanchang	1	750 ,,	3600	2300	60	1929	,,
Nanking Electric Light Co.	Nanking	1	750 ,,	3600	2300	60	1929	,,
Sino British Coal Mining Co. (Mentoukou Mines)	Nr. Peiping	2	750 ,,	3000	5250	50	1928	,,
Wuchang Electric Co.	Wuchang	2	800 ,,	2400	2300	40	——	,,
Kiao Ao Electric Co.	Tsingtao	1	5,000 ,,	3000	3300	50	1930	,,
Chekiang Provincial Govt. (Hangchow Power Plant)	Hangchow	2	7,500 ,,	3000	14000	50	1930-1	,,
Total		23	60,850 K.W.					

DIRECT GENERATION OF ALTERNATING CURRENT AT HIGH VOLTAGES*

By the HON. SIR CHARLES A. PARSONS, O.M., K.C.B., F.R.S., Honorary Member, and J. ROSEN, Member

(1) PRELIMINARY CONSIDERATIONS OF THE FUNCTION OF THE TRANSFORMER.

Engineering history contains several examples of a complete change of procedure brought about by developments in a given field of work.

An example in the field of mechanical engineering is the use of step-up gearing in the early days of marine propulsion by screw propellers—a practice which is to-day reversed, for modern marine steam turbines and some of the latest marine Diesel engines are now connected to their propeller shafts through speed-reduction gears. Since an intermediate period in the development of the triple-expansion engine no gearing has been used.

Just as the mechanical gear forms a link between the prime mover and the driven machine, so in the field of electric power generation and utilization by high-tension alternating currents the transformer has for many years been a necessary link between generator and network, and between network and driven apparatus.

About 1890 the transmission voltage, using underground cables, and been raised to 10,000 by Ferranti and Partridge, by whom much pioneer work was done; and alternators of the same voltage were installed in the Deptford power station. These alternators were of lowspeed design and had revolving armatures, in spite of the high generating voltage. They were probably unique in this respect.

The transmission voltage was transformed down in two steps—10,000/2,500, and then 2,500/100. Here we have the complete antithesis of present practice, where generation at 6,600 volts or 11,000 volts is usual, these voltages being stepped up to 22 k V or 33 k V for the distribution network immediately surrounding the power station, and possibly again stepped up to 66 k V or 132 k V for the grid system of intercommunication of power networks.

---

*Reprinted with the permission of the Institution of Electrical Engineers, England.

In view of these considerations, there is ample precedent for a reversal or change of procedure, if such change is in the interests of modern development. The use of high-voltage alternators is, in fact, proposed by the authors, and with it the abolition of step-up transformers for some part of the power to be distributed.

(2) THE GROWTH OF CONDITIONS FAVOURABLE TO THE INTRODUCTION OF THE HIGH-VOLTAGE ALTERNATOR

Apart from the work of Ferranti, pioneer development in England and America at voltages above 6,000 was not encouraged by the engineers responsible for design and operation of power plants.

In America, in 1899, the 5,000-h.p. water turboalternators at Niagara were built to generate 2-phase current at 5,000 volts. The transmission voltage was 11 000 volts. The alternators had an outer revolving field with a central stationary armature, and were widely known as the "umbrella" type.

In England, the early steam-turbine-driven alternators were designed with a revolving armature, which usually consisted of a smooth core having the windings laid over it and secured by binding wire. This type with a smooth core was used up to 2,000 volts, and up to 4,000 volts with a tunnel winding. They were single-phase machines.

It was found that at the higher voltages, especially where there was more than one phase, the difficulties in manufacture and insulation were great, and the revolving armature was discarded in favour of the revolving field.

Turbo-type revolving-field alternators of 1,500 kW at 1,500 r.p.m., generating at 11,000 volts, were built in 1905, and are still in operation.

While mica insulation was fitted to the end-windings, no mica was used in the conductor-insulating tubes, which were made of varnished fibrous materials.

Many alternators were built subsequently with improved constructional deails, at voltages up to 13,000 volts. It was not until 1921 that the authors' attention was again drawn to the possibilities of highvoltage generators, at a discussion in Newcastle-on-Tyne with an engineer who was responsible for a power supply system where the greater part of the energy was transmitted in bulk at 22 kV to a point some distance from the power station,

TABLE I.

Comparison of Stator Conductors for 1.500-r.p.m., 50-cycle Alternators

Voltage		6 600	11 000	13 400	18 000	22 000	33 000
50,000 kW at 0.8 power factor	Current, in amperes	5,470	3,280	2,700	2,010	1,640	1,095
	Total number of conductors	48	84	96	126	162	240
	Conductors per pole per phase	4	7	8	10½	13½	20
	Number of parallel circuits	4 (2 slots in parallel)	2	2	1	1	1
75,000 kW at 0.8 power factor	Current, in amperes	8,200	4,920	4,040	3,010	2,460	1,640
	Total number of conductors	30	48	60	84	96	144
	Conductors per pole per phase	2½	4	5	7	8	12
	Number of parallel circuits	8 (4 slots in parallel)	4 (2 slots in parallel)	2 (2 slots in parallel)	2	2	1
10,0000 kW at 0.8 power factor	Current, in amperes	10,950	6,560	5,400	4,020	3,280	2,190
	Total number of conductors	24	42	48	60	84	120
	Conductors per pole per phase	2	3½	4	5	7	10
	Number of parallel circuits	8 (4 slots in parallel)	4 (2 slots in parallel)	4 (2 slots in parallel)	2 (2 slots in parallel)	2	1

the generating pressure being 11,000 volts. He expressed a wish that a reliable generator might be designed capable of generating direct at the higher voltage. This change of attitude on the part of a supply engineer led the authors to believe that the problem might be a general one; it came at a time when the authors' thoughts were turning to the design of the largest units, which have now materialized—that is, 50,000- and 100,000-kW units.

With the increase in size of generating unit, the greater were the advantages to be gained in the design of alternators by direct generation at 22 kV or 33 kV, and as there were also advantages to be gained in the power station it was felt desirable to make investigations.

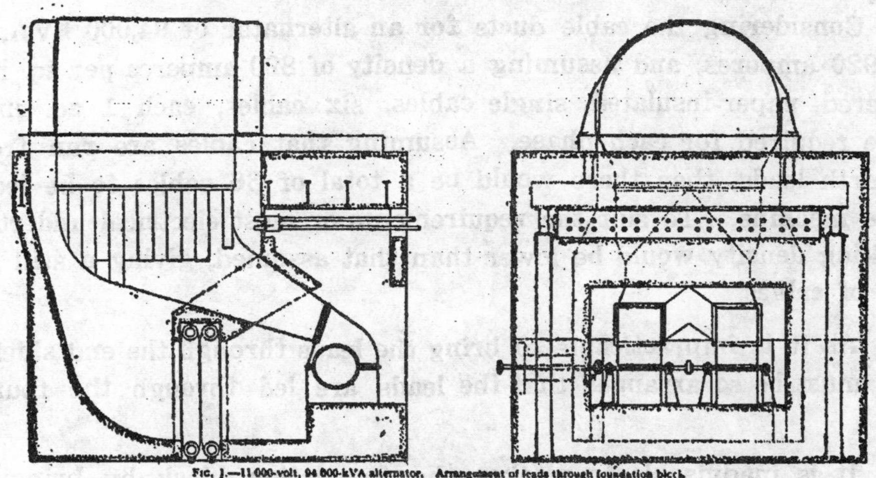

FIG. 1.—11 000-volt, 94 000-kVA alternator. Arrangement of leads through foundation block.

In order to illustrate the difficulties in the design of the largest units, and the lack of flexibility at lower voltages, a list (Table 1) of the current values and approximate number of conductors corresponding to various voltages is given for alternators of 50,000, 75,000 and 100,000 kW capacity.

It is clear that for large units at the lower voltages, to keep the current per conductor low, conductors in two or more adjacent slots would be connected in parallel, thus presenting difficulties in winding, in order that the resultant voltages of each parallel path should be equal in magnitude and also in phase.

The number of conductors in series per phase, upon which the designer can ring the changes, may be as low as 10 or 12. This restriction may impose such limits on the design that an alternator to meet the specified conditions may differ by as much as 20 per cent from the most favourable proportions, with consequent lowering of efficiency.

(3) THE ADVANTAGES OF HIGH-VOLTAGE GENERATION IN THE DESIGN OF THE LARGE ALTERNATOR

On investigation, it was found that in many lay-outs the cost of the leads for a high-voltage machine would be much less and that the cost of the copper busbars and switchgear would also be lower. In fact, in the largest units, one switch only would be employed to carry and break the currents when, at lower voltages, two woul be employed to carry and break the currents when, at lower voltages, two would be necessary.

Considering the cable ducts for an alternator of 94,000 kVA, 11,000 volts, 4,920 amperes, and assuming a density of 820 amperes per sq. in. with lead-covered, paper-insulated, single cables, six cables, each 1 sq. in. area, would be required for each phase. Assuming that cables are run from the stator earth leads, then there would be a total of 36 cables to be led away from the machine. To suit the requirements or most electrical undertakings, the working density would be lower than that assumed, giving a still greater number of cables.

As it is impracticable to bring the leads through the end shields, the winding must be so arranged that the leads are led through the foundation block.

It is inadvisable to weaken the foundation block by bringing the leads through the sides, since the latter form the piers supporting the machine.

The leads cannot conveniently be taken down vertically on account of the ventilation system, and in practice it is found that the best arrangement is to form a cable tunnel in the concrete, running longitudinally from the machine terminals to the exciter end of the block. It will be appreciated that the tunnel has to be of sufficient size for accessibility in fitting the various parts in position.

For the 94,000-kVA, 11,000-volt alternator, much space is required to accommodate the cables and sealing ends mounted below the alternator terminals. Fig. 1 shows the arrangement for this machine. It is seen that the tunnel has to be made nearly the full width of the foundation block, and difficulty is experienced with the girders which reinforce the concrete.

With so many cables grouped together, the maximum output would not be obtained from them.

Most of these difficulties are overcome when an alternator of higher voltage is employed. Considering a pressure of 33 kV, the current is reduced from 4,920 amperes to 1,640 amperes; it will be seen at once that against 6 or 7 cables for the lower voltage machine there are only 2 cables per phase, and these can be loaded up to their full capacity.

A typical arrangement of the leads for the highvoltage machine is shown in Fig. 2. The leads and sealing ends are readily accommodated in the foundation block, the width of the tunnel being reduced from 14 ft. to 8 ft. 6 in., and the machine placed on a firmer foundation. The sizes of 33-kV cable-sealing bells are only slightly larger than those for 11,000 volts, and little extra space is therefore required for them.

Regarding the cost of cables, those for the 33-kV machine cost £20 per yard run, and for the 11,000-volt machine £37 per yard run for the above output.

As an example of the necessity for reducing the current in large units, the General Electric Co. of Schenectady, in designing the 208,000-kW unit for the State Line station of the State Line Generating Co., near Hammond, Ind., found it necessary to increase the alternator voltage for this purpose. There are three main units running at 1,800 r.p.m., comprising a high-pressure turbine driving a 76,000-kW alternator (0.85 power factor), and two low-pressure turbines, each driving a 62,000-kw alternator.

The main alternators were first designed for a voltage of 18 kV instead of the standard voltage of 13,400 volts.

It was later found necessary to raise the voltage to 22 kV.*

The transmission voltages are 33, 66 and 132 kV; it is apparent that the generating voltage has been increased due to the difficulties which arise with the heavy currents.

The authors have been repeatedly reminded that if they can show sound reasons for generating at a higher voltage, then 66 kV, a voltage recognized as one of the standard transmission voltages, would be the most advantageous.

While 66 kV is a generator voltage which may be recalled that there are many conditions under which 22 kV and 33 kV would be considered economical, more especially where the power station is at the centre of, or at only a few miles' distance from, the main consumption of the power, and where underground cables may be employed for transmission. Several such plants may be quoted where the conditions for generating at 33 kV might be considered favourable, for example, Baton, Clye Valley, the proposed site at Carrington, and others. All have surrounding areas which can economically consume power at 33 kV, and enable a saving to be made by direct generation and switching at 33 kV.

**General Electric Review*, 1928, vol. 31, p. 7, and 1927, vol. 30, p. 5.

(4) Financial Savings Effected by Direct Generation at High Voltages.

The authors have worked out the savings which may be obtained with the use of such high-voltage alternators.

Where the generator can be constructed to supply the transmission system directly, without the intervention of step-up transformers, the whole sum, representing the cost of the transformers and their housing, and the capitalized value of their losses, less the additional cost of the high-voltage generator, can be saved.

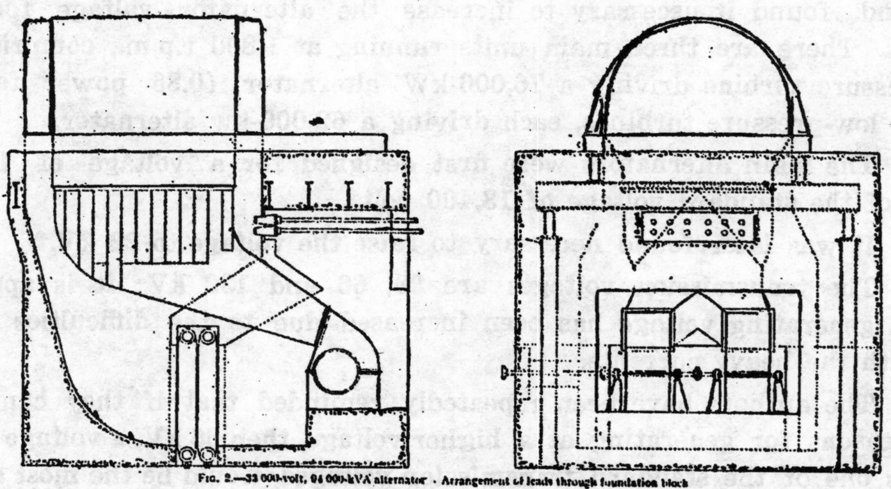

Fig. 2.—33 000-volt, 94 000-kVA alternator : Arrangement of leads through foundation block

Where the generator voltage is stepped up at once to the transmission voltage, there is the advantage to power station designers of freedom of choice in placing the switchgar either on the l.t. or h.t. sides of the transformers.

The various items in which there is a direct financial saving by the use of high-voltage generators may be summarized under the following heading:—

(*a*) Cables.
(*b*) Transformers.
(*c*) Transformer losses.
(*d*) Transformer cooling equipment.
(*e*) Buildings.
(*f*) Switchgear.

(*a*) *Cables.*—As already stated, the cost of the cables between generators and switchgear or transformers increases rapidly with the current and size of unit. The expense of laying the heavy cables will also naturally be greater.

(*b*) *Transformers.*—These will usually be 3-phase units up to 20,000 or 30,000 kVA, and bands of three singlephase units for larger outputs. Artificial cooling, either by water or air, will be required above 15,000 kVA and is included in the prices given below. The figures apply to transformers stepping up the generator voltage to 33 kV.

(*c*) *Transformer losses.*—The capitalized value of transformer losses represents a considerable part of the total capital outlay, and is so much greater than the extra cost of the high-tension generator that the saving of this expenditure is alone sufficient to justify the use of higher generating voltages, when transformers can be eliminated. The correct basis on which to charge the transformer losses is not always easy to determine, and depends on the size and arrangement of the plant and system and the operating conditions, especially the load factor. It may be assumed that the load factor of a large generating unit in a modern power station supplying an extensive network is somewhere between 50 and 70 per cent.

The annual cost of the losses can be obtained by reckoning the actual generating cost of the losses thus obtained, to which must be added a fixed charge per kW of maximum demand, representing the proportion of the fixed charges of the installation which is chargeable at the point at which the transformers are situated; for step-up transformers at the generating station this charge should be lower than for distant distribution transformers, where the losses are supplied through the transmission system.

In order to capitalize the annual cost of the transformer losses, a rate of about 10 per cent per annum may be taken, to include interest on capital, depreciation, obsolescence and insurance.

(*d*) *Transformer cooling equipment.*—The cost of running the transformer cooling-plant motors (oil pump and fan) must be added to the cost of fixed losses of the transformer, as the cooler is usually in serivce all the time the transformer is alive. Where water-cooling is used the cost of water must be considered, although it is only in exceptional cases that this charge is appreciable.

(e) *Buildings.*—Transformer banks are now usually installed out-of-doors, but a certain expenditure is incurred for foundations and accessory structures. There may also be some saving in the construction of cable ducts and switchgear housing.

(f) *Switchgear.*—The cost of switchgear increases rapidly when the current exceeds certain values, and may become excessively high for very large units at low voltage. The cost of maintenance, also, will be higher for the very heavy gear required for large currents.

As a representative plant, a unit of 75,000 kW will be taken, giving its full output at a power factor of 0.8, the equivalent output thus being 94, 000 kVA. It will be assumed that a generator of this output is wound for 11, 000 or 33,000 volts, and has to supply a network at 33,000 volts, so that a transformer bank will be necessary in the first case. It is assumed that the switchgear is on the 33-kV side. The following comparison can then be made:—

	"A"	"B"
Generator output, kVA	94,000	94,000
Generator voltage	11,000	33,000
Approximate additional cost of generat	——	£10,000
Cost of cables (100 yards' run)	£3,675	£2,000
Cost of transformers	£19,000	——
Transformer losses.		
Fixed losses, kW	250	——
Variable loss at full load, kW	400	——
Variable loss at 60 per cent load factor, kW	144	——
Transformer cooling equipment.		
Input to motors, kW	50	——
Reduction in cost of buildings, etc., and transformer foundations	—	£500
Annual cost of fixed losses [at £1 per kW (0. 1d. per unit)]	£1,400	——
Annual cost of variable losses	£ 925	——
Capitalized cost of transformer losses at 10 per cent	£23,250	——

The 33-kV machine thus shows a capital saving of £44,425, from which the extra cost of the former must be deducted, leaving a net saving of of £34,400.

Where the generator voltage is stepped up to 66 kV it might be possible to take advantage of the low step--up ratio and effect a saving by the use of auto-transformers.

	"A"	"B"
Generator voltage	11,000	33,000
Cost of cables (100 yards' run)	£3,675	£2,000
Cost of auto-transformers	£16,000	£9,750
Capitalized value of losses	£20,000	£12,500
Reduction in building costs	——	£1,000
Cost of switchgear	£9,000	£6,200

The total capital saving on the 33,000-volt generator is then £19,225, less £10,000, or a net saving of about £9,000.

The above figures are approximate, but are submitted as representative of the conditions prevailing at the date when the paper was written.

(5) Historical Survey of Past Alternator Construction for Direct Generation at 30kV

The use of a generating pressure as high as 30 kV is not in itself new, since Porf. Mengarini installed two 5,200-kVA, 30-kV allernators, running at 450 r.p.m. and generating 3-phase current at 45 cycles per sec. in the hydro-electric power station of the Societa Anglo Romana at Subiaco, on the upper reaches of the River Aniene. The power was transmitted to Rome, a distance of 34 miles.

The engineers of the Ganz Co., Ltd., constructed these 30-kV alternators in 1905, in addition to others for service in Italy.

Credit must be given to the engineers for this early pioneer work, and the success of these plants shows a thorough understanding of the art of insulation.

The machines have not been repeated in recent times. The amount of power transmitted over long distances from hydro-electric stations has very much increased, and the transmission pressures now employed in Italy have been increased to 100 kV and over. As it is unusual to have a large demand adjacent to a hydroelectric power station situated in the hills, the pressure of 30 kV has fallen into disuse, and the generating voltage was reduced to a lower figure suitable for the design of the moderately large electrical units employed.

The methods by which the engineers succeeded in constructing several successful 30-kV alternators as far back as 1905 are well worthly of study. The precautions which they recommended are now essential in electrically high-stressed materials such as are used in underground cables, etc.

From experience, it was found that the temperatures at which the alternators operated had to be kept at a moderate figure. Any difficuties that were experienced were traced to charring of the insulation. These difficulties emphasized the importance of using mica between turns, where the potentials were low, as well as between phases and to earth.

No attempt was made to grade the conductor insulation, but micanite was used throughout, and attempts were made thoroughly to impregnate the insulation and to expel the air.

The stator end-windings were not clamped, although the plants feeding the overhead transmission lines must have been subject to heavy short-circuits, surges, etc.

The necessary wide spacing of the end-windings, due to the high voltages, with large distances between phases and to earth, no doubt accounts for the remarkable freedom from mechanical failures or movement of the windings.

The authors take the opportunity of mentioning here this explanation of the lower mechanical stresses and forces in the end-windings, as a natural criticism has been levelled at a construction which removes the transformer, which, in the past, has acted as a buffer between the system and the alternator.

The forces on the end-windings are, in fact, much reduced, but this problem is dealt with later in the paper.

(6) THE DESIGN OF THE 33,000-VOLT, 25,000-KW, 3,000-R.PM. ALTERNATOR NOW INSTALLED AT BRIMSDOWN POWER STATION, NORTH LONDON.

The authors' experience as far back as 1905 led them to believe that the voltage with ordinary design could be much increased, but not sufficiently to keep pace with modern developments in 1921. Some better and simpler solution had to be sought, and, in view of these considerations, the authors directed their attention to the design of a high-voltage winding for incorporation in the largest alternators. Several designs were prepared.

In all investigations their efforts were principally directed towards the use of recognized standard insulations, such as micanite, without subjecting the materials to greater electrical stresses or employing greater thicknesses than those which had already proved satisfactory over a period of years.

After considering different schemes, including the grading of the quality and thickness of the insulation, it occurred to the authors that a concentric type of core conductor, of which knowledge was already available

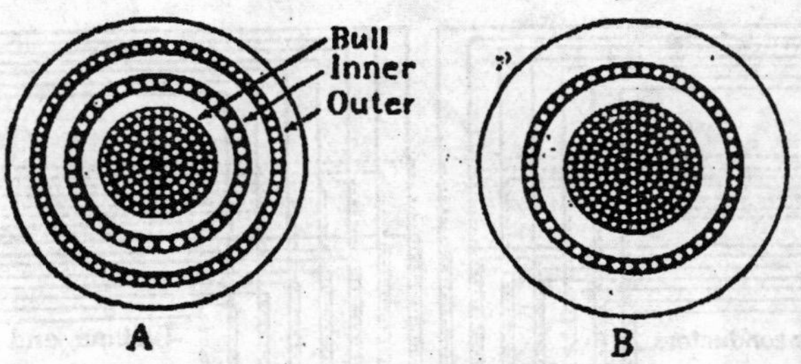

FIG. 3.—Sections through conductor bars.

through its application in other directions, might be adopted. By incorporating this type of conductor it became possible to prepare designs with greatly increased phase voltage without increasing the voltage gradient across the winding insulation.

The concentric conductor, which is the basis of the design and which is to be described, appears to afford a simple solution of the problem. By its use an alternator can be so wound as to distribute the dielectric stress and to lower its mean value at that part of the machine where there is limited area, and the maximum of heat generation at the regions adjacent to the stator bore. In the designs of alternators for voltages of 33 kV and 44 kV between phases, there is sufficient margin to permit insulated conductor bars to be used with thicknesses of insulation not exceeding those of whcih experience has proved satisfactory.

A section through one conductor bar is illustrated at "A" in Fig. 3 and resembles an ordinary concentric cable, with the exception that the insulation between conductors is of micanite.

There are three conductors per slot, nested one within the other, the conductors being wound in such a manner that the voltage is gradually stepped down from the innermost conductor. This formation of conductor is also very strong mechanically—a distinct advantage where the conductor projects beyond the stator core for coupling to the end-connections. For ease in description, the respective conductors in each slot are referred to as the "bull," "inner" and "outer" (see Fig. 3).

The "bull" conductors of each phase are connected in series, and are then connected to the surrounding "inner" conductors which are again connected in series and finally connected to the "outer" conductors, which are

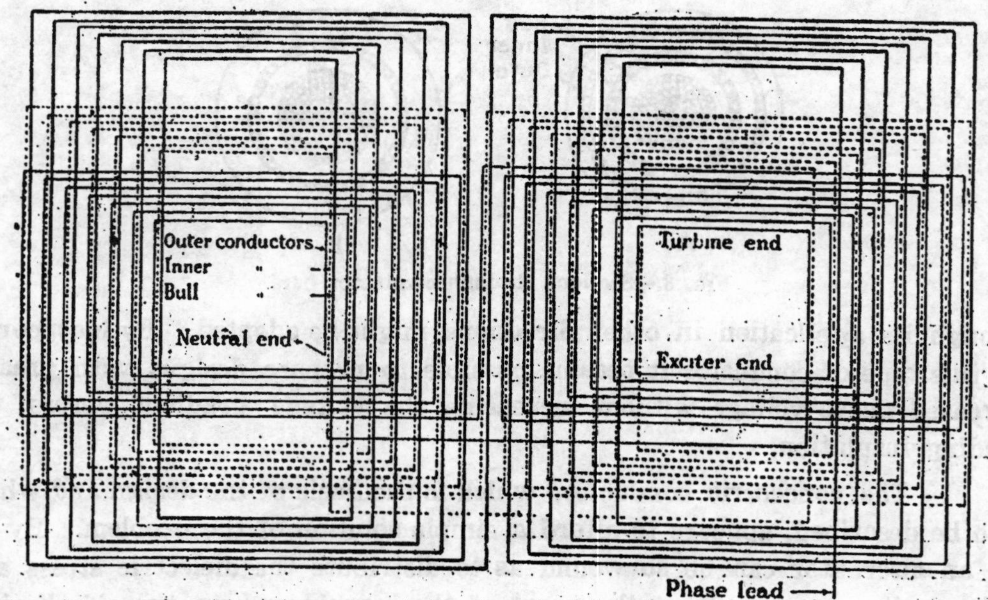

FIG. 4.—Stator winding diagram (one phase only) of a 3-phase, 2-pole winding with 90 slots and 264 conductors, 3-core concentric conductor.

starred to the ends of corresponding cnductors of the remaining two phases, and then connected to earth. A diagram of connections is given in Fig. 4, in which the method of winding is clearly indicated.

Fig. 5 is a vector diagram from which, in the initial stages of design, the voltage difference between conductors in the same or different phases was readily obtained.

By numbering the conductors and using a straightedge on the diagram, it is possible to trace the potentials at the different points round the whole of the windings.

The diagram of connections shown in Fig. 4 is for a 3-phase, 2-pole alternator having 90 slots and three conductors per slot.

The winding of each phase has 88 conductors distributed between 30 slots, the voltage generated per conductor being 217. The phase voltage and the maximum voltage to earth is 19,080.

The "bull" conductor potentials range from 19,080 volts to 13,000 volts to 6 500 volts, and the "outers" from 6,500 volts to zero. It is clear that with such a design there is a substantially constant potential difference of 6,500 volts between the conductors in any one slot, and a maximum voltage from the conductor to eath of 6,500 volts. Such voltages are moderate and are readily dealt with.

The conductors are arranged in three rows as shown in Fig. 6, such an arrangement being found specially suitable for a high-tension machine. By staggering the conductor slots the flux density in the stator teeth is kept uniform, and it was found possible with this design, instead of providing an elongated conductor, as for a low-voltage machine, to use the round form above described and, at the same time, obtain increased internal reactance.

By adopting this arrangement it is not necessary to bend the core conductors where they project from the stator, in order to provide a reasonable leakage gap between the end-windings and the rotor end-caps.

The staggering of the slots gives a uniform distribution of winding, and in effect has the advantages of the smooth-core armature, without the disadvantages of an unduly increased air-gap. The excitation energy is therefore retained at very reasonable figure.

It will also be seen from Fig. 6 that the conductor ends are accessible, giving greater space for sweating and insulating the joints.

Several interesting problems were met in the manufacture of the conductors, but it is unnecessary in this paper to describe the mode of manufacture. It may be sufficient to mention that the "bull" conductor was made in the same manner as an ordinary cable, and varnish-impregnated *in vacuo.* The cable was cut into the requisite lengths, and alternate layers of mica and insulated copper strands were applied. After the application of each thickness of insulation the conductors were re-impregnated *in vacuo.* In no part of the manufacture were the conductors in any way bent.

From Fig. 7 it is seen that two slots per phase contain only two conductors instead of three; a detail of the conductor is shown at "B" in Fig. 3. The conductors of the highest potential are not carried to the slots adjacent to another phase. This increases the distance between regions of maximum potential, and so minimizes the electrical stresses. This reduces to 28 000 volts the voltage between adjacent conductors projecting from the core.

(*To be Continued*)

會針加價啟事

本會會針,向售陸元,嗣因前定之數,近已告罄,亟宜添備,以應會員之需.惟廠方以金貴銀賤,懇請加價,而維血本.當經第五十八次董執聯席會議議决:每枚實售洋八元,郵費照舊.再本會會證業由薛董事次莘担任接洽印刷想不日可以告竣每位會員實贈一紙,合併附誌於此!

開立方捷法

原因　下次之三方廉,如過多位,計算較繁,余於民元研究得一捷法,茲特錄出,乞教明達.

捷法

上次之 { 三長廉面 (A)；一隅面 (B)；三方廉面,三長廉面,一隅面和 (C)；一隅面 (B) } 之總,再乘以 100

即爲下次之三方廉,(D)

即(A+B+C+B)×100=D

劉增甕識於山東建設廳

民19.6.3.

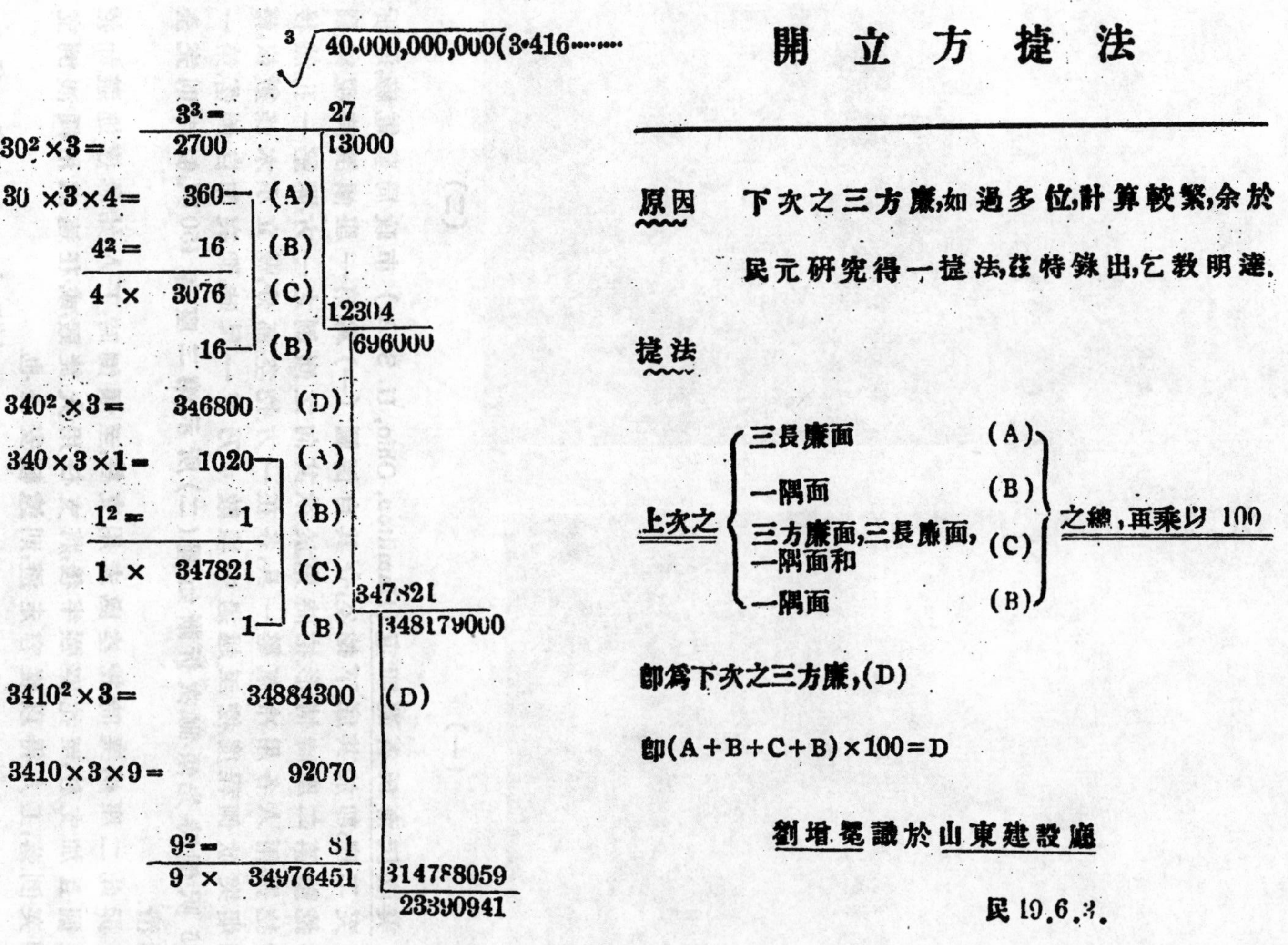

（一）　　（二）

美國沃海沃省漢明屯(Hamilton, Ohio, U. S. A.)市政局發電廠,近拆毀磚烟突二座,即於其底下挖空,令其自倒.圖(一)爲第一座傾倒時現象.圖(二)爲挖鑿第二座時情形.挖時先於欲其向之傾倒之一方,開鑿一孔,然後向兩旁挖掘,兩人各用冷氣鑿一具,各在一方.挖空部分,豎立小木條數根,藉以表測烟突欹側程度.挖至越過中線數寸乃至一尺時,烟突即自倒落.第一座高165呎倒時先傾側,次折斷如圖(二)所示.第二座高150呎,倒時則完全向下崩潰.

拆此二座烟突時,挖空處未用支柱,至傾倒時,工人始急速逃避.如挖掘時且掘且加支柱,至挖空强牢後,將支柱用火焚毀,或在遠處用繩洩倒支柱,令烟突倒落,工人得以從容走避,則遠較安全也.

十八,十二,十二陳宗漢記.

美洲分會附刊

第一期

中國工程學會美洲分會編輯出版委員會編輯
委員會主席　顧毓琇
書記　黃輝

目錄

本會發軔於美洲,後遂移植於本國.有此緣源,故現下該處雖屬分會,而一種迸勃之氣,仍不衰弱,實爲各分會之冠.該會顧君,去年在滬時,曾以專闢一欄見商,本刊自極歡迎,惟以篇幅太多,祗可分期登出,不必限於某期.又凡事倡議易,持久難,深望該會能將佳稿,陸續見賜,則無任感荷矣.

周厚坤誌

引言

美洲分會編輯出版委員會,今年起決定出版刊物,實是了幾年來未償之宿願.分會方面早就感到需要一個『自己的園地』來耕種,國內方面也久已殷切的希望美洲方面——中國工程學會發祥地——能不時給點新的Information,刊物,到現在方應運而出,實在已經延遲了.

因爲分會方面經濟不大充裕,刊物只得暫時『借地造屋』.在總會『工程』上附一欄,由分會編輯稿件.這個『借地造屋』辦法,承蒙總會編輯委員會周厚坤先生答應,並給我們很多鼓勵,十分覺得感激.

其次委員會要感謝這次投稿諸君,並且希望會員指教.

MODERN DEVELOPMENTS IN RAILWAY SIGNALING

By S. V. Chang (張錫蕃)

Introduction

In technical sense railway signaling means the art of regulating and controlling railway traffic by means of signals. The first problem confronted by the railroad was not one involving train rights or train interference. It was mainly a problem of mechanical design, motive power, rolling stock and tracks. Train rights were introduced when there was more than one train on a division of road at the same time. As soon as the number of trains increased and also did the length of runs, the train schedule, which gave each train definite rights, became inadequate. Now the Morse telegraph which led the way to the telegraph train dispatching system gave the answer. Train orders were used, which were in reality telegrams copied on prescribed forms and delivered to the parties concerned. This enabled the centralized control of traffic by a dispatcher to a certain degree. By this method the dispatcher could nullify the train schedule, or he could establish new temporary schedules. Trains classified by right by class and right by direction has lessened the work of the dispatcher, by enabling the train crews to dispatch themselves to a great extent. Then the dispatching trains by signal indication as used in automatic block signal system and interlocking plants has given a new era in modern railway signaling. These systems have greatly the safety of operation in that it checks the dispatching and protects against other hazards which are not detected by any other expedient.

Signaling is primarily divided into two main classes, namely, interlocking and block signaling. Interlocking is of English origin, the first interlocking was made in England in 1859, it was applied to the stirrup frame. The first patents of interlocking plants for manually operated devices were granted between 1856 and 1867. The first installation in America was made in 1874. The first electric interlocking plant embodying the dynamic indication was installed in 1901. The early history of block signaling runs about the same as interlocking as it was a development of the English practice and was first used in England in 1842. In 1863 it was first employed near Philadelphia in America. Wire circuit automatic block signals were installed in New England in 1871. Closed track circuit was invented by William Robinson in 1872 and was first introduced into actual practice in 1879. Not until 1882 the first successful use of directing train movements by signal indication was installed at Louisville, Kentucky. In 1907 this method of operation, following an installation made on a single-track line with track circuits for controlling the manually operated signals, fully gave the efficiency and reliability as a means for directing train movements without train orders. Now the field of signaling is not limited wholly to the use of signals for conveying informations to trains, for the signal scope has been enlarged in recent years to cover the protection with signals of roadway traffic at highway grade crossing, the interconnecting of signals with dispatching systems, the development and installation of automatic train stop and train control devices, and the application of car retarder system.

Modern Developments

As previously stated the most remarkable developments in recent years in railway signaling are remote control of power swithes and signals or centralized traffic control by the dispatching system, the automatic train stop and automatic train control, and the car retarder system or the power operation and control at classification yards.

The dispatching system provides an economical means whereby signals are made use of, not only to protect and space traffic, but also to dispatch traffic. Operating costs are greatly reduced by reducing delay, by dispensing with intermediary operators, by reducing the number of stops. Delays are so reduced by eliminating two out of three stops when trains take siding, by making closer meets and passes due to flexibility of system. In this new system, the signals and switches of a division are operated and controlled from a certral plant by a dispatcher solely by the use of electrically operated signaling devices and without the aid of operators. The dispatcher directs the movements of trains by operating the signals whose indications authorize the movement. By direct operation of the switches as well as signals, the train dispatcher also sets up the route as required.

The dispatcher seated at his desk which is in front of and attached to the control machine. He directs and arranges all meets and passes, and in fact all operations and movements of trains over his territory by merely operating small levers which control the switches and signals. On the control board locates the track model, light indicators, control levers and key switches. The track model serves for the purpose to visualize the road which the dispatcher controls. The dispatcher is kept in constant contact with trains by an ingenious automatic indicating and "OSing" system. A typical installation of which showed that 70 per cent of all meets in a 40-mile section are being made without either train stopping, and that equipment failures, such as broken draw bars and couplers, have been reduced 72 per cent. Furthermore the tonnage per train has been increased from 5,700 to 6,900 tons or 20 per cent.

One of the most important advantages of train control, as reported by the operating officers and enginemen, is the benefit of the cab signal which brings the way side indication into the locomotive cab. The first permanent automatic train control was placed in actual service in 1914. The object of

the system is to enforce safe train speeds by imposing maximum, tapered and low speed limits which are continuously responsive to traffic and track conditions. Under normal operating conditions in a clear block, a green light in the cab indicator is lighted. In case a signal is at stop indication, this cabsignal indication changes from green to red when the engine reaches the braking dstiance from the stop signal. In order to prevent an automatic application of the brakes, it is necessary for the engineman to operate the acknowledging lever and start a manual reduction of brake pipe pressure sufficient to apply the brakes and reduce speed below a low speed limit. If the engineman should be incapacitated, or for any other reason should fail to take action to acknowledge and apply the brakes after the change from green to red in cab indication, then the brakes are applied by an automatic train control device and the train being stopped.

In the continuous train control system receiving coils are mounted on the locomotive just ahead of the leading pair of wheels. Its function is to provide the means to receive alternating current energy from the rails for energizing the train stop equipment, thus controlling its performances. In transmitting the energy from the rails to the locomotive equipment the principle of induction is used. The action between the rails and the receiver acts as a transformer, in which the rails act as the primary and the coil windings as the secondary. The energy induced in the receiver is amplified and then passed through the master transformer to operate the master relay.

The total track mileage installed with train control in America as to Dec. 1928 amounts to 19,728 miles and the locomotives equipped to 7,297.

The invention and introduction of the car retarder have been made by George Hannaner and E.M. Wilcox in 1924. When railroading was in its infancy, freight was simply handled from one point to another by being picked up by one freight train and set off by the same or another freight train. Then the classification yard has become a very important link in this chain of transportation. In this system hump and gravity yards were built where gravity did the work formerly done by the switch engine. In years past, the time gained in running cars faster over the road was often lost in the receiving yard due to slow and inadequate methods of classification. And

now we have the power operated switch where one operator taking the place of a number of switchman and the car retarder where a few operators taking the place of a large number of car riders.

The car retarder itself is, in effect, a car brake, and at its location, performs the same function. It is an arrangement of brake shoes located along and parallel to the track rails, which shoes are forced against the inside and outside of the car wheels by electric energy or compressed air through levers. The retarder, as its name implies, reduces the speed of the car and the effect of which one the car is the same as if either the hand brake or the air brake had been applied.

The classification yard is made to use in parallel with receiving yard and departure yard. The hump engine pushes the train over the hump and is broken up and classified. This movement is controlled by signal indications. With the retarder system switching lists are prepared by the conductor when his train pulls into the receiving yard. With the switching list in front of the operator it is easy for him to direct each car into its proper track, and by means of retarders, to control its speed so that it couples to the cars already on that track as just the proper speed—that is, a speed to insure the coupling being made, but a speed not great enough to cause damage to either car or lading. The chief advantage for adopting the retarder system in classification yard is no doubt the decreasing the cost of yard operation by eliminating car riders and switchmen. Other inherent advantages for this system are reduced delays in the yard and more even spacing of departing trains. More than 17 yards had been equipped with retarders prior to 1929, and several further installations are just under way.

Economics of Modern Signaling

The economic advantages of modern signaling by directing the train movements by signal indication can be summarized as follows:—

(1) Delays are reduced.—By signal indication instead of train orders in directing the train movements, slow-downs or stops for orders are eliminated and trains are kept moving.

(2) Track capacity is increased.—The track capacity is increased when delays are reduced and trains are kept moving. When track capacity of single track line is increased, double-tracking (often prohibitive in cost) can be postponed.

(3) Safety is increased.—Train operation by signal indication provides many facilities that reduce the operating hazards to a great extent.

(4) The costs for operating freight trains are reduced. Operating costs are reduced when delays are reduced. It has been shown that the time lost in making meets under time-tables, train orders and manual block system is 20.3 minutes, while the time lost in making meets under train operation by signal indication is 11.15 minutes. So the saving in time is 9.15 for each meet, that is, about 45% saved of lost time. In a particular installation the freight train speed before the directing trains under signal indication is 13.35 M.P.H., and after that it has been increased to 16.52 M.P.H. For this same case the total cost of installation includes engine and wayside equipment for train control and revision of interlocking is $2,329,227. The total annual saving made possible by this installation is $367,943. Now we subtract an amount of 139,753 dollars for interest on cost of installation at 6% from the total annual saving we get the net annual saving of $228,190.

In conclusion, it has been best known that in addition to the increased safety of train operating, the modern railway signaling makes it possible to reduce or eliminate the written train orders, lengthen the maintainer's sections, reduce the capital investment charges and provide a more efficient and reliable train control system.

中華民國全國廣播無線電網之重要建設

(附工程及經濟計劃)

著者: 楊樹仁

無線電為二十世紀之運幸兒.歐戰時,無線電訊,初次大出風頭.歐戰停止後,國際商業競爭,無線電訊更有成効.一九一九年,巴黎和會開議之際,美國大總統威爾遜氏,深慮遠謀,目睹英倫早執世界海線電訊之牛耳,乃移目光注重無線電之發展.專商美國奇異公司,威司丁電機公司,及波士頓合衆鮮菓公司,互相籌資合作,促成美國合組無線電公司簡稱(RCA)之成立.十年苦心經營,發展國內無線電各種事業,並與歐洲各國,南美,中美,日本,及南洋羣島互通國際電訊,使紐約造成為國際無線電網之中心點.又日本政府,重視無線電巳久,在一九二二年底,日本政府交通部向美國購辦長波收報機全副,訂約與RCA公司互通商報.不幸半年後,即有東京及橫濱之大地震,繼以火災,有線電訊,毀殘殆盡,幸賴無線電台,得國內之急救,並即電告美國,轉知各處,引起各國之同情.無線電,對於吾國去年北伐軍事進行,亦極有成績.龍潭戰事及去年五三濟南慘案,其効用之最著者也.

無線電事業,就其應用之不同,可分為四類:(一)國際無線電訊,(二)國內無線電訊,(三)海岸船舶無線電訊,及(四)國內廣播無線電網.吾國國際無線電訊,倡議甚早,迄未實現.北京腐敗政府,利用電訊借欵,為發財之機會,引起英日美列强之越俎代謀.日本三井公司,在雙橋代造大電台,毫無成績可言,野心有餘,而力不足.當一九二二年間,美國加省電報公司與合組無線電公司協作,向中國建議,在上海設立國際大電台,同時上海,北平,廣州,及哈爾濱,各設較小電台一座,由美商包辦出資代造,並享長期代管之利.美國之獨專計劃,為英國日本政府所反對,外交之抗議,使美商之進行,毫無結果.國民政府成立以來,注重無線電之建設,為保障主權起見,出資向合組公司訂購大電台二座,可與歐美二洲,直接通報.並向德國購小電台四座,分佈於上海漢口,廣州及天津,專作國內通報.將於上海北郊,興造電台,大無線電機,由美運滬,靜待裝置.明夏倘能成立通報,正當一九三〇年,大東,大北海線電訊合同,重行簽訂之時,吾國恃大無線電台作為後盾,有恃無恐,必使不平等之條件,無存在之餘地也.遼寧省政府,因軍事上之關係,向德購大電台,早與柏林通報矣.但國內西南諸省,重山峻嶺,有線電網,設備不全,電報通訊,十分困難,又轉電手續欠佳,時有延遲之弊.上海往四川之電報,因各種關係,或較郵寄快信為遲,此係過去之事.自國府交通部及建委會之努力建設,造成國內短波電訊網,成績極佳.上海往四川之商電,數時內即有回音.又上海拍往雲南之電報,在昔每字須洋一元.(經法國水線轉)現每字降至一角,亦足稱快也.海岸船舶無線電訊,交通部提倡甚力.如招商局等航海商輪,均有長短波無線電機之裝設,使海上生命之安全,多一保障也.

吾國自辦之廣播無線電台,成立於大城市者,已有年,而廣播無線電網之重大建設,無人注意,尚須鼓吹,以促其實現所謂廣播無線電者,乃一地發出電波,而四鄰收受也.在理論方面,其周圍之收音者,並無限數.但事實上,有天時地形之種種關係.中等電力之廣播台,在距離千里以外,即難得可靠及滿意之收音也.當暑期以內,北平廣播電台,播送京劇,(例如梅蘭芳之六月雪)住在上海之熱心收音者,不易收得良好之結果也.全國廣播無線電網建設之目的,乃使優美之節目,由南京,上海,杭州,北平,天津,漢口,廣州,等電台,同時一致廣播,凡重要之新聞,科學之演講,及北平京調,廣州粵曲,上海之崑腔等節目,一瞬間,駕臨國內數十萬之家庭,及萬百之聽衆,亦非難事也.廣播無線電,為美國所發明.威司丁公司所首創.第一廣播台,(KDKA)設在美國之東畢城,於一九二一年夏,作廣播之試驗.半年後,從事正式之廣播,大受各界之歡迎,爭起興造廣播台於各地.未滿二年,全美境內,設有四百五十電台.現有五百八十八台,(一九二九年)從事廣播也.至於吾國境內,最先美人某氏,在上海永安公司屋頂,設立廣播電台,電力極小,嗣受當局之干涉卽停止.但一九二三年後,美商開洛公司上海分店,除電話機外,兼售無線電料,爲推廣生意起見,在徐家匯附近,築造一百瓦特電力之廣播台,(現巳增大五倍)每日廣播滬市商情,蘇灘,及唱片等.滬上人士,好奇心生,爭買收音機,作家庭之娛樂品,滬蘇甬商家,亦意願甚衆.在一九二五年春,收音機之生意爲最盛.嗣後無線電熱,向北流行於華北區域.北平及天津早辦廣播台.北平電台並得京戲院主之合作,廣播好戲之節目,頗博聽衆之歡迎.東北方面,哈爾濱巳向美商購機,設立一千瓦特之廣播台.因其地勢優異,頗少天電阻擾,且隣近日本之大阪,名古屋等廣播台,故無線電之風行東北,理所當然也.民國十七年,南京中央廣播電台,及杭州廣播電台,相繼成立.去年夏季,中央派人在北平,開封,漢口,長沙,南昌,等處,作收音之試驗.但南京電力不大,數千里以外,卽難獲滿意之結果.廣州,南昌,及西安,聞亦有設立廣播台之議,不久亦可實用也.

吾國各地廣播電台,相繼設立,有如雨後春筍之怒發,乃極好現象.惟各台經費,未能十分充裕.以供給良好之節目.北地電台,時有拿手之節目,頗難接近長江流域之聽衆.南方亦有優美之節目,只限於百里以內之收音者.各廣播台不相協助互勵,僅恃唱片之廣播,壽星唱曲,乏味之至,長此以往,聽衆易生厭惡之心理,而無線電營業,將有失敗之危機.欲救此弊,必連絡各方廣播台,互相合作,以完成全國之廣播無線電網?茲將其計劃略述於下:

例以南京電台,爲廣播網之中心點,其電力可增大二十倍.擇適當地段,建立中心電台,計有 20 K.W. 長波廣播機二座,(其一爲備用權)及 1 K.W. 短波廣播機一座,播音室仍在舊址,利用電話線,與電台相通連,同時以長途電線,接通上海之播音室,及杭州之播音室.滬甯間,早通長途電話,倘得電話局之合作,增添電線一對,每晚四小時間,專作滬甯互通廣播節目之用,餘時仍開放商用.甯杭長途電線,亦仿此辦理.(參觀附圖)南京,北平之間,亦可沿津浦線,增設長途電線.北平往奉天,早有平奉電話線之建設.漢口與廣州二地,擬造 2 K.W. 廣播台,及 1 K.W. 短波廣播台各一座,其發電台及播音台亦由電話線相接連.第一步計劃,南京台賴長途電線之連絡,使北平,上海及杭州台之好節目,可傳送至南京播音室,並由短波無線電之收受,漢口台及廣州台之廣播節目,亦可採爲南京電台之播音資料.南京之節目,利用二種不同之波長.(假定爲 380-Meter 及 33-Meter) 同時廣播於四方.北平播音台之京戲,既賴長途線,與南京台同時廣播,廣州方面,恐不能直接收受北平台之電波但可收得南京大電台之 380-Meter 電波,或 33-Meter 之短波.廣州台收到北平節目,即由其電台轉播,則羊城之居民,僅用鑛石收音機,亦可享受京劇之權利也.反之,廣州既有長短波電台各一座,(假定爲 350-Meter 及 45-Meter 二種)亦可同時播送極好之節目.南京台收得之後,再由其大電台轉播送,而福州或南昌等處之收音者.可於四種不同之波長,揀選其一而收受之.或當地有小播音台,(100 Watt)亦再轉播,使近郊之居民,雖用簡易之收音機,可聽受千里外之音樂.南京台除轉播外,並將收得之廣州節目,由長途電線傳達至北平,或奉天電台轉播,使塞外客商,東北農民,亦可聆嶺南飛來之歌曲也.

第二步,增 1 K.W. 廣播電台於各大城,如安慶,南昌,長沙,福州,汕頭,梧州,南甯,雲南,貴陽,常德,重慶,成都,襄陽,宜昌,迪化,蘭州,甯夏,西安,鄭州,太原,綏遠,熱河,錦州,濟南,青島及徐州等處,並附有上等之短波收音機以便收受南京及廣州等節目,作廣播之材料.倘若中央有重要之宣傳,或政府之緊急佈告,由此種廣播網之利器,以傳達於各地之民衆,對於意志之統一,及文化之宣傳,無線電有極大供獻也.或謂各地人民,生活十分艱難,無力購備收音機者甚多,此困難亦可設法解決,即增添公共收音機,於茶館,浴室,飯店,旅社,及學校醫院等處,使無線電成為民衆之娛樂.各地廣播台之電力漸漸增大,使普通之收音機,亦可應用滿意,而國內無線電製造事業,同時進行大量生產制,(Mass-Production) 使價廉物美之收音機,易於銷售也.吾國內地,因數千年內之交通不便,造成各省不同之方言,言語之異歧,易生歧視之心理.例如今歲陝西省旱災之重大,遠地各省頗難引起同情之救助.倘有廣播網之設立,政府即採用標準之方言,作為統一國語之宣傳,並促進中華民族之團結,務使『秦人視越人之肥瘠』之地方主義,根本剷除也.又吾國未實行强迫教育,以致不識字之人民,占大多數,廣播電台,兼作普及教育之演講,及時事之報告,所以補助平民教育之不足也.

廣播網之建設,最初須政府之資助,有欵十萬元,從事進行.至於廣播台之經常費,其籌款,有間接法及直接法之不同.由政府抽收無線電收音機之特稅,作為電台之經常費,英國政府廣播電台,及菲列濱羣島之廣播台等,均採用此種間接法.又美國商辦廣播台,在節目內,雜和廣告之宣傳,以得相當報酬 費若日本及德國政府,則直接向聽衆,抽取收音之照會捐,以供給政府所辦之廣播台,此種辦法,似大麻煩,與吾國情形不相宜.可仿英國之方法,由政府設立廣播局,(Broadcasting Bureau)得財政部之許可,由海關增收無線電收音機進口稅一成,或無線電材料進口稅半成.此種特稅,專用於廣播網之建設,及維持,即可從事工作,使南京,北平,廣州等七大電台,利用長途電線及短波轉電制,(Relay System) 以完全國廣播網之初步,供給優美音樂之節目,及報告確實之新聞.為廣播之持久政策,易得民衆之歡迎,逐漸推廣,使第二步計劃,亦有實現之希望.又音樂對於國民之生活,頗有重大關係.國樂之廣播,亦宜提倡,務使民衆賞聆音樂,成為普及之嗜好也.

SKF

BALL & ROLLER BEARINGS

TRANSMISSION MATERIAL

TEXTILE SPINDLES

鋼珠軸領
羅勒軸領
地軸傳動
各種機件
以及最新
式之紡紗
羅勒錠子

THE

EKMAN

FOREIGN AGENCIES, Ltd.

SHANGHAI

維昌洋行

上海江西路六號

請聲明由中國工程學會「工程」介紹

TELEFUNKEN

全世界的無線電波皆可用

得力風根眞空管

收音或放音

對於每一種用途總有一只最適宜的

得力風根眞空管

總經理——中國西門子電機廠

上海 天津 北平 奉天 哈爾濱 漢口 廣州 香港 重慶

上海顧客請向下列各電料行接洽

湖北路	明遠亮記電料行
南京路抛球場	鴻康電料行
南京路日升樓	大中華電器行
南京路	永安公司電器部
南京路	新新公司無線電部
博物院路	大華科學儀器公司
霞飛路福開森路	中華三極鋭電公司

請聲明由中國工程學會「工程」介紹

△中華郵政特准掛號認爲新聞紙類▽

民國十九年十二月

工程

第六卷第一號

中國工程學會會刊

THE JOURNAL OF THE CHINESE ENGINEERING SOCIETY

VOL. VI. NO. 1 DECEMBER 1930

中國工程學會發行 總會會所：上海寧波路四十七號 電話一九八二四
每册三角預定全年四册定價一元每册郵費本埠二分外埠五分國外一角八分

3669

工程

中國工程學會會刊

季刊第六卷第一號目錄 ★ 民國十九年十二月發行

總編輯 周厚坤　　總務 支秉淵

本刊文字由著者各自負責

插圖：

正文：

附刊：

中國工程學會發行

中國工程學會職員錄

（會址上海寧波路四十七號）

歷任會長

陳體誠(1918—20)　吳承洛(1920—23)　周明衡(1923—24)　徐佩璜(1924—26)

李垕身(1926—27)　徐佩璜(1927—29)　胡庶華(1929—1930)

民國十九年至二十年職員錄

董事部

淩鴻勛　鄭州隴海鐵路工程局

李垕身　南京鐵道部

徐佩璜　上海市教育局

陳立夫　南京中央執行委員會秘書處

吳承洛　南京工商部

薛次莘　上海南市毛家弄工務局

執行部

（會長）胡庶華　吳淞同濟大學

（副會長）徐佩璜　上海市教育局

（書記）朱有騫　上海新西區楓林路公用局

（會　計）朱樹怡　上海四川路215號亞洲機器公司

（總務）支秉淵　上海寧波路四十七號新中公司

基金監

惲　震　南京建設委員會

裘燮鈞　杭州大方伯浙江省公路局

各地分會

上海分會　（會　長）　黃伯樵　上海新西區楓林路公用局

（副會長）　朱其清　上海老北門無線電管理局

（書　記）　王魯新　上海九江路22號新通公司

（會　計）　鄭葆成　上海新西區市公用局

南京分會　（委　員）　惲　震　南京建設委員會　薛紹清　南京中央中學工學院

張自立　南京建設委員會

蘇州分會　（委　員）　魏師達　蘇州吳縣建設局

北平分會　刻在進行重組中

天津分會　（會　長）

（副會長）　嵇　銓　天津良王莊津浦路工務處

（書　記）　顧穀成　天津西沽津浦機廠

（會　計）　邱凌雲　天津法界拔柏葛鍋爐公司

武漢分會　（會務委員）陳彰琯　漢口工務局

（書記委員）孔祥鵝　武昌建設廳　朱樹馨　武昌建設廳

（會計委員）繆恩釗　武昌武漢大學建築工程處

青島分會　（會　長）　林鳳歧　青島膠濟路四方機廠

（書　記）　嚴宏淮　青島公用局

（會　計）　孫寶墀　青島膠濟鐵路工務處

唐山分會　（會　長）　李書田　唐山交通大學土木工程學院

（副會長）　路秉元　唐山北寧鐵路機廠

（總幹事） 王華棠 唐山交通大學土木工程學院
（會　計） 朱物華 仝　上

杭州分會 （會　長） 易鼎新 杭州杭州電廠
（副會長） 朱耀庭 杭州工務局
（書　記） 茅以新 杭州裏西湖三號杭江鐵路局
（會　計） 裘燮鈞 杭州大方伯公路局
（幹　事） 吳琢之 杭州浙江省公路局

太原分會 （會　長） 唐之肅 山西太原育才鍊鋼廠
（副會長） 董登山 山西軍人工藝實習廠計核處
（文　牘） 曹煥文 山西太原山西火藥廠
（庶　務） 郗三善 山西大學校

梧州分會 暫告停頓

濟南分會 （會　長） 王洵才 濟南膠濟路工務總段
（副會長） 朱樹勳 濟南建設廳
（書　記） 杜寶田 濟南膠濟路第五機務段
（會　計） 陸之順 濟南三馬路陸大鐵工廠

瀋陽分會 （會　長） 張潤田 瀋陽北甯鐵路港務處
（副會長） 王孝華 瀋陽兵工廠
（書　記） 余稚松 瀋陽東北大學
（會　計） 胡光燾 瀋陽東北大學

美洲分會 （會　長） 顧毓瑔 Rm 4 West Sibley, Cornell Univ, Ithaca N. Y, U. S. A.
（副會長） 施孔懷 Rm 4 West Sibley, Cornell Univ, Ithaca N. Y, U. S. A.
（書　記） 黃學詩 500 Riverside Drive, N. Y. C.,U, S. A.
（會　計） 龔理河

委員會

建築工程材料試驗所委員會

委員長 沈　怡 上海南市毛家衖工務局

委　員 徐佩璜 上海市教育局
李屋身 上海福熙路成和邨605號
支秉淵 上海甯波路四十七號新中公司
裘燮鈞 杭州大方伯浙江省公路局
薛次莘 上海南市毛家衖工務局
徐恩曾 南京建設委員會祕書處
顧道生 上海福州路九號公利營業公司
黃伯樵 上海新西區楓林路公用局

工程教育研究委員會

委員長 金問洙 江灣復旦大學

委　員 楊孝述 上海亞爾培路309號中國科學社
茅以昇 鎮江江蘇省水利局
張含英 遼甯葫蘆島港務處
周子競 上海亞爾培路205號中央研究院
李熙謀 杭州浙江大學工學院
程干雲 江灣勞動大學
阮介藩 上海蒲柏路平安里6號
[illegible]伯羽 吳淞同濟大學
鄭肇經 青島港務局
陳懋解 南京中央大學
笪遠綸 北平清華大學
徐佩璜 上海市教育局
戴　濟 上海法界邁爾西愛路家麐坊一號
陳茂康 唐山交通大學
梅貽琦 清華大學駐美監督處
陳廣沅 天津西沽津浦機廠
許應期 上海徐家滙交通大學
孫昌克 南京建設委員會
俞同奎 北平北平大學第一工學院
鄒恩泳 上海新西區楓林路公用局
李昌祚 上海西愛咸斯路 55 號
唐藝青 長沙湖南大學
徐名材 上海徐家匯交通大學

3673

會員委員會

委員長　黃炳奎　上海高廊橋申新第五廠

委　員　上海　徐紀澤　南市十六浦荳市街恆吉巷五號　上海　黃元吉　愛多亞路38號凱泰建築公司

南京　徐百揆　南京科巷洪鑫里四號　杭州　朱耀庭　工務局

天津　邱凌雲　法界拔柏葛鍋爐公司　濟南　王洵才　膠濟路工務段

青島　王節堯　膠濟路工務處　武漢　孔祥鵝　湖北建設廳

廣州　桂銘敬　粵漢鐵路株韶段工程局　山西　唐之肅　太原育才鍊鋼廠

瀋陽　張潤田　北寧鐵路港務處　美國　薛楚書　500 Riverside Drive, New York City.

經濟設計委員會

委　員　徐佩璜　上海市教育局　朱樹怡　上海四川路215號亞洲機器公司

胡庶華　吳淞同濟大學　張延祥　南京建設委員會

李　俶　上海徐家滙交通大學

編譯工程名詞委員會

委員長　程瀛章　上海梅白克路三德里639號

委　員　張濟翔　廣州光樓中國電氣公司　尤佳章　上海寶山路商務印書館編譯所

馮　雄　上海寶山路商務印書館編輯所　徐名材　上海徐家滙交通大學

張輔良　上海福開森路378號中央研究院社會科學研究所　孫洪芬　北平南長街22號中華教育文化基金董事會

藍春池　上海膠州路大夏大學　錢昌祚　南京中央陸軍軍官學校航空隊

林繼庸　江灣俞涇廟大南製革廠　鄒恩泳　上海新西區市政府公用局

胡衡臣　南京建設委員　吳欽烈　南京軍政部兵工署

職業介紹委員會

委員長　朱有騫　上海新西區楓林路公用局

委　員　馮寶齡　上海圓明園路慎昌洋行　徐恩曾　南京中央組織部

工程研究委員會

委員長　徐恩曾　南京中央黨部　主任委員　胡博淵　南京農礦部

化工組主任委員　徐鳳石　南京工商部　土木組主任委員　許心武　南京復成橋導淮委員會

電機組主任委員　惲蔭棠　南京建設委員會　機械組主任委員　鈕何受　南京交通部技術委員會

委員　鍾兆琳　上海交通大學　委員　周坤厚　上海福州路一號德士古火油公司

礦冶組主任委員　楊公兆　南京農礦部　建築組主任委員　齊兆昌　南京金陵大學建築部

委員　吳稚田　上海九江路六號沙利貿易公司

建築條例委員會

委員長　薛次莘　上海南市毛家街工務局

委　員　朱耀庭　杭州工務局　薛卓斌　上海江海關五樓濬浦總局

徐百揆　南京科巷洪鑫里四號　李　鏗　上海圓明園路慎昌洋行

許守忠　青島工務局

材料試驗委員會

委員長　王繩善　上海徐家滙交通大學

委　員　康時清　上海徐家滙交通大學　盛祖鈞　上海徐家滙交通大學

科學咨詢處

常務委員長　徐佩璜　上海市教育局　黃炳奎　上海高郎橋申新第五紡織廠

常務委員　胡庶華　吳淞同濟大學　周厚坤　上海外灘德士古火油公司

裘維裕　上海交通大學　支秉淵　上海寧波路四十七號新中工程公司

中國工程學會會所圖案

第六屆萬國道路會議中國代表團攝影

花萊峰 貴州建設廳技正　孫謀 鐵道部技正工務司幫辦　劉景山 前交通部路政司司長　凌鴻勛夫人　陳體誠 浙江公路局局長　凌鴻勛 隴海鐵路工程局長　趙祖康 前安徽建設廳技正

日本鋼鐵業概觀

著者: 胡博淵

(一) 緒言

日本新式製鐵事業之肇端,遠在明治七年(同治十三年).時日本政府,思開發岩手縣釜石鐵鑛,曾購置種種新式製鐵設備,不幸此項企圖,其後竟歸失敗.嗣於明治三十年(光緒二十三年),日本政府復於九州創設八幡製鐵所,中經十餘載之奮鬥,然其成績,猶未大著.及至歐洲大戰,鐵價暴騰,該所營業,始克轉入順境,同時民營諸鋼鐵廠,亦如春筍勃發,盛極一時.迨歐戰告終,鐵價復原,民營各廠,除基礎穩固者外,其他雖多告失敗,然日本　鐵事業,已立有穩固之基礎矣.

(二) 國立八幡製鐵所

沿革及產量　八幡製鐵所,在九州福岡縣遠賀郡八幡市.其創立之始,遠在明治三十年.時日本政府,痛感鋼鐵事業,有國營之必要,爰以年產鋼材九萬噸爲標準,創立該所,前後籌備四年,耗金一千五百八十四萬元.迨三十四年(光緒二十七年)二月,始克舉行第一熔鑛爐點火式,同年五月,開始鍊鋼.在最初數年,以事屬創舉,經濟技術,兩感困難.嗣經日俄歐戰兩役,日本之鋼鐵需要頓增,該所事業,日有起色,生產設備,疊經擴充,現在該所,每年所產之鋼材,已達一百萬噸,前後投下之資本,亦逾一億三千萬圓,實爲日本最大之鋼鐵廠.茲將該所最近產額增加狀況列表如下:

第一表

	大正九年(民國九年)	大正十三年	大正十四年	昭和元年(民國十五年)	昭和二年	昭和三年
生鉄	244,000噸	489,000噸	586,000噸	654,000噸	735,000噸	832,000噸
鋼塊	449,000	685,000	863,000	980,000	1,054,000	1,160,000
鋼材	297,000	493,000	653,000	739,000	830,000	937,000

主要生產設備 八幡製鉄所,分八幡本工場,戶畑作業場,及西八幡工場三處.其化鉄設備:在八幡本工場,計有熔鑛爐六座;在戶畑作業場,則有熔鑛爐二座;此外在本工場,尙有新熔鑛爐二座,目下正在建設之中.至其煉鋼設備:在本工場計有平爐工場三所,轉爐工場,坩堝鋼工場,電氣爐工場各一所;在西八幡工場有平爐工場一所.茲將其主要之生產設備,列表於後:

第二表

熔鑛爐	八幡本工場	熔鐵爐號數	第一號 第二號	第三號 第四號	第五號 第六號	新一號 新二號
		化鐵能力	250 噸	300 噸	350 噸	500 噸
	戶畑作業場	有一百五十噸及三百噸熔鑛爐各一座				
煉鋼爐	八幡本工場	第一平爐工場	二十五噸鹽基性平爐			十二座
		第二平爐工場	五十噸鹽基性平爐			六 座
			六十噸鹽基性平爐			四 座
		第三平爐工場	六十噸鹽基性平爐			七 座
			二百噸 Jalbot 式平爐			一 座
		轉爐工場	十噸酸性轉爐			二 座
		坩堝鋼工場	坩堝煉鋼爐			七 座
		電氣爐工場	弧光式二噸半電氣爐		各	一 座
			誘導式三噸電氣爐			
			傾注式六噸平爐			
	西八幡工場	平爐工場	五十噸平爐			二 座

關於煉焦設備:在八幡本工場有煉焦工場五所,洗炭工場三所,計有煉焦爐四百七十座;在戶畑作業場,計有煉焦爐一百七十五座.茲將八幡本工場煉焦爐之詳細狀況列下:

第三表

名稱	列表	爐數	每爐裝入量	燒成時間	副產物收集設備	蓄熱室
Koppers 式	1	120	4.6 噸	48 小時	無	無
Solby 式	2	150	5.7	30	有	無
第一黒田式	1	50 50	10.5 10.5	24	有	有
第二黒田式	1	100	11.0	24	有	有

此外八幡本工場,尚有鋼材製造場.石炭副產製造場,窯業工場及研究所等,茲不備述.

鋼鐵原料　原料鑛石,分鐵鑛,錳鑛兩種,鐵鑛之半數,係由我國大冶,象鼻山,桃冲等鑛購入,餘由馬來半島之柔佛及朝鮮之安岳,利原,殷栗載甯等鑛供給.其需要量去年爲一百七十萬噸,本年預計將增至二百萬噸.錳鑛由我國湖南之湘潭,湖北之陽新,江西之樂平,江蘇之海州及廣西之欽武等處錳鑛,供給其十分之四,餘由馬來半島及日本四國各鑛供給.所用之焦炭,均由該所自製,其原料以該所經營之二瀨炭鑛所產者爲主.惟二瀨炭粘結力稍差,須配合粘結力較強之炭,助其粘結,藉獲良質之焦炭.現在所用之配合炭,爲日本之松浦炭,高島炭,及我國之本溪湖炭,開平炭,淄川炭等,其配合率爲二瀨炭七成與配合炭三成之比例.石灰石每年約須三十五萬噸,均由門司附近各採石場購入.此外所用各種雜料,則利用本工場所產之鑛滓,輾鋼皮,生鉄屑及熔爐烟塵等.茲將熔爐之原料裝入量,(平均)列表如下:

第四表

原料	大正十二年(民國十二年)度		大正十三年度		大正十四年度	
	裝入量(噸)	對於生鉄之比率	裝入量(噸)	對於生鉄之比率	裝入量(噸)	對於生鉄之比率
鉄鑛	720,602	1.65	683,371	1.610	784,660	1.625
錳鑛	23,736	0.054	15,390	0.036	16,822	0.035
焦炭	455,505	1.05	431,447	1.015	490,411	1.016
石灰石	211,364	0.485	183,860	0.432	199,741	0.414
雜料	20,493	0.047	41,158	0.097	35,001	0.072

作業順序　八幡製鉄所之作業,向係採用銑(生鉄)鋼一貫主義.原料鐵,錳,鑛石,與焦炭及石灰石等共同裝入熔鑛爐,依熱風之作用,使鑛石之鐵分還元,成爲熔銑,其夾雜物則成鑛滓,而於爐內互相分離,每隔一定時間,卽被注出,此種作業,不分晝夜行之.

3679

由各爐注出之熔銑,立卽移入混銑爐,使其品質均一,隨時注入平爐之內.平爐多屬鹽基性者,以上述之熔銑及廢鉄爲主原料.如遇熔銑不足時,特裝入冷銑,並加鐵鑛及石灰石等,而依發生爐瓦斯,以熔解精煉之,然後注入鑄型中,作製各種鋼材用之鋼塊.

鋼塊依其在赤熱之狀態,更裝入灼熱爐內.俟經相當時間後取出,更用轆壓機壓之,截成鋼片,亦乘其在赤熱狀態,移入製品工場,製造各種鋼貨.鋼貨種類大小雖極雜多,但其作業之順序,則均相同.鋼貨之主要者,爲重鋼軌,輕鋼軌,工字鋼,溝形鋼,丁字鋼,山形鋼,球山形鋼,圓鋼,角鋼,扁鋼,線鋼,試力板,矽素鋼板,其他各種鋼板,車軸及外輪,工具鋼,其他特種鋼等.

製鐵所於製鋼材之外,尙行與鋼有連帶關係之副產,加工作業.如對於鑛滓,則注以水,使成水滓後,以鑛滓甎及高爐水泥,或用蒸汽吹之,使成綿狀,以製防熱,或防音之鑛滓棉.又對於熔爐排出之瓦斯,則加以洗滌,以供瓦斯機關之使用.又對於煉焦爐發生之焦炭瓦斯,則製硫酸錏 Tar, Benzol 等化學工業藥品.

生產能力　茲將八幡製鐵所每年之生產能力,及昭和二年（民國十六年）度之實產額,列表比較於下:

第　五　表

(1)──生　鐵	鎔鑛爐	8座
	生產能力	830,000噸
	昭和二年產額	734,839噸
(2)──鐵合金	電力熔爐	2座
(3)──鋼　塊	平　爐	36座
	坩堝鍊鋼爐	7座
	電氣爐	2座
	生產能力	970,000噸
	昭和二年產額	1,054,064噸

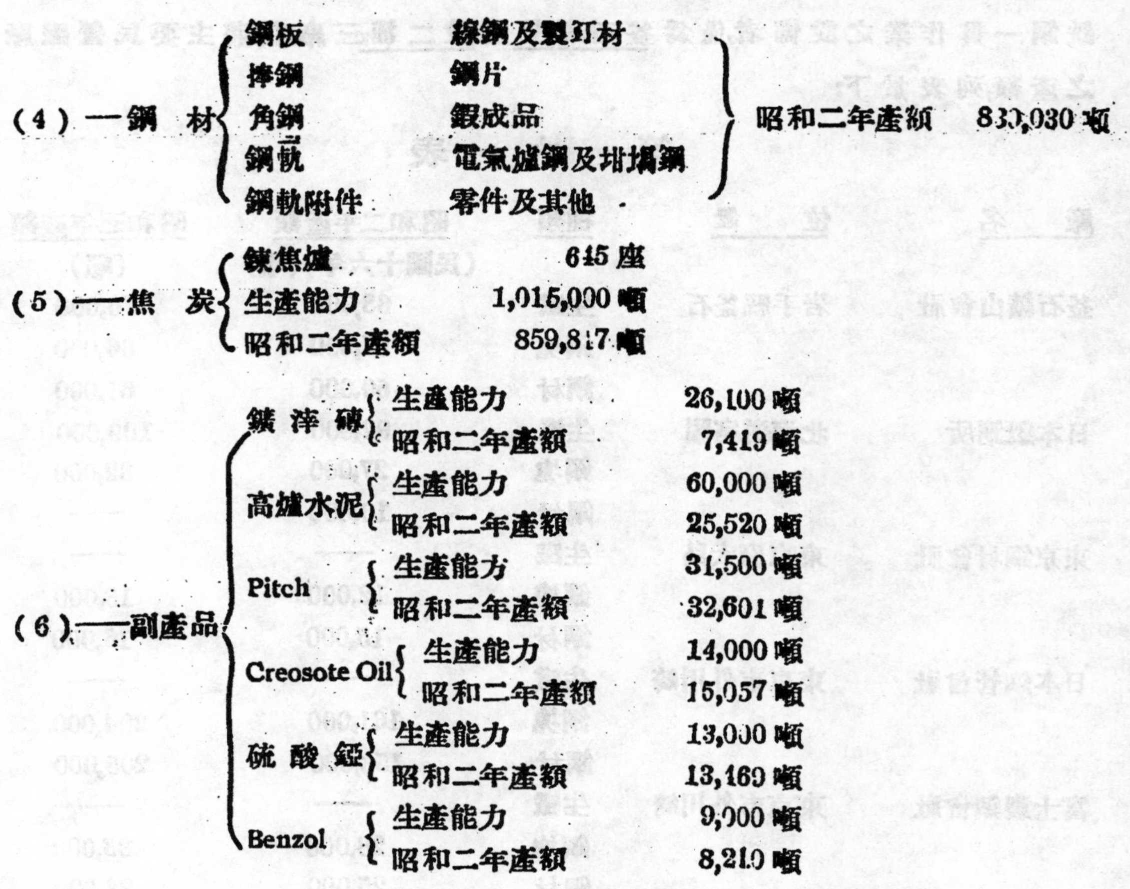

(4)一鋼　材
- 鋼板
- 捧鋼
- 角鋼
- 鋼軌
- 鋼軌附件
- 線鋼及製釘材
- 鋼片
- 鍜成品
- 電氣爐鋼及坩堝鋼
- 零件及其他

昭和二年產額　830,030 噸

(5)一焦　炭
- 煉焦爐　645 座
- 生產能力　1,015,000 噸
- 昭和二年產額　859,817 噸

(6)一副產品
- 鑛滓磚
  - 生產能力　26,100 噸
  - 昭和二年產額　7,419 噸
- 高爐水泥
  - 生產能力　60,000 噸
  - 昭和二年產額　25,520 噸
- Pitch
  - 生產能力　31,500 噸
  - 昭和二年產額　32,601 噸
- Creosote Oil
  - 生產能力　14,000 噸
  - 昭和二年產額　15,057 噸
- 硫酸錏
  - 生產能力　13,000 噸
  - 昭和二年產額　13,169 噸
- Benzol
  - 生產能力　9,000 噸
  - 昭和二年產額　8,210 噸

(三)　民營鋼鐵廠

如前所述,日本政府曾於明治七年,經營釜石鐵鑛,然歷時未久,卽遭失敗.嗣於明治十七年(光緒十年),有田中長兵衛者,獨力接辦此鑛,是爲民營鐵廠之始.迨中日戰役時代,該鑛年產生鐵,已達二萬噸.在八幡製鐵所未成立以前,該鑛實爲日本唯一之製鐵所.嗣後民營鋼鐵廠逐漸發達,及至今日,除國營八幡製鐵所,及專造高級鋼材之吳海軍廠外,其餘民營諸廠,亦有足資記述者.其中尤以北海道之日本製鋼所,川崎之日本鋼管會社,鶴見之淺野製鐵所,神戶市之神戶製鋼所,大坂市之住友製鐵所及大坂製鐵會社,小倉市之淺野製鋼所及朝鮮之兼二浦製鐵所爲最重要.在上述各廠中,其備有

銑鋼一貫作業之設備者,僅爲釜石,鶴見及兼二浦三廠.茲將主要民營鐵廠之產額,列表於下:

第六表

廠名	位置	種類	昭和二年產額（民國十六年）(噸)	昭和三年產額（噸）
釜石鑛山會社	岩手縣釜石	生鐵	63,000	76,000
		鋼塊	57,000	66,000
		鋼材	50,000	61,000
日本製鋼所	北海道室蘭	生鐵	92,000	109,000
		鋼塊	27,000	32,000
		鋼材	16,000	——
東京鋼材會社	東京府大島	生鐵	——	——
		鋼塊	12,000	13,000
		鋼材	15,000	16,000
日本鋼管會社	東京市外川崎	生鐵	——	——
		鋼塊	161,000	204,000
		鋼材	156,000	205,000
富士製鋼會社	東京市外川崎	生鐵	——	——
		鋼塊	28,000	33,000
		鋼材	25,000	24,000
淺野造船所	橫濱市外鶴見	生鐵	22,000	55,000
		鋼塊	——	54,000
		鋼材	43,000	53,000
大坂製鐵會社	大坂市	生鐵	——	——
		鋼塊	39,000	52,000
		鋼材	39,000	48,000
住友諸工廠	大坂市及尼崎	生鐵	——	——
		鋼塊	51,000	51,000
		鋼材	58,000	43000
神戶製鋼所	神戶市	生鐵	——	——
		鋼塊	61,000	73,000
		鋼材	65,000	65,000

川崎造船所	神戶市	生鐵	——	——
		鋼塊	104,000	134,000
		鋼材	77,000	126,000
川崎車輛會社	神戶市	生鐵	——	——
		鋼塊	12,000	8,000
		鋼材	10,000	——
德山製板會社	山口縣德山	生鐵	——	——
		鋼塊	——	——
		鋼材	21,000	27,000
淺野製鋼所	福岡縣小倉市	生鐵	——	——
		鋼塊	44,000	47,000
		鋼材	39,000	51,000
東海鋼業會社	福岡縣若松市	生鐵	——	——
		鋼塊	——	——
		鋼材	3[illegible],000	49,000
兼二浦製鐵所	朝鮮黃海道	生鐵	129,000	146,000
		鋼塊	——	——
		鋼材	——	——
其　　他		生鐵	11,000	不詳
		鋼塊	37,000	仝上
		鋼材	31,000	仝上
合　　計		生鐵	322,000	387,000
		鋼塊	633,000	767,000
		鋼材	687,000	778,000

次將主要民營各廠之生產設備,列為第七表如下;

第七表

名　稱	熔鑛爐座數		煉鋼平爐座數
釜石鑛山會社	200噸　2座	60噸　1座	25噸　3座
	25噸　1座		

日本製鋼所	120噸 3座	100噸 1座	50噸 1座
			60噸, 10噸, 5噸 各2座
東京鋼材會社	——		8噸 1座 10噸 1座
日本鋼管會社	25噸 2座	20噸 1座	30噸 2座
富士製鋼會社	——		25噸 1座 15噸 3座
淺野造船所	150噸 1座		——
大坂製鐵會社	——		25噸 2座
住友諸工廠	——		40噸, 80噸 各1座
	——		25噸 3座 15噸 2座
神戶製鋼所	20噸 1座		12噸, 16噸, 30噸 各1座
			25噸 2座
川崎造船所	——		25噸 10座 20噸 3座
淺野製鋼所	20噸 1座		25噸 3座
兼二浦製鐵所	150噸 2座		50噸 3座
後志製鐵所	20噸 1座		——
大島製鋼所	——		25噸 1座 15噸 2座
三菱造船所	——		25噸 1座
茂里製鋼工場	——		15噸 1座
羽室鑄鋼所	——		8噸 2座

又日人在我國遼寧省,尙有與我國合辦之鞍山本溪湖兩鐵廠,玆將其產額及設備列表於下:

第八表

名稱	民國十六年生鐵產額(噸)	民國十七年生鐵產額(噸)	熔鑛爐座數
鞍山製鐵所	203,000	221,000	250噸 2座
本溪湖製鐵所	51,000	64,000	130噸 2座 20噸 1座
合計	254,000	285,000	每年產鐵能力 272,000噸

此外別種工廠之兼產鋼材者,每年所產亦達三四萬噸,故日本民營各鐵廠每年出產生鐵總量約達六十八萬噸,鋼材約達八十萬噸.

（四）鋼鐵之供求量

明治五年（同治十一年），日本敷設京濱鐵道，始用洋式鋼鐵，迨中日戰役，其國內所產鋼材，年僅一千噸，同時外鋼之輸入，則年達二十二萬噸，故當時日本所需之鋼材，可謂全數仰給於舶來品。迨日俄戰後之明治四十年（光緒三十三年），需要量已增至五十萬噸。其內八成仰給外鋼，餘為本國所產。但至昭和三年（民國十七年），則國內需鋼量更增為二百三十萬噸。外加運至台灣，朝鮮等處者，約十八萬噸，合計鋼材需要總額，為二百四十八萬噸。其內由外國輸入者，僅佔三分之一，即年約八十二萬噸，餘均由八幡及民營各鋼鐵廠供給。茲將最近三年日本鋼材之供求狀況，列表於下：

第八表

	昭和元年（民國十五年）		昭和二年		昭和三年	
	數量（噸）	百分率	數量（噸）	百分率	數量（噸）	百分率
八幡製鐵所產量	658,000	30	713,000	32	841,000	34
民營各廠產量	587,000	27	687,000	31	820,000	33
外國鋼材輸入量	925,000	43	814,000	37	821,000	33
合　計	2,170,000	100	2,214,000	100	2,482,000	100
日本鋼材輸出量	120,000	—	157,000	—	180,000	—
日本鋼材淨需要量	2,050,000	—	2,057,000	—	2,302,000	—

（註）表內八幡產量，與第一表所揭者，稍有不同，蓋因該所之會計年度，與曆年不同而然。

次就錳鋼，鍛鋼，鑄鋼等之普通鋼材，及包括本鋼之高級鋼材，一較其需要量時，則普通鋼材，約佔總需要量百分之九十四，高級鋼材約佔百分之六。

生鐵可依用途而分為製鋼用及翻砂用之兩種，如聯想鋼鐵兩項之總需要量時，則有區別生鐵用途之必要。茲將日本最近生鐵之需要量，列表如下：

第九表

	昭和二年(民國十六年)		昭和三年	
	數量(噸)	百分率	數量(噸)	百分率
國內產量	911,000	61	1,098,000	61
自朝鮮輸入量	103,000	7	140,000	8
自我國東三省輸入量	190,000	13	182,000	10
自印度輸入量	261,000	18	311,000	17
自英美瑞典輸入量	22,000	1	76,000	4
合　計	1,487,000	100	1,802,000	100
自日本輸出量	4,000	—	5,000	—
日本生鐵淨需要量	1,483,000	100	1,797,000	100

(註)　上表生鉄需要量之內,其供翻砂用者,年約四十萬噸,餘可視其係供煉鋼之用.

如前所述,日本現在之鋼鐵總需要量,計在國內年需鋼材二百三十萬噸,翻砂生鐵四十萬噸,合計共達二百七十萬噸之鉅.次就日本鐵業過去之實績,並參酌其他各種要素,而一測其將來需要之增加狀況,則在昭和七年(民國二十一年),預計國內所需鋼鐵,將增至二,七五〇,〇〇〇噸,在昭和十年,更將增至三,三〇〇,〇〇〇噸云.

(五) 製鐵原料之供求量

製鐵之根本原料,自屬鐵鑛,石炭等類.然就最後目的之鋼材生產而言,則生鐵,廢鋼等物,亦可視作製鐵之準原料.茲就日本現有之製鋼設備,一考其所需之生鐵及廢鋼之數量時,則其生產能力,每年至多能製鋼材二百萬噸,外加製鋼消耗約百分之十一(根據昭和元二年度八幡轉爐之成績).故日本每年約需製鋼原料二百二十二萬噸,除每年使用合金鐵二萬五千噸,及儘量收買廢鋼,計在國內年能購用廢鋼二十七萬噸,由海外年可輸入二十二萬五千噸(昭和三年曾輸入三十七萬噸當屬例外)外,每年實需生鐵一百七十萬噸.現在日本百噸以上熔爐之化鐵能力(連鞍山,本溪湖兩廠),

每年約達百五十萬噸.今假定其以全能力化鐵,則日本每年尚差生鐵二十萬噸.此外翻砂生鐵,每年約需四十萬噸.故日本所差生鐵之數量如次:

第十表

日本年製鋼材二百萬噸,如能儘量收買廢鋼,則其所需之

生鐵數量	1,700,000噸
翻砂用生鐵	400,000噸
合計	2,100,000噸
日本現有百噸以上熔爐之製鐵能力	1,500,000噸
每年生鐵不敷	600,000噸

上項之不足量,約與三座五百噸熔鑛爐之製鐵能力相當.現八幡製鐵所已在建設五百噸熔鑛爐二座,以應此項需要矣.次就日本年產鋼材二百萬噸,一檢其所需之鐵鑛總量.其關係如下表:

第十一表

日本現有設備之製鐵能力	1,500,000噸
現在建設中之製鐵能力	550,000噸
合計	2,050,000噸
鞍山,本溪湖兩廠可用當地之鐵鑛	500,000噸
其必需供給鐵鑛之製鐵量	1,550,000噸
製造上項生鐵所要之鐵鑛(每噸生鐵需鐵鑛一·六噸)	2,500,000噸
製鋼用鐵鑛之使用量	180,000噸
故日本國內所要之鐵鑛總量	2,680,000噸

對於上項需要量,每年由其本國,朝鮮,及我國長江沿岸,以及南洋羣島各鑛所能供給之數量如下:

第十二表

日本釜石及俱知安鐵山	500,000噸
朝鮮各鐵山	500,000噸
我國大冶桃冲各鐵鑛	1,000,000噸
南洋羣島各鐵山	1,000,000噸
合計	3,000,000噸

石炭亦爲製鐵重要原料之一.日本石炭,資源雖尙豐富,然以一般石炭,大多缺乏粘結性,故於製造鎔爐所用之硬焦炭時,須配合粘力較强之炭,如我國開平炭等約三成,藉獲良質之煉鐵焦炭.現在日本每年製鐵所需之石炭,約達五百萬噸,其每年所產石炭總額則爲三千五百萬噸,故製鐵所需之石炭,約佔總數百分之十五.錳鑛爲改善鋼鐵品質所不可缺少之物,現在世界年產錳鑛共約二百餘萬噸,其中九成用於鋼鐵事業.惟日本產錳不多,故製鐵所需,殆均仰給於我國湘,桂,贛,粵,及南洋,印度各鑛,其額年達十萬噸.

(六) 日本鋼鐵事業之地位

據昭和三年(民國十七年),之統計,日本鋼材需要量,每年約達二百四十八萬噸,其中本國所產者,佔六成七分,爲百六十六萬噸,其餘三成三分之八十二萬噸則由外國輸入.此外尙有製造鋼材用及翻砂用之生鐵,約三十八萬噸,層鋼約三十七萬噸,鋼塊及鋼片約十萬噸,共計該年所用之外國鋼鐵,尙達一百六十七萬噸.次就舶來鋼鐵類之價格與普通貿易額一加比較,則自大正九年(民國九年),至昭和二年(民國十六年)之八年間,鋼鐵類之輸入價格,達十四億九百萬元.在此期間之入超額,則達二十九億一千七百萬元,故鋼鐵之輸入價格,約佔入超額之四成七分.又昭和三年入超額爲二億二千四百萬元.同年鋼鐵之輸入價格,達一億四千九百萬元,約佔入超額之六成七分.此外以鐵爲主要材料之機械類,及鐵製品之輸入,每年約達一億圓,合計兩項約達二億五千萬元.故今後日本如能確舉鋼鐵自給之實,則其國家經濟,及對外貿易狀況,當更行改善矣.

次就世界總產鋼量而言,最近世界之鋼材產量,每年計達八千萬噸,其內美產三千萬噸,德產一千一百萬噸,英,法各產七百萬噸,餘爲其他各國之總產量.故日本所產之一百七十萬噸,較諸列强尙屬渺無足道.今試一按日本鐵業不能十分發達之緣由,殆因一般企業家,鑑於釜石官廠之首遭失敗,及八幡之連年虧損,毫無底止,以及歐戰時代,投機家臨時濫設,各廠之旋歸幻滅,叠次失敗的事實,故多視鐵業投資爲畏途,此實斯業不能充分發展之最大原因也.況製鐵事業,非有極大之規模,不能合於經濟的採算.今如假定設一年產鋼材三十萬噸,有銑鋼一貫設備之最小限度之鋼鐵廠,則其設備費至少須有四千五百萬元,方合經濟原則,較諸別種企業,投資鉅而獲利不易,此亦斯業不易發達之又原因也.惟近年八幡之製鐵技術,進步極速,其營業成績,亦極順利.如大正六七兩年,該所贏餘,曾各達四五千萬元.即於昭和三年,其贏餘亦達一千五百萬元.其餘民營各廠,亦咸有相當之成績,不但所產鋼鐵之品質,較諸歐,美製品,毫無遜色,即其售價,亦屬不相上下.茲將八幡鋼材,與歐,美製品之價,列表比較於下:

第十三表

		八幡圓	英圓	法圓	德圓	美圓
棒鋼	國內	89.00	79.14	59.25	67.21	90.62
	輸出		79.14	55.27	55.72	至 95.07
工字鋼	國內	87.00	77.82	55.72	65.88	90.62
	輸出		70.74	50.84	50.40	至 95.07
厚鋼板	國內	97.00	85.34	65.88	75.61	90.63
	輸出		77.82	61.45	61.89	至 95.07

(註一)　八幡製品售價,係依去年八月該所發表市場交貨之期貨價格,其自工場運至市場所要之費用,每噸約需四圓餘.

(註二)　歐,美製品,係依去年八月二十二日發行 Iron Trade Review 所載市價之換算爲日金者,表中國內欄,示工場交貨之每噸定價,輸出欄示最近港口交貨之每噸定價.

(七)　結論

查日本製鐵業歷來最感困難之問題:(一)爲國內缺乏鐵鑛,此實日本製鐵事業最大之缺憾,但近年因斯業當局不斷之努力,已能以極廉之代價,輸入鉅量之鐵鑛;(二)爲製鐵技術上之困難,在八幡創業時代頗受此項問題之影響,但近年日本製鐵技術極爲發達,所製鋼材之品質,較諸歐美製品,毫無遜色,故能漸次將外國鑛鐵,自市場驅逐過半;(三)爲生產成本過高,去歲日本之鋼鐵市價,雖與英美市價,大致相仿,惟側聞各鐵廠,均以維持此項市價爲苦,且自本年新正以來,日本實行金解禁,外鐵尤將乘勢湧入,此實目下日本製鐵業者所最感脅威之點也.惟日本一般鑛鐵廠之生產設備,較諸歐美各廠規模,尙覺過小,似應儘量擴充,以期增加生產能率.此外如促進銑鋼一貫作業之實現,及各廠協力避免重複作業,實行單種多產制度等項,均爲減輕成本之要圖.倘能於最短期間,逐漸實現,則日本今後,不惟易於抵制外鐵,而於鋼鐵業前途,不難更有長足之進步也.

國外工程新聞二則

(一)世界最大鐵橋　英國著名鋼廠道曼朗公司承造鐵橋一座,其南北兩部,已於本年八月二十日在中央啣接.該橋橫跨澳洲 Sydney 海港,自南至北,共長 3,370 呎,最大跨度 1675 呎,重 5 萬噸,高出海面 440 呎,故雖在潮水極高之時,船舶亦可自橋下通過.橋面寬 160 呎,有闊 57 呎之大路一條,鐵路四條,10 呎寬人行道二條.預計全部工程,明年可以完竣,重車即可通行,其建築費聞爲英金六百萬磅,可算世界最大鐵橋云.

(二)蘇俄政府計劃填塞間宮海峽　蘇俄政府因欲銜接北樺太與西伯利亞大陸陸路,擬將間宮海峽填塞.經愼密調查,確悉如能將該海峽最狹處拉查列賚附近之海面 2 哩餘填塞,即可達到計劃之目的.因之北方之寒流,不致通過該處,海參威即可成爲不凍港云.

化學工程

著者：顧毓珍

緒言

化學工程之名,始聞于一八八八年,(1)而其能在工程界獨樹一幟者,不過近二十年事耳.蓋化學工業之猛進,始自歐戰.(2)歐戰前之化學工業,乃在技藝時代,無化學工程之可言;歐戰後之化學工業,已入科學時代,實賴於化學工程.故化學工程,實爲科學時代之化學工業之產生品也.

返觀技藝時代之化學工業,多操於化學家或機械工程師之手.化學家缺乏工程智識,機械工程師缺乏化學智識;然以其堅強毅力,不折不囘,經若干之演試修改,(3)而得良好結果.至其所得結果,往往守爲祕密,不輕告人;是以染色業守其染色之祕密技藝,製革業守其製革之祕密技藝.習其業者,由學徒而至技師,由技師而至廠長,均知其然而不知其所以然;所可恃者,其熟能生巧之經驗耳.如是而望化學工業進步之速,亦難矣哉.迨歐戰發生,因原料之恐慌,戰器之製造,而化學工業驟然猛進.(2)感糧食之缺乏,德國有食料代替品之製造,以染色之缺乏,美國有顏料製造業之勃興,餘若橡皮,製紙,酸鹹等業,無不因時代之需要而發達.當需要孔急之時,世人方明技藝時代之化學工業之不可恃,不經濟,不能供其所求;必欲用科學方法研究化學工業,於是化學工程尙焉.蓋昔日所謂之祕密技藝,用化學工程可以解釋其原理;昔日所謂染色之祕密技藝與夫製革之祕密技藝,用化學工程可以歸納其方法而溝通之.去昔日固有之祕密技藝,易以合於科學方法之化學工程,

(1) E.F.Hodgins: Methods of Chemical Engineering Education M.I.T. Tesis

(2) 趙承嘏 歐戰時代之化學工業 科學 八卷七期

(3) Try out Method: 對於質量無從預定,須視試驗如何而定.

宜其能節省消耗,效率增加,而有價廉物美之出品,宜其能將無數實驗室方法(Laboratory Methods),變爲工廠製造法(4)(Commercial Processes).故化學工程學術之倡明,實使化學工業由技藝時代而趨於科學時代.

方今化學工程於歐美工業界中,所占地位,不在任何工程之下,其應用之廣,且在其他工程之上.(5)諺云:『開門七件事,柴,米,油,鹽,醬,醋,糖』是皆重要之化學工業,何一不需化學工程.再觀一國之基本工業,(6)如鋼鐵,(7)精鹽,酸鹹,紙漿,酒精等業,何一不需化學工程.中國欲振工業,不能無化學工程,至少化學工程之於中國,與土木,電機,機械等工程有同等之需要.然而國人習化學工程者有幾?明化學工程果爲何物而知其內容者有幾?國內大學中有化學工程科者又有幾?爲發展實業計,爲富庶民生計,爲建設新中國計,化學工程之在中國,應急待注意而提倡者也.此卽作者介紹化學工程之本意,望國人三注意焉.玆分爲(一)化學工程之意義,(二)化學工程之內容,(三)化學工程與工業,(四)化學工程教育四項,申述之.

(一)化學工程之意義

世之誤解化學工程者曰:『化學工程乃半爲純粹化學,半爲機械工程』或曰:『化學工程卽化學』.(8)此類解釋,非特失化學工程之眞相,抑且給聽者

(4)如 Haber Process, Hydrogenation of Coal 等,均由實驗室方法變成大規模之工廠製造法.

(5) Chemical Engineering Achievements Reported in May Fields—Editorial Chem. & Met. Vol. 36 No. 7

(6)孔祥熙著『基本工業計劃書』工程四卷四期

(7)徐式莊著『鋼鐵與化學工程之關係』. 科學 七卷八號論鋼鐵業中冶礦與化學工程有同等重要,蓋焦煤與煤氣製造,均爲化學工程也.

(8)作者恆見人指習化學工程者曰『彼習化學』,是無異指習土木或電機工程者曰:『彼習物理』.於此可見國人常以化學與化學工程混作一起而不能分別.

以謬誤之映象.美國之有化學工程,不過三十餘年;卽數年前對於化學工程根本觀念,亦頗有誤解與批評.[9] 評者謂『化學工程師係不澈底之化學家與工程師』,並謂『旣有純粹化學 (Chemical Science) 及機械工程兩科,學者應任擇一科習之,不應再有化學工程科之設立』.

物質文明愈進步,專門職業愈多,而其分類愈細;專則精,精則效率增加.考工程業之演進,最初當僅有土木工程,由土木而分機械與電機工程,由機械而分紡織,汽車,航空等工程,由電機而分電力,電報,電話等工程.工程之基礎,莫不曰數學,物理,與化學;[10] 最能以物理致用者,當爲土木,電機,機械諸工程.而最能以化學致用者,惟獨化學工程.工業中應用化學之處日多,故歐,美於四十年前,[1] 有化學工程之添設,美國於三十餘年前,[9] 亦有化學工程之添設,此乃社會之需要使然.至今化學工程能在工程界中獨樹一幟,大放光明,蓋亦專而精之結果;是以化學工程之應否添設,無庸置辯矣.

工程乃科學與社會之橋樑(Engineering is the bridge between society and science.)造橋樑者,工程師也.於物理與社會間,有土木,電機,機械諸工程作橋樑;推而至於化學與社會,其爲橋樑者,非化學工程乎?於此可見化學工程應用之多,範圍之廣,而欲與以一的確之定義頗難.略言之,其定義可謂化學工程乃一種工程,在設計,構造,及管理,工廠與設備,俾化學方法得致用於實業(Chemical Engineering is one which deals with the design, construction, and maintenance of plants and equipment set up for the commercial utilization of chemical processes.) 英國威廉博士[11] (Dr. E. C. Williams) 關於化學工程師之定義曰:[12] 化學工

(9) J. H. James Chemical Engineering Education Journal of Chemical Education Vol. I. No. 7

(10) 欲明工程與數學,物理,化學間之關係,可參閱陳廣沅著工程師一篇,載科學 十卷五期及十一期

(11) 前英國倫敦大學化學工程教授 Former Ramsay Professor of Chemical Engineering University of London, England.

(12) W. E. Gibbs: Chemical Engineering Education and Research in Great Britain. (Journal Soc. Chem. Ind. 1928)

程師之責任在能規劃大規模之化學製造方法,並能設計與處理該項工廠,使所有之化學反應與物理變化見於實用(A chemical engineer is a scientific man whose duty is to plan the large-scale commercial operation of chemical processes and to design and operate the plant required for the carrying out of the chemical reactions and physical changes involved.)

從上列化學工程與化學工程師之定義,可知化學工程與化學及機械工程確有密切關係,而決不能以化學或機械工程替代之.故化工事業,決不可以化學家或機械工程師任其勞.凡化學工程師,必須有鞏固之化學基礎——尤其在物理化學方面——,工程訓練及經濟常識.謂化學工程乃化學與機械工程之結晶品或可,若謂乃化學與機械工程之合併物卽非.(A chemical engineer is more than a combination of chemist with mechanical engineer.)

在試驗室中研究一問題,化學家與化學工程師之觀點不同.化學家研究物性物量,分析之,綜合之,其步驟之簡繁不論,其所用物品之價格更勿論.化學工程師則求最簡易之化學方法,然後研究如何從玻璃管煤氣燈變爲大鍋與煉爐,(13) 旣欲顧原料之價格,又欲知出品之銷路,再欲解決如何運送,如何裝置.所理之質量驟增,處理困難亦驟增.(14) 科學家求眞,工程師求致用;如何致用,經濟,工人,地點等,均成爲重要問題矣.至若身臨化學工廠,機械工程師與化學工程師之觀點又不同.各項化學工業中,物質不同,物性亦異,故於設備機具,非深有化學工程智識者,難於勝任.此種學識,遠在機械工程範圍以外,故若以機械工程師充之,其設備往往非不合用,卽不經用,常爲笑柄.如『腐蝕』(Corrosion)爲化工機械中之大問題,化學工程師研究各物之腐蝕情形,故於設計工廠機具,亦無不顧及.善哉!英國 Hugh Griffiths 之言曰:(15)

(13) 玻璃管等在工廠中之相當物,下書論列甚詳. J. Grossmann: The Elements of Chemical Engineering

(14) 趙承嘏著『化學工程之意義』 科學 八卷四號

(15) Hugh Griffiths: The Elements of Chemical Engineering Design. London: Benn Brothers, Ltd. 1922

『化學工程師在化學家中爲工程師,而在工程師中則爲化學家』『既非尋常工程師又非純粹化學家可以瞭解化學工程,若謂化學家與工程師各有其言論觀點,則化學工程師又另有其態度風采.』("——that the Chemical engineer when in the company of chemists is an engineer, and when in the company of engineers is a chemist. ——it simply serves to show that neither of engineers on the one hand nor the chemist on the other understand the scope of chemical engineering, and whilst it is often said that the chemist and the engineer speak in different languages, it might also be stated that the chemical engineer has still another mode of expression.")

化學工程之範圍極廣,吾人平日所見其應用,僅在化學工業方面,孰知化學工程,實較廣於工業化學(Industrial Cemistry)之範圍.(16) 羅斯福總統曾對美國富源保管委員會言:(17) "Our object is to conserve the foundations of our prosperity; to use our resources, but to use them as to conserve them; not to limit the wise and proper development and application of these resources, but to prevent destruction and reduce waste." C. F. McKenna (18) 聞之則曰:『是乃化學工程師之工作也.(17) 蓋化學工程對於天然原料之處理,如何得最經濟之應用,如何利用其副產物,以及如何減少耗費,最爲注意.例如以燒柴之法燒煤爲不經濟,則用崩解蒸溜法(Distructive Distillation)取其煤氣,利用煤中之揮發液體,並利用蒸溜後所餘之焦煤;如是同爲一噸煤,用化學工程方法處理(19) 後,其代價驟增.

(16) Electrochemical and Met. Industry Vol. 6—Editorial
"The chemical engineer should not be confounded with the industrial chemist. The latter's work is restricted to the chemical industry, and the former's field of activity is wider."

(17) American Institute of Chemical Engineers Transactions Vol. 1 (1908)

(18) C. F. McKenna 爲美國化學工程學會第一任會長

(19) 欲更明燃料之處理法者,可參閱:丁嗣賢著『近年歐美各國對於燃料之研究』科學 十卷十一期化學工程.且為他項工業製造原料.例如汽車工業之於常人,似為純係機械工程之範圍,孰知橡皮業與石油業,實供給汽車工業以最重要之原料.若無化學工程師之努力,設法製造價廉質堅之橡皮輪,一加倫行六十里之汽油,汽車業之不振可必也.

(二) 化學工程之內容

化學工程學說之應用於工業,遠在化學工業產生之前,蓋知其然而不知其所以然也.例如染色與製革,其技藝之倡明,遠在其理論之發現之前.其機械設備多恃不折不撓之試驗精神所得,而所得方法,多視為秘密.故欲致力於化學工業者,必須專習一業,得其秘訣,或製紙或精鹽或炸藥;蓋一業有一業之專枝,無普通理論以領導數業也.

以科學方法治化學工業,則知其方法之應用於造紙者,可應用於造精鹽,亦可應用於造炸藥.故雖以化學工業之不勝枚舉,與其製造法之繁多,而可歸納成少數簡單方法,名曰『單箇處理法』[20](Unit Operation).在每項『單箇處理法』中,自有原理可尋,而適用於各項化學工業.例如石油業中所用之蒸溜法與煤氣業中所用之蒸溜法,其原理一也;製紙業中所用之蒸發法,與製糖業中所用之蒸發法,其原理亦一也.於是在數百化學工業中,[21] 可用數十『單箇處理法』管理之,是乃由至繁而臻至簡之境,此化學工程之功也,此化學工程之所以能使工業進步神速也.觀下列兩表,可明『化學工程之內容,乃為數十『單箇處理法』之工程,而其尤要者不過十一法耳.

---

(20) 美國 A. D. Little, W. H. Walker, W. K. Lewis, R. T. Hastam 均倡用『單箇處理法』研究化學工程

(21) 一項化學工業中,有用六七『單箇處理法』者,有用八九『單箇處理法』者,詳本篇第二圖.

第一表(22)

化學工程中之單箇處理法
(Unit Operations in Chemical Engineering)

(一) 能力之輸送

1. 熱之輸送—— 應用機具(23) 如蓄熱器,換熱器,預熱器,加冷器等.

(二) 物料之輸送

1. 氣體之輸送—— 應用機具如管,唧筒,運風機等.
2. 液體之輸送—— 應用機具如管,唧筒,射揚器等.
3. 固體之輸送—— 應用機具如搬運器,升降機等.

(三) 物料之初步處理

1. 壓碎(Crushing)　2. 研磨(Grinding)
3. 混和(Mixing)　4. 溶解(Dissolving)
5. 沉澱(Precipitating)

(四) 物料之離分

1. 固體與固體
   (1) 檢分(Mechanical Separation)　(2) 水分(Hydraulic Separation)
   (3) 風分(Air Separation)　(4) 浮分(Flotation)
   (5) 浸分(Leaching)
2. 固體與液體
   (1) 積澱與傾注(Sedimentation and Decantation)　(2) 清濾(Filtration)
   (3) 結晶(Crystallization)　(4) 風乾(Air Drying)
   (5) 液取(Extraction)　(6) 凝收(Adsorption)
3. 液體與液體
   (1) 蒸溜(Distillation)　(2) 蒸發(Evaporation)
   (3) 轉分(Centrifugal Separation)

---

(22) 原表見 Report of Committee on Chemical Engineering Education of American Institute of Chemical Engineers. 1922 本表由作者略加增減,表中所舉之專門名詞,係採用中國工程學會新印之工程名詞,不詳者則由作者自譯

(23) 應用機具,可參閱 哈忒著 韓組康譯『工業化學機械』

4. 氣體與氣體

(1) 吸收 (Absorption)　(2) 凝收 (Adsorption)

(3) 液化後之分解蒸溜 (Fractional Distillation after liquefaction)

5. 固體與氣體

(1) 洗滌 (Washing)　(2) 靜積 (Settling)

(3) 清濾 (Filtration)

6. 氣體與液體

(1) 吸收 (Absorption)　(2) 液取 (Extraction)

(3) 減濕 (Dehumidification)

(五) 反應作用與反應方法 (物料之化學處理)

1. 燃燒 (Combustion)　2. 烤燒 (Roasting)

3. 崩解蒸溜 (Destructive Distillation)　4. 電解 (Electrolysis)

5. 接觸作用 (Catalysis)　6. 其他如醱酵,鹹化,硝化等.

(六) 化學工廠之設計及建造

1. 物料分配 (Materials)　2. 機具佈置 (Layout)

3. 經濟較量 (Economic Balance)

上表所列化學工程之範圍,不免仍感繁多,今再擇其尤爲重要之『單箇處理法』列如第二表.

第二表 (23)

化學工程中之重要『單箇處理法』

(1) 流體之流動　(2) 熱之流動

(3) 清濾　(4) 蒸發

(5) 加濕與減濕　(6) 乾燥

(7) 蒸溜　(8) 吸收

(9) 液取　(10) 壓碎與研磨

(11) 檢分

第二表中十一項之『單箇處理法』,當以『流體之流動』及『熱之流動』兩

(24) 每項『單箇處理法』之定義及範圍詳於 Walker, Lewis, MaAdams: Principles of Chemical Engineering McGraw-Hill Book Co., 1927

項最爲重要,其餘九項爲次要.蓋無論何項化學工業,其反應率,其出產之數額,其製法之效率,均繫於原料之分配移送與能力之消減;前者恒爲流體之流動,後者恒爲熱之流動.是以其餘九項之『單箇處理法』亦無不根據於此,試舉『風乾法』(Air Drying)以明之.

風乾法(空氣乾燥法):——冷空氣必先用預熱氣燒熱,將已熱之空氣吹入乾燥機(Drier),而同時必將欲乾之物料送入該機.物料(如濕紙或潮磚)中之水分遇熱空氣而蒸發,而由離乾燥機之空氣帶出.如何運送物料,如何運送空氣,則需應用流體之流動之原理;如何預熱空氣,如何保持乾燥機中之溫度,則需應用熱之流動之原理.此外應用之法當爲加濕法與乾燥法;惟於此可見無論何項製造法,不能脫離『流體之流動』與『熱之流動』兩項『單箇處理法』矣.

化學工程無時不需機具設計與經濟較量[(25)](Apparatus Design & Economic Balance).卽由上舉之例而言,預熱器之大小,乾燥機之構造,均須視空氣之濕度(Humidity)與溫度,物料之質地多寡而定.運送空氣之速率愈大,所需之煽風機亦愈大,則所需之動力(Power)亦多;多用動力,用較大之煽風機則多費金錢.而同時空氣之速率增加,預熱器傳熱愈快,物料中水分之蒸發亦愈快.是以欲設計而施用經濟之空氣乾燥機具,則至少須顧及上列諸點.是則非諳習化學工程不能爲也.

(三) 化學工程與工業

現代工業發達,化學工程之功爲大.本篇緒言中已詳述化學工程,實爲科學時代之化學工程之產生品,而使化學工業由技藝時代趨於科學時代.H. C. Parmelee[(26)]稱化學工程闢工程界之新紀元而爲工業發達之產物.工業

(25) 欲詳知箇中情形,可參閱 C. Tyler: Chemical Engineering Economics. McGraw-Hill Book Co. 1926

(26) H. C. Parmelee: Chemical Engineers for Industry Chem. and Met. Vol. 35 No. 1

界之需用化學工程，美國在二十年前[26]已如此，至近年尤甚。[5] 美國Sir Alfred Mond[27]曾對化學工業社演說謂『以前之化學家，過受工業界之誇獎。蓋化學家在實驗室中，似覺極容易極簡單之方法，執意施諸工廠，化學工程師不知須耗費多少心力，戰勝多少艱難。眞異大工程師能使不能生利之化學方法變爲生利』。於此可見化學工程在工業界之地位，日加重要。下列第一圖，[28]即表明化學工程於美國二十項工業中之地位（此表係美國一九二八年之調查最近調查，尙付闕如）。

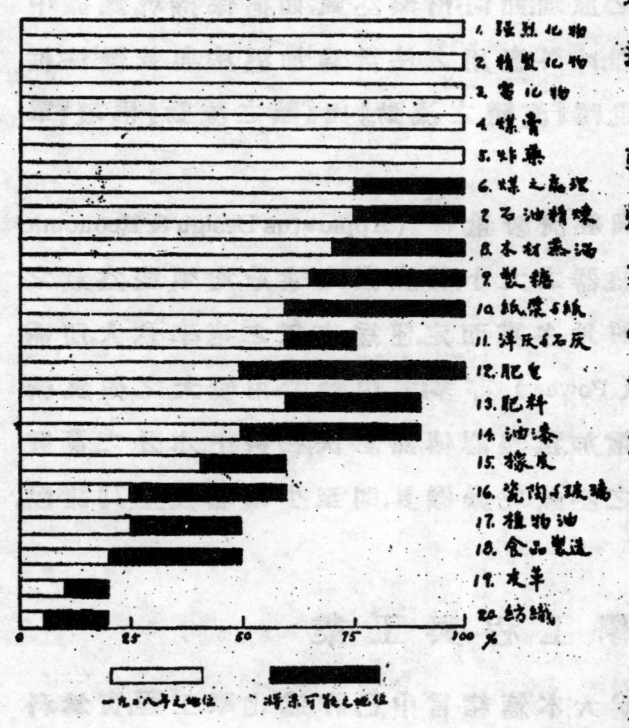

第一圖中之最上五項工業，即強烈化物，精製化物，電化物，煤膏，炸藥，已完全爲化學工程之範圍。其餘十五項中，將來可完全爲化學工程之範圍者，爲煤之處理，石油精煉，木材蒸溜，製糖，製紙，肥皂六種；將來可大有發展者，爲肥料，油漆，洋灰與石灰。將來紡織業中(29)——尤其是人造絲——，應用化學工程之處必多，可以預卜。

本篇前已述及化學工業中之製造方法，可歸納成較爲簡單之

(27) Sir Alfred Mond: Chairman of Imperial Chemical Industries, Ltd.

(28) Industry's Common Bond in Chemical Engineering Chem. and Met. Vol. 35 No. 1

(29) E. C. Bertolet: The Textile Industries need influence of Chemical Engineering. Chem. and Met. Vol. 35, 1

3700

『單箇處理法』(Unit Operation). 每項『單箇處理法』之如何應用,雖因各箇工業而異,然其應用之原則,與及處理之方法則一.所以欲明化學工程與工業之關係,又可觀其『單箇處理法』與各項工業之應用如何而定,列如第二圖[30](圖中空出者,爲應用該項『單箇處理法』).

第二圖

單箇處理法與化學工業

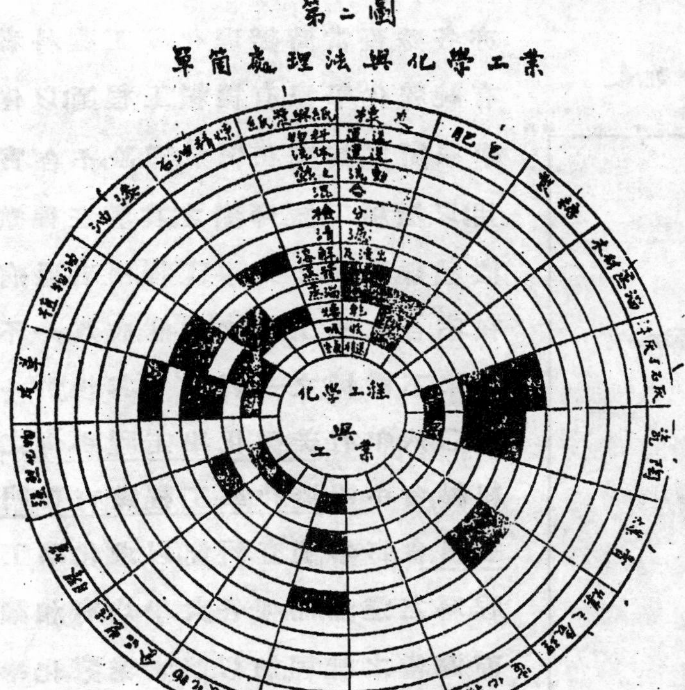

(圖中塗黑者指尚未應用該項單箇處理法)

工業愈發達,分工愈精密,於化學工業尤甚.廠有專製原料者,有處理原料者,有專製商品者;往往一專製商品之廠仰求原料於五六廠者.譬如造紙廠,須向酸鹼廠中購買鹹灰,明礬,硫磺,再須向電化廠中購買氯氣及苛性鈉.各項化學工業之唇齒相依,當無疑義.且『化學品』實爲原料之原料,各項工業無不需用,至其最後出品,吾人覺難辨其固有之化學物料矣.據一九二八年美國工業之調查,[31]凡三十工業,必須用化學工業出品,價額合美金八萬萬元之多,而三十工業中所用之天然原料不過五倍此數耳.

---

(30) 摘錄 Industry's Common Bond in Chemical Engineering Chem. and Met. Vol. 35 No. 1 擇原圖中較爲不重要之『單箇處理法』刪去.

(31) Estimated Consumption of Chemicals by Consuming Industries Chem. and Met. Vol. 35 No. 1

(四) 化學工程教育[32]

今日歐美化學工程之如是發達,皆由於工業與教育之共同合作;[33] 故言化學工程不能不及化學工業,而更不能不及化學工程教育.美國大學之初設化學工程科,設備既不同,課程亦各殊.蓋當時辦理化學工程科者,有純粹化學家,有機械工程師;以化學家辦之,其弊在偏於理論,不合實用;以機械工程師辦之,其弊在偏重機械課程,缺少化學課程.兩者通病,在不合化學工程之本意,成爲一不化學不機械之一科.[9] 其後於一九〇八年,有美國化學工程學會之組織,會中附設化學工程教育專門委員會,每年調查研究,以求化學工程科之進步,而使各大學均有相類而適當之課程.加以諸實業家,化學工程師,化學家之提倡,始克臻現在完美之境.現今有化學工程科中凡

第三圖

美國化工畢業生之百分分配表

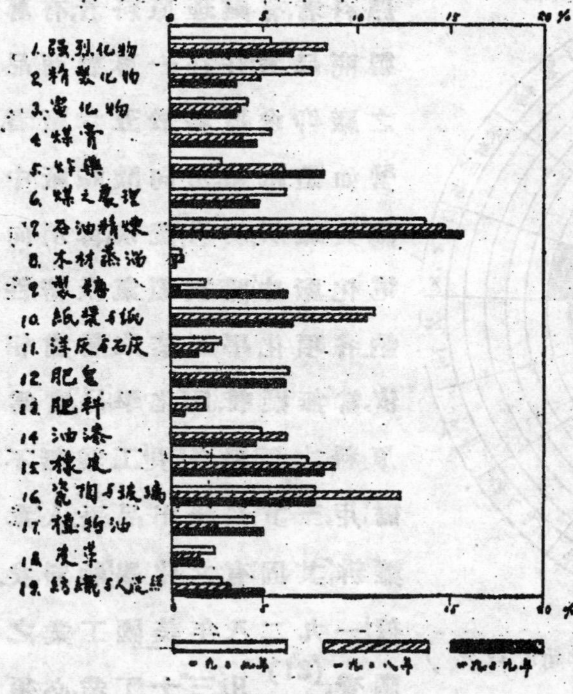

(32) 本節所述,多限於美國之化學工程教育.關於英國者,可參閱 W. E. Gibbs: Chemical Engineering Education and Research in Great Britain. Jour. Soc. Chem. Ind. 1928

(33) H. C. Parmelee: "The Development of Chemical Engineering in this country has been fortunate in many respects, not the least of which has been the close co-operation for 20 years between industry and education in expressing and meeting, respectively, the need for the chemical engineer." Chem and Met. Vol. 35 No. 1

八十校,[34]每年化學工程科畢業生有八百人.一九二八年之化學工程科學生較前一年增百分之二十.每年畢業生之入工業界者之分配,列爲第三圖.[34]

觀第三圖,可知近三年來,美國化學工程科畢業生於工業界之分配.讀者須知美國化工科畢業生,年多一年,原圖旣系根據百分計算,故於許多工業中,一九二九年之百分計算,或較弱於一九二八或一九二七年,惟其人數則未必較少,或竟過之.然分配之大致趨向,可以立見;以入石油煉業者爲最多占六分之一),其次爲橡皮業與炸藥業.十九項工業中之需要化學工程師如此;然無完善之化學工程科,良好之化學工程教育,以培植人才,使致用於工業,則工業決不能繼續發達,此化學工程教育之所以爲重要而工業界之必須與合作也.最重要者爲「化學工程實習學校」之設立;使習化工者自臨化學工程,實地試驗.研究製造方法.考察化工原理,以及機具之構造.工廠之佈置.是乃以現代之化學工廠.供作化學工程科之試驗室,紙上談兵等弊盡行去矣.美國麻省理工大學(M. I. T.)河海河省立大學(Ohio State University)均有此類設立,前者之化工實習學校,作者曾身歷其間,殊覺實地經驗之可貴,遠非書本之所能盡述,更非於實驗室中所能得也.

結 論

由上而觀,可見所爲化學工程者,並非一種工業之技藝,又非半爲化學半爲機械工程之學科,而爲設計,構造,及管理工廠與設備,俾化學方法得致用於實業之一班獨立工程也.於近代工業中,應用之廣,範圍之大,遠勝任何工程;而於國家之富源,更負重大使命,中國果欲振興工業也,則化學工程不能後於任何工程,國人而不欲注意於化學工程,則中國雖有其他諸工程之發展,吾可斷中國之工業,依舊不振,而其他工程之發展亦必有限量也.故中國欲振工業,必須以化學工程與任何工程並重.方今國內洋貨充斥,利權外溢,愛國者痛心疾首,高呼『抵制』『抵制』;然徒呼『抵制』,不自製造,是束手待斃之策.應用化學品之仰仗於外國者,觸目皆是;望國人急起直追,從事化學工程.以圖化學工業之發達也可.

(34) How Industry Absorbs College Graduates in Chemical Engineering Chem. and Met. 36 No. 7 1929

INTER-COMMUNICATION BETWEEN AUTOMATIC AND MANUAL TELEPHONES

自動電話與非自動式通話之研究

By B. J. Yoh (郁秉堅)

General.— In order to carry the traffic from both manual to auto matic and automatic to manual in a telephone network or interconnected offices, when one of the exchanges has been changed into automatic working, several systems have been developed in each case. Of course, they will not be in existance as soon as all exchanges in the network have been changed into automatic.

From Manual to Automatic Systems. When the automatic exchange is introduced into existing manual networks, one of the problems is naturally to have a scheme for handling traffic from manual subscribers. The methods used can be generally segregated into two classes.

Class 1. Methods where the manual "A" operator completes connection without the aid of a "B" operator. The "A" operator, as shown in Fig. 1, goes on in an order wire to the desired automatic exchange and then controles the establishment of the connection in the automatic exchange by means of either a dial or key-sending equipment located on the "A" position of the manual exchange.

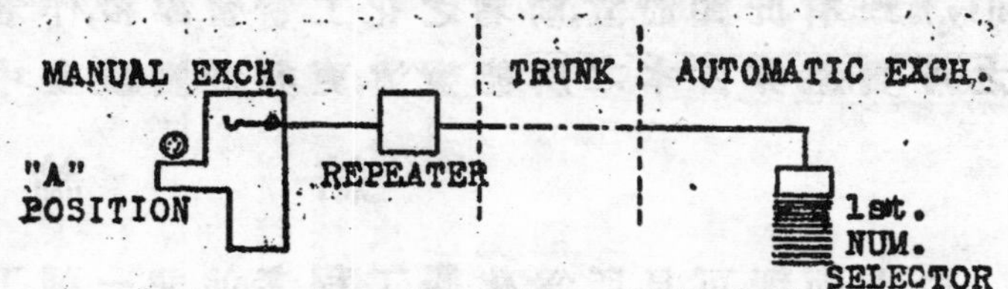

Fig. 1. *Manual Operator "A" Completes Connection Without The Aid Of "B" Operator.*

Class 2. Methods where the manual operator "A" completes connection with the aid of a "B" operator via an order wire. As shown in Fig. 2, the "A" operator goes in on an order wire to the desired automatic exchange and gives the order to a "B" operator, as usually referred to as a "Semi B" or "Cordless B" operator. This operator assigns a trunk and completes the call in the automatic exchange either by means of a dial or a keyboard. It may

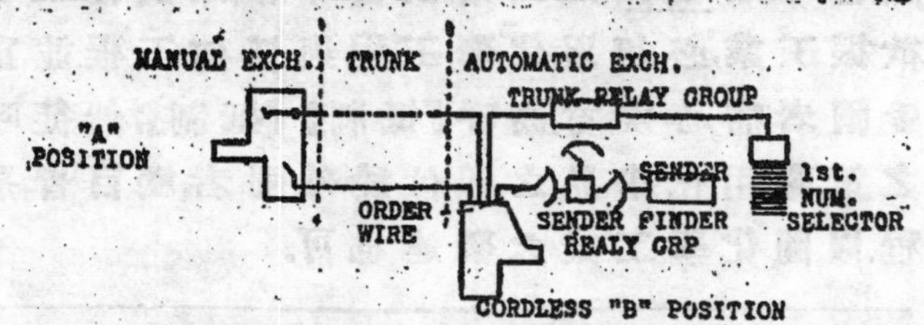

Fig. 2. *Ordinary Order Wire Scheme From "A" Operator To Cordless "B" Operator.*

be noted that in so far as the "A" operator is concerned, this method of completing a call from a manual to an automatic subscriber is identical with that used for the completion of a call from a manual to a manual subscriber.

The order wire method of trunking contains several inconveniences, principally during heavy loads in confusion on the order wire. The tendency is, therefore, to-wards the application of methods in Class 1. They have the feature of "direct" or "straight-forward" trunking and no "cordless B" equipment or operators in the automatic exchange. There are, however, cases where a "cordless B" operator is preferred, so, the methods in Class 2 have still been used.

In order to get the merits of "direct" trunking and to do away with the order wire but retaining the "cordless "B" position, Mr. Ostline of Automatic Telephone Manufacturing Company has suggested a method, which is represented by Fig. 3.

The volume of manual to auto natic traffic at any one exchange is at a maximum when the exchange is just under conversion and will gradually decline until the time when all offices in the network become automatic, then, it will be a minimum. The traffic remaining at that time may be that incoming from the trunk and toll exchanges, and a small amount from exchanges outside the area.

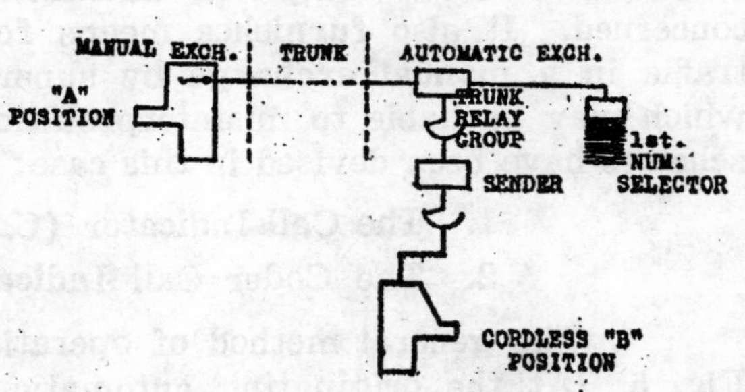

FIG. 3. *Ostline's suggested Scheme.*

From Automatic to Manual Systems.— The traffic outgoing from auto-going from automatic exchanges to manual exchanges may be divided into matic exchanges to manual exchanges may be divided into four types, namely: Junction, Trunk, Toll and Special. The junction traffic which is the inter-area traffic, and by far the largest proportion, forms one of the most difficult problems when converting a large area to automatic working. There are three possible methods of working automatic to manual calls:—

Method. 1. Automatic subscribers dial exchange code only and inform "B" operator verbally, as shown in Fig. 4. The objection of this method of working would be the bad effect on service due to the automatic subscribers sometimes having to dial the code and also numerical digits and at others dialling the code only.

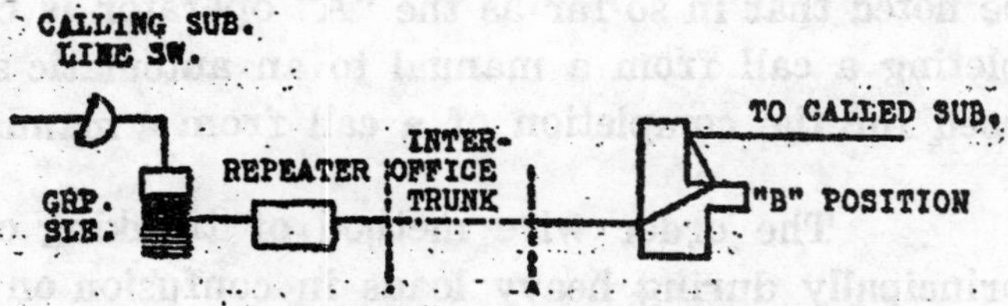

FIG. 4. *Automatic To Manual Call, Method 1.*

Method 2. This method fits automatic switches at the manual exchanges. It seems very attractive, as if each manual exchange were equipped with first numerical, second numerical and final selectors, incoming calls from automatic exchanges could be directed to these switches as thus proceed as for an ordinary straight automatic call. There are, however, several drawbacks, chiefly due to the limitted space and the different operating voltage in manual exchanges.

Method 3. Automatic subscribers dial full number as the ordinary practice, and numerical portion displayed on lamps in the front of "B" operator. It renders calls from subscribers in an automatic exchange to those in a manual exchange completely automatic so far as the calling subscriber is concerned. It also furnishes means for greatly facilitating certain classes of traffic in a manual exchange by superseding the usual order wire method which may be liable to misinterpretation during busy hours of traffic. Two schemes have been devised in this case:

1. The Call Indicator (C. I.) System.
2. The Coder Call Indicator (C. C. I.) System.

The general method of operation of the call indicator may be seen in Fig. 5. At the originating automatic office the dialing by the subscriber of the first or first and second digits of the number determines the routing of the call to the required manual office through a repeater. The first disengaged trunk is seized, immediately engaging a trunk relay group and associated cord circuit at the manual position. Each incoming trunk on the position is directly associated with a trunk relay groups which is

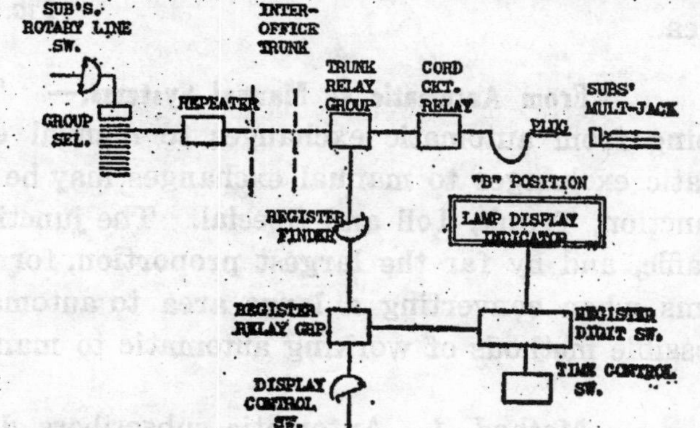

FIG. 5. *Call Indicator System, Schematic Showing Method of Operation.*

particular to that circuit. A rotary line switch, individual to the trunk, is now engaged and functions as a finder to select the first idle register relay group and register digit selector.

The register relay group is now ready for received the impulses representing the number and are received and stored via the register relay group on the register digit switch, which consists of controlling relays associated with single action step-by-step Strowger minor switches responding to the regular impulses from the dial. Having received all the impulses, the wipers of the minor switches are in the position determined by the numerals dialled, and on the completion of the unit digit, if the display board is not already showing a number, the call will be displayed. At the same time, an assignment lamp in front of the associated cord and plug will be lighted, informing the operator which plug to use.

If the subscriber's line is disengaged, the plug is inserted in the multiple and the number shown on the display-board disappears and the next number stored appears in its proper sequence. The assignment lamp associated with the trunk is also extinguished and the supervision of the call is taken care of by the supervisory lamp. Should the called number be already engaged, the operator plugs into a busy jack, putting the usual busy tone on the calling subscriber's line.

In the Coder Call Indicator System, the calling subscriber at the originating automatic office, as shown in the Fig. 6, dials in the usual manner, the first digit or first two digits seizing an outgoing C. C. I. repeater to the manual office. A rotary line switch associated with the repeater finds and connects the repeater to a disengaged coder. The numerical digits in the shape of regular Strowger impulses are now dialled into the coder, and there translated into code impulses, and stored ready for transmission along the junction until the last numerical digit has been received.

Upon the seizure of the trunk, the relays in the incoming trunk relay set operate and a marker controlled from a marker distributor, immediately starts rotating and connects itself to the trunk through the repeater. Depending on the sequence of the calls, the display controller will next connect the trunk through the marker to the decoding and storing relays. As soon as the decorder is connected to the trunk, the number stored in the coder at the originating office will be transmitted in code along the junction. The code impulses are then transferred to the storing relay group and the coder and decoding rclays are immediately released for other calls.

As soon as the number is stored, it is also shown on the position display board. The operator now takes up any plug on the position and tests the multiple jack of the number required and if disengeged, inserts the plug. This causes the rotary switch associated with the cord to act as a finder, proceeding to connect itself with the trunk displaying the number. If the called line is engaged, the operator depresses a key common to the position, which causes a busy tone to be put on to the calling subscriber in the usual manner. The number shown on the display-board is also automatically wiped out and the next number displayed if stored ready.

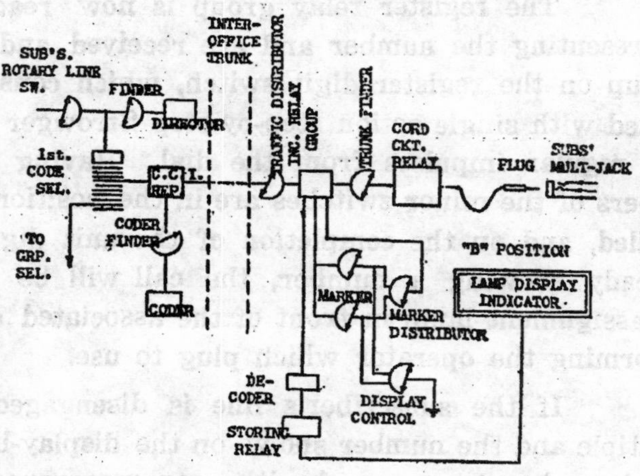

FIG. 6. *Coder Call Indicator, Method Of Operation.*

Immediately on the storing and displaying of a number, and before the connection is complete, the decording relay group is connected to the next trunk on which a call may be waiting, and is available for immediately receiving another in code form.

The chief differences between Call Indicator and Coder Call Indicator may therefore be briefly summarized. In the Coder Call Indiacator System, regular Strowger impulses are converted into coder impulses by means of positive, light negative and heavy negative working for transmission along the inter-office trunk and are translated back into numerals for display at the manual end. The code impulses are transmitted along the junction in a minimum of time. In the Call Indicator System, all the equipments are rigidly located in the manual office, while in the Coder Call Indicator, the Coder is located in the automatic office and the decording equipments are located in the manual office. Thus, the methods of trunking and grading in both cases are different.

The economical and operating advantages which attend the use of the coder call indicator are considerable. The method of call distribution embodied in this system enables economics in the apparatus to be made compared with ordinary call indicator system as the amount of equipment can be apportioned on a traffic basis. The flexibility of coder call indicator for handling the traffic is ideal and far superior to the usual call indicator, a uniform load being distributed to all manual operators.

The total amount of automatic to manual traffic in the area will steadily increase from the time when the first automatic exchange is installed until a maximum is reached about the middle of the transition period, and then it will steadily decrease as more and more exchanges become automatic.

Mechanical Tandem Exchang.— Chief reasons on the introduction of the tandem junction working in telephone networks are:

1. Saving inter-office trunks, and
2. Offering higher trunking efficiency.

Ordinary tandem working on a strictly manual basis, however, has limitations due to the increased time required to complete a connection and the liability of error caused by the introduction of the additional operator. With the introduction of Mechanical Switching in the tandem exchange and the projection of the call to a call indicator position in the terminating exchange thereby increasing the speed of completion and reducing the liability of error, the limitation of tandem working are largely removed and the full benefits realized.

In the network having both automatic and manual exchanges, the Mechanical Tandem Exchange is required to route the traffic as follows:

From manual exchange to automatic exchange.

From manual exchange to manual exchange.

From automatic exchange to manual exchange.

As soon as all the manual exchanges in the network are converted into automatic working, the tandem exchange, then, becomes a Junction Center.

The method of handling traffic in the mechanical tandem exchange is illustrated in Fig. 7. Traffic from "A" positions of a manual exchange will be dealt with in a similar manner to the traffic incoming to a cordless "B" position at an automatic exchange. The "A" operator will "order wire" the call to the cordless "B" operator at the tandem exchange, who will set up the call to the automatic switches via her key sending equipment. If the call is for a manual exchange, it will be directed to a call indicator position at that exchange, and the numerical portion of the required number with effect the required display on the lamp panel at the call indicator position. If the call is for an automatic exchange, the numerical digits will go out to the switches at the automatic exchange in the regular manner.

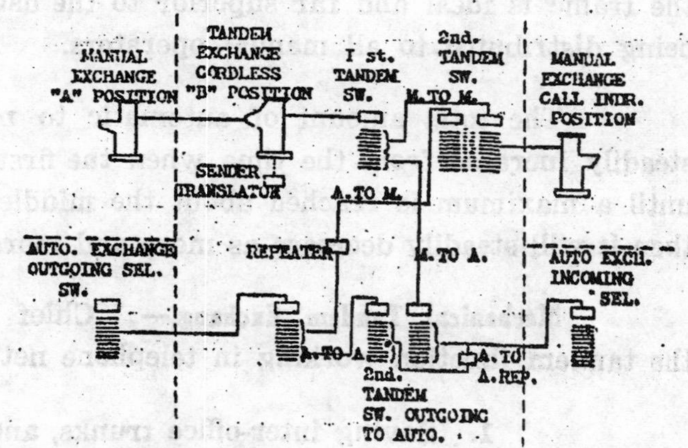

FIG. 7. *Mechanical Tandem Exchange Routing Scheme.*

Traffic originated in an automatic exchange will be carried direct from the levels of the outgoing switches to the first tandem switches at the mechanical tandem exchange. In the case of a call for a manual exchange, the first and second tandem switches will be operated and the numerical portion of the required number will be displayed at the call indicator position of the manual exchange. A call for another automatic exchange will pass out via levels on the first and second tandem switches and will be routed through repeaters to the required automatic exchange.

Conclusion. — This paper has mainly concerned itself with the application of the different methods for the inter-communication between automatic and manual telephones. The requirements of these methods during the period of transition may be briefly summarized as follows:

1. To ensure smooth operation between the two systems.
2. To give efficient service between the two.
3. Full automatic method of calling from automatic subscribers to manual ones is preferable.
4. Both initial and operating costs should be as low as possible.

複筋混凝土梁計算捷法

著者：趙國華

(一) 緒言　(二) 複梁在計劃,複核上所起之問題.
(三) 公式之說明及列題之說明.　(四) 結論.

(一) 緒言

鋼筋混凝土梁內之鋼筋,恒安置于斷面之下部.此係鋼筋混凝土性質上當然之事實.但在特殊情形之下,有時于上下二方各置之者,則前者稱之曰單梁(Simple R. C. beam)後者稱之曰複梁(Double R. C. beam).關于單梁之計算甚簡易,而應用範圍亦較狹,惟複梁之計算,用通常方法,甚爲繁雜但其用途甚廣.本篇即就該項計算上加以改良,並利用最便利之方法曰 Nomograph* 製成圖表三種以資應用,使之充分簡單,並可免除大部分之計算,因而勞力與時間,亦可節省多多矣.惟事屬初創,雖經作者試驗至再,惟恐尚有差池,深幸海內同志俯而教之.

本篇之作原因向先吾國所通行關於鋼筋混凝土工學之書籍如 Holl's R. C. Construction," Tayor & Thompson's "Concrete, plain and Reinforced" 以及新近通行之 "Urguhart & O'Roucke's "Design of Concrete Structures" 等書,恆用冗長之公式及理論以充教材,缺少實用上之例題.且處處用繁雜而須經驗之假定,已使初學者茫無頭緒,更因各書所附之表格,範圍較狹,往往不能直接讀得,而須用判讀法讀出之,以致時生差誤,且所用之單位制度,俱爲英制,但本國現下情形又須積極改成公尺制,在此過渡時期,至少限度亦須英公

---

*nomograph 一字擬譯成『圖表學』或有用 nomogram 者原字之意義謂 "The graphical presentation of laws", 在美國所通用曰 Alinement Chart, 英國則用 Alignement Chart.

二制參用,好在中國之工程事業,尙未十分發達,能在數年之內,努力改革用英制之習慣,猶未晚也.因以上種種之原因,欲以棉力補救於萬一,故不揣翦陋,將數年來研究所得,分別錄出以供諸先進之指正.今先就關於複筋梁計算之一部,先行供獻,俟他日另有機會,再將關於T梁及彎曲與直壓應力同時作用之部材(member)之計算方法,逐一發表之.

(二) 複梁在計算,複核上所起之問題

由已知彎曲力率,及材料之應力而定梁之斷面,此項步驟是曰計劃.由已定之斷面而求其抵抗力率或材料之應力,此項步驟,是曰複核.計劃與複核二種尙可分成若干問題討論之.今分別言之如次:—

歸於計劃上所起之問題有二:

(一) 上下鋼筋量及梁寬一定時,由彎曲力率以定其深度.

例如連梁(Continuous beam)支點處,受負彎曲力率時,有一部分鋼筋向上彎曲,而過支點,有一部分則在下部通過,此時鋼筋量與梁寬爲已知,而所須求出者僅爲其深度.

(二) 彎曲力率大於該梁所能任負,而混凝土斷面尺寸,又爲其他條件所限止者,(如建築,裝飾,採光,換氣等關係)則增加鋼筋用量以求其平.

例如懸臂梁(Cantilever beam)之弧度已定,及外彎曲力率已知,而求懸臂內所需鋼筋之量,如首都之中山橋,外觀爲一拱形,實爲二懸臂梁所兜成,該懸臂梁內鋼筋用量之決定,卽用下述之方法.

歸於檢算上所起之問題有二:

(一) 由已知之斷面而求其最小抵抗力率.

(二) 由已知之斷面而求其最大之應力.

此項檢算上所起之問題,由於混凝土斷面之尺寸,在計算時往往帶有另星小數,而在實際施工或製圖時,常作爲整數,又所用鋼筋之大小,往往與市

場上所供給者不符,致生差異,又斷面上部鋼筋之位置,往往不能與其所假定者符合,因以上之種種原因,故須有檢算之必要究竟該項最後決定之斷面,能負多大之抵抗力率及其材料所起最大之應力.

(三) 公式之誘導及例題之說明

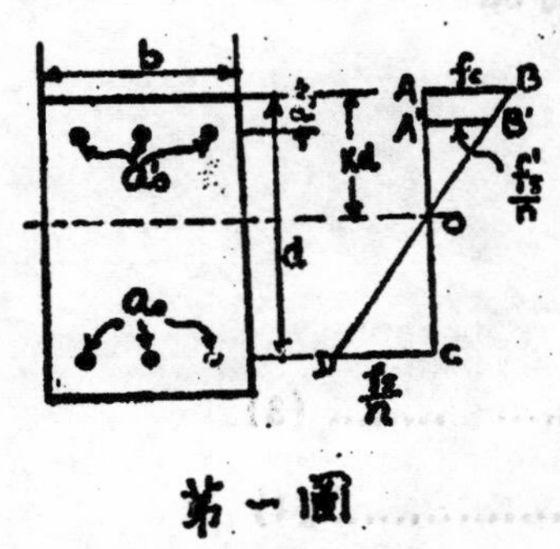

第一圖

(一) 第一問題之解法.

依第一圖之斷面.

假定　梁之有效深度 (Effective depth) $=d$

梁之寬度 $=b$

抗壓側鋼筋百分比 $=P_c$

抗張側鋼筋百分比 $=P_t$

抗壓側鋼筋中心至邊線距 $=d'$

中軸距比 $=k$

彈率比 $=n=15$

$d'/d=\beta$

求該斷面之混凝土及鋼筋關於中軸線所起之一次率而置之為零即

$$bd^2\left\{\frac{k^2}{2}-n\left[(1-k)P_t-(k-\beta)P_c\right]\right\}=0.$$

$$\frac{k^2}{2}-n\left[(1-k)P_t-(k-\beta)P_c\right]=0.$$

$$k^2+2n(P_t+P_c)k-2n(P_t+\beta P_c)=0.$$

解上列 k 之二次式則得

$$k=-n(P_t+P_c)+\sqrt{2n(P_t+\beta P_c)+n^2(P_t+P_c)^2}\text{..........(1)}$$

上式中,如 P_t, P_c, β, n 四值為已知,則 k 值即可求得.

又求該斷面混凝土及鋼筋二項關於中軸線所起之二次率 (moment of Inertia) 應為

$$I=bd^3\left\{\frac{k^3}{3}+n\left((1-k)^2P_t+(k-\beta)^2P_c\right)\right\}$$

將上式代入材料力學上之基本公式

$$M=\frac{If}{c}=\frac{If_c}{kd}$$

$$=\frac{f_c\ bd^3\left\{\frac{k^3}{3}+n\left((1-k)^2P_t+(k-\beta)^2P_c\right)\right\}}{kd}$$

$$=f_c\ bd^2\left\{\frac{k^2}{3}(1-k/3)+\frac{nP_c}{k}(k-\beta)(1-\beta)\right\}$$

$$=R\,f_c\ bd^2 \qquad\qquad (2)$$

卽 $R=\frac{k^2}{2}(1-k/3)+\frac{nP_c}{k}(k-\beta)(1-\beta)=\frac{M}{f_c\ bd^2}$ ……(2)'

又由相似三角形AOB,COD中可得

$$f_s=\frac{n(1-k)}{k}f_c=\frac{f_c}{Q}$$

卽 $$Q=\frac{k}{n(1-k)}=\frac{f_c}{f_s}$$

$$M=R.\ Q.\ f_s.\ bd^2 \qquad\qquad (3)$$

$$\therefore \quad d=\sqrt{\frac{M}{Rf_c\ b}} \qquad\qquad (4)$$

$$d=\sqrt{\frac{M}{R.\ Q.\ f_s\ b}} \qquad\qquad (5)$$

今由第一式可知,因 P_t, P_c, 值之變更,而使 K 值同時亦起變化.而所求之 d 值.亦因之發生問題,究用(4)式抑用(5)式以定之乎?此點在通常之書籍上都未論及,使初學者莫衷一是.但依研究所得,凡由(1)式求得之K值大於 $\frac{15}{15+\frac{f_s}{10}}$ 者,則用(4)式決定其深度.若小此數者,用(5)式以決定之.今舉一例以明之.

例一 已知之項目

$b=12''$, $P=P'$, $as=3$ sq. in.

$M=750{,}000$ in. lbs.

$f_c=750$ lb/sq. in. $f_s=16{,}000$ lb/sq. in.

求梁之有效深度.

解. 今假定 $j=7/8$*

*如用於本國現在各市所規定之容許應力範圍之內,j 值恆可假定爲 ⅞ 以計算之.

則　$d=\dfrac{M}{asf_s j}=\dfrac{750{,}000}{3\times16{,}000\times\frac{7}{8}}=17.85$ inches

而　$P_t=P_c=\dfrac{3}{12\times17.85}=.014.=1.4\%.$

假定　$d'/d=0.10$

今　$P_t+P_c=2.8\%.$

$P_t+\dfrac{d'}{d}P_c=1.54\%.$

由圖表（1）得　$k_1=0.38$

但　$k=\dfrac{15}{15+\dfrac{16{,}000}{750}}=.414$

∴　$k_1<k.$

故用（5）式以求 d 令

$$Q=\frac{k_1}{n(1-k_1)}=\frac{0.38}{15(1-.38)}=0.0408.$$

由圖表（1）得

$$R=\frac{M}{f_c bd^2}=0.302.$$

$$\therefore\quad d=\sqrt{\frac{750{,}000}{0.302\times0.0408\times16{,}000\times12}}=17.75 \text{ inches}$$

即爲所求梁之有效深度,苟若不察,誤用（4）式以求之,則其結果又復不同,此點不可不加以注意,今就用（4）式以求 d 値之結果

$$d=\sqrt{\frac{750{,}000}{.302\times750\times12}}=16.6. \text{ inches}$$

又關於 j 値一項先由假定而算出 d 値復將此 d 値代入次式

$$j=\frac{M}{f_c bd^2}\cdot\frac{Q}{P_t}$$

而得 j,此値是否與假定相同,此亦檢算之一法.惟在本例中之⅞,及依本圖情形而用⅞,往往適合,故此項手續可省.惟因其他之關係,則須有檢算之必要.

如斷面已經決定,因在實施時之便利,不能吻合計算所得之尺寸,則由此

已決定之斷面以求其最大應力或最小抵抗力率.但應用何式以決定之,則在通常書藉上又復含糊莫辨,往往將二值,俱行求出而比較之,但亦可視K_1值之大小而立卽判斷之,如K_1值大於$\frac{n}{n+\frac{f_s}{f_c}}$時,則梁之抵抗力率及材料之最大應力以抗壓側爲準,卽用(a)式決定之,若小於時,則以抗張側爲準或卽用(3)式決定之,今舉例以明之:

例二 若於例一所得之結果中爲實施時之便利,其斷面應如第二圖圖內所示.

$$d=18\ \text{in.}$$
$$as=as'=3\ \text{sq.. in.}$$
$$b=12\ \text{in.}$$
$$d'=2\ \text{in}$$

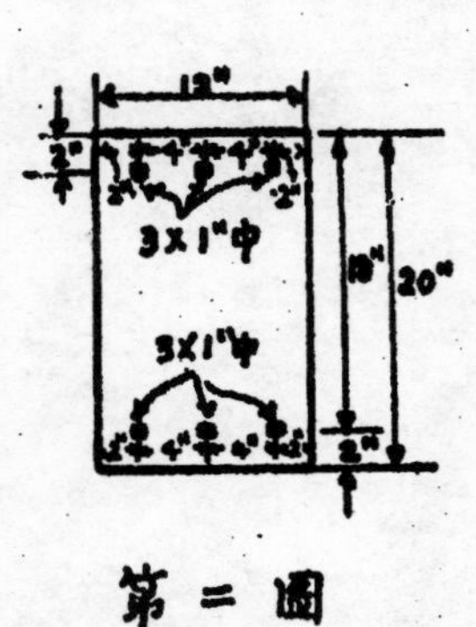

第二圖

求其最小抵抗力率及材料之最大應力.

解. 令. $P_t=P_c=\frac{3}{18\times12}=1.39\%$

$$\frac{d'}{d}=\frac{2}{18}=0.111.$$

$$P_t+P_c=2.78\%$$

$$P_t+\frac{d'}{d}P_c=1.55\%.$$

由圖表一可得

$$k_1=0.385<\left(\frac{n}{n+\frac{f_s}{f_c}}\right)$$

故可逕由(3)式求之

由圖表一可得 $R=\frac{M}{f_c\,bd^2}=0.295.$

而 $Q=0.0417$

故 $M_s=0\ 295\times0.0417\times16{,}000\times12\times\overline{18}^2=766{,}000$ in. lbs.

而 $f_s=\frac{750{,}000}{0.0417\times0.295\times12\times\overline{18}^2}=15.700$ lbs./sq.in.

如用(2)式求之則其結果如下

$$M_3=0.295\times750\times12\times\overline{18}^2=861{,}000\ \text{in. lbs}$$

$$f_3=\frac{750{,}000}{0.295\times12\times\overline{18}^2}=654\ \text{lbs./sq. in.}$$

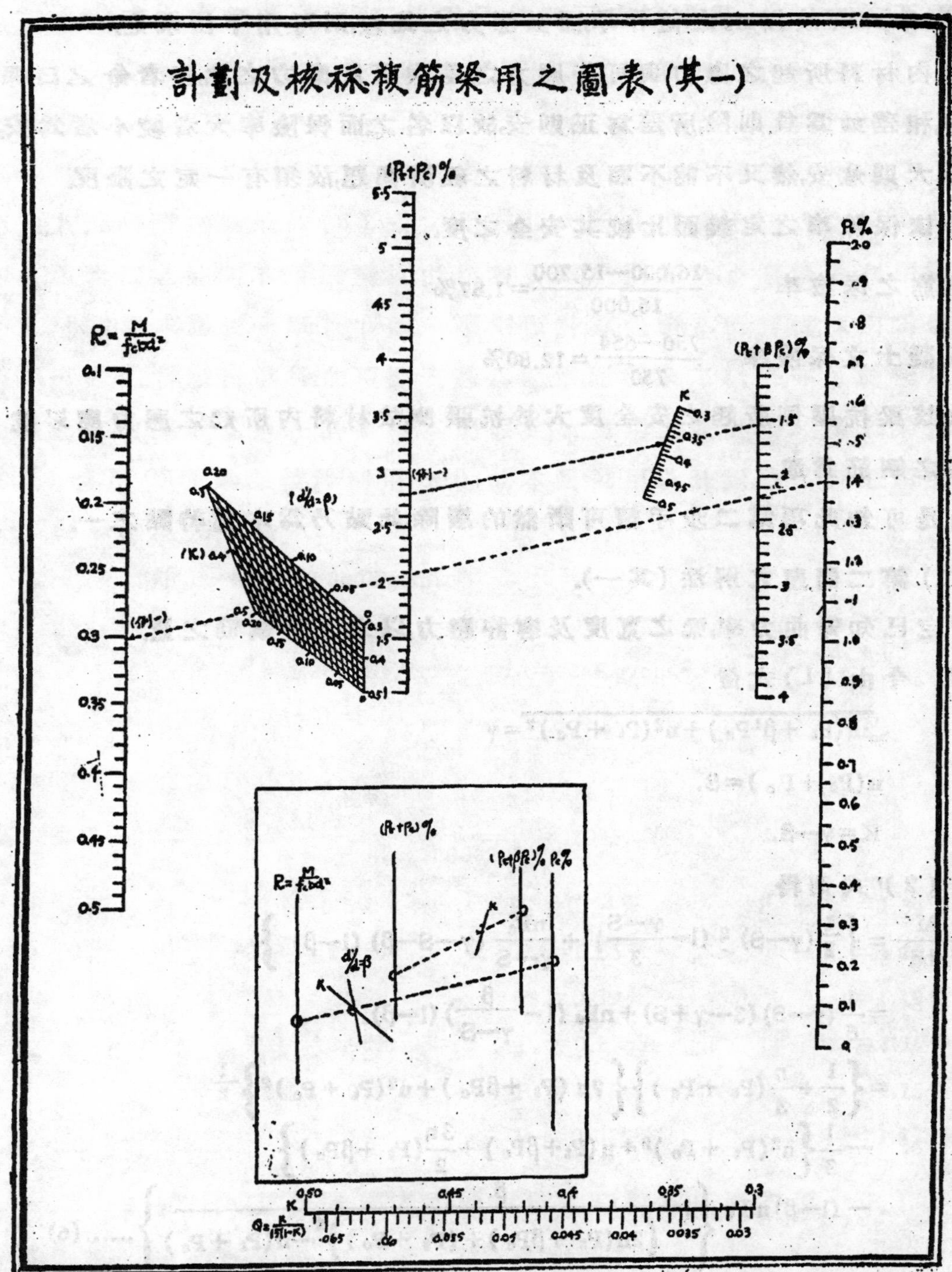
計劃及核袮複筋梁用之圖表(其一)

今 $M_s > M_c$ 故以 M_s 之值作準,而其應力之比較,則可用下法求之.

梁內材料所起之應力,與可許應力之差,與可許應力之比作者命之曰保險率,相差如為負則險,所差為正則安,故以名之而保險率大者較小者為安,值愈大則愈安,然又不能不顧及材料之經濟問題,故須有一定之限度.

今依保險率之定義而比較其安全之度.

鋼筋之保險率 $\frac{16,000-15,700}{16,000}=1.87\%$

混凝土之保險率 $\frac{750-654}{750}=12.80\%$

故該梁抗壓側所起之安全度大於抗張側,故材料內所起之應力應以抗張側之鋼筋為準.

由是可知此項第二步手續可斷然的廢除,此點乃為本篇特點之一.

(二) 第二問題之解法 (其一).

假定巳知彎曲力率,梁之寬度及容許應力,以求上下鋼筋之量.

解. 今由 (1.) 式置

$$\sqrt{2n(P_t+\beta^1 P_c)+n^2(P_t+P_c)^2}=\gamma$$

$$n(P_t+P_c)=S.$$

則 $K=\gamma-S.$

代入 (2)' 式則得

$$\frac{M}{f_c bd^2}=\left\{\frac{1}{2}(\gamma-S)^2\left(1-\frac{\gamma-S}{3}\right)+\frac{nP_c}{\gamma-S}(\gamma-S-\beta)(1-\beta)\right\}$$

$$=\frac{1}{6}(\gamma-S)(3-\gamma+S)+nP_c\left(1-\frac{\beta}{\gamma-S}\right)(1-\beta).$$

$$=\left\{\frac{1}{2}+\frac{n}{3}(P_t+P_c)\right\}\left\{2n(P_t+\beta P_c)+n^2(P_t+P_c)^2\right\}^{\frac{1}{2}}$$
$$-\frac{1}{3}\left\{n^2(P_t+P_c)^2+n(P_t+\beta P_c)+\frac{3n}{2}(P_t+\beta P_c)\right\}$$
$$-(1-\beta)nP_c\left\{1-\frac{\beta}{\left(2n(P_t+\beta P_c)+(P_t+P_c)\right)^{\frac{1}{2}}-n(P_t+P_c)}\right\}\cdots\cdots(6)$$

於（6）式之中若P_c，P_t，β三值爲已知，即可直接求得$\frac{M}{f_c bd^2}$之值而不必再由（1）式求k_1代入（2）式求R.如若R，β二值已知則由該式可定P_t，P_c之值.但P_t，P_c二值可得任定，而成種種之配合，故如先知斷面內一例之鋼筋量，即可求得他方應需之量，又於若干種配合之中比較鋼筋用量之多寡，而取其最經濟之結果，故（6）式之用處甚大.但在事實上該式之繁，一望而知之.因此法之長處不可放棄，故由作者費數星期之久，不避繁雜，利用圖表學，將該式製成圖表一紙.惟β值一項，則因計算過繁，不能詳備，深引爲憾.如將來有暇，定當繼續進行.此項圖表甚爲簡單，而所求之結果，亦可於極短時間找出之.而關於β值在計劃時，本屬假定，故實用上在計劃較薄之梁或坂(Slab)時用0.15，梁深較厚則用0.10，故本圖表用以計劃，已足敷應用矣.茲舉二例以明其用法如次：

例三. 已知 $M=320,000$Kgcm.

$b=100$ cm, $d=20$ cm

$f_s=1,260$ Kg/cm^2 $f_c=42$ Kg/cm.^{2}

$d'/d=\beta=0.10.$

求上下鐵筋之百分比.

解. 今 $R=\frac{M}{f_c bd^2}=\frac{320,000}{42\times100\times20^2}=.1905.$

由圖表二可得

$P_t=0.6\%$	$P_c=0.74\%$	$P_t+P_c=1.34\%$
$P_t=0.7\%$	$P_c=0.55\%$	$=1.25\%$
$P_t=0.8\%$	$P_c=0.39\%$	$=1.19\%$
$P=0.9\%$	$P_c=0.27\%$	$=1.17\%$
$P_t=1.0\%$	$P_c=0.15\%$	$=1.15\%$
$P_t=1.1\%$	$P_c=0.075\%$	$=1.175\%$

由以上之結果，可見抗張側之鋼筋用量大時較爲經濟.本項中以1.15％爲最經濟之鋼筋百分比.

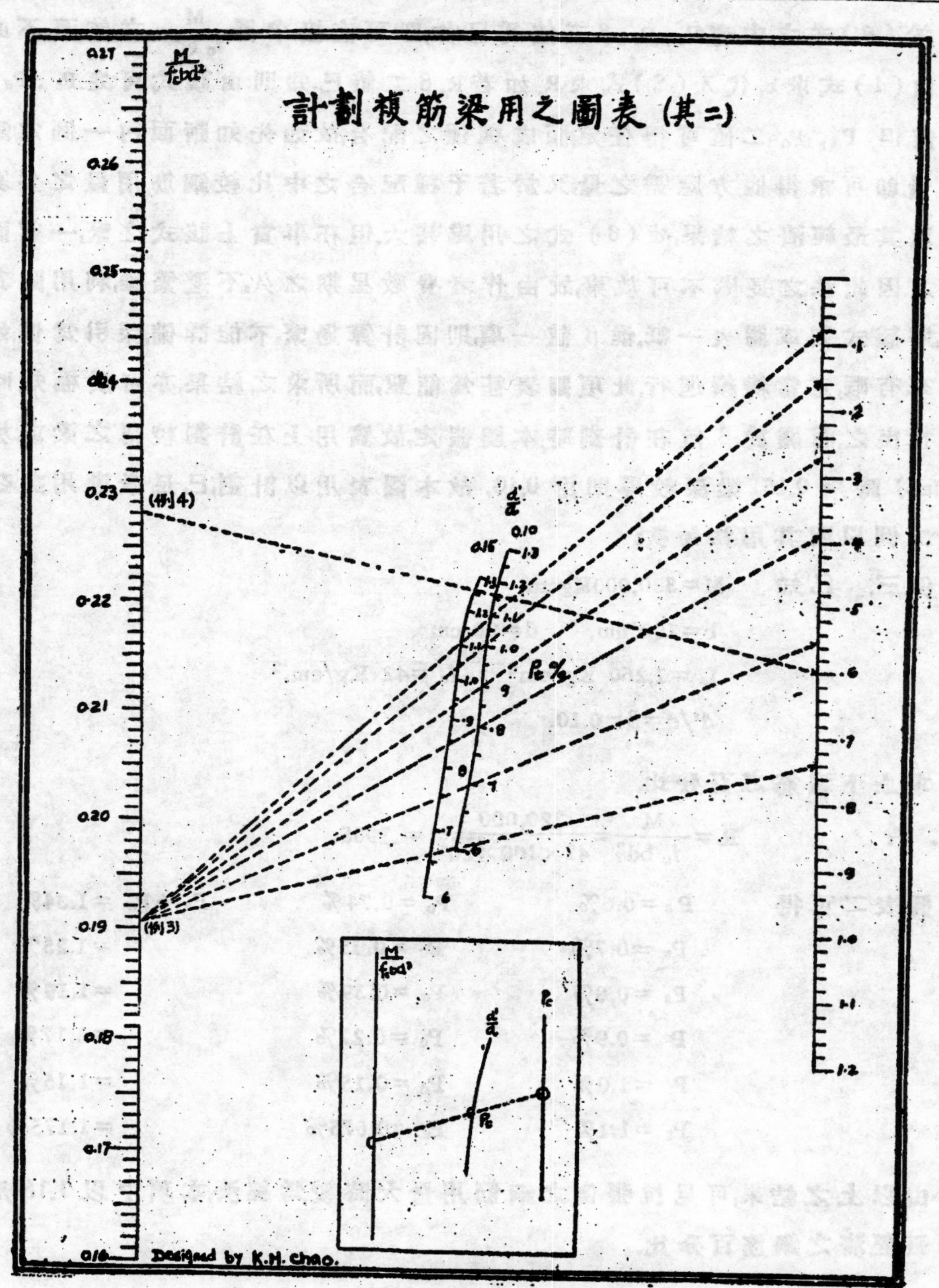
計劃複筋梁用之圖表 (其二)
M/fc bd²
0.27
0.26
0.25
0.24
0.23
(例4)
0.22
0.21
0.20
0.19
(例3)
0.18
0.17
0.16
d'/d
0.10
0.15
Pc %
Pc
Designed by K.H. Chao.

例四．已知　$M=860,000$ in. lbs.

$b=12$ in. $d=22$ in.

$f_s=16,000$ lbs./sq. in.　$f_c=650$ lbs./sq. in.

$d'/d=\beta=0.10$,

$as=4\sim\frac{7}{8}''$中$=3.06$ sq. in.

求抗壓側應須鋼筋之量

解．今　$P_t=\frac{3.06}{12\times 22}=1.16\%$

而　$R=\frac{860,000}{650\times 12\times 22^2}=0.228.$

由圖表二可得

$P_c=0.6\%$.

故　$as'=0.6\%\times 12\times 22=1.584$ sq. in.

用　$2\sim\frac{7}{8}''$中　稍不足其所需，

(三) 第二問題之解法 (其二)

今假定鋼筋梁斷面bd及經濟鋼筋比P，則此斷面所起之抵抗力率爲

$$M_1'=f_s P\left(1-\frac{k}{3}\right)bd^2$$

如此項斷面不足勝任外來之彎曲力率，其所不足之彎曲力率，應由上下二方各增鋼筋用量，使上下鋼筋所起之隅力，適足以抵抗其不足．荷鋼筋加於該斷面之內而不變其二側之最大應力時(卽使k值不變)，此項增加鋼筋之量所起之偶力，應依原斷面所起之應力計算之．故在抗張側所增之鋼筋百分比，應由下式求得之：

$$M_2=M-M_1=f_s(d-d')as,=f_s\left(1-\frac{d'}{d}\right)P_1bd^2.$$

或　$$P_1=\frac{M-M_1'}{f_s(1-\beta)bd^2}\quad\cdots\cdots\cdots\cdots(7).$$

故抗張側鋼筋百分比之總和應爲

$$P_t=P+P_1\quad\cdots\cdots\cdots\cdots(8)$$

又由第一圖中之相似三角形 A'OB',COD,而得

$$f'_s = f_s \frac{1-k}{k-d'/d} \quad \cdots\cdots(9)$$

故抗壓側所用之鋼筋百分比 P_c 可由下式求得之

$$M_2 = M - M_1 = f'_s (d-d')as_2 = f'_s (1-d'/d) P_c bd^2.$$

或
$$P_c = \frac{M-M_1}{f'_s (1-\beta) bd^2} \quad \cdots\cdots(10)$$

將(7),(10)二式中消去 $M-M_1$,而以 $f'_s = f'_s \frac{1-k}{k-d'/d}$ 代入而化簡之則得

$$P_c = \frac{1-k}{k-\beta} P_1 \quad \cdots\cdots(11)$$

今因
$$M = M_1 + M_2 = f_s [P(1-k/3) + (1-\beta) P_1] bd^2.$$

$\therefore$
$$M' = \frac{M}{f_s bd^2} = P(1-k/3) + (1-\beta) P_1.$$

或
$$P_1 (1-\beta) = M' - P(1-k/3).$$

$\therefore$
$$P_1 = \frac{M' - P(1-k/3).}{(1-\beta)}$$

而
$$P_t = P_1 + P = \frac{M' - P(1-k/3)}{(1-\beta)} + P \quad \cdots\cdots(12)$$

$$P_c = \frac{M' - P(1-k/3)}{(1-\beta)} \cdot \frac{1-k}{k-\beta} \quad \cdots\cdots(13)$$

故若已知 M, P, β, b, d, f_s 等值卽可由(12),(13)式直接求得其上下所需之鋼筋百分比.

茲將以上各式,求得一適合本國現在各市所規定之建築條例,假定 $f_s/f_c = 30$ (因 f_s = 18,000 lbs/sq.in f_c = 600 lbs./sq.in f_s = 1,260 Kg./cm², f_c = 42 Kg/cm²)

依以上之假定,則斷面內之經濟鋼筋百分比

$$P = 0.565\%$$

而中軸距比　$k = ⅓$

故
$$P_t = \frac{M' - 0.565\% \times (1-1/9)}{1-\beta} + 0.565 \quad \cdots\cdots(14)$$

$$= \frac{M' - .00503}{1-\beta} + 0.565\%$$

$$P_c = \frac{M' - 0.00503}{1-\beta} \cdot \frac{2/3}{1/3-\beta} = \frac{2M' - 0.01006}{(1-\beta)(1-3\beta)} \quad \cdots\cdots(15)$$

今將(14)(15)二式,利用圖表學製成圖表一種,以資實施應用.今舉一例,以明其用:

例五. 已知　$M=320,000$ Kg cm.　$\beta=0.10$

$b=100$ cm,　$d=20$ cm,

$f_s=1,260\text{Kg/cm}^2$,　$f_c=42\text{ Kg/cm}^2$,

求 P_t, P_c 之值

解. 用(12)(13)二式之解法.

(1)　$k=\frac{1}{1+nf_s/f_c}=1/3$

(2)　$P=\frac{f_c k}{2f_s}=0.565\%$

(3)　$M_1=f_s p(1-k/3)bd^2=253,300$ Kg cm.

(4)　$M_2=M-M_1=320,000-253,300=66,700$Kg/cm.

(5)　$P_1'=\frac{66,700}{1,260(1-0.1)\times100\times20^2}=0.00147,$

(6)　$P_t=0.00565+0.00147=0.00712.=0.712\%$

(7)　$P_c=P_1\frac{1-k}{k-\beta}=0.00147\frac{1-1/3}{1/3-0.1}=0.00421.=0.421\%$

用圖表之解法.

先算出　$M'=\frac{M}{f_s bd^2}=\frac{320,000}{1,260\times100\times20^2}=0.00635$

由圖表三卽讀得　$P_t=0.71\%$

$P_c=0.42\%$

由是可知用圖表以求解答,其便捷較諸計算,何啻倍蓰而所起之差誤,又復極微.

(四)　結　　論

本篇所附之圖表,其長處除能減省冗長之計算及樽節時間之外,尙俱有普遍之性質,無論所用之單位制度爲英制或公制,或爲市制,本圖表俱可應

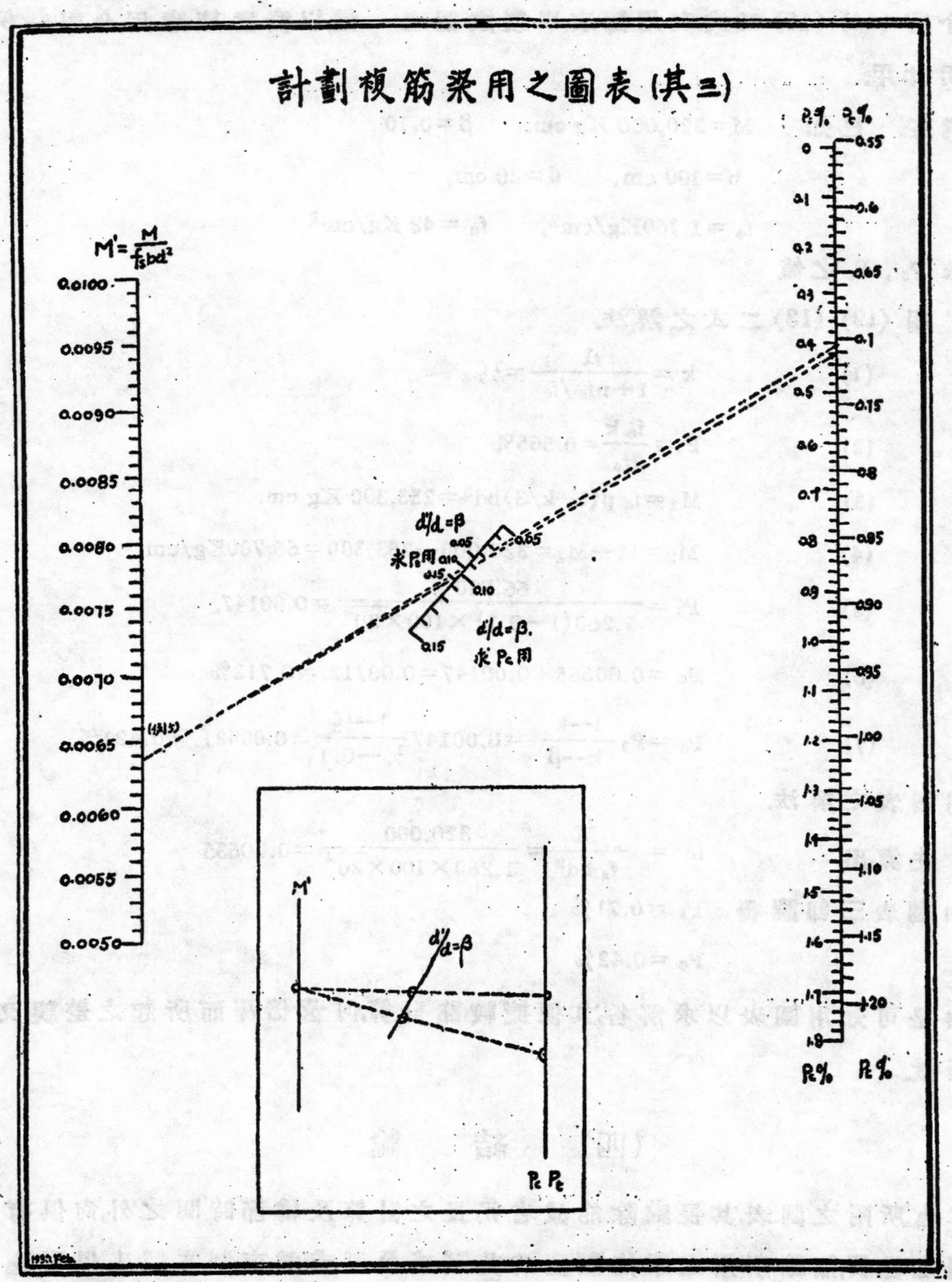
計劃複筋梁用之圖表(其三)
$M'=\frac{M}{f_s bd^2}$
0.0100
0.0095
0.0090
0.0085
0.0080
0.0075
0.0070
0.0065
0.0060
0.0055
0.0050
d'/d=β
求Ps用
0.05
0.10
0.15
d'/d=β.
求Pc用
$P_s\%$ $P_c\%$
0
0.1
0.2
0.3
0.4
0.5
0.6
0.7
0.8
0.9
1.0
1.1
1.2
1.3
1.4
1.5
1.6
1.7
1.8
0.55
0.6
0.65
0.7
0.75
0.8
0.85
0.90
0.95
1.00
1.05
1.10
1.15
1.20
$P_c\%$ $P_c\%$
M'
d'/d=β
P_c P_c

用,今示一例如下:

例六. 已知 $M=64,000$寸斤(市制) $\beta=0.10$.

$b=10$市寸 $d=6$市寸

$f_s=28,200$ 斤/市寸^2

$f_c=940$ 斤/市寸^2

求上下鋼筋之百分比.

解 $$M'=\frac{64,00}{28,200\times10\times6^2}=0.00644.$$

由圖表三讀得 $P_t=0.715\%$

$P_c=0.45\%$.

故作者以為欲決心改革單位制度,實非難事.如本篇所述之方法歸於單位制度所起之困難問題,消滅殆盡,無論何種制度均能適用,此點亦為本篇之特點.而圖表一舉,實為促進改革單位制度之急進先鋒,深幸我國工程界諸先進,起而圖之.

本篇因專就關於複梁之計算方法上着點,故不克附有良好而完善之實例,以明其步驟,深引為憾.但對於該項計算上所起之問題,大致俱可解決.故如能明瞭以上之步驟及方法,則在實施應用時,已足敷應用矣.又歸於施用圖表之解法,則每一例題,卽在該圖表上用虛線表示其徑途,此項步驟,可不辯自明,故不贅述.惟該項圖表製成之方法,則因限於篇幅,不克序述,俟他日另有機緣,再行另立題眉,專述此事可也.

本會編輯部啟事

下期(六卷二號)工程,專載本會在瀋陽東北大學舉行年會時,所集各種論文,稱為年會論文專號.該稿付印在卽,約於明年二月杪,可以出版.此啓.

十九年十二月

整理東北水利芻議

著者: 朱重光

(一) 概 論

東三省——遼寧,吉林,龍江,——位於我國東北隅,東控沿海州,以出日本海,南附朝鮮半島之背,以瞰日本,北扼黑龍江,以阻俄人東下之路,西控蒙古,跨長城,以制中原,西南據遼東半島,以控黃渤,隱握東西兩洋之關鍵,此地理上之位置也.若以幅圓論之,東西廣約二千八百餘里,南北長約三千餘里,面積三百六十餘萬方里,較日本三島,幅圓約大二倍有半.其境內有最大之山嶽,——內興安嶺——,最大之川流,——黑龍江——,最大之平原,——內蒙古——,最良之海港,——大連,營口,葫蘆島——,其礦產約計六百處.林產有面積六萬方里,約占三省總面積六分之一.農產計十餘萬方里,約占三省總面積三分之一.他若水產,畜產等,亦無一不備.苟開發而利用之,即此一隅之經濟,足以抵抗日,俄帝國主義之侵略,而況擁有百餘州之富乎?故論東北建設事業,凡我技術同志,亟應起而圖之,本各人之專長,盡個人之職責,抱互助之精神,謀國力之富強,豈東北一隅之福哉?重光服務東北水利事業有年,因就管見所及,草擬整理東北水利芻議,以就正於同志諸公,幸辱教焉!

(二) 河流之性質—國際河流,國內河流

東三省當蒙古高原之東,擁有三百餘萬方里之面積,其南北二部有最大之山脈,盤繞其間,故川河交錯,流域綿長,為我國第三大河流.查東省與日,俄接壤,多以河流為界,故東北河流有國際及國內之分.今將其較大河流,而有關於交通農利者,略述如下:

(甲)國際河流　(1)黑龍江流域——額爾古訥河,黑龍江,混同江,烏蘇里江(2)圖們江(3)鴨綠江

(乙)國內河流　(1)松花江,嫩江,牡丹江(2)遼河,柳河,渾河,太子河(3)大淩河

以上國際河流中,除鴨綠江外,均無海口,航行我國本部之輪船,不能駛入上述之河流.而該河流內之我國商輪,亦不能駛至中國本部.故東北貨物之運輸,非假道於中東鐵路,由海參威出口,即借路於南滿鐵路,由大連出口.每歲運費損失之鉅,不問可知,此就商業而言.若論及國防,危險尤甚.查東北江防軍艦九艘,困守松花江內,不能越三江口一步,黑龍江之江防,從未過問,此何故耶?蓋缺少海口,萬一與俄發生事故,我海防各艦,不能援助,孤軍無援,不得不示退讓也.

(三)整理國內河流

東北國內河流之犖犖大者有三,即松花江,遼河,大凌河.後二者均有海口,故遼甯省貨物之運輸,可不假道於人.陸路由平遼鐵路直達關內.海道由營口以達四方.吉黑兩省,則不然.陸路無直接關內之鐵道,水道缺直接入海之海口,貨物運輸,不假道於南滿,即借路於中東,利權外溢,年以千萬計.若不挽救,兩省必至受經濟之壓迫,至於民窮財盡而後已.挽救之道何如?實現　總

理之實業計劃,開鑿松花江及遼河間之運河,爲治本之維一善策.蓋將來鑿成之後,其利益之點甚多,略述如後:

(1)黑龍江流域各河流,均有直接入海之海口.

(2)吉,黑兩省之貨物,均可由水道運輸,運費比鐵道爲廉.

(3)可以振興營口之商業,與大連,海參威競爭.

(4)可以打倒日,俄之鐵路侵略主義.

(5)可以聯絡海防江防,鞏固我邊陲.

至於運河之路線,參考日人濱井松之最新滿洲全圖,在東遼河與伊通河之間:(一)由東遼河流域內之十屋地方起,經懷德縣城至大嶺鎮,與伊通河合流約長三十六公里.(二)由十屋上流雙城堡起,經三道崗,至大嶺鎮下流萬家橋附近,與伊通河會合,其長亦不過三十六公里.其工程費假定每公里十九萬元,(據孫哲生部長去歲雙十節發表之建設大綱草案新開運河經費預算每英里三十萬元).共需國幣六百八十四萬元.區區建設費,或由兩省分擔,或由中央建設經費項下支出,想易爲力也.對於黑龍江流域之河流,如松花江,嫩江,牡丹江等開濬工程,均歸東北水道局辦理.遼河,柳河,渾河及太子河等之整理工作,均歸遼河工程局辦理.分工合作,三五年後,行見最重要之運河,最流長之幹河於東北,豈不快哉?至大凌河之工程,將來可歸開關葫蘆島商埠工程局辦理,蓋該島在大凌河口,有直接關係在也.

(四) 整理國際河流

吾人對於國際河流,偶不注意,強鄰將佔爲己有,譬如黑龍江,乃中,俄兩國之國際河流.今俄艦向兩岸可以自由行動,而我艦不但不能至俄地,併我國河流不能駛入,不平等何如?不自由何如?况整理黑龍江之經費,中,俄各半,由黑河中國海關支付.工程人員均係俄人,而中國方面河流,是否年年修濬,向無報告,中國官廳亦未過問,權利旁落,宜乎俄人之野猴自大矣.今爲爭自由平等計,首先停給修濬經費,工程歸東北水道局負責,且將修濬經費,以補該局工程經費之不足,是則望於東省當局者也.黑龍江問題如能解決則混同江,烏蘇里江,及額爾古納河均可迎刃而解矣.

鴨綠江爲中韓之國際河流,上流有極大森林,下流有不凍港口,日人將在安東添築商港,以壟斷鴨綠江.今爲抗制日本侵略主義保護我國領土計,依照　總理之實業計劃,將安東闢爲漁業港,爲唯一補救之策.

圖們江爲中韓及俄韓之國際河流.上流爲中韓合流,下流爲俄韓合流.我國竟被俄,韓封鎖入海自由,喪失殆盡,此誠外交上之傷心史也!此江爲吉林入海最近之道,且與吉會鐵路平行,北滿貨物之輸出入,均可由此江轉運,商業之發達,可與俄之海參威及韓之清津爭霸.今欲達到此目的,非將江之左岸(自江口溯上三十里)之地恢復不可,如我國以革命外交爭之,或有一線之望,此則非我技術同人之責,乃執政諸公之任也.

(五) 結　論

整理東北水利,既如上述,但玆事體大,實行談何容易.其最困難之處,有下述數點:(一)東北介居兩大之間,外交棘手,一舉一動,易起衝突;(二)因政治關係,經濟更受影響,建設進行,遂被阻止;(三)東北人士,素抱門羅主義,不論人之才學如何,不顧本地有無人才,非我者去,爲客者逐,技術同志,每多却步;(四)主持行政者,往往爲腐敗官僚,工程經費,藏而不用.有此種種原因,故東北建設事業之前途,整頓實未易言也.愚意東北河流,或屬國際,或屬兩省,整理之權,應歸中央管轄,由中央建設委員會主持辦理,則事權統一,既可免兩省之爭,(遼吉爭辦東北水道局甚劇),而　總理實業計劃,亦於此可以完成矣.敢貢區區,尙希諸同志有以教之,幸甚!

THE DESIGNING OF HIGHWAY CONCRETE FRAME STRUCTURES FROM A PRACTICAL POINT OF VIEW

BY WM. H. F. WOO, M. S.

Synopsis. — Compared to other types of structures used in highway constructions either to bridge streams, or for grade separation, the concrete frames are little used, and judging from my personal knowledge, such a structure is not classified as a standard structure. Like railroad turnouts, no two frames are of exactly identical design, this being on account of the difference in physical and geological conditions in the locations where they are built, difference in the loading they are designated to carry, and several other factors affecting the design.

This paper deals with and discusses the methods commonly employed in the design of frames and the arrangement of members and reinforcement, attempting to throw some light on the practical side of the design. Discussions in this paper are written primarily to apply to frames when such are used as highway structures, but, as will be found, a great part of them can apply with equal force to similar structures designed for other purposes.

Generol Considerations. — A frame is a structure consisting of a girder or girders and columns, the ends of the girder being so rigidly connected with the columns supporting it, that deflection of one will cause deflections of all other members thereby inducing stresses in these members. The columns may be designed as, and built with fixed, free ends or any state intermediate between these two limits. Like a simple structure, the girder is the first member to take the load the structure is designed to carry, which then traverses through the joint, pass the footing, and then reach the foundation.

The use of frame structure as a highway bridge type is so far limited to those appearing in the forms of viaducts and overhead structures used in grade spearation. For bridging over streams of normal magnitude, it has not been much adopted due to its unsightliness and difficulty of its construction at such locations, and danger of being disturbed for being near the water.

Frames as overhead structures are very common, and in fact some railroad companies in the United States specifically desire that only such be the type built over their tracks for gr ade separation.

Frames as viaducts are also common. They are adopted not only due to the saving and economy it offers as compared to the similar concrete structures, but also due to their stiff resistance to longitudinal and lateral deflections. This latter property owned by some types of structures is gradually receiving attention and appreciation of highway engineers on account of the great number of fast moving vericles travelling on highways at the present time, whose speed constitutes a better agent of attacking on less rigid structures.

Whether a frame is an economical structure or not depends to a considerable degree upon the way in which the floor system is arranged. The types of floor system commonly adopted to such structures are three: (a) slab type in which the beams are omitted, the load being directly transmitted from the slab to the girder, (b) slab and beam construction in which the floor beams are arranged perpendicular to the roadway supporting the slab and and resting on girders, (c) like (b) only stringers are added. In several of the actual designs, the writer had occasions to investigate each of these arrangements and compare their relative economical value and reach the following conclusions:

(1) The use of (a) system is limited to those frames with short spans in which the dead load alone does not form a great part of the total moment from which the section is determined. The omission of floor beams necessitates that a thicker slab be used and therefore brings more dead load to the girder than in the case when the floor beams are used.

(2) Under any circumstance, type B is the most economical arrangement, and with skillful arrangement of the girders and floorbeams, the panel of the slab enclosed on the four sides by girders and floorbeams can be made to approximate a square, and therefore a doubleway reinforcement can be used. Any load that comes on the slab is distributed among and carried by the two systems of reinforcement. Thus a thinner slab may be used, and therefore a deduction in dead load is the result. It should be added, however, that by using the double system more steel is required, but the thickness of the slab is greatly reduced. Considering the prices to be paid for each, the amount saved, expressed in terms of money, through the adoption of thinner slab under ordinary circumstance is more than that to be spent for the additional steel resulted from the use of double reinforcement. The thinner slab contributes a lesser dead load to the structure, thus indirectly affecting a further saving in the cost of entire structure, a factor demanding

notice for long spans frames, say over 50 feet. (c) The use of stringers in addition to the floor beams system that was mentioned above is not general on account of the added expense to pay for the complication of form that result. The sections of the stringers with the lengths that are commonly designed are generally governed by the allowable shear at the ends and when this is the case, it explains itself that much concrete is wasted at sections where the allowable shear is not exceeded.

In general it may safely be concluded, that the floor system of a frame structure unlike that of a simple structure cannot be economically designed without a consideration of the arrangement of the other members with which it acts as a single unit. Under some circumstances the type of floor system cannot be choosen at randem to fit economy or to suit the wish of the designer and is dictated by the local conditions. Thus lack of head room, unfavorable foundation conditions are some among the governing conditions.

The designing of a highway frame structure is a problem requiring sound judgment in addition to precision in computations. The arrangement of the structure, that is the manner in which the columns, girders and floor systems are connected to form a unit, is very much determined by local physical requirements and geological conditions, and the success of the design depends upon the degree of thoroughness with which these matters are investigated.

Good subsurface foundation is one of the chief elements favored to the erection of a frame structure. It is a fatal policy to try to build a frame on foundation whose bearing value is low or nonuniform. Poor foundations cause settlement of the structure and if the settlement is anything but uniform, it is detrimental to the structure, for it causes tremendous stress in the entire structure even for a slight unequal settlement, the magnitude being dependents upon and varies with the rigidity of the members connected to form the structure.

As a foundation for frames and for every kind of structure there is nothing that is better than sound rock. The allowable bearing value of this kind of foundation is usually no less than the working stress of the concrete of the columns resting on it. In such a case, spreaded footing is theoretically unnecessary. If the distance of the rock below the ground level is not too great to prohibit the arrangement, on account of economical reason, the columns should rest directly on top of the rock or be keyed into it according to whether they are designed as with hinged or fixed ends.

As has been said, the foundation conditions dictate to a great extent the locations of the columns. As we all know, there is seldom a case in construction where the foundation under the ground is so uniform in character that it does not offer an economical problem in connection with the design of the structure. Rock, for instance, has sharp drops and steep slops, and it is entirely common that two borings located only several yards apart find the difference in elevation of the rock to be several tens of feet. Therefore a profile giving full details of the information of the subsurface condition should be in the hands of the designer who after careful consideration of it should determine the places where the foundation alone permits the location of the columns.

The columns should be placed at points so that they will satisfy every designing and physical requirement. When the frame is a grade elimination structure, the columns should be located outside the boundary lines of the right-of-way of the tracks over which the structure spans. As for a viaduct structure the columns may be located in such a way as to produce greatest economy.

The arrangement of the columns with regards to the girders depends a great extent on how the floor system is arranged. In the usual type of construction where the girders and floor beams are used, there is placed one line of columns to every line of girder. Longitudinally, the girders and columns are connected to form the frame to carry the vertical load. Transversely, the tops of the columns should be connected rigidly by a cap, thus forming another set of frames to take wind and eccentric force from vehicles running on it. This arrangement is the most economical serving the purpose of transmitting such loads to the foundation.

Frames may be built of a single span or of multiple spans. The multiple span frame gains advantage over the single one in that the maximum positive moments at the center of the girders are less than that of the corresponding point of the single spanned one of similar design and therefore its girders require a smaller cross-sectional area. On the other hand, the former requires a better foundation, and if the foundation falls short of satisfaction, they should not be adopted.

Frames may be designed and built with fix-end columns or hinged end columns. The use of hinged-end columns concrete frames as highway structures is very limited on account of (1) the difficulty involved in making the hinge that will function as a hinge, (2) little economy that will derive from its use when compared to a simple structure (3) difficulty of construction and (4) greater deflection both horizontal and vertical as compared to a similar structure with fixed columns ends. However, when the foundation is subtle and if ever a frame is to be built on it the use of a hinged end column frame is preferred to a frame with fixed end column frame, for to the former structure, the effects of unequal settlements of the foundation are not as seriously as to the latter. This rule however only aplies to single-span structures.

Temperature stresses sometimes play such an important part in the entire structure that it may need careful consideration of the designer. Like in arches, the temperature stress is a function of and varies with the rigidity of the members and connections, and with the overall length of the structure. Temperature variation exerts greatest stresses at the connections of the end columns and girders, where the structure is the weakest, and therefore where it should receive most careful consideration. (It is not uncommon to have the temperature stress at these points some 20% to 30% of the d. & l.l. stresses combined). When the temperature stresses at these section are so high as to endanger the safety of structure, some means of reducing it is necessary. The most effective method of doing this is to reduce the rigidity of the end columns. By this is meant either to reduce the cross-section area of the end columns (or more exactly the moment of Inertia of the end columns), increase the lengths of the columns, thus making it more slender, making the ends of the end columns hinged if the original design is of fixed ends, or a combination of these. If these remedies still do not help to reduce the temperature stress as low as is desired, then the final step should be resorted to, namely that of disconnecting the joints of end columns and girders, thus making the girders simply supported at these points and reducing the temperature stress to absolute zero.

Preliminary design and dimensions. — After the arrangement of the frame and floor system is determined upon a rough design should be made next on the floor system to get the approximate thickness of the slab. In this connection, concentrated loads, instead of uniform, usually govern the design. The value of the thickness of the slab will be used in calculating the approximate dimensions of the girders.

The preliminary dimensions of girders may be either interpolated from a structure of like design and dimensions and carrying like loads, or determined through employment of approximate calculation. For this, uniform loads are to be employed for live load. For the coefficient of moment, a value of 1/12 to 1/16 may be used, depending upon the degree of the rigidity of columns. The width of the stem is usually governed by the number of bars desired to be placed in a row in it. Thus for a span length of 50' say, a width of 2'-0" is usually adopted, allowing seven bars to be placed in a row.

There is no frame with a span length of say 50' and economically designed that can safely resist the load without the addition of haunches at the ends of the girder. The depth of the haunch is determined primarily by considering the total shear at the section and the unit shearing stress allowed. The value of total shear referred to above may be obtained by assuming the girder to be simply supported.

The width of column is usually made the same as that of the girder. The depth of this column is another uncertain matter requiring judgement. As will be noted later, the deeper the column the more close the girder acts as a fix-end beam, and therefore a more economical combination is produced.

If a multiple span frame is determined upon, it is advisable to have the spans of approximately same lengths. This rule is established on the assumption of course that the geological and physical conditions permit it.

If the span lengths in a multiple span frame are approximately the same, it is advisable to make all the girders of the same depth. By so doing, the process of designing is greatly shortened, and a great deal of labor required to detail the steel is reduced.

Methods of Analyzing the Frame Structure. There are in use several methods to compute analytically frame structures. The following are the ones more often used:

1. Slope and deflection method
2. Area moment method
3. Method of least work
4. Method reciprocity.

All of these, however, are based on the same principle, namely the principle of Elasticity, and for any given problem, they should yield the same result.

It is not within the scope of this paper to discuss the theories involved in each of the above methods. Neither should there be such an attempt, for almost every method mentioned above is well discribed in standard text books dealing with indeterminate structures. Technical periodicals sometimes also carry descriptions as to how these methods applied in the solution of actual problems.

Of all the methods suitable for the solutions of frame problems the slope and deflection method deserve special mentioning and recommendation as it is this method that has been developed to such a practical form and involves so many mechanical fool-proofing features that it earns itself the reputation as being the easiest to apply.

Nearly all the methods used in the solution of statically undeterminate structures involve the assumption that moments of inertia of the members composing the structure are of constant value. With a reinforced concrete frame, the moment of inertia varies from section to section in any individual member, on account of the variations in reinforcement and different effectiveness of the concrete in taking tension. Many argued that since the values of the moments of inertia are used only as a means of determining the relative rigidity of the members composing the frame, the reinforcement may be omitted from the consideration and that plain concrete sections may be used for the calculation of such values without introducing serious errors. This opinion finds the favor of most of the designers, and this method is greatly followed.

After the values of I, the moment of inertia, are ascertained, actual computations may be started, using any of the methods that a designer is in favor of. In order to determine the maximum stresses at different sections, influence lines should be drawn. The method called the "Fix point method," "Feste Punkle," is recommended to the designer for expediting the drawing of the influence line. Maximum moments and shear should then be calculated at regular intervals of the girder and curves should be draw connecting the extreme values. Such curves are of great service to the designer, and in important structures are an absolute necessity.

Details of Design. (1) *Column.* The cross sectional shape of columns most used in frame structures is rectangular. As has been said, the width of the column is generally made the same as that of the girder resting on it. By so doing, the form work is simplified and the steel can be more accurately placed.

For columns with fix ends the cross sectional area of the column is usually maintained constant throughout the length of its column. The fixity of the column at the ends is usually produced by keeping the footing into the rock foundation when it is within reasonable reach or by other means: such as capping on piles, etc. The depth of the column should be such that it will give a section great enough to carry the load. In general the more deep the column the greater is the economy, provided by so doing the stresses due to the dead and live loads and especially from temperature change are still kept within allowable limit, that it will not spoil the look of the entire structure and that the column is not too high.

The reinforcement in the column should be arranged as symmetrical about the gravity axis as is practical, so as to reduce the eccentricity to a minimum. For intermediate columns, this is possible, economically, for the stress there are mostly in the form of direct loads, the moment is generally small compared to the outside columns. In end columns, the moment is generally high and the direct stress low as compared to those of the intermediate columns. In such columns, a symmetrical arrangement of steel generally indicates a waste of steel, although a better structure is resulted. It is the job of the designer to determine that arrangement which is best to suit the local conditions.

The moment diagram of a column with fix ends due to dead load and live load is a straight line varying from + to —, with a zero value at a point 1/3 up the base of the column. The moment at the top of the column has the maximum value and is twice that at the base. The temperature change casues the maximum moment at the top of the column. Therefore at this section, the steel area should be more than at any other section of the column. At a certain distance downward from the section above referred to, a part of the steel may be stopped, to conform with steep slope of the moment diagram. By so doing, a great percentage of steel is saved.

When a column is hinged, the cross section area may be made constant or variable, increasing with the distance between the base of the column and the section to conform with the moment diagram which in this case, is a straight lino varying from 0 to the maximum at the top of the column. If the latter method is followed some concrete is saved but it is handicapped by the resulting complication of form work. A careful designer should adopt the one that is on the favorable side of economy.

The hinge of the hinge end columns may be either a mechanical provision or merely the result of manipulation of the steel in the column. Mechancial hinges are expensive for concrete frame structure, and are therefore little used. The latter method is generally followed. It is accomplished by stopping the steel in the column above the hinge section, and leaving only few to pass through the center of the column to connect it with the footing. The steel being at the center of the section, the section is unable to take moment, for concrete of the section, the section is unable to take moment, for concrete is not able to take tension, as is commonly assumed. After the structure is loaded, the concrete at the hinged end is cracked leaving the steel alone to take the shear. Such an arrangement is open to objection because the cracked concrete absorbs water easily which may do damage to the structure in time of freezing. To eliminate this danger, a curved plate may be used at the base of the column, through the center of which a hole is punched. The plate is concreted with the columns and rests on the prepared footing. The reinforcement in the column is allowed to run through the hole of plates and is then securely fastened in the footing. It will take the shear; but no moments. In this way, a much better arrangement is resulted. It is necessary that that part of the reinforcement exposed to the weather should be coated to prevent errosion.

With the size of column usually employed the unit shearing stress in the column is usually so low that it will not modify the spacing of the hoops commonly adopted in simple columns of similar sizes.

The joint between the end columns and girders should be well protected with reinforcement on account of the high moment there that result from the dead and live load and temperature variation. If every part of the structure is designed with equal care and if ever a sign of weakness appears after the structure is built, such a sign most likely appears first at the corner above mentioned. The failure of the corner to take the load is not a local problem, for it disconnects the continuity of the girder and columns, thus causing the positive moment in the girder to increase and may finally be a cause of the collapsing of the entire structure. It is a job of a good designer to arrange the steel at this part of the structure to take care of the mo-

ment and direct load fully and economically. The following rules may be used as a guide to produce a good design:—

(a) Try to arrange the steel as symmetrically about the gravity of the columns as possible.

(b) Do not use too many short bars, for short bars usually produce undesirable stresses at the places where its ends are and where the stresses are high.

(c) Extend the bars far enough from the points where they are unnecessary and hook the bars at their ends. Steel spent this way is usually justified.

(d) Do not attempt to run he bars from the girder to the column for such bars are hard to place in the field and hard to hold accurately in place in the form.

(e) Be a little extravagant in using hoops and stirrups at this part of the structure. They prevent the main reinforcement from buckling when they are under stress.

(f) Be not liberal to let the stress exceed that allowed.

(g) Girders. The design of the reinforcement in the girder is similarly to that of a fixed end beam if the frame is a single span frame and to that of a continuous beam if it is a multiple span frame.

Before any actual design is started, it is best and advisable to have the maximum moment diagrams drawn, using any one of the methods mentioned in the foregoing. These diagrams give one the notion as to the amount of steel area required at different section and give indications as to where the steel may be bent, after the steel at the important sections is designed.

To save reinforcement, advantage must be taken to the limit in making use of the slab above the girder which two form a T-Beam. This is possible only in the design of positive steel, that is steel taking positive moment. For negative moment, the slab is on the tension side, and therefore

only the rectangular section should be considered in designing the reinforcement.

As has been said before, haunches are necessary in almost every frame used as a highway structure to take care of the negative moment and reduce the unit shearing stress to within allowable. The depth of the haunch at the springing is usually governed by the latter. The bottom of the haunch may be a straight line or a conical arc, depending upon the desire of the designer. Although it enhances the looks to some extent, the using of curved bottom haunches is hardly justified. It complicates the formwork, and the bending and placing of steel, and therefore is more expensive than the straight haunch.

The design of shear reinforcement in the girder is the same as that for simple structure after the total shears at various sections are known. Bars of the main reinforcement may be bent up or down to take shear at points where they are not needed to take the bending stress. The ends of the bars disposed in the way are thus utilized in a most economical manner. A word or two, however, should be mentioned with regard to the bending of the bars in the view of making them to do the service as is expected. In bending the bars of simple beams, the bends are usually made long enough so that it will pass the neutral axis for a certain length, say a certain number of diameters, so that it will develop the tension through bond in the compressive area along that length. In the girder of a rigid frame, however, the positions of neutral axis are not definitely located and they change with different loadings. It is therefore reasonable that a greater length of imbedment than that mentioned above should be allowed to assure that the bars will work under any loading. A still better structure will be produced if the bars are extended to meet and then run side by side with the steel on the other side of the axis for a distance say ten times the diameter of the bars bent. With bents of this shape such bars will not only take the shear effectively but will function as stiffners in plate girders, and therefore are in one way a safeguard against distruction caused by the impact of the modern wheel loads.

DIRECT GENERATION OF ALTERNATING CURRENT AT HIGH VOLTAGES

By the Hon. Sir Charles A. Parsons, O.M., F.C.B., F.R.S., Honorary Member, and J. Rosen, Member

(Continued from Page 544 Vol. V, No. 4)

The end-windings illustrated in Fig. 8 are composed of flat copper strip formed on edge and having a full radius on the edges. This formation gives a very rigid construction, enabling it to withstand the stresses set up on short-circuit. Additional mechanical support is given to the end-winding strips by fitting them into recesses formed in the impregnated-wood supporting clamps.

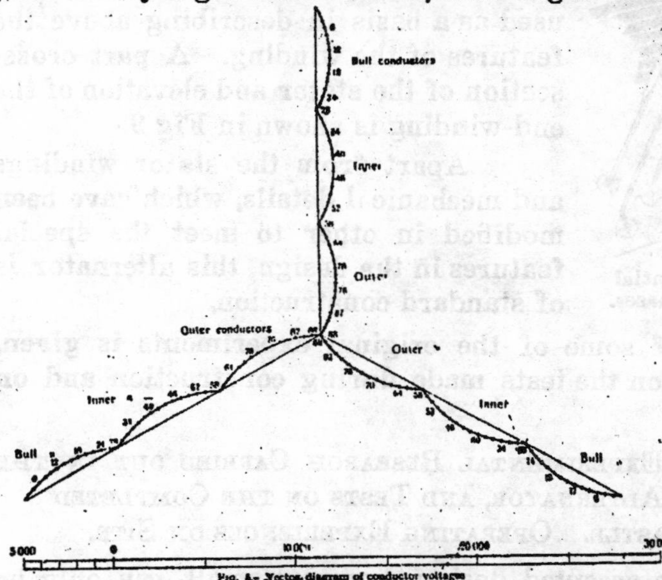

Fig. 5.- Vector diagram of conductor voltages

There are three banks of end-connections in each phase, corresponding to the "bull," "inner" and "outer" conductors. The cross connections between each bank are provided with removable links to enable each third of the phase to be pressure-tested separately. The link forms a ready means to fit between-turns protection if desired.

Ample distance is provided between phases, so that it is unnecessary to provide insulating shields between the end windings, and adequate leakage surface is also provided over the impregnated-wood packings between the phases.

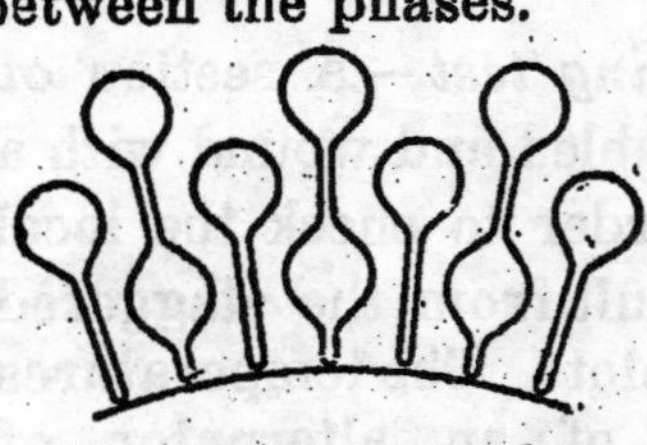

Fig. 6.—Conductor slots arranged in three rows.

In a normal design of alternator the full phase potential exists between banks in the end-windings. In the high-voltage alternator design the end-connections for the "bull" conductors of one phase are adjacent to the end-connections joining the "outer" conductors of another phase. The difference in potential between the phase banks is therefore less than the normal voltage between phase terminals. Thus the maximum potential difference in the end-windings between the adjacent conductor No. 1 in any phase and No. 58 in another phase is, from Fig. 5, referred to above, only 23,000 volts.

A 25,000-kW, 31,250-kVA, 33,000-volt, 3,000-r.p.m. alternator incorporating these principles was built for the North Metropolitan Electric Power Supply Co., Ltd., for installation in their new power station at Brimsdown, North London. It has been used as a basis in describing above the features of the winding. A part cross-section of the stator and elevation of the end-winding is shown in Fig 9.

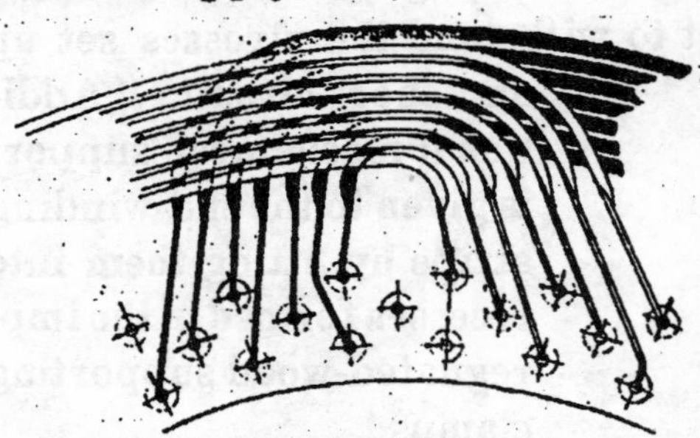

Fig. 7.—Position of highest-potential conductors relative to adjacent phases.

Apart from the stator windings and mechanical details, which have been modified in other to meet the special features in the design, this alternator is of standard construction.

A short account of some of the original experiments is given, together with a few notes upon the tests made during construction and on completion.

(7) An Outline of the Experimental Research Carried out in the Development of 33-kV Alternator, and Tests on the Completed Alternator, at Newcastle. Operating Experiences on Site.

Any departure from accepted design, however small, can only be accomplished successfully after extensive research. This is particularly so in large electrical apparatus, where, in addition, the proof of actual operation must be applied. A few of the tests previous to and during the assembly of the alternator are outlined below.

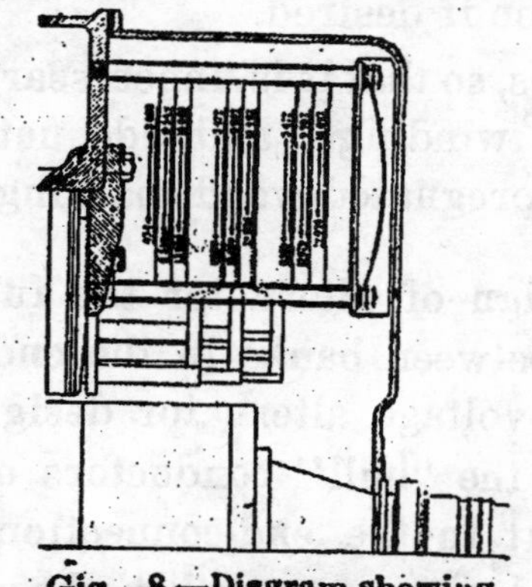

Gig. 8.—Diagram showing voltages of end-winding connections of 33000-volt altnator.

Stator tooth-heating test.—A section of the core plate was assembled and wound with a temporary winding in order to check the local heating which might result from the staggered disposition of the stator slots. The temperatures did not exceed those of an alternator of low voltage.

Pressure-testing apparatus.—A single-phase transformer with a voltage ratio of 440/110,000 was used for all pressure tests. Tappings were brought out at one-third and two-thirds of the maximum voltage.

In order to overcome the liability to flash-over at the pressure test of 10,0000 volts, individual bars while under test were immersed in a bath of varnish.

Test conductor bars were constructed, one set being un impregnated and a second set being impregnated after the application of each insulating tube. The latter bars, as seen from Fig. 10, had 30 per cent lower dielectric loss.

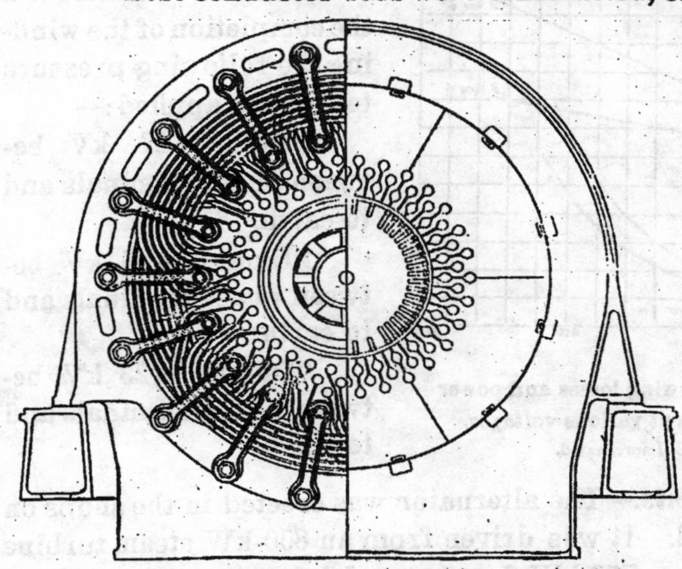

Elevation of stator end-windings. Part cross-section of stator.

FIG. 9.—33 000-volt, 3-phase, -pole, 3 000 r.p.m., 31 250-kVA alternator

The distribution of electrostatic capacity between the three conductors is given in Table 2.

The corresponding figures for a single conductor bar of an 11,000-volt alternator of similar output is 0.00024 μF per foot run obtained from test, the calculated value being 0.00028 μF per foot run.

Reliability tests extending over several months were made on the test bars, and an extract from the log is given in Table 3.

TABLE 2.

Capacity per Foot Run.

Position	Concentric bars	
	Measured capacity	Calculated capacity
	μF	μF
Between "bull" and "inner"	0.00018	0.00025
Between "inner" and "outer"	0.0003	0.0005
Between "outer" and "sheath"	0.00034	0.00046

The potential gradient across the in sulation in single-core and 3-core cables, with the same sectional areas, is given in Figs. 11 and 12. It will be seen that the maximum gradient is considerably less in the 3-core concentric cable. Flaws, such as cracks, or air pockets, are less liable to occur in the concentric cable, as the layers of insulation are much thinner.

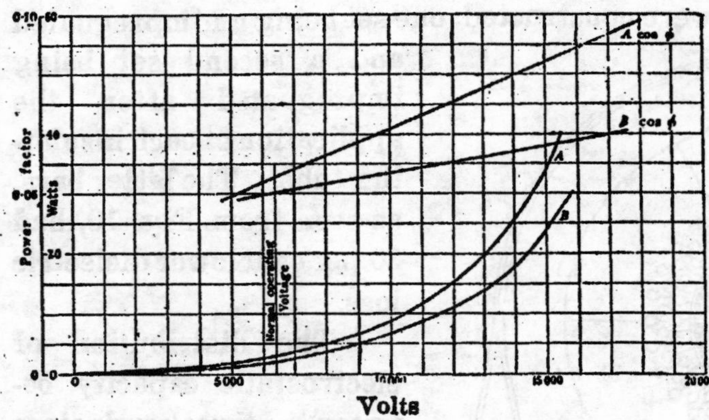

FIG. 10.—Comparative curves showing losses and power factors of concentric conductors at various voltages.
A. Un-impregnated. B. Impregnated.

Pressure tests.—On the completion of the winding the following pressure tests were applied:—

"Bulls"—67 kV between phase terminals and to earth.

"Inners"—45 kV between phase terminals and to earth.

"Outers"—23 kV between phase terminals and to earth.

Testing arrangements.—The alternator was erected in the shops on a specially designed test-bed. It was driven from an 800-kW steam turbine and coupled in paralled with a 750-kW d.c. motor driving through gearing. Either machine was capable of driving the alternator at full speed when fully excited. A completely enclosed air system, with a fan delivering 40,000 cub. ft. of air per minute, and surface air cooler, was provided for cooling the alternator.

TABLE 3.

Average Dielectric Loss in the Bar.

Test conditions	Voltage		Bar temperature		
			20°C.	60°C.	110°C.
33 kV on "bull" ... ... 22 kV on "inner" ... ... 11 kV on "outer"	174 per cent	Dielectric loss Leakage current	30 watts 0.008 amp.	105 watts 0.016 amp.	235 watts 0.03 amp.
19 kV on "bull" ... ... 12.8 kV on "inner" ... ... 6.4 kV on "outer"	Normal	Dielectric loss Leakage current	1 4watts 0.001 amp.	18.8 watts 0.004 amp.	61 watts 0.008 amp.

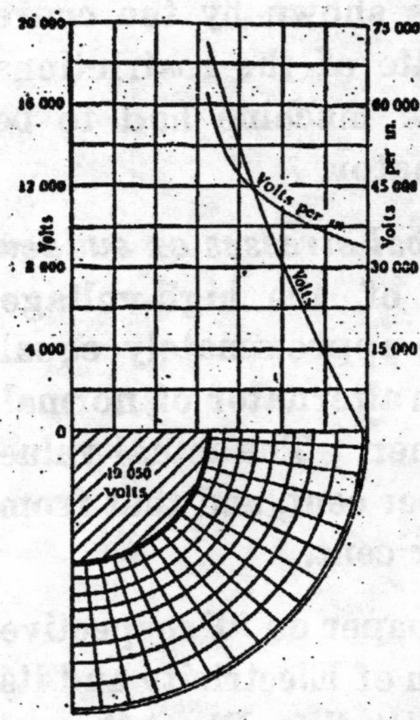

Fig. 11—Curves showing gradient and voltage-drop across insulation for a single-core conductor.

Temporary air ducts were constructed, to re-circulate the air. The excitation was provided by a d.c. generator.

The open-circuit and short-circuit characteristics are shown in Fig.13, and voltage oscillograms in Figs. 14 and 15.

Wave-form.—The voltage wave-form between phase terminals and between phase terminals and earth departs less than 1 per cent from a pure sine wave, and is free from ripples: no difficulty due to harmonics when operating on a cable network has been experienced.

Sudden short-circuit test.—The alternator was suddenly short-circuited by an ironclad, oil-immersed switch, and an oscillogram was obtained of the current in three phases, the curves being shown in Fig. 16. It should be recorded that after the test the end-windings showed no sign of movement.

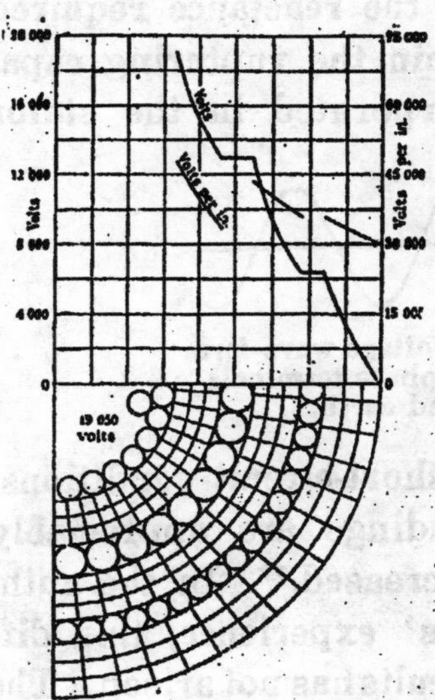

Fig. 12.—Curves showing gradient and voltage-drop across insulation with concentric conductions.

Heat runs.—As it was impracticable to dissipate 25,000 kW at 33 kV, the windings were rearranged so as to circulate current, i.e. with the three sections of each phase in parallel. The difference in voltage between the "bull" section and the other two sections causes a circulating current which is controlled by a choke in the circuit. While the conditions do not represent the actual conditions of operation, they give a very good indication of the results to be expected on load.

The temperature-rises were moderate and very satisfactory, the figures on load being well within the estimated values.

During the open-circuit tests, a record of the leakage current per phase was taken. This figure was 80 mA at 33 kV at 60° C.

Efficicncy.—The efficiency of the alternator is shown by the cruve in Fig. 17. The efficiency is high at all loads, in spite of the restrictions imposed on the design owing to the fact that the machine had to be interchangeable as a whole with the low-voltage alternator.

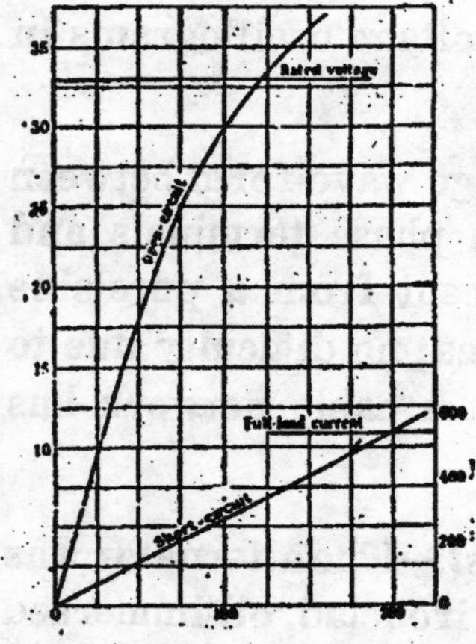

Fig. 13.—Actual open-circuit and short-circuit characteristics.

Reactance and mechanical stresses on sudden short-circuit.—The reactance of the high-voltage alternator proved on test to be approximately equal to the combined reactance of an alternator of normal design and a step-up transformer. The actual value obtained by calculation is 22 per cent, and that from the short-circuit tests is 21 per cent.

Dr. S. L. Pearce, in his paper on "Prospective Development in the Generation of Electricity and its Influence on the Design of Station Plant," read before the Engineering Conference of The Institution of Civil Engineers in June 1928, draws attention to possible increased forces on the stator windings in a statement reading as follows:—"In the absence of step-up transformers, or external reactance coils, the reactance required for limiting the short-circuit currents to values within the rupturing capacity of the switchgear has necessarily to be incorporated in the stator

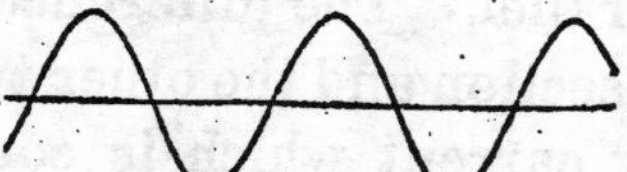

Fig. 14—Voltage wave-form between phase terminals.

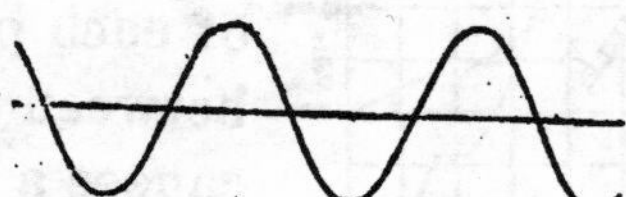

Fig. 15.—Voltage wave-form between phase terminals and earth.

windings, with the natural consequence that under short-circuit conditions, the mechanical forces on inherently weaker windings are appreciably increased." In the authors' experience, this difficults has not arisen. The short-circuit tests carried out on the 25,000-kW, 33-kV alternator have proved that the windings are quite as robust as those of a low-voltage machine.

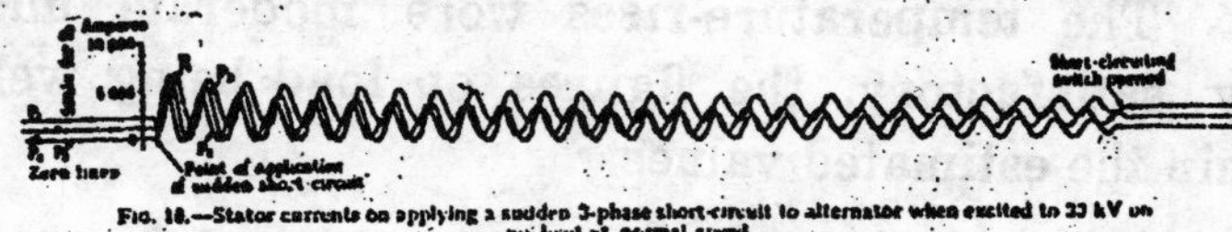

Fig. 16.—Stator currents on applying a sudden 3-phase short-circuit to alternator when excited to 33 kV on no load at normal speed.

The arrangement and spacing of the end-connections of the 11-kV and 33-kV 25 000-kW alternators, respectively, are shown in Figs. 18 and 8. It will be seen that greater space is provided for the accommodation of the windings of the latter.

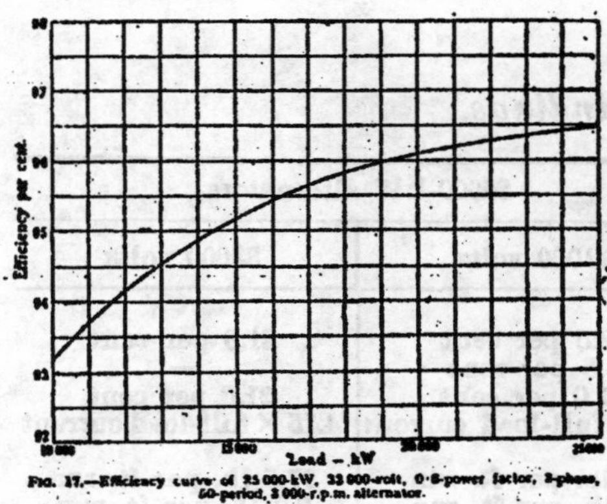

FIG. 17.—Efficiency curve of 25 000-kW, 33 000-volt, 0·8-power factor, 3-phase, 50-period, 3 000-r.p.m. alternator.

In making a comparison of the forces between conductors under short-circuit conditions, it is assumed that the short-circuit takes place at the terminals of the high-voltage alternator, and at the secondary terminals of the transformer connected to the low-voltage alternator.

If a short-circuit occurs at the terminals of the low-voltage alternator, the stresses in the windings are increased nearly three times.

The number of ampere-conductors is approximately the same in each design, but the intensity of the leakage flux surrounding the end-windings is smaller in the high-voltage design, due to the longer magnetic path.

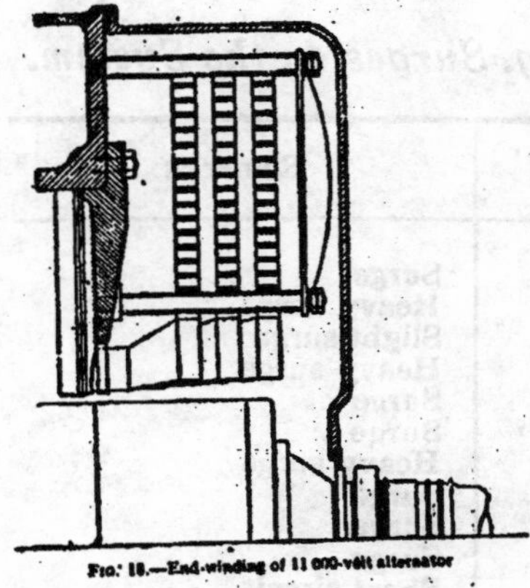
FIG. 18.—End-winding of 11 000-volt alternator

The conductor slots are circular and much wider than the reactance slots, and the intensity of the leakage field is low in the former. For these reasons, and owing to the lower current per conductor, the stresses between conductors both in the core and end-winding are lower in the high-voltage design than in the low-voltage design.

A comparison of the reactance and stress in the two designs is given in Table 4.

On site.—Since the plant was installed in August 1928, it has operated continuously up to its maximum load at voltages varying between 34,000 and 35,000, and it has withstood without any sign of distress the most severe faults on the large overhead and underground network to which it is coupled.

The control and regulation have proved eminently satisfactory in every way.

Some extracts from the station log are given in Table 5.

TABLE 4.

Forces on End Windings.

	25000-kW alternators	
	11000 volts	33000 volts
Inherent reactance of alternator...	12.5 per cent	21.0 per cent
Reactance of transformer	8.5 per cent	—
Total reactance	21.0 per cent	21.0 per cent
Current on short-circuit	4 75 × full-load current	4.75 × full-load current
Forces in windings:—		
(a) End-connections between phases ...	75 lb. per ft. run	17.5 lb. per ft. run
(b) End-connections between turns ...	14 lb. per ft. run	2.0 lb. per ft. run
(c) Between core conductors...	250 lb. per ft. run	140 lb. per ft. run

TABLE 5.

Extract from the Brimsdown Power Station Log. Surges on the System.

Date	Time	Generator on load	Remarks
31 August 1928	01.00	No. 2	Surge
31 August 1928	01.35	„ 2	Heavy surge
31 August 1928	18.25	„ 2	Slight surge
19 September 1928	08.30	„ 2	Heavy surge
12 October 1928	10.35	„ 2	Surge
2 November 1928	21.15	„ 2	Surge
5 November 1928	00.06	„ 2	Heavy surge
6 November 1928	19.41	„ 2	Surge
6 November 1928	20.14	„ 2	Surge
6 November 1928	21.01	„ 2	Surge
10 November 1928	16.15	Nos. 1 and 2	Short-circuit
16 November 1928	17.15	„ 1 and 2	Heavy surge
19 November 1928	03.25	No. 2	Heavy surge
19 November 1928	05.57	„ 2	Heavy surge
21 November 1928	06.09	„ 2	Slight surge
29 November 1928	17.29	Nos. 1 and 1	Heavy surge
6 December 1928	14.38	No. 2	Surge

Generator No. 2 is the 33000-volt alternator.

(8) Conclusion.

Only the fringe of the possibilities has been touched; it certainly seems in the natural course of design that a reliable high-voltage alternator is essential to the rapid increase in size of power systems and their interconnections.

The paper is confined to generation at higher voltages, but other units of ever-increasing size, such as motors, motor-generators, synchronous condensers, etc., may be economically designed and coupled direct to the network without the use of transformers.

Although the purpose of the authors has been to give information and to promote discussion on the high-voltage machine, their primary object is to gain from engineers an expression of opinion on the possibilities, advantages and disadvantages of generating alternating current direct at voltages higher than those now recognized as customary.

The authors hope that they have shown that a definite advance has been made in the generation of electricity and in the design of large generators which may be used for connection to the "grid." If their work contributes to the problem of providing means for the more efficient and economical generation of electrical energy, and thus to some saving in the consumption of coal, they will feel that their efforts have not been wasted.

They wish to take this opportunity of paying tribute to, and to acknowledge the courage and foresight of Sir James Devonshire, K. B. E., Chairman of the North Metropolitan Electric Power Supply Co., and the chief engineer, Captain J. M. Donaldson, M.C., of the same Company, in installing the pioneer high-voltage unit, and without whose co-operation it would have been impossible to undertake the work.

They also wish to acknowledge the assistance given by the staff and officials of Messrs. C. A. Parsons and Co., Ltd., who have made unstinted efforts in support of all investigations and in the compilation of this paper.

AERODYNAMICS AND AIRPLANE DESIGN

Paper presented to the Chinese Engineering Society in its Eleventh Annual Convention, Sept. 6, 1929.

By GEORGE C. CHOU, AERO. E. (周 [illegible] 璋)

PART 1. INTRODUCTION

As is generally understood, aeronautics is a common name under which, aerodynamics, airplanes, airships, landing fields, airways, technical flying, air transportation, commercial applications, industrial application, military application, and so forth, each occupies a special field for specialization. Today it is more convenient to deal with only two topics in aeronautics, Aerodynamics and Airplane Design.

Before any subject is treated on its high spots, a general review or survey is always advisable. Let us survey the general mode of operation of an airplane. Figures 1, 2, and 3 are the usual drawings to express the main features of an airplane while the chief operation characteristics have to be tested in flight. Suppose the airplane is flying level like figure 1 which is technically known as in its cruising attitude. Attitude is term itself under which the airplane displays all its activities. The lift is equal to its weight so that airplane will stay in that particular level, the equation is,

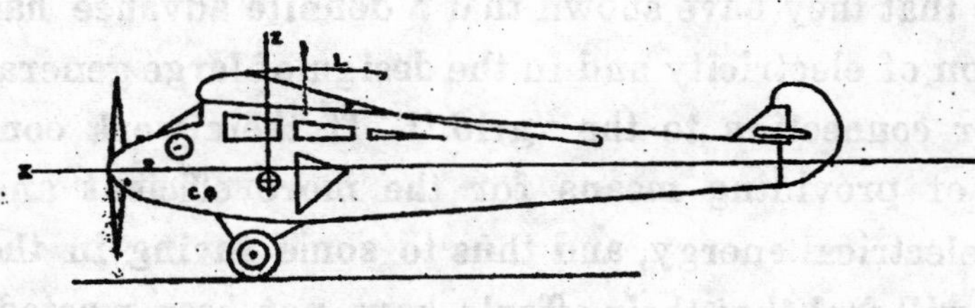

FIG. 1

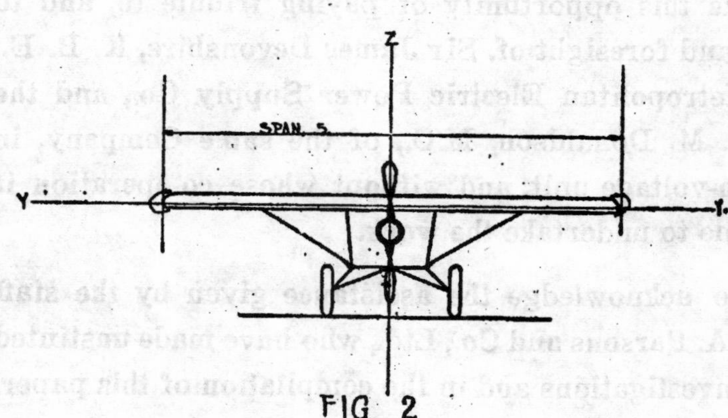

FIG 2

$$L = W = k_y A V^2,$$

Where L is the lift on the wings, W the gross weight of the airplane, K_y is the specific lift per

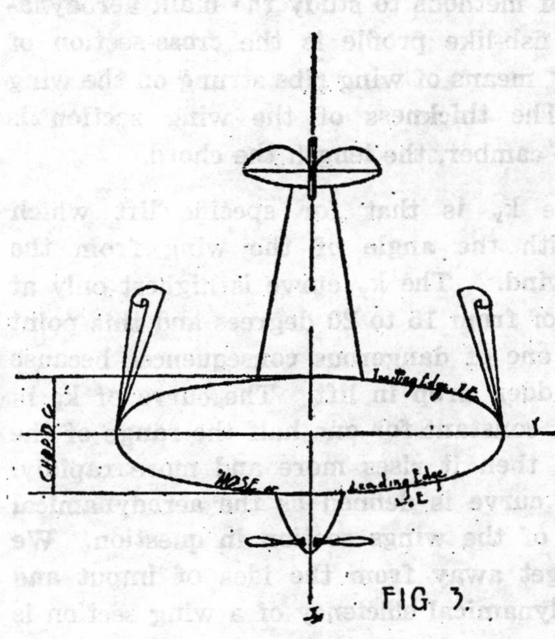

FIG 3.

unit area of the wing surface at unit velocity per unit time, A is the total area of the lifting surface, and V is the velocity[1] of the airplane. The head resistance, technically known as the **Drag** D, is overcome by the thrust of the propeller, The drag of the wing alone is expressed by

$$D = k_x A V^2$$

where k_x is the specific drag coefficient of a certain wing. The thrust of the propeller must be equal to the entire drag of the airplane at uniform velocity. The development of the thrust will be dealt with later.

The level attitude of the airplane is maintained by the predetermined balancing elements expressed in the following equation,

$$M = o = M_w, M_t, M_b, M_p, \text{etc.}$$

where m is the total moment about the center of gravity of the entire airplane and must be equal to zero so that nothing will turn, M with subscripts w, t, b, p, etc., means the moments due to wing, tail plane, body, propeller, and so forth. A thorough explanation will be given later. For simplicity the total moment about the c.g. contains that of wing and tail only, which is accurate enough to determine the flying features of a machine. The above equations satisfy the law of motion at constant speed, as

$$\text{sum } V = o,$$
$$\text{sum } H = o,$$
$$\text{sum } M = O.$$

Now more information is to be sought in lift L, drag D, thrust T, and moments M of the various parts of the airplane.

Figure 4 is a crystallization of methods to study the main aerodynamical characteristics of a wing. The fish-like profile is the cross-section of the wing whose shape is maintained by means of wing ribs strung on the wing beams. The thickness of the wing section is called the camber, the length the chord.

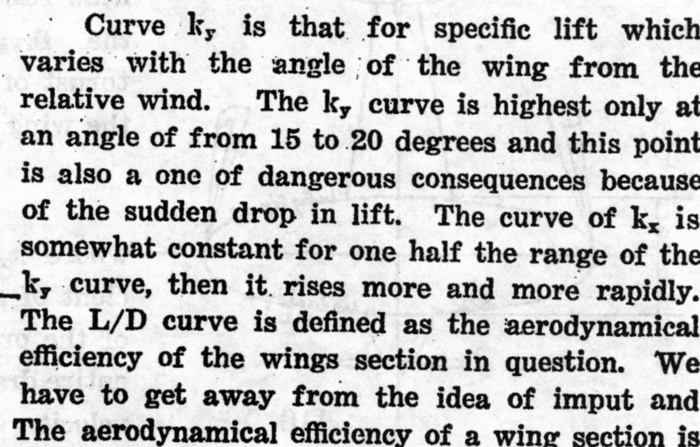

Curve k_y is that for specific lift which varies with the angle of the wing from the relative wind. The k_y curve is highest only at an angle of from 15 to 20 degrees and this point is also a one of dangerous consequences because of the sudden drop in lift. The curve of k_x is somewhat constant for one half the range of the k_y curve, then it rises more and more rapidly. The L/D curve is defined as the aerodynamical efficiency of the wings section in question. We have to get away from the idea of imput and output ratio for efficiency. The aerodynamical efficiency of a wing section is the ratio of lift over drag L/D.

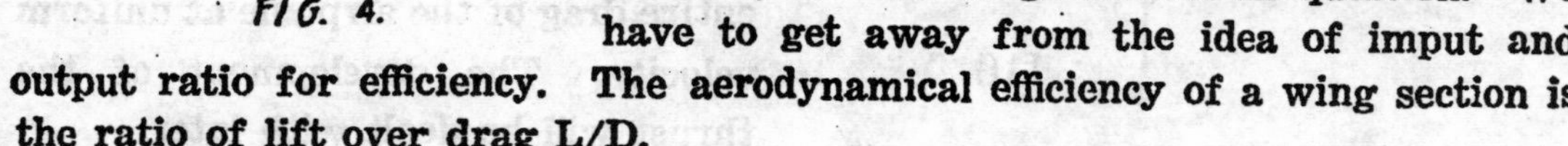

FIG. 4.

The curve C. P. is the travel of the center of pressure of the wing section along its chord line (c in fig. 2 & 4). It is assumed a straight line along the length of the wing called the span. The curve is plotted in per cent of the chord line c either from rear edge (Trailing Edge T. E.) towards the front edge (Leading Edge L. E.), or vise versa. It is to be noted that the travel is forward on high angles and rearward on low angles of the wing against the relative wind.

These curves are of prime importance to the airplane designer. The k_y and k_x curves are sometimes plotted one against the other like figure 5 which is called the polar plot of the wing section. Curve a is a theoretical one parabolic in nature and curve b is the experimental curve.

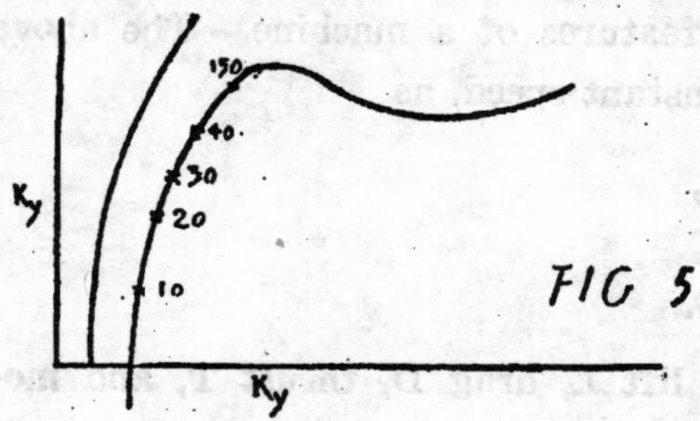

FIG 5

To illustrate the importance of these curves we take the airplane in its level flight condition. Since efficiency is the aim of flying, one must fly at the top of the L/D curve where gasoline will be comsumed the least and a long distance can be flown over. The C. P. at that point will assist the designer to divide the load onto the wing beams. The drag k_x at that point will give the designer how much power is needed to go. The size of the wing is determined by the maximum value of k_y and the equation,

$$A = \frac{W}{K_y \ V_m^2},$$

where K_y is the maximum value of lift and V_m is the minimum value of flying speed. Higher k_y gives smaller wing and less construction work. Smaller V_m insures safety in landing or gliding down to land. Next we will find how thrust is developed.

The development of the thrust of a propeller is similar to the lift of the wing. Figure 6 is my perspective drawing to show how the thrust is developed. Here we have the air velocity components, that due to the motion of the airplane and that due to the turning of propeller. The resultant velocity is represented by V_r. The large angle ϕ is the blade angle which the chord of the propeller blade section makes with the plane of revolution YY. The actual angle of flying corresponding to that of a wing section is designated by i and which is never to be too large. The actual section of the propeller should be cylindrical since the propeller revolves in a cylinder. But for practical purposes, the section can be assumed to be plane. Resolving the lift L to the OX direction will give the desired thrust for pulling the airplane through the air. The other components will be resolved to the OY direction which is nothing but a waste or a necessary evil. The ratio of thrust to the other energy wasted will be the efficiency of the propeller, at that section. By summing up a number of these sections the total thrust of the propeller can be found. The propeller is designed for that required thrust usually. Before I go any further, I shall sum up what has been said before in a brief manner.

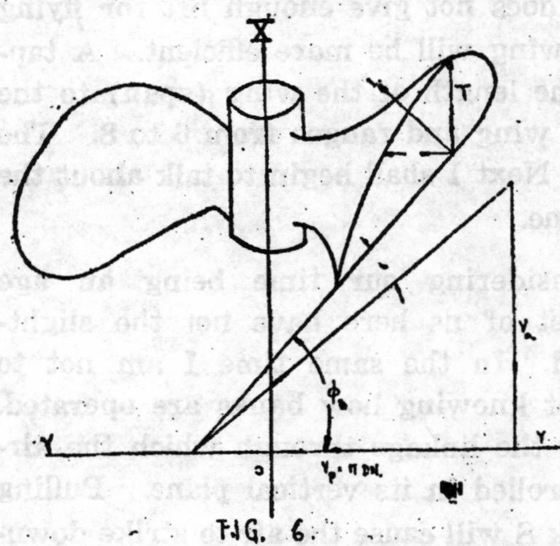

Figs. 1 & 2, some more items.—Referring back to fig. 1, some more elements of aerodynamics still exists but that's not all. Beginning from the propeller, the two tips of it are forcing the air to form what is called tip **Vortex.** Behind the propeller, the air is forced to go faster than the speed of the airplane and is called the **Slip Stream.** The slip stream does not go in a straight manner and the twist is called the **Rotational Mass.**

On top of the wing is a curve describing the air path which is smooth and then gradually develops into a turbulant form. By improperly elevating the wing, the turbulance will be excessive and called upon to drop the wing down (this is the burbling of the wing) Just behind the wing and tangent off from the trailing edge, the air is forced downward. On the tariling egde it is called trailing **Vorticity** and further back is called the **Downwash** This has some effect of the tail plane.

Fig. 3 has some arrows showing the directions of the air at the wing tip which is called the wing **Tip Vortex** similar to that of the propeller. This offers the reason why a short wing does not give enough lift for flying because the tips are drowned. A longer wing will be more efficient. A tapered wing is just as good. The ratio of the length of the wing (span) to the width (chord) is the **Aspect Ratio** of the wing and ranges from 6 to 8. The larger the better if construction permits. Next I shall begin to talk about the turning and maneauvering of the airplane.

Rise & Dip of the Airplane.—Considering our time being an age of specialization, I don't wonder if most of us here have not the slighthtest idea of how airplanes move around. In the same time I am not to blame for not knowing how banks are operated. Fig. 7 shows the linkage through which the airplane is controlled in its vertical plane. Pulling back the stick S will cause the air to strike downward on the tail T. Thus the airplane will go upward through the pull of the propeller. The airplane will slow down somewhat. The pilot P is shown in outline. In a similar manner, pushing the stick forward will cause the air to strike the tail from below and push it upward affecting a descending attitude. The airplane will go faster due to the pull of the propeller and gravity. You can well imagine what a change is going in the balancing moments of the equation and curves.

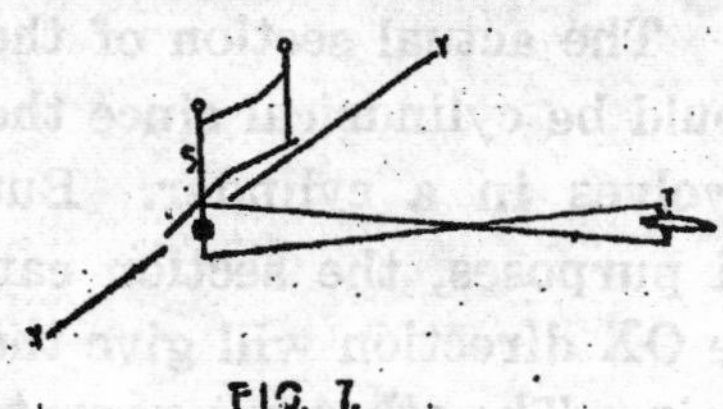
FIG. 7.

Bank & Turn. Fig. 8 is the inside view, of an airplane to show how the contols work for a turn, and fig. 9 the motion of the parts necessary to turn. The arrows show the directions the parts should move to. Looking at the aileron which is a flexible part of the wing, the upward motion of which will force the wing downward as the elevator tail does the tail of the airplane, and vise verse. Attention is called to fig. 8 that the airlerons are drawn in a plane of the controls in the cockpit. Moving the stick to its right will cause the right airleron to go up hence the wing down. The left wing will do the opposite. This is to get ready to turn to the right and providing for the plane to act against the centrifugal force. Fig. 10 is an instantaneous view of the airplane during a turn. The rudder control is shown below it which is very simple. Next you will be ready to be introduced to the more difficult subject of maneuvers of the airplane.

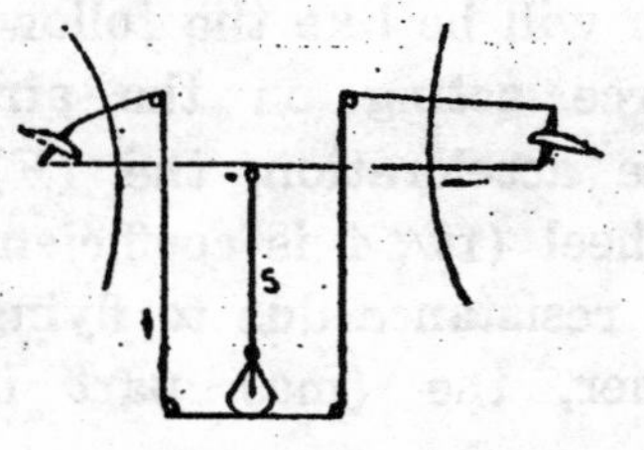

FIG. 8.

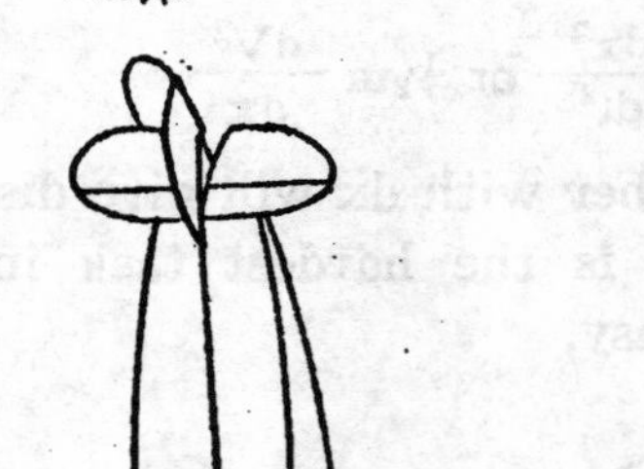

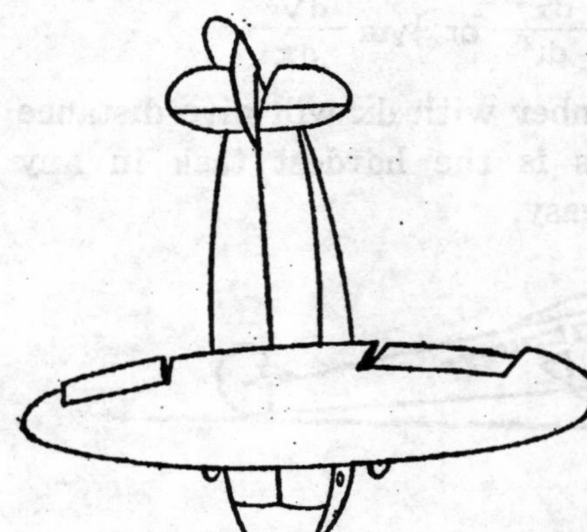

FIG. 10.

After understanding the motion of the airplane in level attitude, we shall touch some of the other more technical movements. It is easy and logical to follow the events, from start to finish.

Take-Off

Fig. 12 shows an elementary almost childish picture of an airplane ready for take-off. Forgetting other precautions and technical point, we suppose that the engine is turned full power on. The tail must be pushed down to receive the slipstream for lift. The slipstream will lift the tail up till the plane is in a level

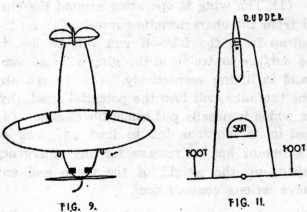

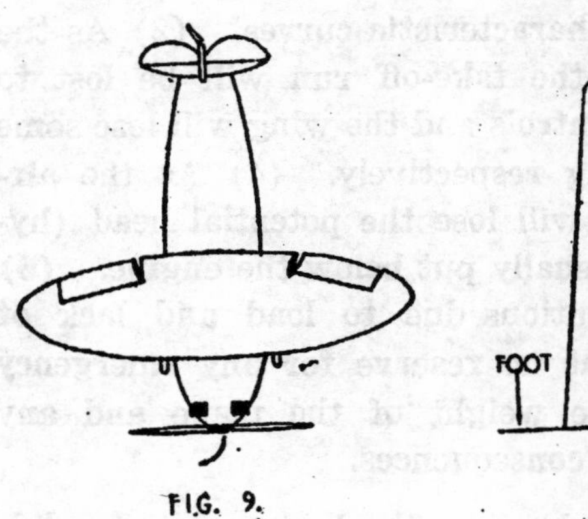

FIG. 9. FIG. 11.

attitude. This will be in a position offering least resistance to gather up speed. As the speed of the airplane gather up, the wing will begin to be responsive to the air. The equation of motion will be like the following, from F=ma, where F is the unbalanced force acting on the airplane, b is the mass of the airplane, and a the acceleration, the (F) part consists of thrust T, ground resistance on the wheel (fw, f is coefficient of friction and W the weight of plane) and the wing resistance due to flying which will develop drag in an aerodynamical manner, the (ma) part is $m(dx^2/dt^2)$,

$$T - f(W - k_y AV^2) - k_x AV^2 = m\frac{dx^2}{dt^2} \text{ or } \frac{1}{2}m\frac{dV^2}{dx}.$$

The member with dt will give time and the member with dx will give distance of the airplane in taking-off. Gathering terms is the hardest task in any mathematical problem. The solution is then easy.

FIG. 12.

Climbing. To leave the grouund is so easy and so simple, as described in the previous page. Next thing is climbing. Climbing is one of the most dangerous stages in flying. (1) The wing is operating around the burbling point as can be remembered from the characteristic curves. (2) As the plane noses up, the speed gathered up from the take-off run will be lost to some extent. (3) Thus both the surface controls and the wing will lose some of their effectiveness in control and in lifting respectively. (4) As the airplane noses up, the gasoline in the tantanks will lose the potential head (hydraulic) relative to the carburetor which is usually put below the engine. (5) The propeller will lose of its speed in revolutions due to load and lack of sufficient head. Therefore (6) the motor has no reserve for any emergency since it is fully used up for raising up the weight of the plane and any attempt to do maneuvering will have serious consequences.

In case the engine stalls due to overloading, the best cure is to glide down immediately without any attempt to turn or side-slip which necessitates the expenditure of additional energy.

Craisirg Granted that getting into the air did not entail any difficulties, the next move it to aim at cruising. Cruising. means to fly at the most economical speed at the service ceiling of the airplane. This is done by leveling out the plane from its climbing attitude. During this motion, the center of pressure of the wing has moved backwards and the tail must supply a down load to keep the plane in trim. So the tail-plane is sifted by moving down the stabilizer which is in front of the elevator. In this respect, the downwash of the air due to the wing decomes an asset.

Judging from the fact that any unnecessary inclination of the parts of the airplane will produce drag which reguires power to be overcome, crusing will need a very good balancing on all parts of the airplane. This is the job of the designer. It also involves the utilization of air currents to gain height and tail to save fuel or cover more resistance for a given amount of fuel. That is the job of the flier.

The pilot must also know what is the cruising r.p.m. of the engine so that he fly at that rev. Fig. 13 suggests the idea. The engine makers must supply the test data for brake horsepower against r.p.m. and fuel consumption against r.p.m. R.p.m. is an indirect measure of the airplane speed. The lowest point on the c curve (pounds of gasoline per horsepower per hour) indicates the most economical condition. The brake horsepower (b. hp.) will decrease due to lack of oxygen to burn the gas. The engine maker must test the engine in vacuum to give such complete data to the buyers. Here finishes the introduction to my talk. In the next few minutes we shall study what aerodynamics is composed of and landing will also be taken up.

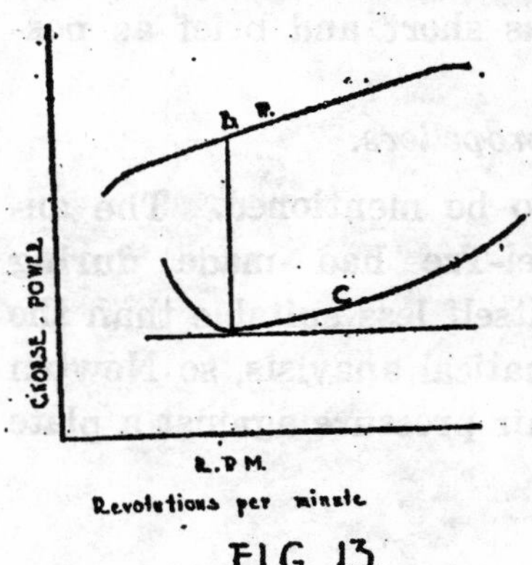

FIG. 13.

PART II. AERODYNAMICS

The Previous introduction had enabled us to follow a straighter channel on the subject of aerodynamics proper which is very wide and very old subject in its nature. Aerodynamics had been studied since time has begun. In a broader sense aerodynamics can be considered as that branch of science that deals with all possible contacts of any object with air. It is easy to imagine how wide and immense the field will be, perhaps wider than any magnetic field any radio station can set up. Coming down to specific things, aerodynamics deals with the following definite topics.

(1) general aerodynamics including hydrodynamics which is the basis for any aerodynamical analysis of skin friction, law of resistance, viscosity of air, standard atmosphere,

(2) theory of the wing sections, of dirigible shapes, of propellers, and so forth.

(3) Stability and dynamics of the airplane,

(4) Performance of the airplane,

(5) Maneuvers of the airplane.

(6) Controls of the airplane, and

(7) Miscellaneous topics.

Scientists have spent years of time with concentrated effort to study these subjects, engineers have put out enormous amount of energy to make apparatus for tests, and fliers paied their lives to help out the scientists and engineers in testing, only finding still more new phenomena and elements to be smoothed out for the benefit of mankind. A bibliographical list of the findings was prepared and occupies many a feet of paper in the governmental organs of various nations. Here I shall make it as short and brief as possibly can.

(2) *Theory of wing sections, of propellers.*

The wing theory is the first logical thing to be mentioned. The anscestor of the wing is the kite which philosopher Mei-Tze had made during Confucius's time. As our Chinese language lends itself less suitable than the Greek or Phoenician language in figure and mathematical anaylsis, so Newton had the first chance to formulate the equation on air pressure against a plate or plane surface in a moving stream of air,

$$P = \frac{d\,A\,V^2 \sin^2 \Theta}{g},$$

where P is the total pressure, d the density and V the speed of air steam, A the area of the plate, g the gravity, and Θ the angle between the plate plane and the direction of air. New ton had no experimental data nor paper to verify this law. So Euler, Raleigh, and Lanchester all had a chance to develop more exact equations. Now the equation stands like.

$P=0.0011\ A\ V^2$ for plate perpendicular to air, and

$P=0.00003\ A\ V^2$ for small angles of attack.

These are the rules for testing flat plates. The results will be used to compare with the resistnace of any wing section. In fact the resistance of complete airplane are usually reduced to that of a equivalent **Flat Plate.** Next we shall survey the field of modern wing theories.

There are three 3 famous wing theories based on pure mathematics the most marvelous of which is that of J. Joukowski, then von Mises, and Max. Munk (also Glauert). The best wings are developed by actual trials.

(a) Joukowski Wing. This great mathematical accomplishment as well as aeronautical engineering starts first with streamline function and velocity-potential function which is nothing but Lagarnges' equations. All of us are familiar with magnetic fields which exists around a wire carrying a steady current in one direction. We are familar to with electrical fields which radiates outward from a wire with high voltage. Streamline function will be used on magnetic and potential function on electrical fields. The next step Joukowski took was to apply the methods in complex variables and then conjugate function (Greene's theorem). The final step was the application of conformal transformation which enabled him to change circles into ellipses and vise versa. The details of making such a wing section is simply geometry. This is not all. The air velocity around the Joukowski wing, the pressure the moment, and the drag can be all calculated and the results are verified by experiments.

(b) Von Mises Wing. The Joukowski wing has this disadvantage that it has large C. P. travel and thin trailing edge which is hard to manufacture. The von Mises wing is an improvement that the trailing edge is turned up (fig. 14.) This turning up of the rear edge will neutralize some of the C. P. travel the other wing has. But the difficulty of making one is

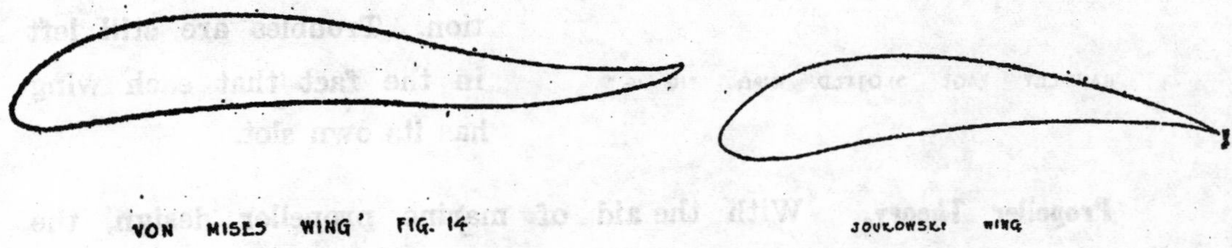

VON MISES WING' FIG. 14 JOUKOWSKI WING

intensified. The mathematics of the wing is similar to that of the other save one more term is added during the development of a Fourier series.

(c) **M. Munk Theory** Munk's wing (also Glauert) differs from the other two in that this wing is based simply on the properties of straight or curved lines expressed in Fourie's series and representing the mean thickness of the wing (camber), fig. 15. The properties of the section are left to wind tunnel tests. In fact all the wing sections, when to be applied in making airplanes, have to be tested in order to have certainty. Therefore the theoretical developments does not put aside any of the wing sections developed by tests like that of Col. V. E. Clark, a friend of Mr. H. K. Chow. They really encouraged the use of wing sections based on tests as the latter will greatly facilitate manufacture.

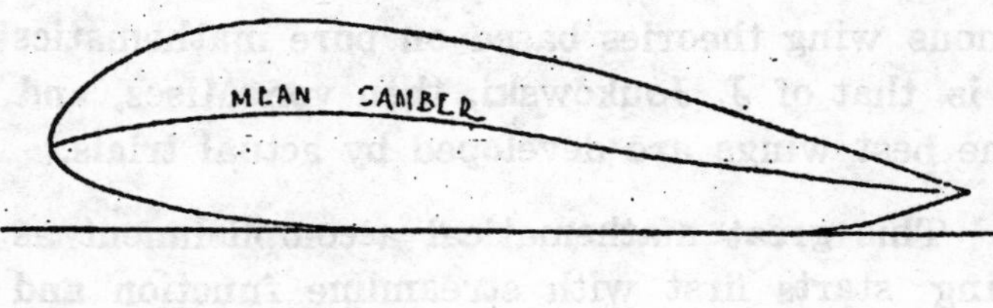

MUNK DEVELOPMENT. FIG. 15

The slotted wing of Handle Page is entirely empirical, fig. 16. Knowing that the air will get disturbed when the wing is elevated to too high an angle, he put a winglet in the nose of section to deflect the air down so as to make air flow smooth and graceful. It has been rated as the first great step forward in aviation. Troubles are still left in the fact that each wing has its own slot.

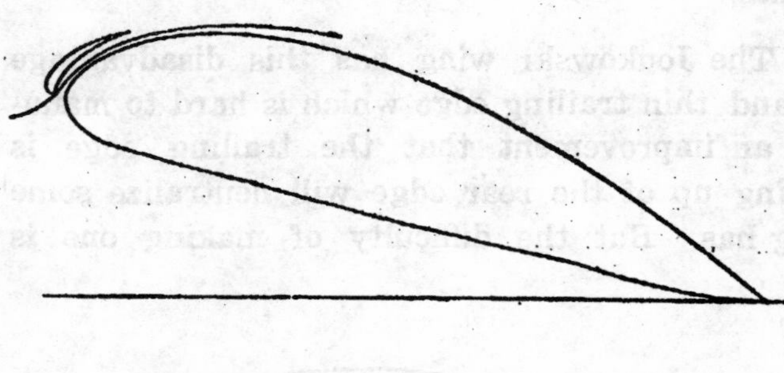

HANDLEY PAGE SLOTTED WING FIG. 16

Propeller Theory. With the aid of marine propeller design, the propeller theory has been thoroughly treated somewhere else. Here it is is intended to give a very brief survey just as what was done to other parts

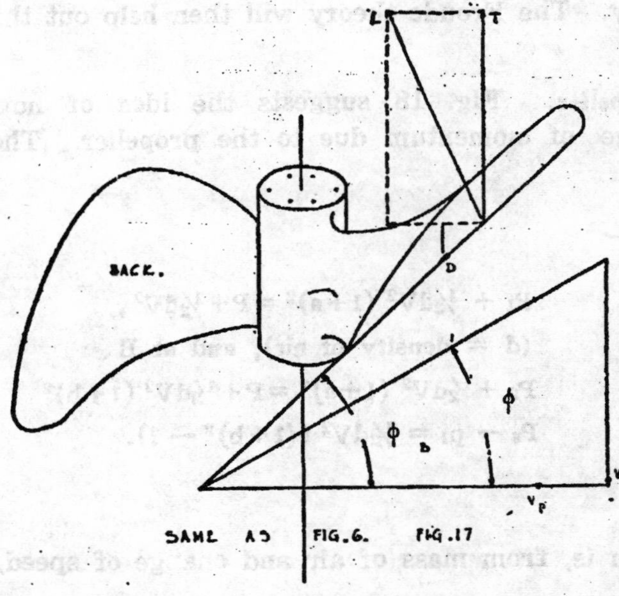

of the subject so as to make way for further study if you are interested in aeronautics. Some definitions:—

Aspect ratio, radius width, is blade angle, Viewpoint from cockpit, is $\tan^{-1}(V/nD)$, V is air velocity, D is diameter, Back of propeller is cambered, Plan view of propeller is its back, A section cut anywhere is assumed plane (actually cylindrical), Pitch is forward distance of one revolution of propeller, Geometrical pitch is pitch at 2/3 the radius, Experimental pitch is forward distance in one revolution resulting in zero thrust, (this is like a screw going into wood). Constant pitch is one whose sectional pitches has varying angles so that the forward advance is constant, Varying pitch propeller can have sections making any angle desirable.

The above definitons will greatly simplify the study of the propeller. The simple theory is developed by Drzweicki Fig. 17. The figure is almost self-explanatory as to how the thrust is developed by resolving the lift of the wing section to the forward direction of the propeller hub which is in line with X-axis of the airplane. The other components will be resolved into direction perpendicular to the propeller hub and is wasted, as a useless torque Q.

$$\text{Efficiency } e = \frac{\text{useful work of } V\ T}{\text{imput of } 2\ \pi\ N\ Q} \text{ of each}$$

section of the propeller. By studying the various sections set in arbitrary intervals, the complete thrust and imput needed can be determined by graphical integration.

The Drzweicki theory of the propeller is not complete because it neglected the change of air velocity after the propeller strike it and sends it flying back very fast. Also the rotational velocity of the air is not considered in finding the imput necessary. The Froude theory will then help out the situation.

Froude Theory Propeller. Fig. 18 suggests the idea of how the air is under going change of momentum due to the propeller. The Bernoulli equation is, at point A.

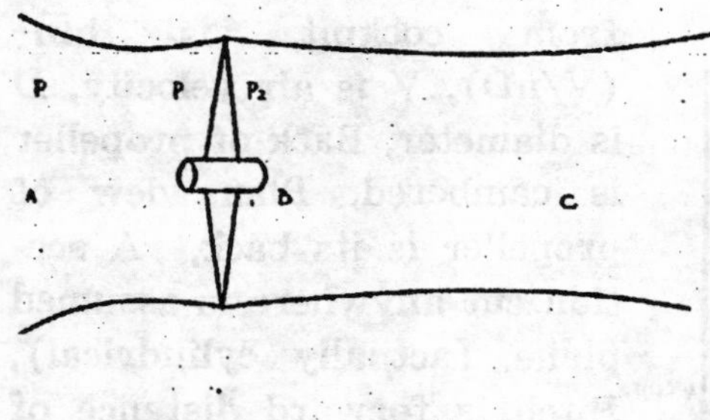

FIG. 18

$$P_1 + \tfrac{1}{2}dV^2(1+a)^2 = P + \tfrac{1}{2}dV^2,$$

(d = density of air), and at B

$$P_2 + \tfrac{1}{2}dV^2(1+a)^2 = P + \tfrac{1}{2}dV^2(1+b)^2$$

$$P_2 - p_1 = \tfrac{1}{2}dV^2((1+b)^2 - 1).$$

The momentum equation is, from mass of air and change of speed,

$$d\,A_a V_a = d\,A_b V(1+a) = d\,A_c V(1+b).$$

where A is area and V is velocity of air at sections A, B, & C, a, b are change of air speed,

$$A_A = A_B(1+a) \text{ and } A_C = \frac{A_B(1+a)}{(1+b)}.$$

The thrust T is a force of mass times acceleration,

$$T = A_B(p_2 - p_1) = d\,A_b(V + v\,a)\,b$$
$$= \tfrac{1}{2}\,d\,A_b V[(1+b)^2 - 1], \qquad b = 2a.$$

But sometimes acceleration b is assumed as 3a. Few knows the intricacies of the actual condition existing in air. From the above considerations, it is seen that the efficiency is

$$e = \frac{1}{1+a}.$$

Propeller design is a mighty complicated subject. It is wise to stop here.

(3) *Stability and dynamice of the airplane.*

.. More fortunate are the mathematicians who worked out the stbility of airplanes than those who worked on the wing theories. There are two kinds of stability about bodies moving in fluids, the statical and dynamical stability. The stability equation I put out at the beginning of the talk was that for statical stability during level flying. Instead of going into mathematics I shall briefiq state the conditions. Since the balancing of the airplane requires the interconnection of the wing force and the tail force, there is differential calculus in the development of that inter-relationship. There is also differential calculus in treating the thrust and its relation to the center of gravity of the entire airplane. The slip-stream and downwash play an important part too. Fortunately, when the airplane is balanced in these statical conditions they will be dynamically stable at certain range of speeds.

Dynamical Stability. The dynamical stability deals with two phases of the problem, the longitudinal and the lateral. Since the former is more important, I shall briefly state the problem. The following table is convenient for a ready reference in visualizing the directions and forces,

Axis	Moment	Motion	Ang. vel.	angle	velocity	Force	Torque
x	L	roll	p	Θ	U	X	M_p
y	M	pitch	q	Θ	V	Y	M_q
x	N	yaw	r	ψ	W	Y	M_r

Since the airplane is free to move in hree dimensions, it can go forward and backward (not easily), right and left, up and down. It can also climb and descend, roll and yaw like a ship in sea. When translation is made there is developed an air resistance to oppose the translation. When rotation is made, there is resistance due to air developed to resist the motion. The forces thus developed are called the resistance derivatives. There are three translation motions and three rotational motions. So there will be six forces acting on the airplane when it is moving freely in all directions, and the derivatives will have reference vectors of six. Thus thirty six simultaneous equations will result. Among the thirty six unknowns seven can be measured in the windtunnel. So the problem is still unsolved.

However expert mathematicians have spent plenty of time in succession to study the motion by taking small changes of velocity called the step-by-step method and calculate the motion for a few seconds. During the process of analysis, some terms will be ignored on account of being higher orders of derivative. Aided by the method of determinants, the complete motion is reduced to a bi-quadratic equation with proper coefficients of twenty three. The form is like.

$$Ax^4 + Bx^3 + Cx^2 + Dx + E = 0.$$

Routh has made known of a factor called Routh's Discriminant,

$$R = BCD - AD^2 - EB^2$$

which must be positive for good stability. The solution of such an equation is rather complicated. But Bairstow has factored out the equation into two quadratic equations one of which represents a fast oscillating motion, the other a slow oscillating motion The equation is like the following,

$$(x^2 + B/A + C/A)\ [x^2 + (D/C - BE/C^2)\,x + E/C] = 0.$$

The first factor represents a quick oscillating motion and the second a slow one.

Similar equations are derived for the lateral motion of the airplane but will be omitted for convenience.

(4) *Performance.*

The performance of the airplane varies with the purpose of the design. Commercial airplanes require long voyage economy while racers needs speed and climbing capabilities. I shall make known the list of items usually calculated from equations in standard text books,

Take-off time and distance,
Rate of climbing at various altitudes,
Maximum speed attainable at service ceiling,
Cruising speed at service ceiling,
(Service ceiling is place where airplane will climb 100 feet per minute),
Calculation of absolute ceiling where airplane cannot climb any more due to decrease of power.
Stalling speed when airplane will not lift the weight,
Landing speed (similar to stalling speed),
Gliding speed with power on and power off.
Landing distance and time after wheel touches ground,
Endurance (time to stay in air),
Range (maximum distance the airplane can go).

Engineers working on performance usually occupy a special department of considerable size. Their job is to predict performance to avoid waste for the company.

(5) *Maneuvers of the Airplane.*

Maneuvers of the airplane is a subject of the aeronautical engineer but the carrying out depends upon the flier. The measure of maneuverability is the time to make a 180 degree turn, in other words to reverse the direction of flight, at a certain given speed. Commercial airplanes are hard to maneuver because they are made too stable and smooth flying. Military airplane are made short, small, and fast, so that the maneuverability is very high. High maneuverability means extra strength required. While the excessive wear of racing automobiles is on the tire and track, the excessive force on the maneuverable airplane is on the wings. Diving down the airplane at very high speeds, and suddenly pull it out by making the wing nearly perpendicular to the relative wind, the airplane will slow down but will suffer a sudden deceleration. Thus wings often came off due to this load. A list of the names of maneuvers seems interesting, for reference,

Pulling out of dive,
Loop, inside and outside loops,
Inverted flying,
Immelmann turn,
Roll, barrel, Dutch, spiral,
Wing over,
Falling leave,
Flat spin,
Tail spin,
William's turn.

Explanation of each can be asked any body in the airports, so I shall pass on to the next topic.

(6) *Controls of the Airplane.*

With the rise of commercial airplane industry, control of airplane became a special branch of study, because slower speeds need better controls. The first designs of control in the commercial airplanes are based on military results. Very soon it was found the rule does not work. At present there is no data for the proper design of airplane controls. The candidate aeronautical engineer has to gather data from various successful airplanes. The tail area, the rudder area, and the aileron area, are usually expressed in percentages of the wing area with special reference to their center of loading from the center

of gravity of the entire airplane. This is very similar to the proper location of the center of pressure of the wing from the c.g. of the plane. Rough figure are this,

aileron........8—12% of wing area,
Horizontal tail area, 14—20%,
vertical „ „ 4— 5%.
20% increase for flying boats and amphibions.
20% increase for any special slow speed airplane.

Controlling big airplanes must be done by one man and the load on the control surfaces are very great. So balanced controls are devised. This is done by making the surface well distributed on either side of the turning hinge, leaving just a little over balance for the pilot to work on. Automatic controls have been designed in Germany already but the use of mercury electrodes and electric motors are rather heavy.

Another difficulty with controls is when the control is moved the air is disturbed sometimes to dangerous conditions. Then comes the slotted controls where the motion of the control will not disturb the air so badly as to make other parts of the airplane lose their effects.

(7) *Miscellaneous Topics.*

A list of the topics will be enough to incite interest,

wind tunnel studies,
air motion around large erections like chimneys, buildings, etc.,
air above rivers and lakes,
air around cliffs and surrounding canyons,
meteorology,
air in clouds,
winds around hill-sides,
and so forth.

PART III. DESIGN OF THE AIRPLANE

Like any other designing work, airplane design involves a lot of changes on account of the sad limitations. There are points best in aerodynamics but impossible of design like great aspect ratios and other things. The wing can never be designed like that of bird. Still the airplane industry is prospering if we go within those limits. The following steps will guide any designer from going astray.

(1) Purpose of the airplane, to be commercial, military, racing, messenger, farm dusting, photographic, or others.

(2) Free hand drawing to carry out the main feature of patentable specialities.

(3) Weight estimates for the empty airplane and the contents, and the balance about the entire center of gravity.

(4) Scaled drawing in three views with outlines of contents like pilots, passengers, baggage, gas tank, oil tank, and other important items.

(5) Study of wing sections and decision on one.

(6) Decision of the engine to be used.

(7) Preliminary performance estimates.

(8) Preliminary stress analysis.

(9) Arrangement of controls.

(10) Stability estimates.

(11) Changes of design to suit aerodynamics or vise versa, to avoid bulky members, and so forth.

(12) Control of weights, not t be over the estimated value to any too great extent.

(13) Final aerodynamics and tress analysis and completion of design.

Since all airplanes look alike, if one just draws up the outline of an airplane from common sense, it is most probable that it will fly after it is designed and built. So in this part of my paper, I will just point out what kinds of loads are coming on to the airplane, so that the plane will be designed to stand those loads.

The most important groups of the airplane that take up the loads are the wing group, the body group, the controls, the tail, the engine mount, and the landing gear. For simplicity, a monoplane will be taken up which has only one wing with the body hanging down from its middle.

Loads on Wing. Since the airplane is supported by the wing, the air load on the wing is distributed. So the wing is like a cantilever beam carrying air load and concentrated loads due to the body and contents.

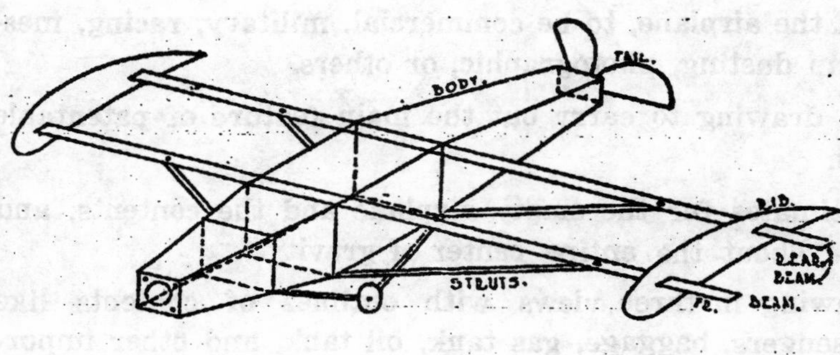

FIG. 19

To secure lightness and keep the wing section in comformity with the theory, above figure shows roughly how the skeleton of the airplane is built up. The ribs keep the wing section while the wing beam transmits the load on the ribs to itself together with the body load. Two wing beams are the least we can use, one front beam and one rear beam. How the air loads come onto the wing is analized as follows,

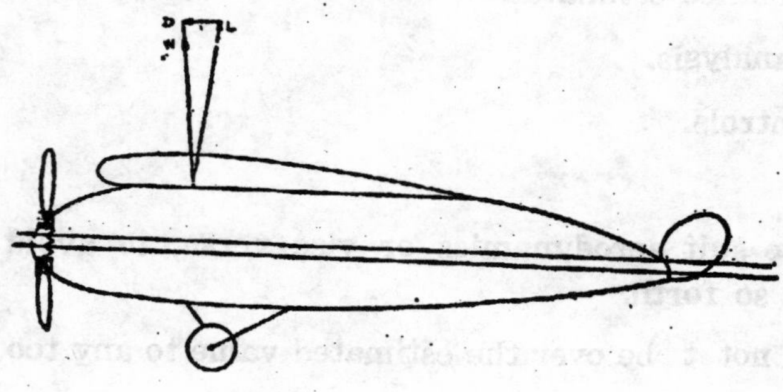
FIG. 20

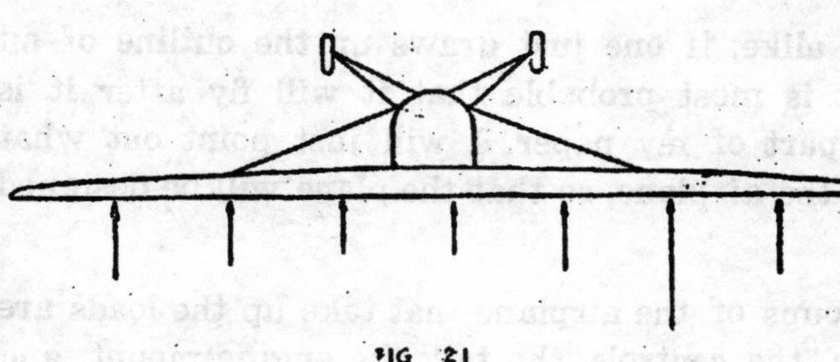
FIG 21

(1) High angle (or climing) flying when the load is near the front beam.

(2) High speed flying when the load is near the rear beam.

(3) Inverted flying when the load changes direction.

(4) Nose diving when the wing is twisted somewhat.

Figures at left are self explanatory.

The wing loads are expressed in pound per square foot of wing area. For stress analysis, the load is in pound per in. of wing span. This load is multiplied by a load factor for the four conditions,

condition (1) 6.5 x load per inch, (i.e. design load),
(2) 65% of (1),
(3) 40% of (1),
(4) below is the ex planation.

If the strength of the wing beam is stronger than the design load, the extra strength is called the margin of safety.

Wing Beam Design. In designing the wing beams, a loading curve curve for one pound per inch is the proper start. Figure at left shows the idea where w is the load per inch, s the shear at different sections, m the bending moment at the various sections.

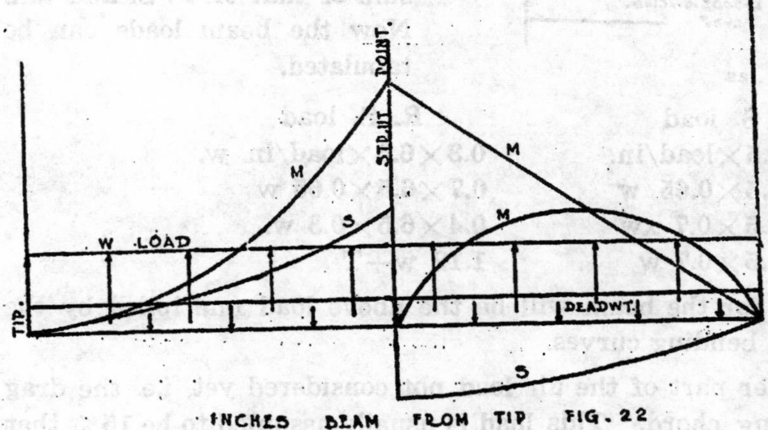

INCHES BEAM FROM TIP FIG. 22

But the design load of the wing has to be divided between the two beams. The division is made by the following method. Refering to figure at left, Front spar F. S. will take the larger part of load at high angle while rear will take the larger load at high speed, since the c.p. is moved, fig. shows F. S. takes 70% of condition (1) and rear spar takes 70% of condition (2). For inverted flying, the high angle C. P. is used. For nose diving condition fig. 25 will be used to illustrate. The L. E. of wing is supposed to be under down load (referring to c.g. of airplane) and rear edge of wing is under upward load. The balance is taken care of by the down load on the tail to keep diving straight. The drag loads must be equal that weight of

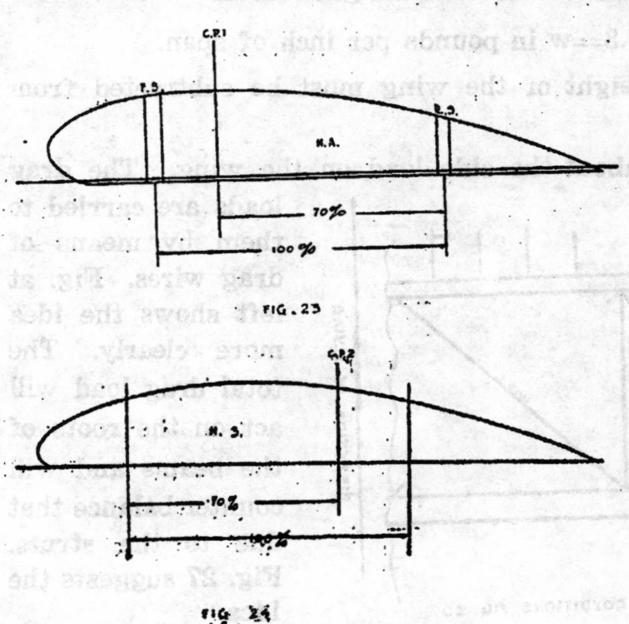

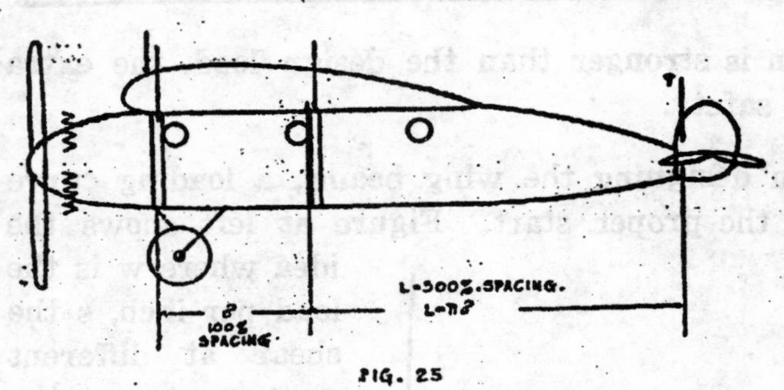

entire airplane. The load will therefore be that of inverted for f. S. Tail load is gotten by taking moment about Rear Spar. Load on R. S. will be sum of that on F. S. and tail. Now the beam loads can be tabulated,

Condition	F. S. load	R. S. load
(1)	0.7×6.5×load/in.	0.3×6.5×load/in. w.
(2)	0.3×6.5×0.65 w	0.7×6.5×0.65 w
(3)	0.4×6.5×0.7 xw	0.4×6.5×0.3 w.
(4)	0.4×6.5×0.7 w	1.12 w+T

The bending and shear for the beam will be the above load multiplied by the values on the shear and bending curves.

There is another part of the air load not considered yet, i.e. the drag loads acting on the wing chord. This load is usually assumed to be 15% that of the beam load for condition (1) and (2). For (3) it is zero. For condition (4), the chord load C is the sum of condition (2), added 30% and modified by a constant k to be equal to the total weight. It is

$$K\times0.15(4,225)\times1.3=w \text{ in pounds per inch of span.}$$

It is to be noted that the dead weight of the wing must be subtracted from the air loads.

The previous talk was about the side load on the wing. The drag loads are carried to them by means of drag wires. Fig. at left shows the idea more clearly. The total drag load will act on the roots of the beams and will counter balance that due to the struts. Fig. 27 suggests the idea.

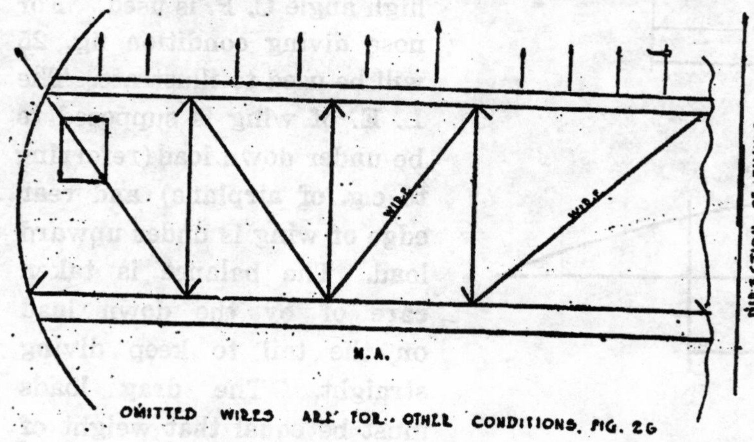

OMITTED WIRES ARE FOR OTHER CONDITIONS. FIG. 26

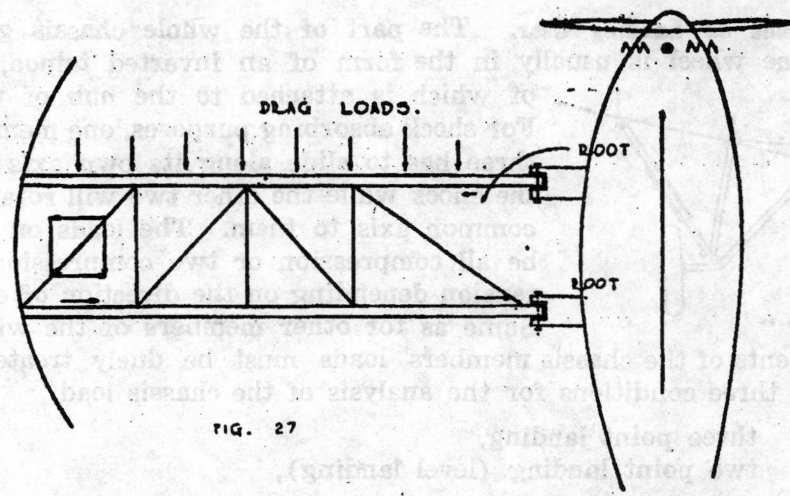

FIG. 27

It can now be summarized that there are the following loads on the beams,

the side loads, acting up and down,
the side loads action back and forth,
the axial compression due to wires or structs,
the aial tention due to wires on compression.

It is always good to define the sign of the loads before going too deep.

+ denotes load acting, up, back, out,
— denotes loads acting down, forward, in.

The nature of the loads have to be visualized for particular designs. After the loads are collected, then the beam can be designed for the most severe cases of loading. The components of the loads in all members must receive due attention. Thus the load component of a drag wire acts in three direction, one along each beam, one along the chord, and one in the wire. The components of the struts will have four, one along the struts, one along the beam, one along the chord, and one parallel to the Z—axis of the plane.

The rib design is a complete engineering unit itself and the usual manner of the rib design is hown as Fig. 28.

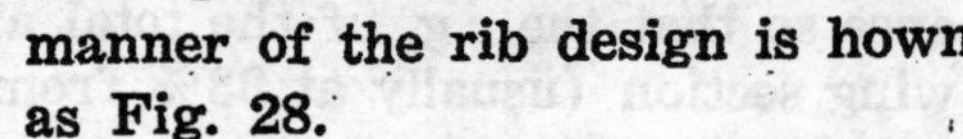

The beam and strut design are finally the same for airplanes as for other structural members like that of a building, or rail, and so forth. Next we pass to the loads on the landing gear.

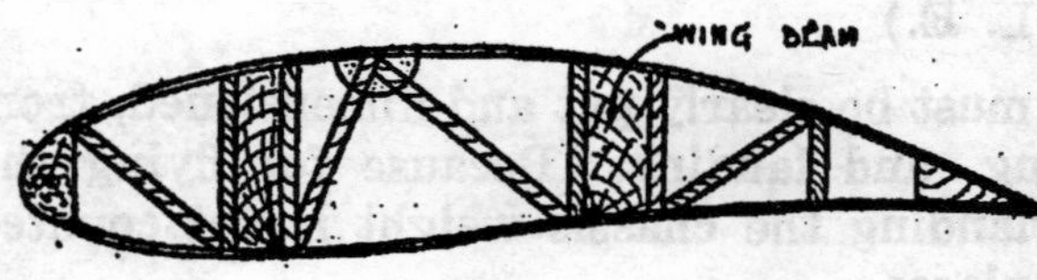

FIG. 28

Loads on Landing Gear. The part of the whole chassis group that supports one wheel is usually in the form of an inverted tripod, the apex of which is attached to the hub of the wheel. For shock absorbing purposes, one member of the three has to slide along its own axis to absorb the shock while the other two will rotate about a common axis to them. The loads on them will be all compression or two compression and one tension depending on the direction of outer load. Same as for other members of the wing group, the components of the chassis members' loads must be duely treated. There are usually three conditions for the analysis of the chassis load,

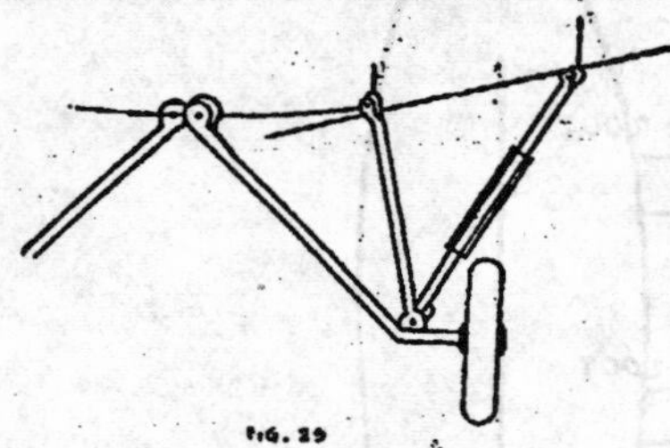

FIG. 29

three point landing,
two point landing (level landing),
one wheel landing.

The loads can be determined either graphically or analytically. The loads from wing analysis and chassis analysis are most important because the body analysis has to be based on these loads which are attached to it.

LOADS ON BODY

(1) First things to be done on the body are to plot the truss top, bottom, and side showing clearly where the joints are by center lines only.

(2) On these joints are placed the loads of the contents and the body's own weight.

(3) Assume the side truss to be in one plane and the loads on the joints are ½ (load factor) (unit loads).

(4) In distributing the loads in (2) the total weight must check the weights at the joints and the moment of each joint weight about the preliminary C. G. must check to zero. The preliminary C. G. was found in the preliminary balance by taking moment about a convenient point of reference so that the c.g. of the total airplane will balance at certain point of the wing section (usually at 33% from L. E.)

(5) C. G. of the airplane must be clearly put and dimensioned, from each joint, for all conditions of flying and landing. Because for flying the wing weight is not counted and for landing the chassis weight is not counted thus changing the C. G. to different places.

(6) Tabulation of the joint loads stating clearly the design loads for all conditions in flying and landing.

(7) Now the forces (or loads) due to flying and landing as found in the wing and chassis analysis have to have zero moment about the C. G. too. Usually they do not because of various arbitrary factors.

(8) In order to have (7) balanced, three simultaneous equations sum V, sum D and Sum M from the joint loads, and the wing or chassis forces must be solved and a coefficient k introduced for balancing. The factor k is used to modify the wing or chassis forces.

(9) This step is to check the forces so that

in flying,
 the lift of wing equals weight of plane,
 the drag of wing equals thrust of propeller,
 and the moment about c.g. be zero.
in landing,
 the ground force up will equal weight down,
 the ground resistance will equal inertia of plane,
 and the moment about c.g. be zero.

(10) After the force systems are checked, a graphical solution of the side truss can be easily done.

After the above steps are taken the forces in the members will have various values. Only the greatest will be chosen for design.

(11) So a list of the members must be made and the loads due to flying conditions (1), (2), (3), (4), and landing conditions (5), (6), (7) to be filled in.

(12) But the rear part of the body will receive larger loads than is thought of. The rudder load usually give a large twist to the body's rear, and the elevators give a large bending force, that the top truss has to be analized for maxrudder load and the side truss under maximum elevator loads. These will be condition (8) and (9) for body.

(13) Also the nose of the body where the engine is usually put, will receive a torque when the engine starts. The nose is usually under severe load in high angle flying. So the side truss has to be analized for high angle load plus the load due to torque, Condition (10) for body.

(14) There may possible be special conditions for the body like hoisting into a steamer about some particular joints. This will occur for special designs.

The air loads that occur to the other parts of the airplane like the tail, and the airleron will be analized similar to that of the wing. The loads on the controls like the stick, the cables, the rudder bar and supports, are very simple and need no further explanation.

紡紗機之拖動法

著者:費福燾

第一節　變速馬達單獨拖動法

自紗廠採取能變換速度之單相或三相更流電馬達拖動紗錠機以來,於今已二十年,該種制度之原理,係紗錠機紡錠子之時,馬達速度,能同時自動調節,俾維持紗線相當之拉力,紗線斷裂之弊,因之減少,而出品於以增多.方今各國大紗廠,採用該種制度者甚多.且應用於製蔴及紡毛機之拖動,亦極相宜.今將該種制度包括紡紗機(Ring Spinning Frame)馬達,及自動齒輪之關係,及互相配合法,述之如后.

自動調節速度之機關,為兩個桃子盤(Cam),其動作則用適當之聯合,傳自紗錠機.該兩個桃子盤之動作,復傳到馬達之調速柄.綠上述之傳動,馬達整流子上電刷之地位,可以移動,而得所需要之紡織速度.至該兩桃子盤之責任其中之一,務使錠子始紡(Copping)及紡畢(Doffing)之時,速度較平常為低.故在該兩種時間內,線紗之拉力減低,而線紗斷裂之虞,亦因之減少.其另一桃子盤則在紡紗機之鐵圈軌(Ring rail)每一起落之時,轉動一次.馬達之速度,因該桃子盤之作用在紗線紡在錠子上最小直徑之時,其速度減少.俟錠子之直徑,紡到最大時,馬達速度,亦增到最高點.

因上述之傳動法,紗線始終能維持其適當之拉力.用該種傳動,紡軸之平均速度.亦較一般用固定速度拖動之法為高.其結果則紗錠機上每日之出品增加.當紡同樣質地之紗線,用變速傳動法則纏繞(Twist)較少,而所紡紗之堅久力,能與須纏繞較多之固定速度法相等.今請舉一例,有時用質地較軟之紗線,其纏繞須較少.若用固定速度拖動法,因紗線質地較軟之故,紗軸之速度,須減到極低度,以免紗線在紡織時之斷裂,倘用卜郎比變速拖動法,則其速度與出品均皆較高.附圖一爲 *瑞士卜郎比變速馬達,在歐洲某紗廠之攝影.今將其緊要項目述之.每馬達爲六八輪製馬工率,四百伏而特,五十周波,每分鐘一千轉.該項馬達係單相制,但平均分佈於三相線上.其速度調節,可自每分鐘六百轉增至一千二百轉而

第　一　圖

並無電損.每紗機裝五百錠子.其所紡線紗支數,自四•五至廿四支.(Counts)因紡該項粗紗之時間甚短,故除自動調節齒輪之外,尚裝一完全停止紗軸之機關(Full Bobbin Stop Motion)直接傳到馬達之起動柄而使馬達停止.當該廠採用該種變速馬達第一批共十部,結果甚爲圓滿.前後復添辦二十部.至用該種馬達之成績,則出品較該廠所用之別種拖動法,增加百分之七.除單獨馬達拖動紡紗機外,(見圖一)該種馬達製造上稍事改變,應用於他種工業甚多.如雙架紡毛機,常裝有兩個滚筒,故應用雙馬達拖動法.(見圖二)又馬達位於一底座之上,而每個馬達之速度,各被其自動齒輪所調節.

有時雙架紡機之滚筒相距太近,而仍欲採用雙馬達拖動制度;則可裝置

*瑞士卜郎比電機廠在中國獨家經理者爲上海九江路二十二號新通貿易公司

第 二 圖

第 三 圖

高效率之變速齒輪.該齒輪在一油缸內轉動,故永得滑潤.(見圖三)

倘紡機之錫滾筒(Tin roller)較馬達之普通速度爲低.可利用變速齒輪配合;俾適合任何紡織機之速度;而同時又得最適宜之馬達特性.倘紡紗機速度之調節更須增廣,可加裝固定子變換鑰.(Stator Change Over Switch)(見圖四)該圖並表明變速齒輪之裝置至該種馬達之冷却法係在廠房內地板下裝設風道.新鮮空氣自廠外引入,經過馬達內部之後,其熱空氣復導至廠外.有時熱空氣散在廠內亦無妨礙.如圖五所示,係三相整流子變速馬達拖動紡蔴機.馬達裝在鐵橙之上,而所通過之空氣,經過馬達後;散在廠房之內.

第 四 圖

第五圖

第二節　價値較廉之松鼠籠式馬達單獨拖動法

紡織廠中,有時因所紡紗線之性質,對於自動變速馬達制不甚合算,則可採取價格較廉之松鼠籠式馬達單獨拖動制.如圖六所示,該種馬達係松鼠籠式亦名短接式(Squirrel cage type)裝設鋼珠軸領,馬達兩端有孔,其內則裝設特製風扇,故馬達之繞線得以冷却,而同時因該種特製風扇之轉動,將黏在馬達上之棉塵吹去,而不至鑽入馬達內部,馬達之壽命及安全於以保障.(作者曾參觀瑞士魏廷根紗廠,試將小棉團塞入馬達孔內,立時爲風扇吹出)如圖六所示,卽係該項馬達裝在紡紗機之端,而位在一鐵架之上.占地旣省,視察亦易,較之用架空拖動法,所省地位甚多.且廠內光線亦因之不受影響.該項馬達,因用特種皮帶盤(Jockey pulley)傳動,故皮帶雖短而傳動之效率甚高.該種皮帶盤尙有一利,卽錠子始紡之時,可將皮帶暫時扳鬆,俾錠

第 六 圖

子軸之速度在此時稍減;而紗線之拉力亦因之稍低;故紗線斷裂之處得以減少.故用該項拖動制較之用架空傳動法等之平均速度爲高,而出貨因之較多.松鼠籠式馬達拖動制雖不能如第一節所述法之盡善盡美,但價格較廉,且勝於他種拖動法實多.至所述兩種拖動制之採用,須權衡出品及經濟情形而後取決之.

本刊職員易人

九月杪,本會舉行年會於瀋陽東北大學,總務一席,由黃炎,姚長安,楊錫鏐三君依次當選.旋三君因故先後堅辭,由本會商請支秉淵君擔任.工程季刊自本期起,凡印刷,廣告,發行等事務,概由支君負責.

SMALL REFRIGERATING PLANTS FOR MAKING ICE CREAM

The manufacture of ice cream cannot be carried on with any prospect of success unless a fairly large daily turnover can be reckoned with. This is an essential condition, because of the relatively high cost of acquiring the necessary plant, and because only a very small quantity of ice cream can be sold in winter. The allowance for annual depreciation must therefore be earned in a comparatively short time, i.e. principally during the few hot months in summer. The problem is therefore in the first place a purely commercial one, depending solely and absolutely on the prospects of selling the manufactured product, so that for the moment the only plants which have any likelihood of being profitable are those situated in districts, where a sufficiently large daily demand for ice cream can be reckoned on.

A plant for making 165 gals. of ice cream per day has been put in service by the Industrie du Froid S.A., Cairo (fig. 1). The cold required for the plant is supplied by a vertical Sulzer compound ammonia compressor rated at 48,000 B. Th. U. per hour at an evaporating temperature of 22 deg. F. and a condensing temperature of 96 deg. F. The compressor can be seen to the left in fig. 2; the other compressor seen in the illustration serves for ice-making and the two Sulzer Diesel engines of 40 and 120 B.H.P., installed in the same room, drive generators which supply the light and power required. In contrast to other plants for making ice cream, no special store rooms are provided for keeping the ice cream, but there are two hardening rooms, kept at a constant temperature of 11 deg. F. The two freezers, fig. 1, and also the refrigerating systems in the hardening and other rooms, work with direct evaporation, whilst the cream cooler and the ice cream ageing vats are worked with cold brine taken from an existing ice tank.

Ice cream is essentially a sweettend frozen cream. In addition to the cream and sugar, about ½ of 1 per cent. of gelatine is nearly always added in order to prevent the formation of large ice crystals; aromatic substances are also added. Here there is endless opportunity for variety, as not only fruits can be added in the form of extract or juice or grated, but also all the other various kinds of flavouring matter hitherto adopted in making sweets and cakes. Nevertheless a certain number of standard makes have been evolved, of which the principal are vanilla, pine-apple and lemon ice creams.

In particular kinds of ice cream other substances are added, such as eggs, for example, the mixture being heated up to 175 deg. F. before being frozen, in order to curdle all the white of egg; starch-flour and other substances are also employed.

After being prepared, the raw materials are mixed and pasteurised, the mix being passed through the homogenising machine and then over a cooler to the ageing vats which are fitted with a stirring mechanism. In these vats the mix is allowed to stand generally for 24 hours at a constant temperature of 40 deg.F. in order to "age." The function of the homogenising machine is to break up the fat globules into still smaller globules that no cream can form on the surface. The homogenising machine works on the principle of forcing the warm milk and cream at a high pressure, and therefore at a great velocity, through a very narrow aperture. This causes strong eddies, which break up the fat into very small globules. The homogenising process works the ice cream into a uniform mass and also prevents the mix turning into butter under the subsequent churning action in the freezer.

FIG. 1.
Refrigerating apparatus for making ice cream.

After the mix has aged sufficiently, it is run into the freezer and then the aromatic and colouring substances are generally added, when such are to be used. In the freezer, which is fitted with a cooling jacket, the mix is simultaneously beaten and frozen, the final temperature being 27 to 25 deg.F. This operation requires abt. 20—30 min., depending on the size of the freezer. It results in a great frothing, to such an extent that up to 2 gallons of ice cream can generally be obtained from 1 gal. of mix. This increase in volume is technically known as "overrun" or swell. Its amount is important, as on it depends to a great degree the quality of the resultant product and also the commercial efficiency of the manufacture. Ice cream should be "light," but there is also a limit beyond which it becomes soggy and unpalatable. This limit lies with an overrun of abt. 100%; it is not advisable to go beyond that.

The mix is now left in the freezer until it is absolutely hard, but shall still be in a semi-liquid state when it leaves the machine. It is then run into pre-cooled cans, or immediately into moulds, which are then placed in the hardening room. This is a room kept at a very low temperature, and there the ice cream becomes consistent, so that it can be cut with a knife. The temperature of this hardening room should be at least +5 deg.F., but better still is —4 to —13 deg.F. The ice cream generally remains here for 24 hours after which it is transferred to the general storage rooms, where it is kept at a temperature of abt. 14 deg. F until distributed. The purpose of the hardening is to solidify the water contained in the ice cream as quickly as possible. It is necessary to do this quickly, in order to keep the ice crystals as small as possible. The slower the rate of freezing, the larger will be the ice crystals, and the coarser the texture of the ice cream. The hardness of the ice cream depends on the completeness with which it is frozen.

Fig. 2. Refrigerating plan of the Industrie du Froid S.A., Caira, comprising tmo Sulzer ammonia compressors; in the background, to the right, two Sulzer Diesel engines diving generators for supplying electric light and power

Certainly even the adoption of very low temperatures will not cause all the water to turn into ice, as the sugar solution becomes more concentrated as the water is frozen out. But, for the reasons given, it is desirable to use the lowest possible temperature, and in the United States the temperature of the hardening room is being always more and more lowered. About 20 years ago +14 to +15 deg.F. was thought satisfactory, but now —5 deg.F. is demanded and there are even hardening rooms in existence with temperatures of —20 deg.F.

Ice cream is generally dispatched in containers packed in a freezing mixture of ice and salt and, when well insulated, can be sent to a comparatively great distance. The best containers are those in which the ice cream is packed in tightly closing metal boxes. The temperature of the freezing mixture can be adjusted as desired down to a temperature of abt. —10 deg.F. by varying the amount of salt.

The question of keeping ice cream in retail shops deserves a chapter for itself. It is a costly matter to supply the retailer daily with ice cream, and for this reason it is always advisable to get him, whenever possible, to acquire a cold cupboard fitted with mechanical refrigeration.

國外工程新聞

德國柏林市之道路概況

柏林市現有之道路約八千條,共長約2,900公里,總面積約計爲23,500,000平方公尺,總價估計約330兆馬克,平均每市民應攤76.3馬克.其中61%爲磚石路,29%爲瀝青與柏油路,6%爲碎石路,其餘爲他種道路.

柏林市築路經費頗爲支绌,其原因,大半由於該市每年汽車捐稅收入,約23.8兆馬克,大部分爲德政府支用,能分配於該市築路之經費,僅2.34兆馬克,不過總數十分之一.故雖有建築新路,便利交通之議,曾於數年前計劃建築放射式道路十九條,環繞式道路三條,及其他重要聯絡道路多條,因限於財力,未克舉辦.其尚能於去年(1929)以18兆馬克之經費,築成道路165條,亦可謂極盡其力矣.在此種困難情形之下,柏林市現在所定之築路方針,僅以完成交通幹路爲限,其次要交通道路,及居住道路之修理,雖亟待舉辦者,亦不得不暫緩從事焉.

柏林市修築之道路,除行人道外,其路面擬分劃爲三部,卽電車道一條,與車馬道兩條.至於交通廣場,其通過之道路在四條以下者,採用直接穿過制;在四條以上者,則採用環行交通制.廣場之中央,仍主張設立寬闊之隔台;例如新築之 Alxander 廣場,中央亦設寬闊之草地,四週車道比較狹隘,雖對於交通,有不甚適宜之點,然亦不之顧也.

凡天雨時道路,往往發生泥濘,車輛往來,易生危險,交通極感不便.柏林市當局,有鑒於此,設法避免天雨時道路之泥濘,減少行車之危險,飭工務局改良道路之建築.築路方式,僅以瀝青柏油砂,混凝土,石塊等數種爲限.又爲減少掘路起見,已擬就埋設各種管纜之統一辦法,庶掘路工作得分段分期進行,道路交通,不致多受妨礙.

柏林市內之運貨汽車,裝載重量往往超越規定之數,甚至二倍以上,殊於道路之維持有礙,當局將釐訂嚴厲之罰則,以取締之云.

美洲分會附刊

第二期

中國工程學會美洲分會編輯出版委員會編輯

委員會主席　顧毓琇
書記　黃輝

目錄

鐵道電化及鐵道與人口

著者：劉乾才

(一) 鐵道電化新聞二則

(1) 紐約與華盛頓間之鐵道電化

本雪物泥亞鐵道公司,於十月一號宣佈,與巴迪摩爾城議妥,完成紐約與華盛頓間之鐵道電化.

此項議約曾醞釀數年.其重要問題爲分站計畫,鐵軌改良,及巴城之雙軌山洞.此項問題一決,該公司遂可發展電氣鐵道,先從紐約至豫民頓.再至華盛頓首都.一切客車貨車皆可通行.預計五六年之間,山洞,新站,貨棧,交軌,以及電氣裝設,皆可竣功.

該公司已定造一百五十輛電氣機關車,作拖客車之用.價值合美金一千六百萬元.造完期限兩年.此項機車,由該公司及西屋奇異之工程師規定,其最高速度爲每小時九十哩,馬力六千匹(兩機頭合併,以一小時爲限)大過於現有之蒸汽機車馬力半倍.將來往返紐約與華盛頓之時間,由五小時縮短爲四小時或四小時以下;自紐約至費勒德爾費亞城之行程由二小時減爲一小時半而已.

(2) 英國政府電化大不列顛鐵道之先聲

英國政府擬將全英領土之鐵道電化,藉以救濟失業工人.該項計畫,已於九月七日公佈,派定一委員會審查.委員會主席爲威爾(Lord Weir of Eastwood)該會將來促英國中央電氣部之成立,以資統一.威爾曾聘請英國電氣專家多人,襄助一切.現任美洲加拿大國有鐵道局之總理湘頓(Sir Henry Thorton),亦有被請回國,任政府鐵道顧問之說.

(二) 歐美之鐵道與人口

國名	面積	人口	每方哩人口	國有鐵道哩數	總線哩數	每千方哩哩數	每萬人哩數
奥國	32,368	6,535,363	201	3,608	4,128	128	6.3
Czechoslovokia	54,206	14,388,000	265	2,289	8,239	153	5.8
法國	212,736	40,960,000	193	6,907	26,872	126	6.6
德國	180,972	62,592 000	346	33,320	35,597	197	5.7
意大利	119,744	40,799,000	341	10,300	13,355	116	3.3
尼士蘭	13,213	7,726,072	577	......	2,254	171	3.0
西班牙	195,040	22,128,000	113	......	9,705	50	4.4
瑞典	158,525	6,087,923	38	3,877	10,110	64	16.6
瑞士	15,944	3,959,000	248	1,828	3,492	219	8.9
英國本洲	94,278	44,170,241	469	......	21,165	225	4.8
美國	2,973,774	120,073,000	40	......	249,131	84	20.8

HIGHWAY LOCATION

BY T. Y. YU (余宰揚)

Not until the advent of automobile in the closing years of nineteenth century has the highway development become one of the most important problems of the country. The romance of railroad location engineer has been played up so impressive that its glamour has overshadowed the highway work. Yet the annual expenditure on American highways is already in excess of annual sums spent on railroad work in its palmiest days. Of course any knowledge of railroad engineering is helpful to highway work if the engineer can correlate that knowledge with the requirements of highway. But ignorance of any of peculiar features of highway work would result in material improvement without adequate return. For instance railroad has spent large sums in acquiring long easy grade. In highway work, the consideration of grade is given less weight, and in ordinary design we expect to confirm to natural topography as much as possible, because the return due to excessive cut and fill for improving the highway grade is intangible, and the modern automobile is so designed as to be able to ascend steep grade. This is only one of the marked differences between railroad and highway location which should not be overlooked. The object of this article is to present these fundamental points to be considered in highway location.

Survey of Highway. The complete survey of a highway, quite similar to that of railroad, may be divided into following four steps:—

1. Reconnaissance
2. Preliminary survey
3. Location survey
4. Final survey

The former three involves the location and design of a highway, while the last one, to be made after the highway is completed, serves only to check engineer's construction estimates and needs not to be discussed here.

The reconnaissance is rough preliminary investigation of possible routes between two terminals of a highhway project immediately after reconnaissance, the preliminary survey are made for the purpose of collecting all informations along the routes as may be considered feasible in the reconnaissance. The informations collected are to be mapped, and a rough estimate of cost prepared. As soon as best lines selected on the paper, it is laid on the ground subject to minor changes if necessary, which procedure is said to be location survey.

Importance of prepliminary survey. The negligence of preliminary survey too often results in the expensive resurveys and readjustments after the construction is started. The trouble would be multiplied in the case of concrete pavement; its hard and fast lines can be hardly altered. The crooked locations, especially in pioneer district, are almost unsoluble withuot the help of preliminary survey. Therefore it is always a good policy to have painstaking preliminary surveys, and map in the field with contours and topography laid to the convenient scale. All items such as grade, alignment, drainage, foundation etc., should be analyzed in order to provide a sound basis for the paper location. These items are to be briefly discussed in the following paragraphs:

Maximun grades and grade reduction. Grade selection depends upon many factors such as safety, conveniencs, cost of construction and maintenance, and traffic operating cost. Not one of these is absolutely dominant without considering the others. The whole question of determining maximum grade depends upon experience and lies in the dicision of what is most reasonable for a specific case. For these roads where there are large percentage of heavy trucks or horse drawn traffic, the maximum grade should be kept under 6%, while for lighter traffic, a short grade up to 8% may not be objectionable. The soil condition, and the types of pavement and ditch have also great bearing on the grade, for the maintenance of shoulders, ditches, macadam or gravel road, increases in cost rapidly on grades over 6%. From the standpoint of maintenance, 6% can be said as logical maximum grade.

It is also important that the maximum grade should be correlated with other features of location. In securing lower grade, the engineer should notice that neither poor alignment, nor excessive cut have been introduced. If there are existing heavy grade between two terminals which can not be reduced. it will be unwise to spend large sums in securing less intermediate grades. In a word, the reduction of grade requires comprehensive preliminary survey and careful analysis of alignment and cost.

Aligament. Alignment affects safety, speed, ease and hauling power of the traffic, as well as construction cost of the highway. Mere theory is not sufficient in fixing the alignment, unless it is tempered by some practical considerations.

In well-settled communities, alignment is practically controlled by existing right of way except some minor relocations. In sparsely-settled communities, it is less handicapped by right of way difficulties. When the right of way can only be secured by paying heavy compensation, it is advisable to find a new location.

As a rule, curves of radius of less than 500ft. should be eliminated, and a minimum sight distance of 500 ft. be secured in both horizontal and vertical alignment, A curve concave toward a hill will preclude its use, unless a bench cut out of the slope can be provided at reasonable expense. The outline of the bench is shown in the following figure.

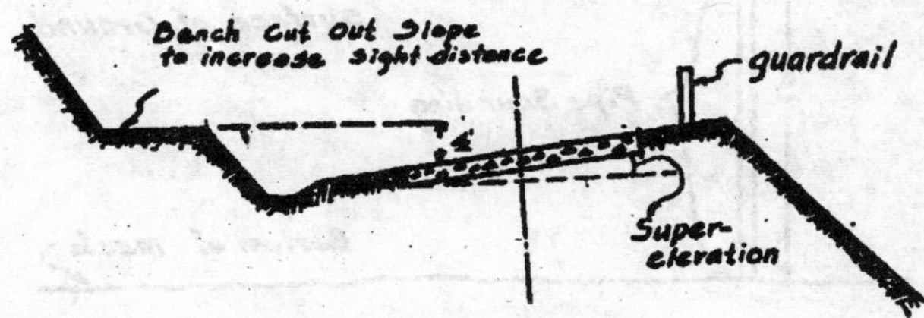

If the road has long tangents the curve should have longer radius than those crooked roads with shorter tangents, because in the former case traffic will attain higher speed. It seems to be a curious thing that more accidents happen on long straight road than the crooked. The reason is that people are apt to drive faster, and feel monotonous on the straight grades. Therefore an alignment composing smooth and safe curves, and yet straight enough to be free from undue amount distance, will be the most desirable one.

Foundation Experience has shown that poor foundation and improper drainage are responsible for the failure of many highways. Special attention must be paid to investigate the condition of soil over which the highway rests, as the design of future road depends greatly upon the results obtained. Due to the fact that soil condition changes abruptly, many engineers advocate station-to-station analysis. It is not infrequent that soft soil develops where road surface prevents fair appearance. Therefore as a matter of precaution, whenever there is suspicion of unstability, the sounding bars must employed, and a length of pipe is driven into the bottom of hole; the earth retained in the pipe showing what kind of soil it is. In short, any serious soil condition should be studied so that the projected line will not pass through any unstable place unadvisably. Sometimes a change of line to avoid the bad foundation is justifiable.

The rod used for subsoil investigation is ½ inch or ¾ inch gas pipe jointed in five-foot lengths. In case the rock underlies the road surface within short distance of surface, the road should be driven down to the bottom, and the elevation of the rock surface is ascertained as shown in the following figure, care should be taken to avoid the misleading due to presence of logs and stones. The results of soil investigation are to be tabulated under general soil classi-

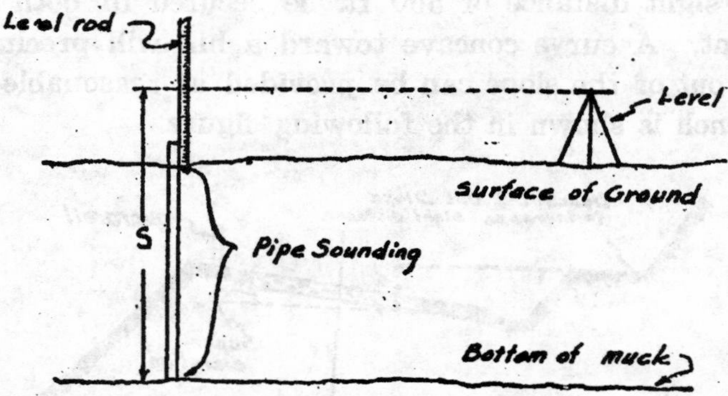

fication, that is gravel, sand, sandy loam, loam, clay, wet heavy clay and quicksand. The elevation of rock surface can be plotted on cross-section sheets.

Drainage. Adequate drainage is essential to the success of the highways. There are many factors which should be investigated in the preliminary survey, such as topography of stream crossing, height of banks, foundations, length of spans, distance of haul and cost of materials etc. Only the most important ones are to be discussed on account of lack of space. The foundation condition is determined by test pits and gas pipes in similar way described in the preceeding paragraph. If the soil condition seems to be not firm enough to support the structure, the piles will be recommended. Information concerning high and low water elevation is also extremely helpful, because these piles above the low water elevation should be either concrete, or creosoted to prevent decay.

Since the size of opening and elevation of bottom of superstructure depends upon the volume and velocity of stream flow and on clearance for ice and floating debris, a careful analysis of drainage area is indispensible, but details of which are too numerous to be mentioned here.

As to location of bridges and culverts, the common fault is the use of right angle location in very unnatural way, the capacity of drainage structure is thus reduced, and the flow of stream checked, resulting in excessive scour and silting. The following diagram shows the relative merits of two locations.

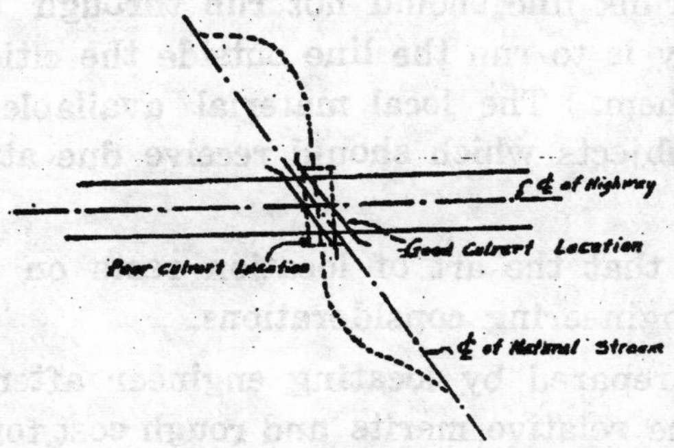

It is true that right angle location saves the length of the culvert, but the saving can hardly balance the above mentioned disadvantages. Considering the maintenance cost, the right-angle location for streams crossing the road on skew angle is unfavorable unless the stream channel can be changed for considerable distance when the skew of stream centerline is too big, or the stream approaches the road in an undesirable manner, a suitable plan of channel change should be worked out. A careful study of all informations gathered in the preliminary would lead to the recommendation of suitable types of structure. Then rough estimates are made for different types before final dicision is reached.

The underdrains also play important part in field of drainage. The engineer should bear in mind that, in case of poor soil condition, it is necessary to get the road bed three feet above the water table. This can be attained by raising the grade or by lowering the general level of water table. An open ditch, or tile drain will serve the purpose. Such ditches are to be surveyed and cross-sectioned for preliminary estimates. Special attention should be paid to locate the springs and seepages which are common sources of drainage trouble. The only remedy is relocation, or to divert them by ditch or tile.

Conclusion Due to broad aspect of the subject, this article barely covers some main points among which special emphasis is laid upon foundation and drainage. Every year the increasing traffic makes the railroad crossing a menace to safety of travel. Every effort should be made to reduce the number of crossings. If the crossing is unavoidable, grade separation and clear vision near the track should be secured. Another problem, which becomes greater each year, is the avoiding of streets in congested cities. Although the highway should be so located as to serve the greatest number of communities, the trunk line should not run through the center of large cities. The best policy is to run the line outside the cities with short branch roads running into them. The local material available and local environment are other two subjects which should receive due attention in locating a route.

It is obvious that the art of location rests on good common sense, and not entirely on engineering considerations.

The report prepared by locating engineer after preliminary survey should show clearly the relative merits and rough cost for the different routes. It should convey full informations, yet it is so concise that even the non-technical readers will grasp the situation easily.

——o——

FROM BOSTON TO DETROIT

BY C. T. CHWANG (莊前鼎)

Early in last March I conceived the idea of taking an inspection trip through the United States. I sent a circular to various concerns; which I selected from the Condenced Catalogue of A.S.M.E. and the Chemical Engineering Catalogue. I received many letters of permission and not a few letters of refusal. In general, the manufacturing plants in the fields of mechanical, electrical, and civil engineering gave me a cordial welcome; while those in the chemical and industrial fields, which are very specific in nature, refused my inspection, for the sole reason that they never allow any visitors.

On account of my financial conditions, I was only able to see the industrial plants near the large cities that I passed. My trip was not complete; but it was successful. With about sixty dollars in my pocket I travelled over two thousand miles in three weeks.

Mr. J. T. Hu and I started from Boston. In Boston, we visited the Sturtevant Co., where all kinds of fans used in power plants and for heating and ventilation are manufactured; and the Walworth Mfg. Co., where all kinds of valves and pipe fittings used in power plants are cast out in a big foundry of cast steel, cast iron, and brass. In the field of fan-manufacturing, the Sturtevant Co. has its world reputation. This company usually admits five Chinese students to work in the Drafting Department for a period of two years. In my opinion, they should be allowed to work for a while in the erecting floor and in the outside installation work. Perhaps we Chinese students perfer the white collar jobs to any work outside the Drafting Department. But it would be a mistake to think that a successful engineer can be made in office only.

We first drove to New York. The places of general interest such as the Woolworth Building, the Statue of Liberty, the Holland Tunnel, the Metropolitan Museum of Fine Arts, need no recount here. We visited the Worthington Pump in Harrison near New York. It is a plant, which manufactures all seizes of pumps, gas engines, and compressors. The company usually gives a training course, starting from June. Some-time ago a few Chinese students worked there. But the lack of trade relations between the company and China in the last few years and the dull prospect in the future serve as a good reason for the company to refuse to admit any Chinese student.

We went to the famous Edison Laboratory in New Orange near New York. To our disappointment, it is not wholly open to visitors. We were only able to see the office desk and chair in the room, where Mr. Edison used to work. We saw, also, on Mr. Edison's desk, the first phonograph and the photos of Mr. Coolidge and Mr. Hoover. We were guided through the Laboratory and the Edison Storage Battery Co., where the storage batteries using iron oxide and nickel oxide in potassium hydroxide electrolyte are manufactured for uses in the electric trucks.

From New York we proceeded to Bathelham, where is located the largest steel company in the world, the Bethelham Steel Company. This company employs 30000 to 40000 workers during the War and is now employing about 20000. There are sixteen blast furnaces, the largest of which takes 800 tons of ore a day and the smallest 400 tons. In making the structural steel, I-beams, girders, columns, channels, and angles, the open hearth process is used. We spent almost a whole day in the plant. The assistant sales engineer accompanied us and explained very clearly the whole process of manufacturing structural steel. To our surprise—and joy—we met a Chinese fellow, who has been working there since his graduation from the University of Illinois in 1925. This company gives a student course caled the "Probation Circuit," for a period of ten weeks. Usually it is very hard to get in, unless one has some help from the Chinese government.

We then went to Philadelphia, where we met the Nanyang men of the University of Pennsylvania. We had a good time. We visited the Leeds and Northup Company, where the electrical measuring and recording devices and meters are made; and the Cochrane Corporation, where the heaters and condencers for the powers plants are made.

We then passed to Washington D. C. In the capital we had the opportunity the see the Senate in session, and hear the eloquent discussion of a senator, charging President Hoover for allowing the capital to be "wet." We visited some places of general interest. By the way, one should never Miss Washington D.C. A trip through the chief cities can never be said to be complete without seeing the capital. A well-planned trip, including a visit of Washington D. C., can be carried out with the smallest expenditure by a group of interested people, who know how to drive a car carefully and how to repair it, in case any slight trouble should happen. A good partner in a car is most desirable. I was lucky to have Mr. Hu for my partner. But a friend of mine is more lucky: he has his bride for his partner. As I am writing this, they were "honeymooning" in their machine. Let us wish them "Bon Vovage!"

From Washington we went toIthaca and back to our Alma mater Cornell University. We met our friends and visited our old professors. Then we went to the Eastman Kodak Co., Rochester. There the cutting and rolling of photo films were clearly explained to us by the guide. However, we were not admitted to the so-called "No-man-room," where the film is manufactured in the automatic machine, designed and built by the Company. Since that machine has nevor been patented, it is not open to the visitors. The guide told us about the life of Mr. Eastman, the inventor of the film. The story is by no means new; it is the story of a man of patience and hard work.

On our way to Buffalo, we passed the Niagara Power Co., the largest hydro-electric power plant in the world. It generates power for the use of the industrial plants within its area of 300 miles. In Buffalo Mr. Hu took the train for California on his way to China. I, now left alone, visited the American Radio Comnany, and saw the wonderful process of enamelling and the officient methods of making radio castings. At Erio, I visited the Union

Iron Works, where large steam boilers are made. The process of rivetting, rolling, and forging were clearly explained.

The Westinghouse Electrical Company was the next place of my visit. It is as large as G.E. in Schenectady. But all the buildings of Westinghouse are connected, whereas those of G. E. are scattered. Both these companies employ from 18000 to 20000 workers. At present only one Chinese student is working at Westinghouse. He came from China direct.

Pittsurgh is a dusty city. But it is the real industrial capital of U.S.A. Around the city there are many iron and steel works, to which I did not have time to go. I visited, however, the American Bridge Co., at Ambridge. All the designing work of the company is done in the office in the Frick Bldg, Pittsburgh. In Ambridge, which is twenty miles from Pittsburgh, all the fabricating work on the structural steel, and the bridge work are done with the multiple punchers, drillers, rivetters. There, also the bridge members are put together in a horizontal position, and subjected to test before they are shipped out for erection.

From Pittsburgh I proceeded to Akron, the world's largest manufacturing center of rubber. (62% rubber supply in the world comes to Akron) The B. F. Goodrich Company is the first rubber company in the Midwest and the largest in Akron. It employs 16000 workers. It manufactures rubber tires, tubes, footwears, and household necessaries. The most efficient method of making footwears according to the wonderful conveyor system, which is used in the manufacturing of the Ford automobiles, interested me most.

Thirty-four miles away from Akron is Cleveland. At Cleveland I made hurried visits to the Machine Tools Exposition in the Public Auditorium, and the Warnor and Swasey Machine Tools Works. Then I hurried to Detroit to find a job in the Detroit Edison Company. I got it. Thus my trip came to an end.

* * * * *

I have been deeply impressed by the fact that the army of workers has made the United States as she is to-day. It is not the soldiers, nor the diplomats, that may be considered as the backbone of the nation. For where comes mass production? There comes power and industry? One word is the keynote; that is work.

For a long time, China, while encouraging scholastic studies has despised the workmen. And that is, as we all know, the main reason why we are lagging far behind other nations in the development of industry. Before her industries are fully developed, China must experience three stages. The first one is the improvement of communications: the railway, the automoblic, the telephone, and the like should be more properly managed and widely used. With a strong central government, these can be done, let us say, about fifteen years. The second will be a stage of power industry. That is the development of the hydro-electric and steam power for various uses. The third will be the development of the industries of iron and steel, chemical acids and alkalies, and cement and structural materials.

Now is the time for China to prepare her students of engineering to come here to work in the different companies of various field. If our government can ever be induced to make some contract with the companies here for the sake of business relationship and co-operation, it will be easy, I think, to get their consent to admit the Chinese students to work. China needs her engineering students to come here to work for practical experience, but not her students to work for the vain titles of Ph.D., D.Sc., and M.S. The real value of research in the highly specialized field of engineering in this country lies in the fact that U.S.A. has got her industries fully developed and needs research to carry on further development. How about China? It will be a long time for her to come to such a state of development as we find here, I would say I shall send my grand-children here to work for Ph.D. or D.Sc., because at that time, let us hope, China would be in a better position to use such highly specialized men of engineering. If these friends of mine, who are candidates for the highest honor, should feel themselves offended by what I have said above, I should be very sorry. I regret that I spent too much time at college. It was against my will to do any research in a specialized field of engineering. Perhaps every engineering student will remember that a good engineer here is made outside school he is made from the real bottom. China has had enough engineering students; but she still needs real workers.

民國二十年四月

工程

中國工程學會會刊

第六卷第二號

THE JOURNAL OF THE CHINESE ENGINEERING SOCIETY

VOL. VI, NO. 2 APRIL 1931

中國工程學會發行 總會會所：上海寧波路四十七號 電話：
每册三角預定全年四册一元每册郵費本埠二分外埠五分國外二角三分

工程

中國工程學會會刊

季刊第六卷第二號目錄 ★ 民國二十年四月發行

總編輯 周厚坤　　總務 支秉淵

□論文號 十五次年會宣讀□

插圖:

中國工程學會發行

3797

中國工程學會會章摘要

第二章　宗旨　本會以聯絡工程界同志研究應用學術協力發展國內工程事業爲宗旨

第三章　會員

(一)會員　凡具下列資格之一由會員二人以上之介紹再由董事部審查合格者得爲本會會員

(甲)經部認可之國內外大學及相當程度學校之工程科畢業生并確有二年以上之工程研究或經驗者

(乙)曾受中等工程教育并有六年以上之工程經驗者

(二)仲會員　凡具下列資格之一由會員或仲會員二人之介紹並經董事部審查合格者得爲本會仲會員

(甲)經部認可之國內外大學及相當程度學校之工程畢業生

(乙)曾受中等工程教育并有四年以上之工程經驗者

(三)學生會員　經部認可之國內外大學及相當程度學校之工程科學生在二年級以上者由會員或仲會員二人之介紹經董事部審查合格者得爲本會學生會員

(四)永久會員　凡會員一次繳足會費一百元或先繳五十元餘數於五年內分期繳清者爲本會永久會員

(五)機關會員　凡具下列資格之一由會員或其他機關會員二會員之介紹並經董事部審查合格者得爲本會機關會員

(甲)經部認可之國內工科大學或工業專門學校或設有工科之大學

(乙)國內實業機關或團體對於工程事業確有貢獻者

(八)仲會員及學生會員之升格　凡仲會員或學生會員具有會員或仲會員資格時可加繳入會費正式請求升格由董事部審查核准之

第四章　組織　本會組織分爲三部(甲)執行部(乙)董事部(丙)分會(本總會事務所設於上海)

(一)執行部　由會長一人副會長一人書記一人會計一人及總務一人組織之

(三)董事部　由會長及全體會員舉出之董事六人組織之

(七)基金監　基金監二人任期二年每年改選一人

(八)委員會　由會長指派之人數無定額

(九)分　會　凡會員十人以上同處一地者得呈請董事部認可組織分會其章程得另訂之但以不與本會章程衝突者爲限

第六章　會費

(一)會員會費每年國幣六元入會費十元　(二)仲會員會費每年國幣三元入會費六元

(三)學生會員會費每年國幣一元　(四)機關會員會費每年國幣十元入會費二十元

交通部

國際大電台真茹發報台

東北鋼鐵廠地點之研究

著者：胡庶華

東北跨有遼甯，吉林，黑龍江，熱河四省及灤河以東之地，物產富饒，地利未闢，黑龍江之金，熱河之銀，皆爲我國第一。而遼甯煤鐵之富，尤爲東南各省所無。煤鑛儲量約一千五百兆噸，鐵鑛儲量約四百五十兆噸。（中國鐵鑛儲量約九百七十兆噸，遼甯幾占其半）。所惜者，强鄰逼處，本溪湖，鞍山，弓長嶺等處豐富之鐵鑛，皆爲日人所據，不復爲漢家寶物。於是基本工業之原料，如鋼鐵焦煤及石油等，皆操於他人之手，瞻念前途，不寒而慄。茲將遼寧煤鐵之已經開採者，表列於下：

(一) 本溪湖煤鐵鑛　現由省政府與日商大倉喜八郎合資開採，名曰中日合辦本溪湖煤鐵公司。

(二) 牛心台紅臉溝煤鑛　由華商楊春元與日商石本鑽太郎合辦，名曰復興煤鑛公司。

(三) 田什夫溝煤鑛　由商人孟凌霄獨資經營。

(四) 雷霹砬子煤鑛　現由那晏浦經營。

(五) 王官溝無烟煤鑛　由商人于文漢獨資經營。

(六) 艾家墳無烟煤鑛　由商人白樂奎經營。

(七) 柳樹排子無烟煤鑛　由商人白佩珩經營。

(八) 賽馬集烟煤鑛　由王亘川獨資經營。　　(以上本溪)

(九) 弓長嶺鐵鑛　由省政府與日商滕田延太郎合辦，名曰中日官商合辦弓長嶺鐵鑛公司。

(十) 蛤蟆坑盧家屯無烟煤鑛　現由官辦名曰天利煤鑛公司。

(十一) 磐山堡無烟煤鑛　由商人張潤科獨資開採。

(十二) 大孤山鐵鑛共有八區　由于冲漢與日商鎌田合辦,名曰振興鐵鑛公司. (以上遼陽)

(十三) 蛤蟆山煤鑛　官商合辦,名曰愛商煤鑛公司,承辦商人爲沈成茂.

(十四) 大紅石粒子煤鑛　由于雲章集股開採,名曰强業公司. (以上錦西)

(十五) 半截河孟亮河等處煤鑛共十區　官商合辦,名曰西安煤鑛公司;中日商人合辦者,名曰秦信煤鑛公司.

(十六) 鴨子圈煤鑛　由趙恩普集股開採. (以上西安)

(十七) 老虎洞山與歪頭山之錳鑛　由王正黼集股開採. (以上興城)

(十八) 新開嶺煤鑛　由佘永泉獨資興辦. (以上西豐)

(十九) 五湖咀子煤鑛　由東北鑛業公司承辦,係王正黼經理. (以上復縣)

(二十) 彬松崗煤鑛　由商人劉建元開採. (以上輝南)

(廿一) 小王大溝硫化鐵鑛　由孫世榮集資開採.

(廿二) 紀家崴子硫化鐵鑛　由陳和瑞獨資開採 (以上鳳城)

(廿三) 楊木橋子煤鑛　官辦歸採鑛局經理. (以上通化)

(廿四) 紅旗杆溝煤鑛　由史玉英獨資開採. (以上柳河)

(廿五) 金溝煤鑛　係由瀋海鐵路公司與商人周文富合辦,名曰官商合辦金溝煤鑛公司.

(廿六) 搭連咀子煤鑛　由周文富與日商合辦,名曰大興煤鑛公司.

(廿七) 得古土口子煤鑛　由姚銘勳與日商合辦.

(廿八) 蛇窩屯煤鑛　由薛永來集資開採.

(廿九) 小瓢爾屯煤鑛　由吳尙忠集資經營. (以上撫順)

觀上列煤鐵各鑛,凡鑛質良好之區,莫不有日本資本在內,而尤以鐵鑛爲最甚.民國四年,日本以二十一條件强迫吾國承認,其中一條爲「與日本以開採奉天九大鐵鑛之特權」.其後遂有本溪湖鐵廠,鞍山鐵廠之設,爲供給

日本鍊鋼原料之源泉.去年滿鐵公司擴充之計畫,更足驚人,錄之於次:

日本南滿洲鐵道株式會社,自明治三十九年由日政府創立以來,迄今二十餘年.據本年份滿蒙年鑑載該社資本總額已達四四〇,〇〇〇,〇〇〇元;服務人員多至三四,三七四人.民國十六年度營業收入爲二三〇,五五八,五二四元,支出計一九四,二八四,二〇一元,收支兩抵,純贏利竟達三六,二七四,三二三元之巨.現擬擴充現有資本四萬四千萬元爲十萬萬元,資本增加後,亟應興辦幷擴充之重要事業如下:(一)鞍山製鋼計畫,(二)硫酸製造計畫,(三)撫順煤炭生產計畫,(四)沙達計畫等.又日本大藏省當局關於昭和製鋼所設立問題,現正在調查製鐵獎勵金及輸入稅等.據仙石總裁意見,由滿蒙之經濟的發展上攷察之,此項設立,置於滿鐵管理之下,雖有多少不便,比較置於朝鮮總督府之管理下有利益.於是在鞍山設立鋼廠之說逐漸有力,其理由有三:(一)生產原價低廉;(二)滿蒙開拓以來,二十年間日人之滿蒙移民不過二十萬人,如昭和製鋼所設於滿鐵管理下,即更可移住二十萬之日人;(三)計畫經濟的開拓滿蒙之今日,如設於滿鐵管理之外,則違反滿鐵本來之使命.

觀此,則日人不僅將遼寧鐵鑛完全據爲已有,且欲進而設立昭和製鋼所,以爲侵略滿蒙之工具.吾人欲於其勢力範圍以內設立鋼鐵廠,殆爲不可能之事.吉林黑龍江雖有鐵鑛發見,尙未有詳細調查,而按之交通情形,亦不宜於設立大廠.

然欲抵制外人侵略,非自行開發東北不可.開發東北,尤非在東北範圍以內,設立規模較大之鋼鐵廠不可.若國防,若農鑛,固無論已.即以交通而論,除已築成之南滿,中東,安奉,四洮,瀋陵,天圖,吉敦,瀋海,呼海,齊昂鐵路外,其尙待建築者如下:

路別	長度(公里)	起訖點
洮熱	八八八	洮南至熱河

長	大	二一二	長春至大賚
敦	會	——	即吉敦路延長至會寧
新	林	——	新邱至林西
盤	大	二四〇	盤山至大虎山
臨	安	二七五	臨江至安東
安	拜	一七三	安達至拜泉
新	邱	四八	新立屯至新邱煤礦
敦	五	——	敦化至五常
濱	黑	——	哈埠至黑岡
海	鏡	一三八	海林至鏡泊湖
密	富	二八八	密山至富錦
密	虎	——	密山至虎林
開	林	三四五	開魯至林西
開	扶	一四八	開通至扶餘
扶	哈	二一八	扶餘至哈爾濱
穆	牡	——	穆陵至牡丹江口
公	伊	五七	公主嶺至伊通
鐵	法	五六	鐵嶺至法庫
海	嫩	二七〇	海倫至嫩江
齊	嫩	——	齊齊哈爾至嫩江
小	林	二七	小喬站至林甸
呼	鶴	四六三	呼蘭至鶴立岡
齊	黑	——	齊齊哈爾至黑河
齊	扶	二四七	齊齊哈爾至扶餘
黑	安	一一二	黑河經林甸至安達
朝	安	三五八	朝安鎮至安圖
濱	依	三〇八	哈埠至伊蘭
吉	五	一六二	吉林至五常
吉	寧	——	吉林至寧古塔
吉	密	——	吉林至密山
吉	呼	二〇八	吉林至呼蘭
窰	德	二二	窰門至德惠
滿	青	九二	滿溝至青岡

滿	肇	三五	滿溝至肇東
赤	林	二七〇	赤豐至林西
滿	興	八八	滿溝至興隆鎮
遼	厲	八六	遼陽至厲家窩鋪
四	西	——	四平街至西安
阜	厲	——	阜新至厲家窩鋪
石	楡	六四	石頭城子至楡樹
延	琿	九五	延吉至琿春
海	索	四八〇	海蘭爾至索倫
達	大	——	達家溝至大和莊
同	五	一六八	同賓至五常
一	五	九七	一面坡至五常
一	依	二二五	一面坡至依蘭
三	一	——	三姓至一面坡
德	九	四五	德惠至九台
穆	三	二六五	穆陵至三姓
甯	海	二三	寧古塔至海林
朝	濛	——	朝陽鎮至濛江縣
興	臨	三二〇	興江至臨江
臨	長	一八四	臨江至長白
打	鄭	二八	打通路彰武站至鄭家屯
瀋	遼	八〇	瀋陽至遼陽
瀋	熱	——	瀋陽至熱河

上列之六十三線,乃急應敷設者.其餘應計畫建築之線當有數十,合計各線之長不下兩萬公里.假定每公里鐵道需鋼軌十六萬一千磅,約合七十二噸,合計共需鋼軌一百四十四萬噸.若以年產六萬噸之鋼廠供給之,約需二十四年,是東北之急應設立大鋼鐵廠不待辭費矣.

又以遼甯未採煤鐵礦言之,約略如下:

瀋陽　懿路煤鑛　劉通士屯煤鑛　四千戶屯煤鑛　宛家溝煤鑛

遼陽　鷄冠山鐵鑛　沈家溝鐵鑛

本 溪	馬鞍山煤鑛 達子嶺鐵鑛 喜雀嶺煤鑛 柴子窪東山煤鑛
撫 順	茨兒溝煤鑛 齊家嶺煤鑛 營盤北溝煤鑛 楊木溝硫化鐵鑛
蓋 平	裂縫山鐵鑛
復 縣	長興島鐵鑛 王家屯煤鑛 三稜山鐵鑛
鳳 城	茅兒山鐵鑛
新 民	楊木林子鐵鑛 馬架子南山鐵鑛
通 化	大羅圈溝鐵鑛
臨 江	四道溝裏峪子溝煤鑛 大粟子溝鐵鑛
莊 河	歪頭砬子鐵鑛
錦 西	大窰溝 秋皮溝 龍灣山等煤鑛
興 城	小鷄冠山鐵煤鑛 窰溝煤鑛
海 龍	廟溝煤鑛
輝 南	杉松崗北山及大洋河等處煤鑛 鞍子河等處鐵鑛
桓 仁	牛藏河鐵鑛
岫 巖	徐家溝之鐵鑛
法 庫	印牛堡朝陽洞煤鑛

以上各鑛區既未開採,而鑛量之多寡,亦未有切實之調查.然集中各處之煤鐵,足以供給一東北大鋼鐵廠之用,可斷言也.東北需用鋼鐵之必要既如彼,東北原料之充足又如此,而臥榻之側,竟容他人鼾睡.若不急起疾追,爲東北工業原料謀一根本之解决,將有噬臍無及之悔.顧鋼鐵廠之設立,以選擇地點最爲重要.其條件不外原料之來路,出品之去路而已.東北各鑛多密集於腹地,無水道運輸之便利.本溪湖,鞍山兩製鐵所,即感受此種痛苦者也.今欲脫離特殊勢力,求一與東北各鑛相距不遠,而交通又極便利者,舍秦皇島及葫蘆島莫屬也.

秦皇島三面環海,爲渤海沿岸不凍之港.西距天津四百六十里,東距山海關三十里,又有鐵道聯絡北寧.開灤之煤鑛(約七萬萬噸),灤縣,司家營,張莊家之鐵鑛(約三千萬噸)爲其最近之原料.歐戰時有人建議在此設立鐵廠,後以龍烟鐵廠成立,而此議遂罷.其實龍烟鐵廠之交通,遠不若秦皇島也.

參看拙作「龍烟鋼鐵廠之將來」載中國建設雜誌鑛冶專號),然以秦皇島與葫蘆島比,則葫蘆島更勝一籌.

葫蘆島爲東北不凍之良港,居遼東灣之中樞,當渤海之要衝,距秦皇島約三百零七里,距瀋陽約五百十里,與北甯路有一支線相連,(連山灣支線)由連山站直達港口約十七里有半.港水極深,即最大海船亦可入口.現已開工築港,將來發展,未可限量.若於連山支線以北之白馬石或以南之玉皇閣附近設一大規模之東北鋼鐵廠,不僅使港內之商務益臻發達,即東北之農工商業,亦將受益匪淺,較之秦皇島優點有二:

(一) 秦皇島常有風災,葫蘆島絕無暴風之危險.

(二) 秦皇島設廠須用開灤之焦煤,易受外人之牽制;葫蘆島設廠可用北票之焦煤由北寧路支線錦朝鐵路(長二百三十里),直達港口,不過三百里左右.錦西煤亦可煉焦,距離更近.

至於原料,關於焦煤一方面者,除熱河北票之煤約一萬萬噸,錦西煤八千萬噸外,有黑龍江之鶴崗煤鑛約一萬萬四千萬噸,吉林之穆稜密山煤鑛約五萬萬噸,可爲補充原料.關於鐵鑛一方面者,海城有二百萬噸,復縣有五十萬噸,臨江有一百二十萬噸,通化有一百二十萬噸,灤縣有三千萬噸.而熱河境內最近有大鐵鑛之發見.錦西一帶,石灰石亦多,頗爲便利.

說者謂葫蘆島設鋼鐵廠有二弊.

(一) 濱海之地,易受敵艦或水上飛機攻擊,且距南滿鐵道及營口大連太近,尤易受人操縱或壓迫.

(二) 復縣海城之鐵鑛均在南滿鐵道線上.而臨江,通化,尚無鐵路可通,卽有亦須經過南滿鐵道.熱河鐵鑛究有若干,亦無調查.灤縣鐵鑛距葫蘆島四百餘里,運費恐亦不輕.

關於第一點,係整個的國家強弱問題.如果國家無獨立之能力,則一切建設問題可以不談.縱然將鋼鐵廠設在內地,亦是毫無辦法,漢冶萍卽其明證.

假若中國能自強自立,卽鋼鐵廠辦在海邊上,亦自有保護之能力.譬如英國東境海濱,鐵廠林立,而日本八幡製鐵所亦距海不遠,則海濱不能辦廠之說,不攻自破.至於受人壓迫與操縱,固屬可慮,然亦不能因噎廢食.吾人自辦鐵廠以抵制日本鐵廠,正與吾人移民東北,以抵制日本之移民東北同一用意.葫蘆島之開闢,亦所以打破大連營口安東之壟斷.吾人當積極奮鬥以工業促其繁盛,而以實力固其捍衛也.

關於第二點亦不成問題.大凡水道便利之鐵廠,不虞原料之缺乏.德國用瑞典及西班牙之鐵鑛,日本用吾國大冶繁昌之鐵鑛,皆不嫌其遠.昔秦皇島設廠計畫,亦以安徽鑛石為重要之原料,今葫蘆島何獨不然.蓋水運每噸鑛石每里運費五釐至一分,鐵道每噸每里運費四分五至八分七不等,(北寧路六分二)平均水運僅值陸運八分之一.而出品去路,四通八達,較之全恃鐵路運輸之龍烟鐵廠,誠不可同日而語.

由是觀之,葫蘆島不僅為商業良港,亦可為工業要區.苟能設大鋼鐵廠於其間,吾知東北之興盛繁榮,可立而待也!

會員介紹本刊告廣酬謝辦法

凡本會會員代招廣告,每期在二百元以上,由本刊贈登該會員有關係之公司廣告二面.每期在一百元以上,贈登該會員有關係之公司廣告一面.每期自五十元至九十九元,得登該會員有關係之公司廣告半面,不另取費.每期自三十元至四十九元,得登廣告半面,本刊僅收成本.三十元以下者,贈登題名錄一格.以上所稱有關係之公司,以完全由華商組織,該會員係公司內股東或職員為限.

總務支秉淵

ASPHALT PENETRATION MACADAM and ASPHALT SURFACE-TREATED MACADAM

BY H. K. CHOW (周厚坤)

Asphalt on account of its many desirable qualities has become the standard surfacing material for all urban roads in large American cities. Its low abraisive factor, resiliency, water-proofness, extreme adhesiveness, ease of opening up for public utilities, and its availability in large quantities, all account for its popularity.

There are in America three general methods of utilizing the material. In the order of costliness, they are the sheet asphalt, the asphaltic concrete and the asphalt macadam. Sheet asphalt is a mixture of asphalt, sand, and mineral filler. The mixing is done in a stationary plant with material and temperature carefully controlled. The asphalt content is about 10-12%. It wears longest under the hardest kind of traffic service. Cost with concrete base is about G. $4.76 per square yard.

Asphaltic concrete is a mixture of asphalt, broken stone and sand. The mixing is also done in a stationary plant with material and temperature under careful control. The asphalt content is about 7½-8½%. Cost with concrete base is about Gold $4.46 per square yard.

The asphalt penetration macadam is the ordinary macadam with its surface voids filled with asphalt. In this type of construction, the asphalt is applied in situ and no mixing plant is required. Amount of asphalt per square yard is about 2-2.5 gallons. Cost is about G. $2.30 per square yard.

Although asphalt penetration macadam is the cheapest kind of asphalt construction in America, yet in China, (with the exception of such metropolis as Shanghai), it is still considered too expensive for general adoption. The prevailing type of construction, where asphalt has been found necessary, is the surface-treated asphalt macadam. The asphalt requirement is about 0.5 gallon per square yard, or 40 pounds per fong for first application.

This paper, while describing briefly the first two types as above, is intended to give a detailed discussion of the two latter types. Under the present day financial conditions in China in general, any discussion on sheet asphalt and asphaltic concrete can have only academic value, while a thorough knowledge of asphalt penetration macadam and asphalt surface-treated macadam is of immediate benefit to those who are responsible for city's roads.

(1) ASPHALT PENETRATION MACADAM

For moderate city traffic, the following specification will suffice:

Subgrade

1. Excavate roadway to proper depth so that the finished surface will be 12" above subgrade, having a crown of 1/4" to 1 foot.

2. Before placing foundation stone, roll subgrade thoroughly. Any weak or soft spots should be filled with suitable materials such as gravel, earth or sand. **No clay should be used for fill.**

Foundation Course

3. Foundation course to be constructed of rubble, field stone or boulders 6" to 8" in size. Larger stone to be broken with sledge hammers to maximum size of 8".

4. After the foundation course has been laid to grade, consolidate with a roller. The finished grade of foundation course should be four inches below finished grade.

5. The voids of foundation course should then be filled with gravel 1" to 3" in size and then rolled thoroughly.

ntermediate Course

6. When the foundation has been constructed to the satisfaction of the engineer, spread the intermediate layer of stone 2.5" loose measure composed of stone ¾" to 2.5" in size. Roll this layer of stone thoroughly, and then fill voids with clean sand or stone dust. Roll, sweep and sprinkle this layer until voids are completely filled. Allow to dry at least one day; then sweep off all excess sand.

Top Course

7. Spread layer of stone 2.5" loose measure, composed of stone 1.5" to 2" in size. Roll thoroughly.

Asphalt

8. Apply 1.5 to 1.75 gallon max. of asphalt per square yard. Asphalt to be applied with 8" spout pouring pot. The asphalt when applied should show no foam on the surface and should be 300°F. to 350°F.

Sealing Course

9. When asphalt has been applied, cover with sufficient stone 5/8" to 1" in size to fill the voids of stone already covered with asphalt.

10. Roll this layer of stone thoroughly and when rolling has been completed, sweep off all surplus stone.

Sealing Coat and Top Dressing

11. Apply 1/2 to 5/8 gallons of asphalt per square yard using road brooms to spread out as thin a coating as possible. Cover this layer of asphalt with clean 1/2" stone (1/4" to 5/8" in size). Roll thoroughly.

Curb

12. Where there are no walls or ground gutters to confine the road material, a substantial curb 4" × 12" is necessary in order that the road material may receive sufficient lateral support and disintegration may not proceed from the shoulders inward.

Comments

Subgrade.— Note that clay is not allowed in this specification even for filling weak or soft spots in the subgrade. For reasons, see under "Asphat Surface-Treated Macadam."

Foundation Course.— The latest practice in America is to use stone of moderate size for this purpose. There labor being expensive, setting foundation stone by hand is economically impossible. The foundation is usually laid down by dumping from motor trucks. Stones of moderate size naturally give much evener contour without much raking or hand shifting. But incidentally it was found that it gave a more flexible foundation that stood up better under service.

Intermediate Course— This course is inserted to give a better contour for the top course to rest on, because as a rule the foundation course even after consolidation and spreading of 1"-3" stones is somewhat uneven. Its second function is to afford a resilient cushion to absorb shocks of the traffic. Note that the filling material is again sand instead of clay for reasons that appear under "Asphalt Surface-Treated Macadam."

Top Course.— This course is the wearing surface of the road. The broken stones should be clean and dry before any asphalt is applied.

Asphalt.— The success or failure of this type of road depends upon the correct application of asphalt. When properly applied, the surface will last for years. On the other hand, bad workmanship will see the surface deteriorate in surprisingly short time.

Asphalt should be applied in wide strips of 8" each. No overlapping is permitted. Any overlapping makes that spot too "fat," and asphalt will creep.

Temperature control is important, because the natural inclination of the contractor is to use less fire-wood. But an under-heated asphalt will neither flow nor adhere well.

Again, an overheated asphalt will carbonize and lose its ductility. This is however more of result of workman's carelessness than contractor's deliberate intentions.

Sealing Course.— The top course, after the application of asphalt, still leaves a honeycomb of voids. These are now filled up with smaller stones.

Sealing Coat and Top Drussing.— The above sealing course is now rendered waterproof and made to adhere to the top course below by a scanty application of asphalt, thus presenting a voidless surface uniformly tough and water shedding.

Top dressing in the form of stone chips is now applied to protect the asphalt as well as to prevent it from adhering to the wheels of vehicles when the road is immediately opened to traffic.

Curb.— Formerly the edge or shoulder of a roadway was not given sufficient attention. They were made thin and, as a result, failed at the shoulders. Then disintegration gradually worked towards the crown. A curb or some other substantial shoulder is absolutely necessary to prevent such occurrence.

(2) ASPHALT SURFACE-TREATED MACADAM

By far the great majority of so called asphalt roads built in China are of this type. It is cheaper than the full penetration, as the asphalt does not go down or between the stones but just covers the surface forming a blanket. The quantity of asphalt used per square yard is much less, say 1/2 to 3/4 gallon. There are right and wrong ways of constructing this kind of road. If properly done, the surface will last 3 to 5 years, with traffic of above 1,000 vehicles a day. On the other hand, if it is improperly done, it may last only a year, or even less on bridges when vibrations are present.

The wrong, yet the prevailing, way of constructing this type of road, is to use clay as a binder for the surface stones. The result is not satisfactory, because clay when dry on the surface of the stone is dirt pure and simple.

and asphalt will not stick to dirt. Clay under ordinary weather conditions also absorbs and holds great deal of moisture which is a deterrent for asphalt to adhere. But for road construction it is worse as the clay is thoroughly mixed with water before application, thus forming colloidal clay from which the moisture can not separate and escape. The consequence is that the voidfilling material is always wet, and asphalt can not adhere to it. Moreover the upper stone surface is also thoroughly coated with this colloidal clay thus preventing the direct contact between the stone and asphalt.

If this type of construction must be persisted in, the only improvement under the circumstances is to sweep the road surface clear of superfluous clay as far as possible with a view to exposing the greatest amount of clean stone surface. In fact, slight depressions, as a result of hard sweeping, between stone surfaces are much preferred. Unfortunately the clay sticks so well that it is doubtful whether by sweeping alone the desired effect of exposing clean stone surface can be achieved. However, the result will be better than if the stone surfaces are completely covered with clay.

Asphalt surfaces prepared like the above, may still tear up during hot weather. The only remedy when this does happen is to spread 1/2″ stone chips over the soft surface. Corners of these chips will project above the asphalt and prevent its adhesion to vehicle tyres, etc.

As an alternative to sweeping off the surplus clay, the surface may be covered with surfacing compound which is a cut back liquid asphalt. The solution penetrates into the clay particles, coating the clay and stone at the same time. When the volatile content disappears, a heavy coating of asphalt is left behind, over which subsequent applications of solid asphalt (melted) can be made. As the surface of contact in the second application is asphalt to asphalt, the union will be perfect.

The right way, however, is to do away with clay altogether. For filling up voids, use coarse sand (1/4″ down to ordinary sand), instead of clay. Sand is of course devoid of cohesion, but when wetted it has been found to compact quite well and easily. The function of a binder in a broken stone course is not so much an actual cement as a compacted filler, because the inter-attrition between the broken stones is so severe under impacts of traffic load that even more cementitious materials than clay will break down.

Spread this sand one inch thick over the rolled surface consisting of 1.5″-2.5″ material, and allow the traffic to run over it for sufficient time to drive the sand into voids as far as possible. Sweep away the surplus sand to such an extent as showing slight depressions between individual stones. Pour melted asphalt over this surface. The asphalt will adhere because in this case both the sand and stone present clean surfaces.

If the void filling material available is stone chips only instead of sand, the same can be used provided it is graded and only 1/2" sizes are used for upper filling. The fine and intermediate sizes, however, need not be wasted as they can be used in the lower filling next to 1/2" size.

Economically, the only difference in cost is between clay and sand as everything else is the same. The difference is so slight when compared with the total cost that the method if worthy of adoption. When traffic is light, any kind of surfacing will do, but when traffic is heavy such as on Chun San Road (中山路), Nanking, a wrong method or material will show up at once. A surface that is well cemented to the broken stone underneath will stand up much better than if the same merely on it. A sand-filled upper course gives an asphalt wearing surface that is an integral part of the road section, while a clay-filled one is separate from the matter below and can be easily ripped up.

In concluding this paper, it will not be out of place to say a few words about asphaltic surfaces on bridges. On Chun San Bridge in Nanking, the surface is poor, but, on the wooden bridge near the City Electric Plant, is bad. The cause of all this is vibration. Where the structure is comparatively rigid, vibration is less, and disintegration is slight. But when the structure is non-rigid like a wooden bridge, the vibration is severe, and disintegration is both rapid and great. Under such circumstances, a better type of pavement is needed. Granite stone blocks set in asphalt paving filler is excellent; but a cheaper alternative would be to pave it with asphaltic concrete of the cold mixed kind at least 3" thick. Since there are few bridges in a city street the extra expenditure need not be a heavy one. With a good bridge surface, the appearance of a street will be uniformly good; without it, a fine street will be marred by the interposition of a broken-up bridge.

Finally, it should be pointed out that though the above discussion is detailed, it can not describe every little step in the construction of the road. Such technique must be gained from practical experience. For satisfactory results, close attention to detail and careful supervision by reliable foremen are absolutely necessary in these two forms of road construction. For while in sheet asphalt and asphaltic concrete, the mixing operations are under careful and expert control, in penetration and surface-treated macadams, this advantage is absent; human equation becomes an important factor and should be carefully watched.

隴海隧道之過去與現在

著者：李儼

1. 緒論　2. 設計通論
3. 汴洛隧道　4. 隴海觀陝段隧道
5. 硤石驛四號隧道　6. 隴海靈潼段隧道

(一) 緒論

隴海鐵路橫亙蘇豫秦隴四省,路線長二千四百公里(Km.).其在鄭州以西,時時通過高原峻嶺.故在汴(開封)洛(洛陽)一段,鞏縣前後十英里(Mile)間.已有隧道十一座,共長二千八百五十一公尺(Meter),在洛陽潼關一段.又有隧道十七座,就中以在硤石驛長一千七百八十公尺(約一英里四分之一)者,爲國中長隧道之一.而靈寶潼關間者,今尚在建築中.至西安蘭州間路線,前經踏勘計由北路行者,有隧道五十四座,共長 1180) 公尺,由南路行者,亦有五十一座,共長 9230 公尺.其北路由平涼至瓦亭,經過六盤山,即增加坡度,亦應鑿一長六公里至七公里之隧道.隴海鐵路所建築與計畫之隧道,旣如是之多,則其資料甚足爲工程界之參考,自無疑義.著者服務隴海十餘年,上述工程,多躬與其役.硤石驛四號隧道初成之日(1923),曾作 Drilling the Sia-Shih-Yi Tunnel 一文,刊於美國 The Over-Seas Highway Magazin Vol. 1, No. 10 中.亡友陳寶驪,又搜採當時資料,作 Construction d'untunnel sur la ligne du Lung-Hai 一文,刊於震旦學院工科雜誌第五期(1923).玆更綴成此篇,藉以貺政當世.圖(1)爲隴海路線圖.

(二) 設計通論

隧道經過之處必先詳細測量該處地面,地底情形,爲他日施工之準備.此項勘測,不厭求詳,藉易得良好之線路.且同時應選二,三路線,依地質,工費,地方情形之比較,最後採取一線.路線之地位,既已決定,坡度灣道之膈,亦應事前商可.因工程師一時爲地勢與經費所限,設爲高峻之坡度,與極小之灣道,

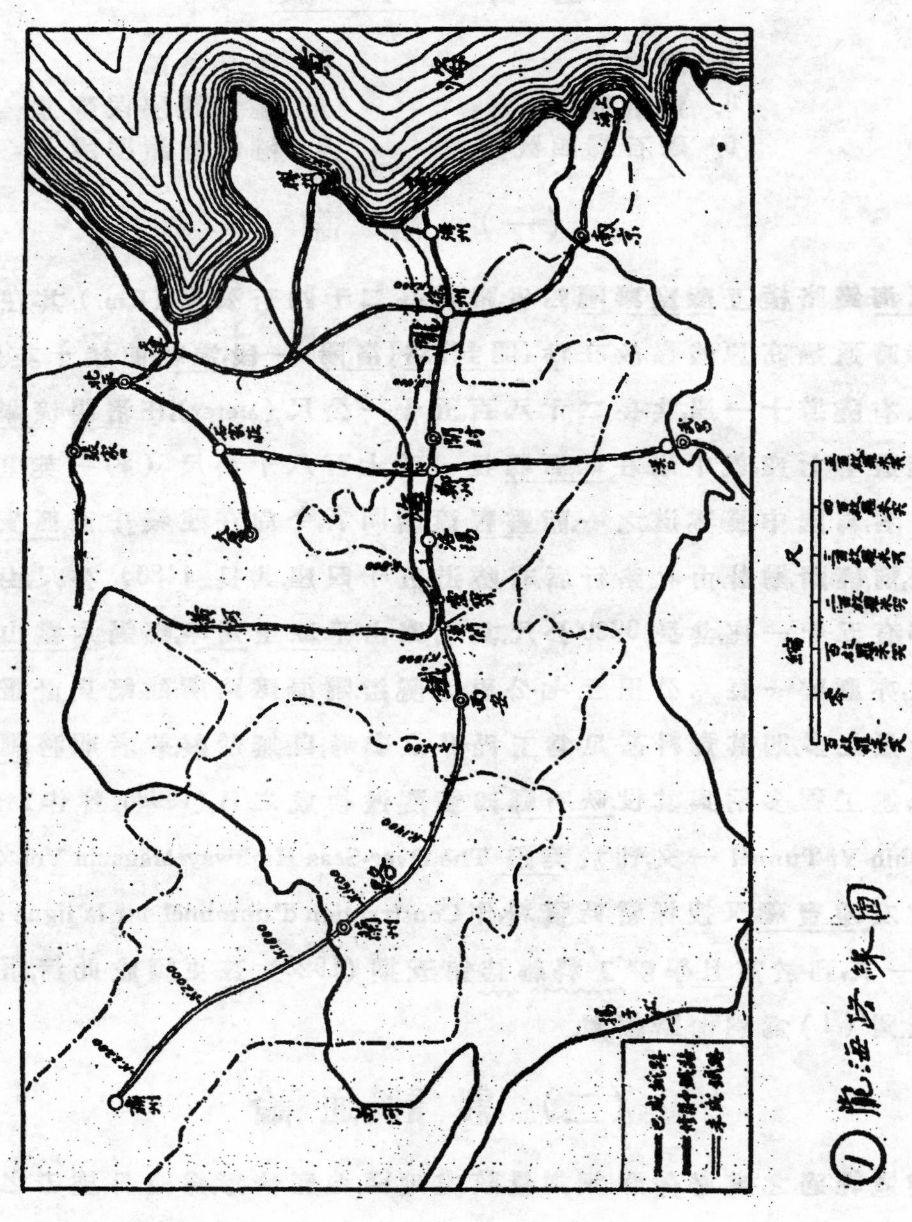

(1) 隴海路線圖

因此害及將來之發展者,往往有之,故事前不可不愼.

開工之先,須詳測所選定之路線.較短之隧道,固可以尋常之方法測量之,若較長者則應有良好之儀器,精密之量度,期他日地底坑道鑿通時,可以洽合相會.其次則隧道之大小,開鑿之次序,及木料支撑之方法,土石工作手續,均須先爲規定,俾作工者有所遵守.復次則作工之器具,及其人物,幷當先爲羅致.

上述設計通論,僅舉大意,其詳則世有專書.此外各處隧道,時有詳細報告,足備參考.此篇所述,則僅限於隴海鐵路.

(三) 汴洛隧道

隴海鐵路汴洛段,係民國前七年(1905)開始測量動工.民國前二年(1910)一月一日,新路行通車禮.全線共長 115 英里.建築計畫,大都取法平漢鐵路.惟平漢武勝關隧道,係備雙線鐵道之用,汴洛路則僅備單線之需耳.今將汴洛段十一隧道情形,列表記述於次.其中公里數概由鄭州向西計算.

號數	位置 (公里)	長度 (公尺)	坡度	灣度 (公尺)	距路基高度 (公尺)
1	48+907——48+957	50	0.000	R=500	28.92
2	49+695——49+938	293	0.010	400	46.23
3	50+215——50+544	329	0.010	400	41.23
4	51+725——52+015	290	0.000	0	59.13
5	52+240——52+493	258	0.004	600	61.11
6	53+320——53+796	476	0.004	400	67.02
7	54+567——54+770	203	0.010	0	40.72
8	54+846——55+065	219	0.010	0	33.54
9	64+234——64+489	255	0.000	1,000	33.20
10	66+023——66+258	230	0.006	800	43.47
11	66+795——67+043	248	0.000	0	39.90
	共計	2851 公尺			

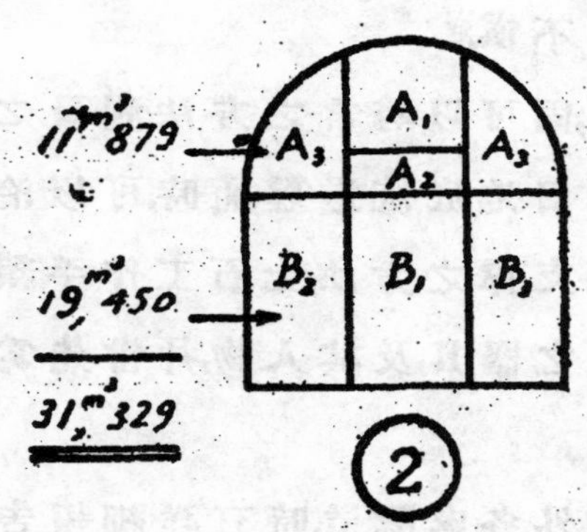

汴洛段隧道,用比國法開鑿.其開鑿次序,如(2)圖所示,首 A_1,次 A_2,A_3,A_3,又次 B_1,B_2,B_3. 每公尺土方,上部圓拱計 11,879 立公方 (m^3),下部幷水溝在內,計 19,450 m^3,共合 31,329 m^3.而砌衣如(3)圖所示,上部圓拱計 3,927 m^3,下部幷水溝(0,090 m^3)在內,計 4,215+0,09) 共 4,305m^3,兩共 8,232m^3.亦有例外者,如接長三號隧道則用(4)圖,頂厚 0,80m. 砌衣面積,共 30,70m^2.其詳細計算法如下:

③

$$\frac{9.58\times5.49}{2}=26.30\ m^2$$

$$2\times\frac{4.58+0.96}{2}\times4.18=23.16\ m^2$$

$$2\times\frac{4.58\times0.47}{2}=2.15\ m^2$$

$$\text{Total}\qquad=51.61\ m^2$$

$$\frac{3.14\times2.25^2}{2}=7.95\ m^2$$

$$3.40\times4.50=15.30\ m^2$$

$$\text{Total}\qquad=23.25\ m^2$$

$$\text{砌衣面積}=51.61-23.25+2\times2.6)+0.45=30.70m^2$$

汴洛段全爲黃土地層,故支撑木材所需極少,每公尺約爲 1,50 m^3.其分配則如(2)圖:

A_1	= 0,600 m^3
A_2, A_3, A_3	= 0,500 „
B_1	= 0,100 „
B_2, B_3	= 0,300 „
	1,500 m^3

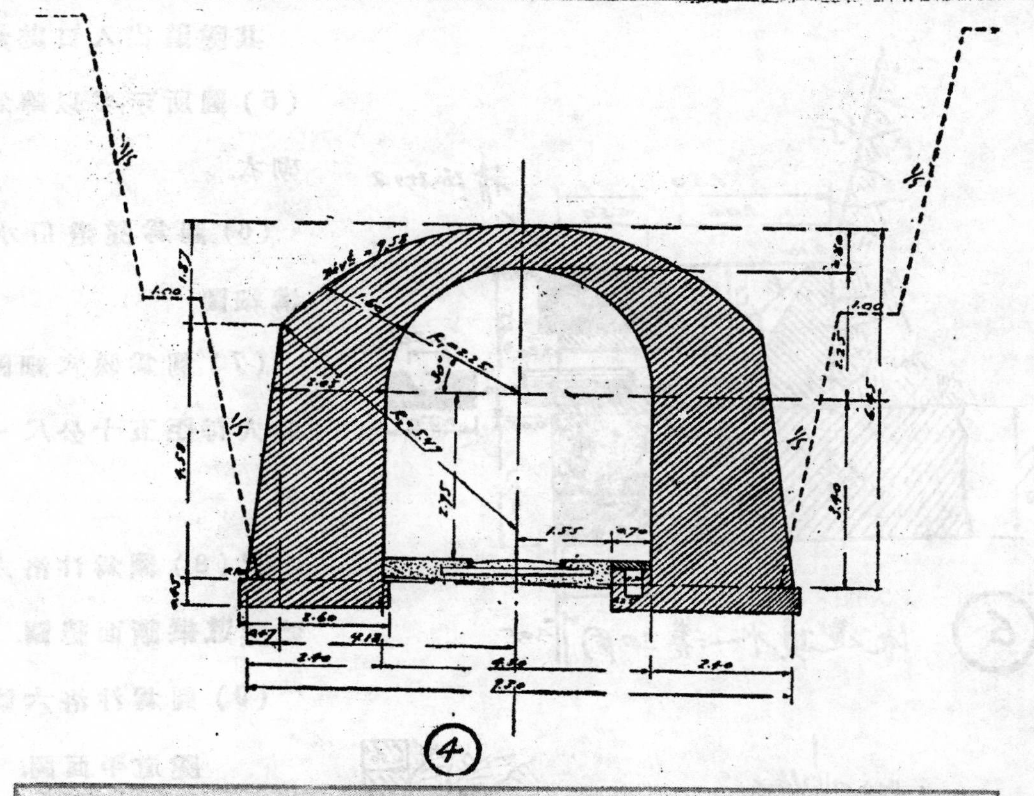

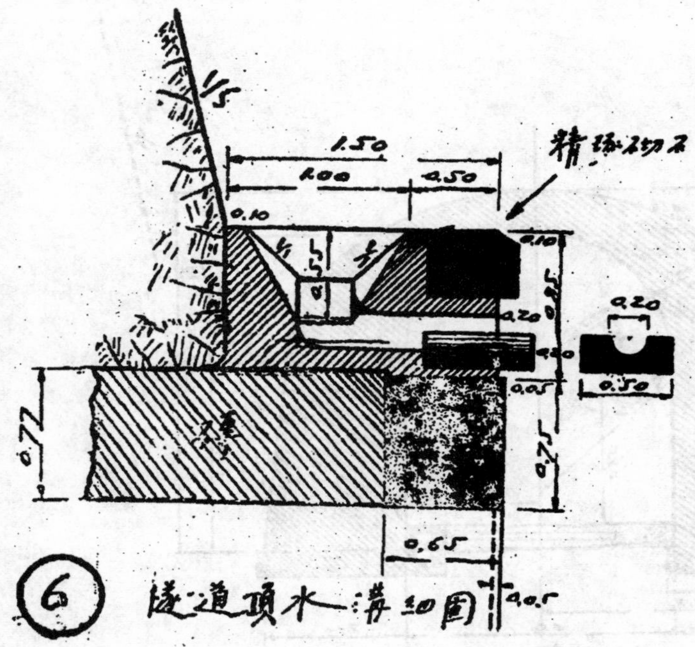

其隧道出入口處如(5)圖所示,多以磚爲砌衣.

(6)圖爲隧道頂水溝細圖

(7)圖爲躲穴細圖.躲穴每距五十公尺一個.

又(8)圖爲汴洛六號隧道縱斷面總圖.

(9)圖爲汴洛六號隧道平面圖.

(10)圖爲汴洛六號隧道縱斷面圖.

(11)圖爲汴洛六號隧道東口正面圖.

(12)圖爲汴洛六號隧道西口正面圖.

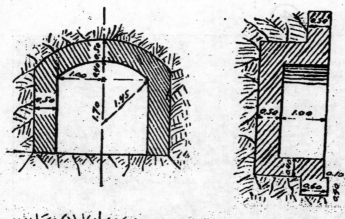

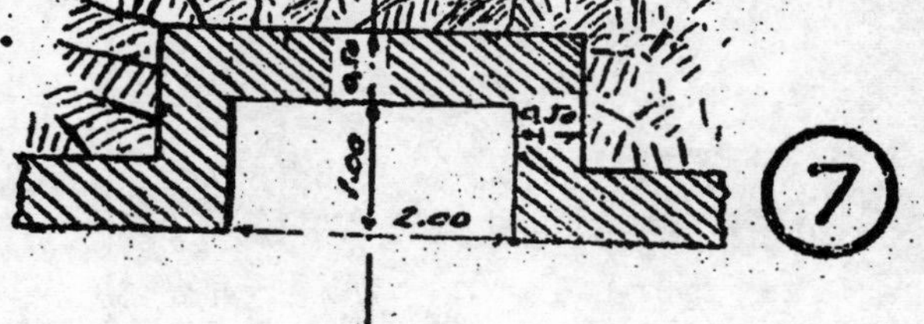

⑦ 躲穴細圖

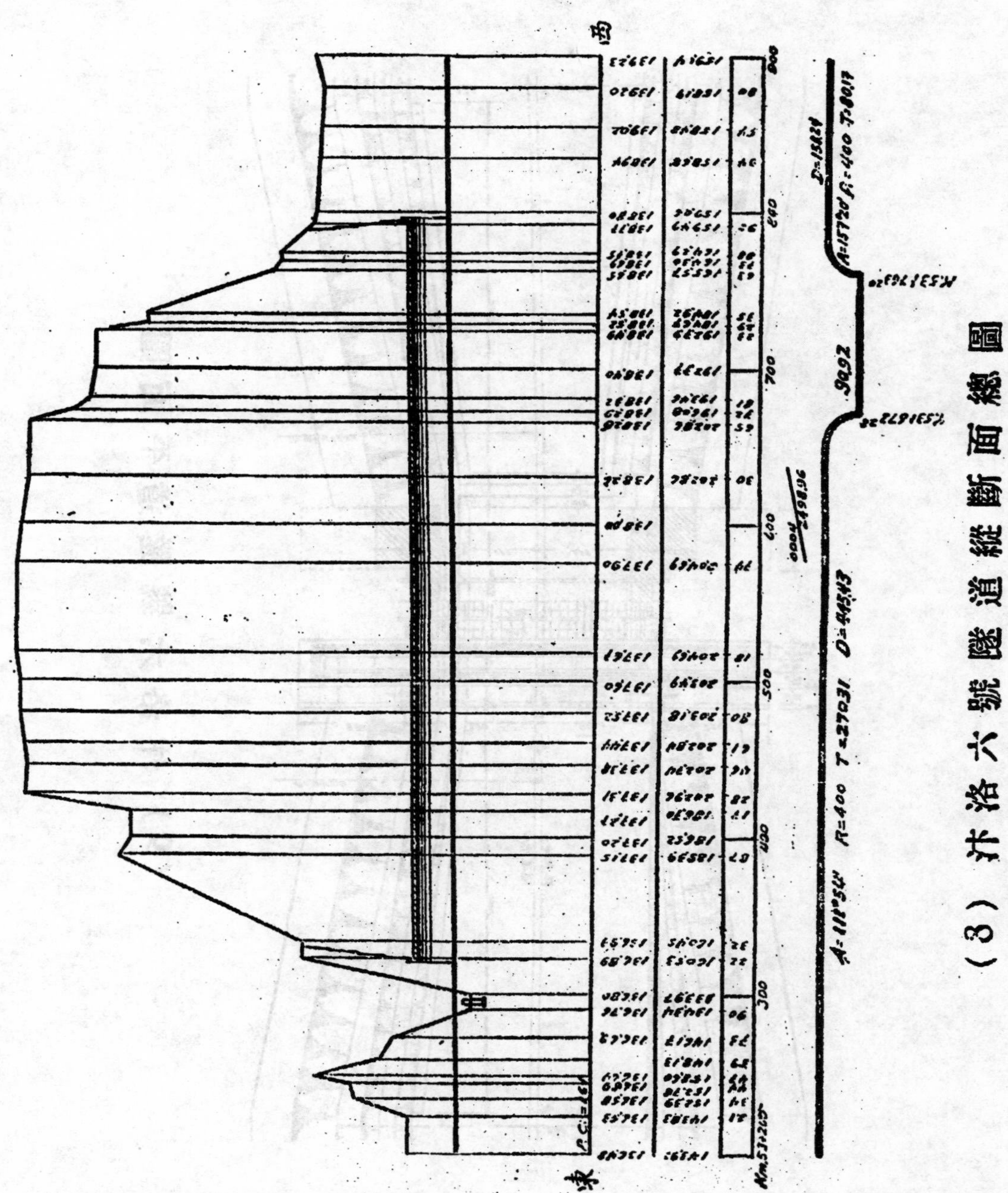

（3）汴洛六號隧道縱斷面總圖

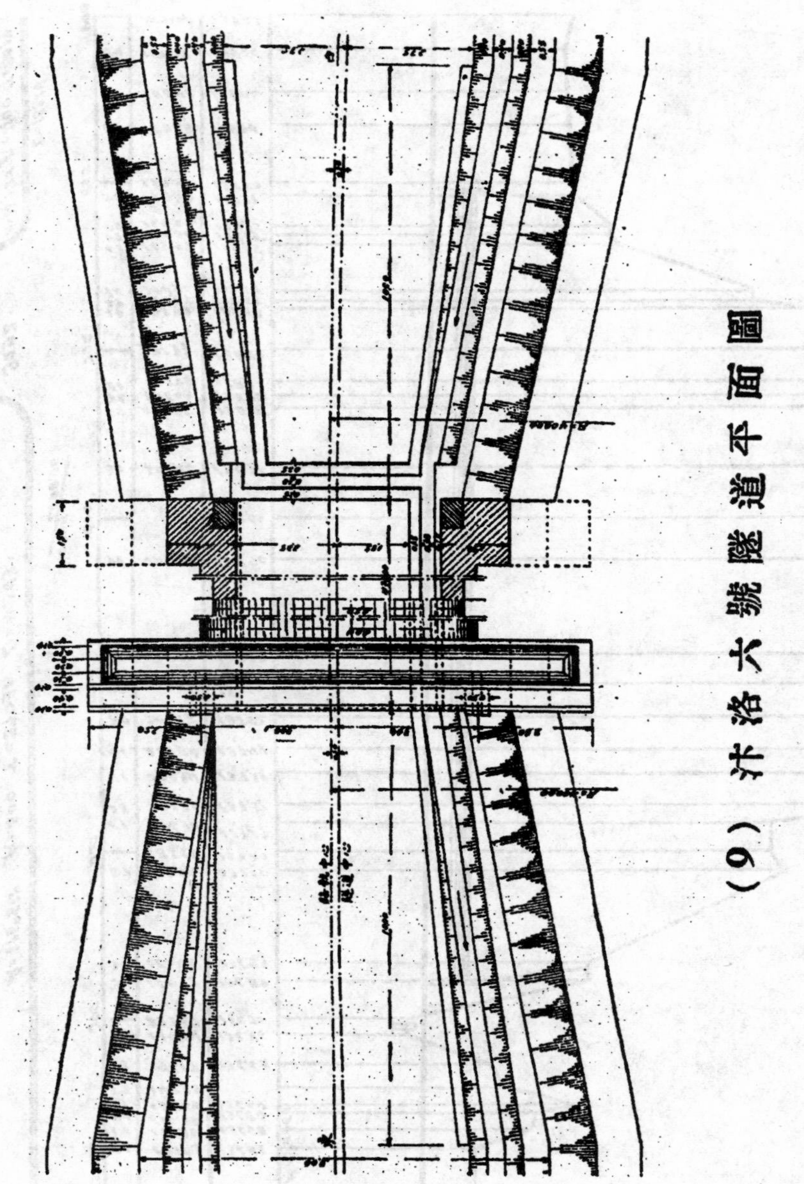

(9) 汴洛六號隧道平面圖

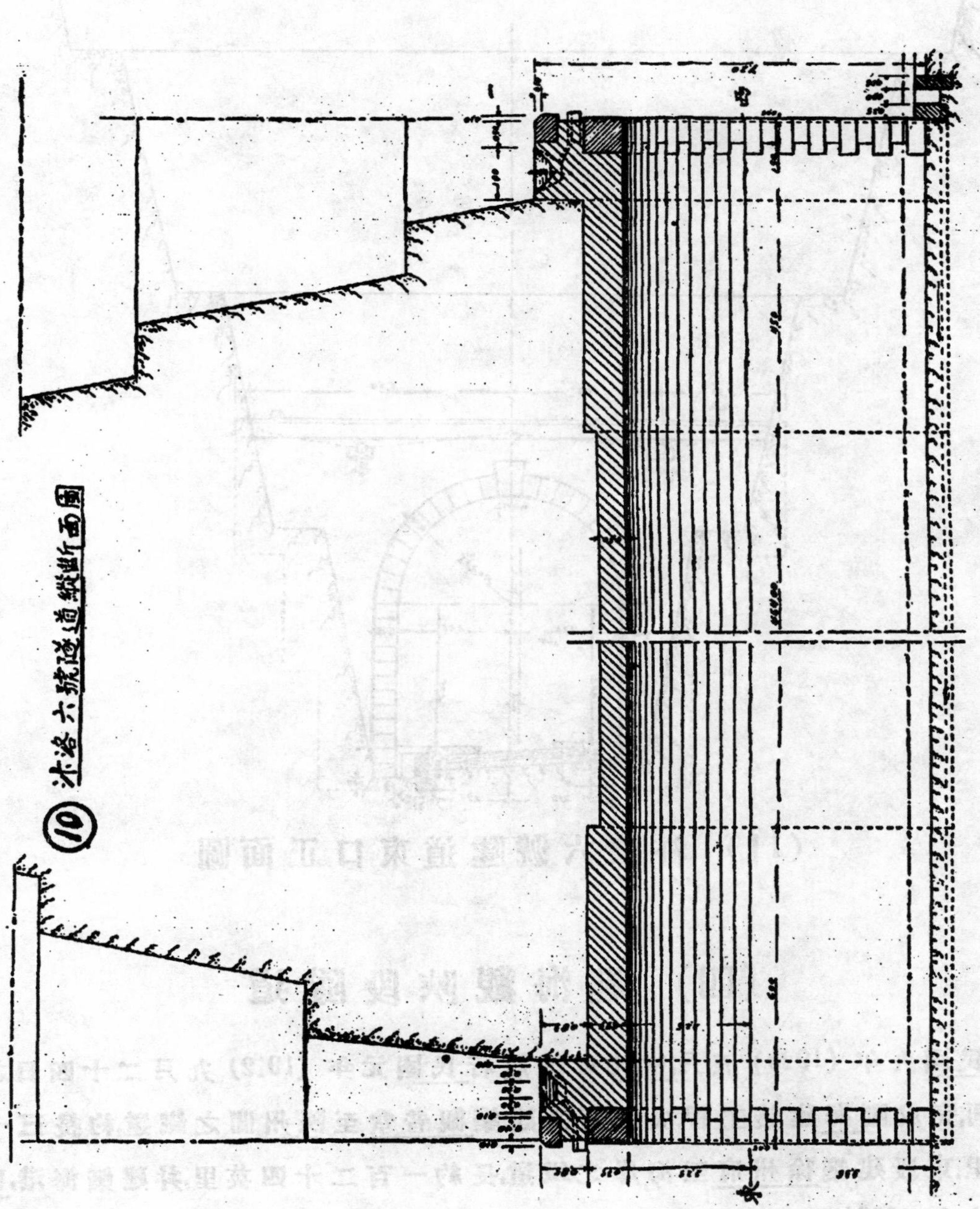

⑩ 汴洛六號隧道縱斷面圖

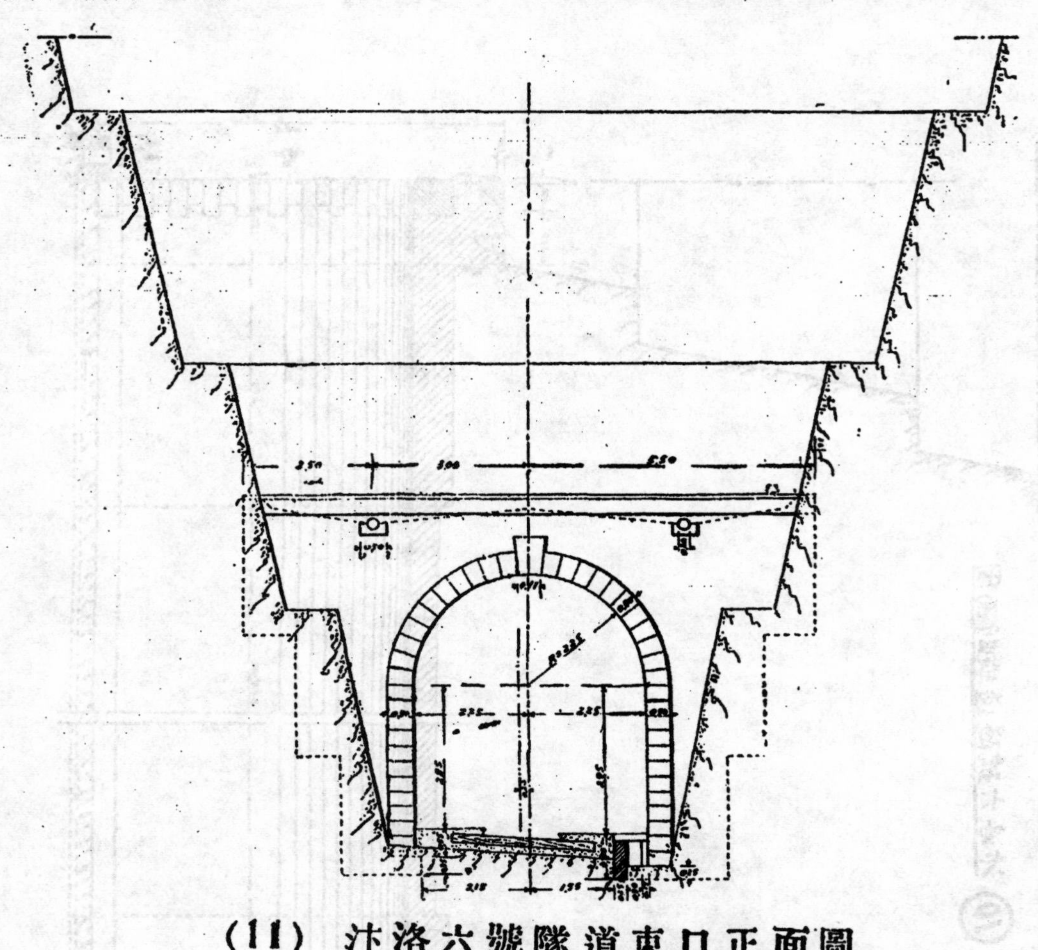

(11) 汴洛六號隧道東口正面圖

(四) 隴海觀陝段隧道

民國九年(1920)五月,中,比會議履行民國元年(1912)九月二十四日之合同,完成隴海建築工程.決議西段建築觀音堂至陝州間之鐵道,約長三十英里.東段建築徐州府至海岸之鐵道,長約一百二十四英里,并建築海港,民國十年(1921)春間,東西段同時進行.觀陝段於民國十三年(1924)通車.此段共有隧道五座,情形如下表所述:

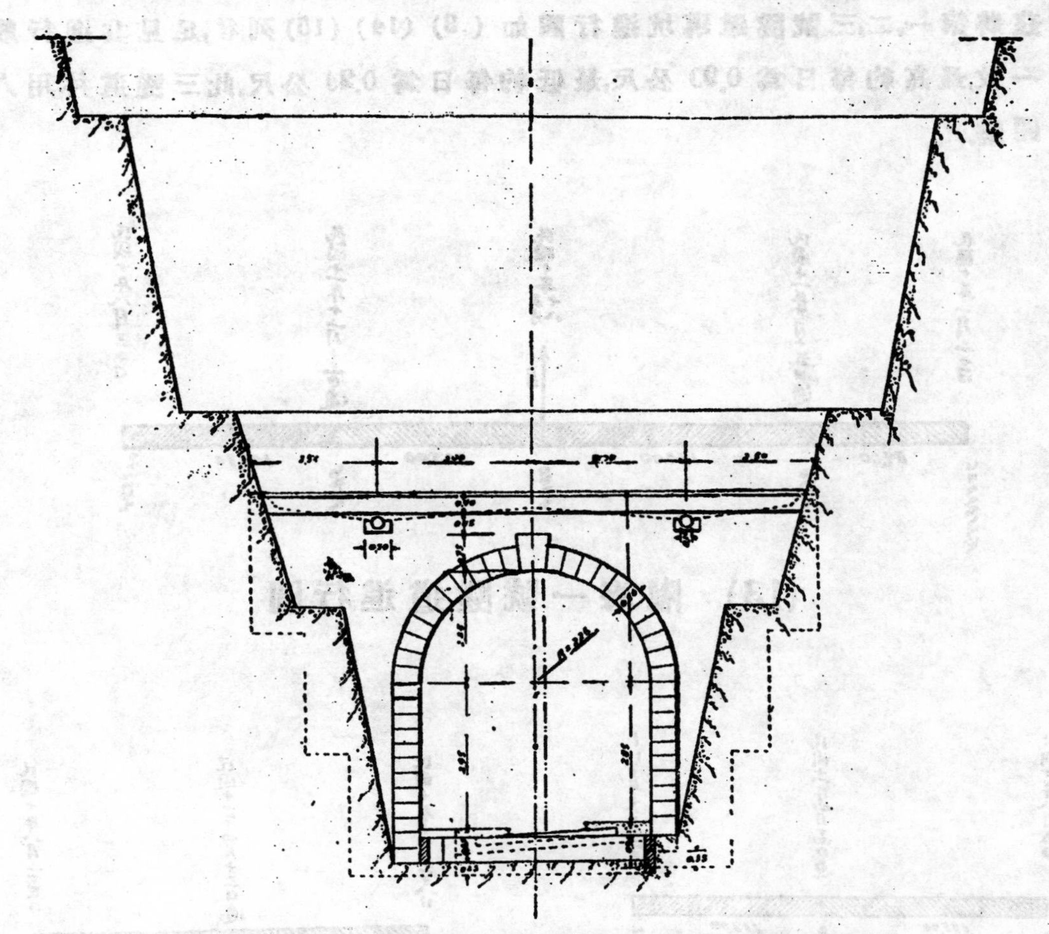

(12) 汴洛六號隧道西口正面圖

號數	位置 (公里)	長度 (公尺)	坡度	灣度 (公尺)	距路基高度 (公尺)
1					
2	213+226—213+652.70	426.70	0.015	R=1000	32.62
	215+785—216+032	247.00	0.005	0000	34.50
3	216+251—216+489	235.00	0.005	0000	54.57
4	218+850—220+630	1780.00	0.002	0000	139.19
5	226+829—227+414	585.00	0.003	500	40.10

兹將第一,二,三號隧道導坑進行圖如(13)(14)(15)列後,足見其進行頗不一致,最高約每日爲0.90公尺,最低約每日爲0.20公尺.此三隧道并用人工開鑿.

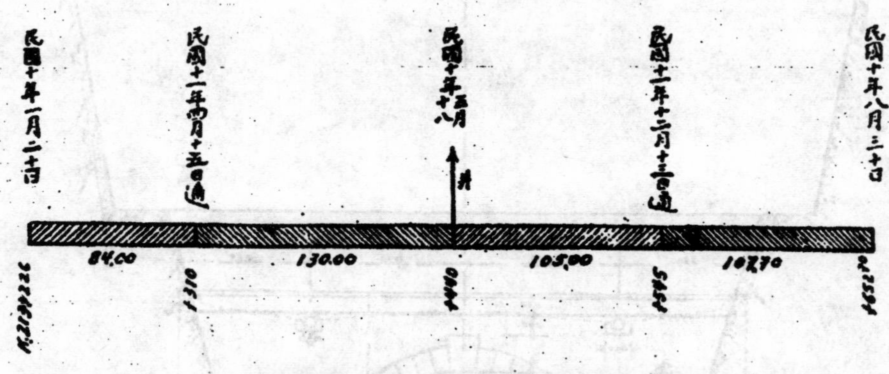

(13) 隴海一號隧道進行圖

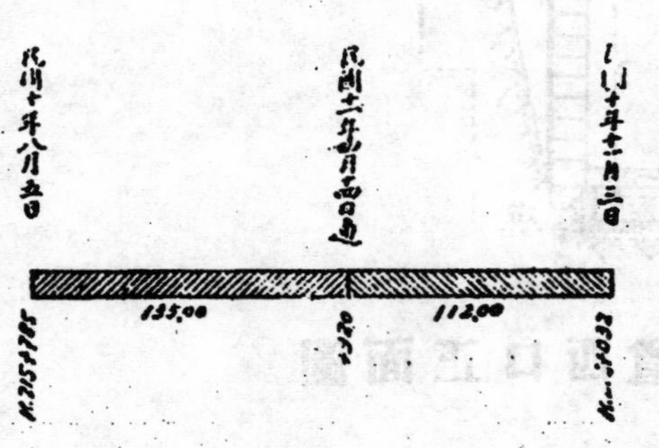

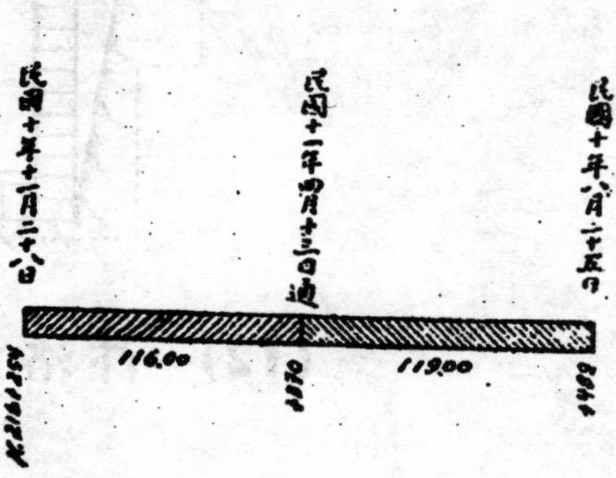

(14) 隴海二號隧道進行圖　　(15) 隴海三號隧道進行圖

隴海鐵路係採用包工制度.今將此三隧道開鑿費用,分爲鐵路公司用費,及包工公司用費,列表比較如下:

(A)一號隧道開鑿費用.(隧道長426.70公尺)

1. 鐵路公司用費

類別	單位	數量	價格	總價
土	m^3	48.00	1.70 元	81.60 元
軟石	m^3	7,400.871	3.35	31,492.92
硬石	m^3	6,664.285	4.20	27,990.00
由井上出土加費	m^3	2,782.363	0.80	2,225.89
			共	61,790.41 元

2. 包工公司用費

類別	單位	數量	價格	總價
本地木	m^3	446	25.00 元	11,150.00 元
黑火藥	斤	18,870	0.15	2,830.50
黃炸藥	磅	10,100	1.20	12,120.00
器具	——	——	——	1,960.00
人工	日	95,240	0.30	28,572.00
包工公司開銷	10 %	——	——	5,663.25
營業損失	——	——	——	(－) 505.34
			共	61,790.41 元

(B)二號隧道開鑿費用（隧道長24[illegible].00公尺）

1. 鐵路公司用費

類別	單位	數量	價格	總價
軟石	m^3	5,000.000	3.35 元	16,750.00 元
硬石	m^3	4,176.000	4.20	17,539.20
			共	34,829.20 元

2. 包工公司用費

類別	單位	數量	價格	總價
本地木	m^3	360	25.00 元	9,000.00 元
黑火藥	斤	21,700	0.15	3,255.00
器具	——	——	——	1,080.00
人工	日	56,230	0.30	16,869.00
包工公司開銷	10 %	——	——	3,020.40
營業純益	——	——	——	(＋) 1,064.80
			共	34,289.20 元

(C)三號隧道開鑿費用(隧道長235.0)公尺)

1. 鐵路公司用費

類 別	單 位	數 量	價 格	總 價
軟 石	m^3	435.619	3.35 元	1,459.32 元
硬 石	m^3	7,684.223	4.20	32,273.74
			共	33,733.06 元

2. 包工公司用費

類 別	單 位	數 量	價 格	總 價
本地木	m^3	98	25.0) 元	2,450.00 元
黑火藥	斤	44,570	0.15	6,685.5)
器 具	——	——	——	1,381.00
人 工	日	47,646	0.30	14,293.80
包工公司開銷	10 %	——	——	2,480.03
營業純益	——	——	——	(+) 6,452.73
			共	33,733.(6 元

上述三隧道,鐵路公司用費及包工公司用費,相差甚微,足見隴海所定工價,頗與當時情勢相符.至全隧道用費,約爲每公尺三百元,例如:

$$一號隧道每公尺用費 = \frac{127,799.\ 9\ 元}{426.70\ 公尺} = 299.5)\ 元$$

$$三號隧道每公尺用費 = \frac{67,104.33\ 元}{235.00\ 公尺} = 287.00\ 元$$

玆將一號隧道詳細用費列表如下:

(D) 一號隧道隴海公司用費(隧道長426.70公尺)

類別	單位	數量	價格	總價
土	m^3	48.000	1.70元	81.60元
軟石	m^3	9,400.871	3.35	31,492.92
硬石	m^3	6,664.285	4.20	27,990.00
由井上出土加價	m^3	2,782.363	0.08	2,225.89
				61,790.41
E號西門土	m^3	48.350	14.00	676.90
D號西門土	m^3	117.609	11.20	1,317.22
G號西門土	m_3	2.476	20.00	49.52
普通砌石用D號灰漿	m^3	13,757.767	8.60	11,831.60
粗琢砌石用D號灰漿	m^3	2,801.713	13.40	37,542.95
細琢砌石用D號灰漿	m^3	65.521	16.40	1,074.54
精琢砌石用F號灰漿	m^3	6.797	19.00	129.14
磚砌石用D號灰漿	m^3	5.866	10.50	61.59
圓拱砌石加價	m^3	1,616.018	2.00	3,232.04
隧道內砌石加價	m^3	4,103.024	1.40	5,744.23
圓拱上敷塗西門土	m^2	1,860.232	0.80	1,488.19
卸除井內支木	m^1	25.770	1.80	46.39
轉運并安置井內圓管	m^1	25.000	2.25	56.25
井內空隙填塞鐵屑	m^1	24.200	1.50	36.30
井蓋用鐵條	Kg^5	823.000	0.50	411.50
用小鐵道轉運坑內土	m^3	4,994.0[illegible]2	0.1175	586.80
轉運石料	—	——	——	1,724.42
				66,009.58
			共計	127,799.99元

此處砌衣,因該地產石至富,價甚廉,每公尺約一百五十五元,其D號西門土以量計,約合1:5:8.[illegible],茲再將隴海一號隧道各圖詳舉如次:

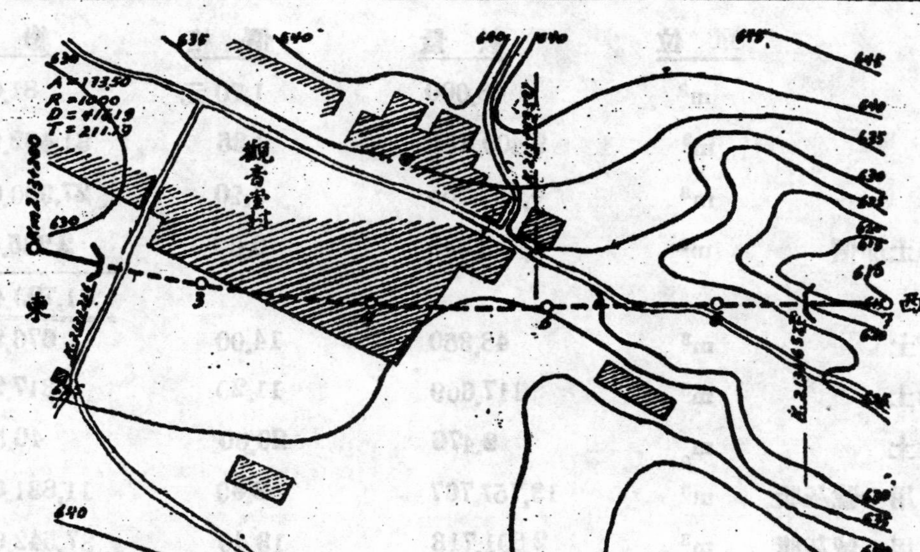

(16)　隴海一號隧道平面總圖

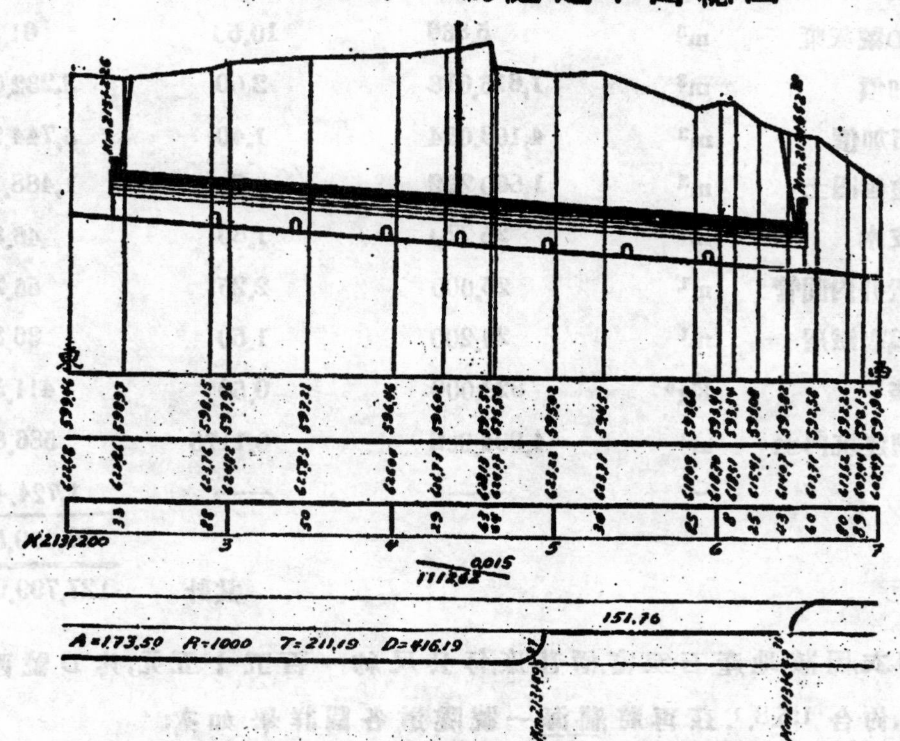

(17)　隴海一號隧道縱斷面總圖

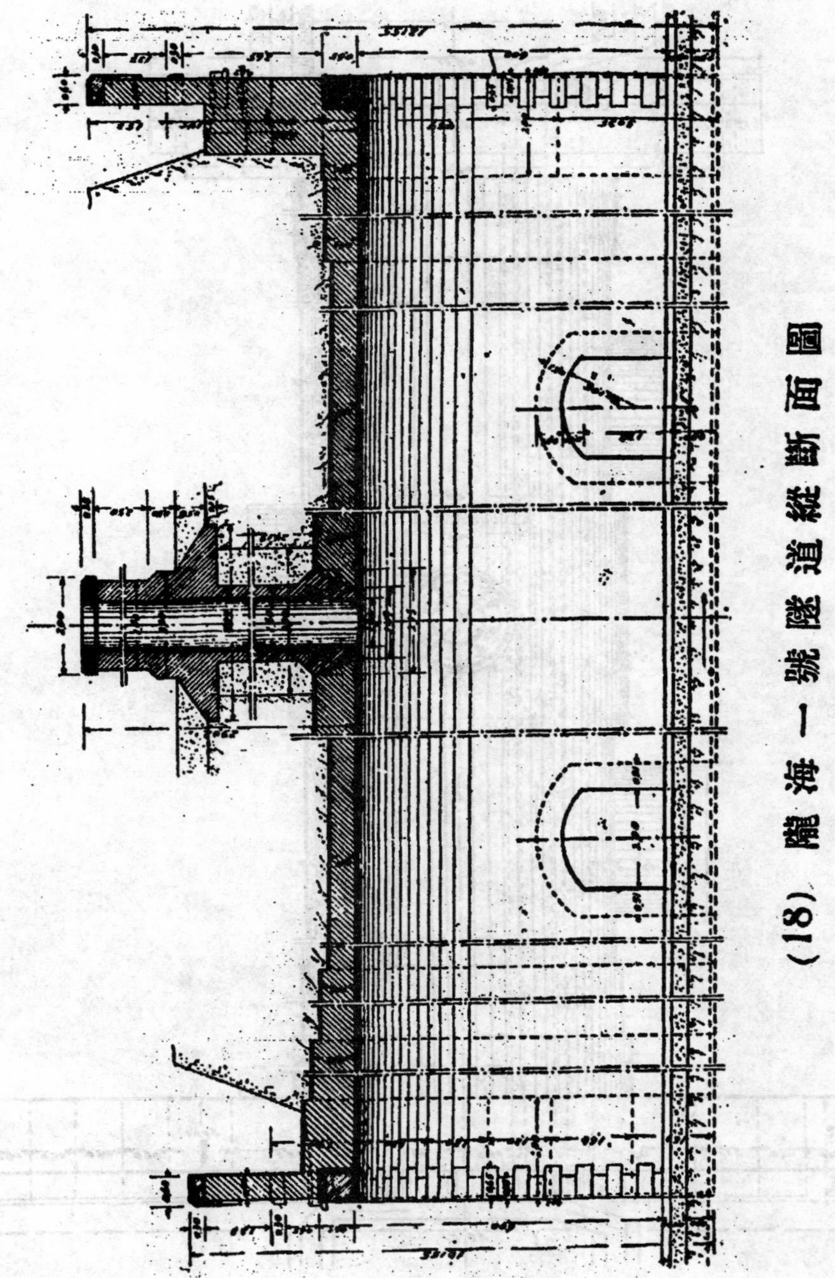

(18) 隴海一號隧道縱斷面圖

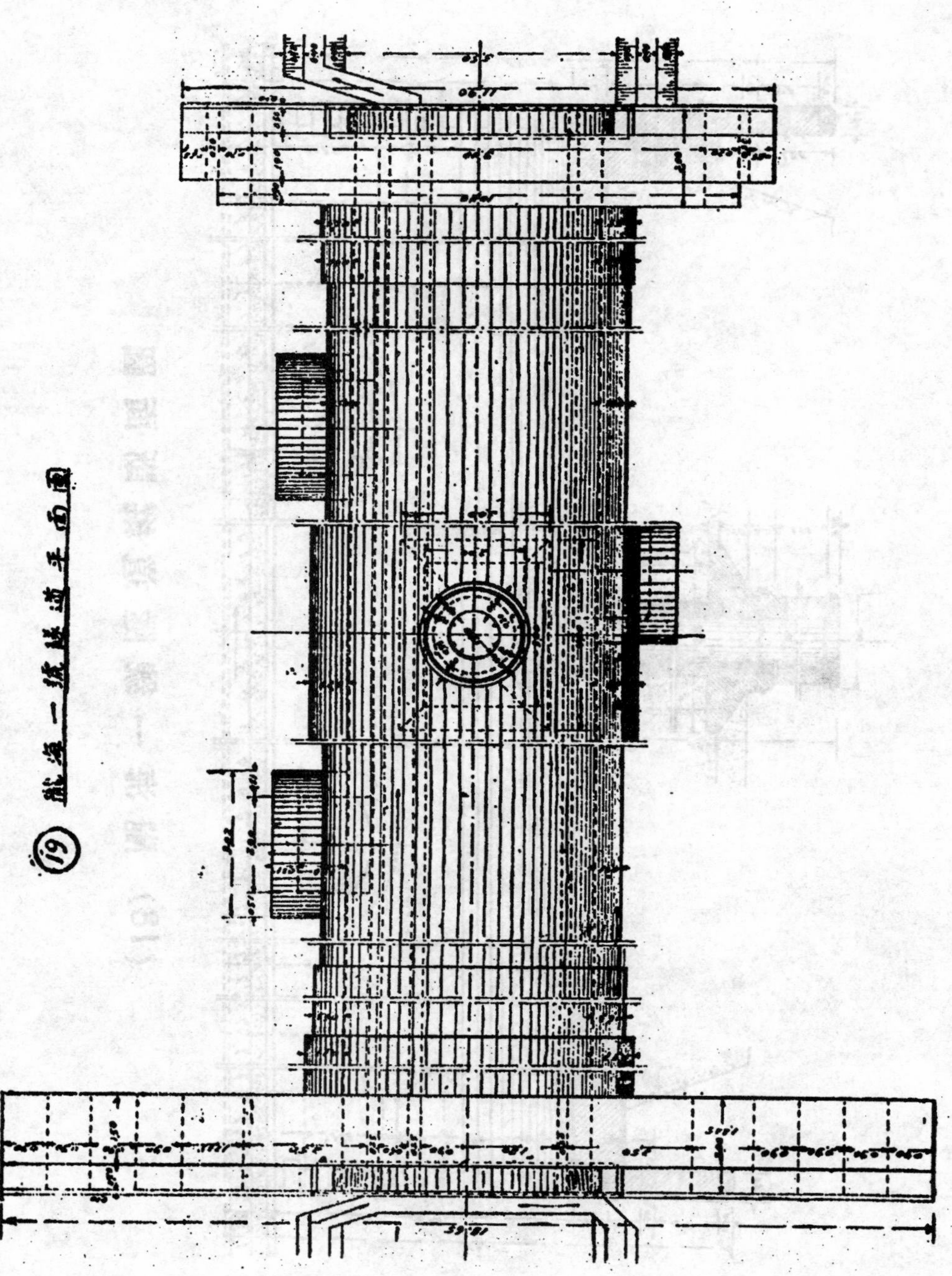

⑲ 龍華一號隧道平面圖

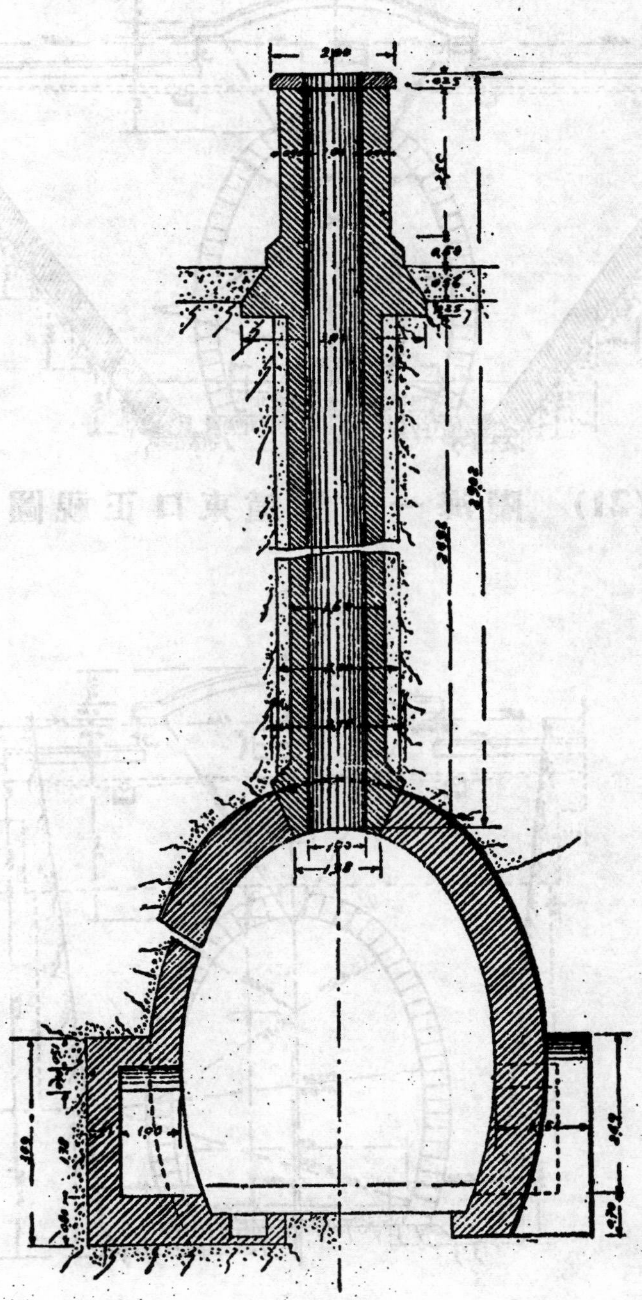

(20) 隴海一號隧道橫斷面圖

(21) 隴海一號隧道東口正視圖

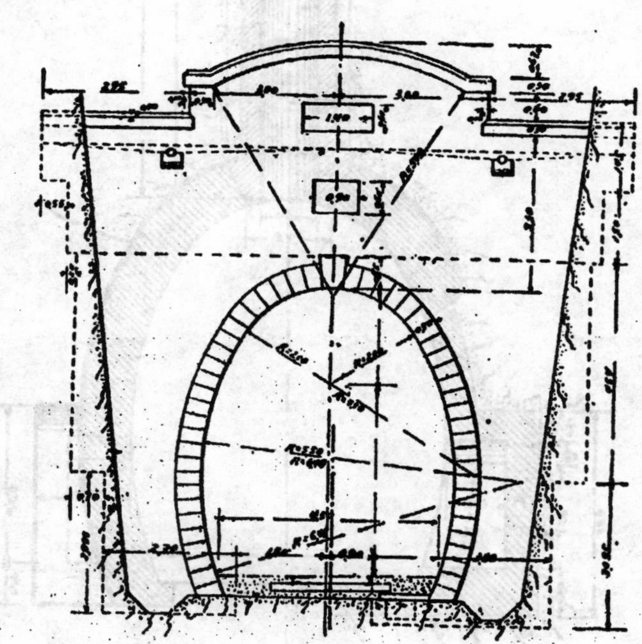

(22) 隴海一號隧道西口正視圖

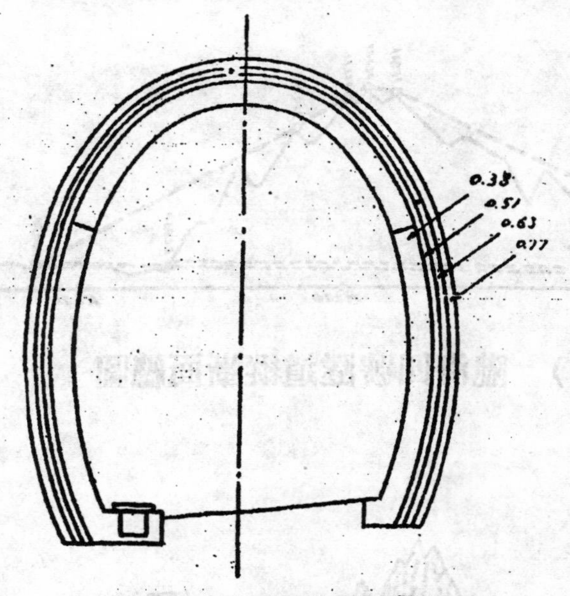

(23) 隧道砌衣圖

隧道砌衣之厚薄,原無由計算,惟有就經驗而定.彼時所採者有厚

0.38 公尺

0.51 公尺

0.63 公尺

0.77 公尺

者數種如圖(23).其土方面積及砌石面積,則如下表:—

種　類	土方面積	砌石面積
無切衣隧道	28.62 1m	0.355 m² 水溝在內
圓拱砌衣厚 0.38 公尺	31.302 ,,	3.036 ,, ,,
全隧道砌衣厚 0.38 公尺	35.062 ,,	6.796 ,, ,,
全隧道砌衣厚 0.51 公尺	37.377 ,,	9.104 ,, ,,
全隧道砌衣厚 0.66 公尺	39.517 ,,	11.281 ,, ,
全隧道砌衣厚 0.77 公尺	43.222 ,,	14.956 ,, ,

(五) 硤石驛四號隧道

隴海鐵路硤石驛四號隧道共長1730公尺.係民國十年(1921)開工.原擬全用人工開鑿,先開五個導井,藉以更番工作.後以全隧道大部爲堅硬石灰岩,運輸遙遠,開鑿爲難,乃於翌年(1922)改用機械開鑿.(24)(25)(23)爲此隧道之平面,斷面及地質圖.其(25)圖中小圈爲經緯儀標點.

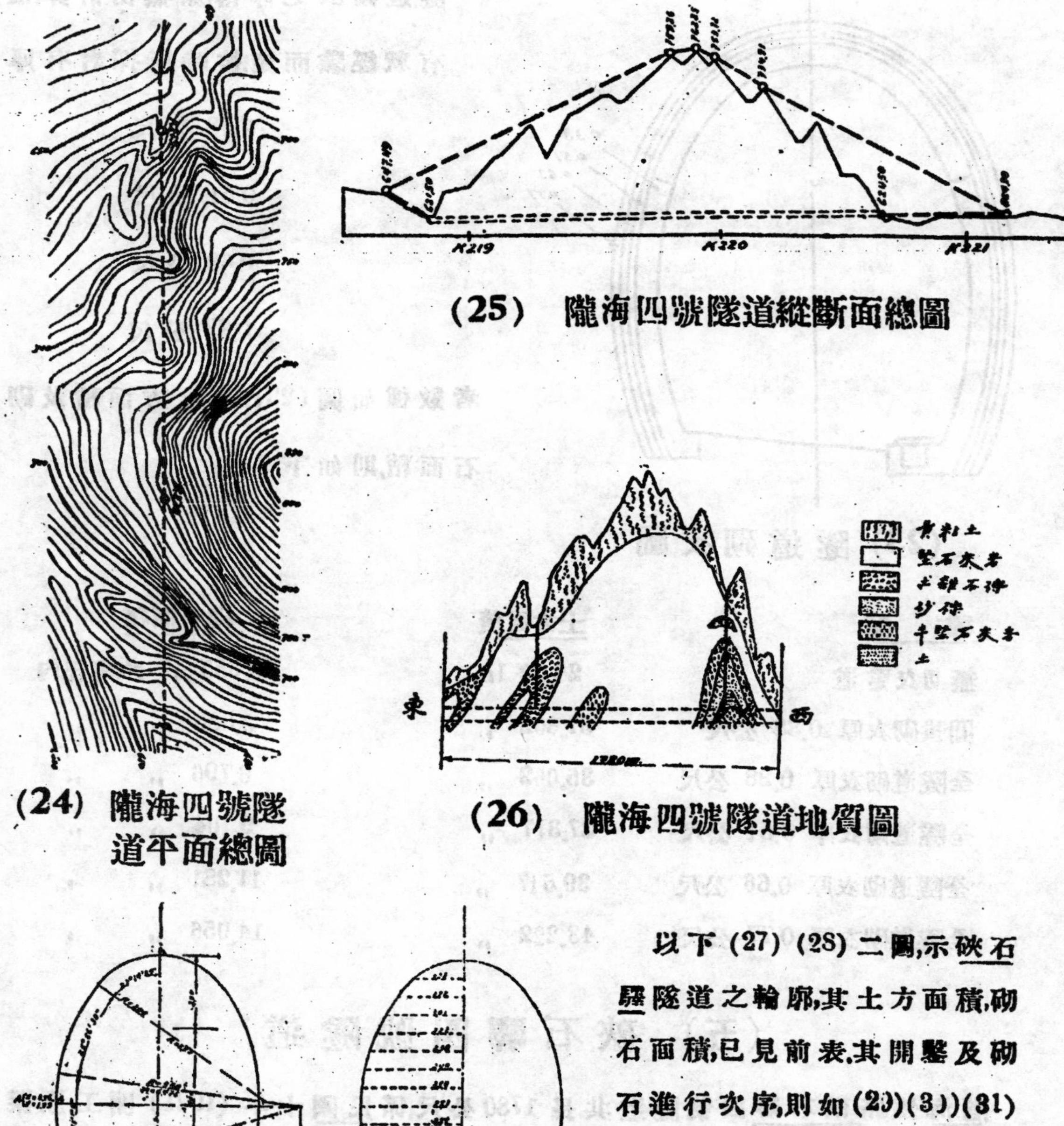

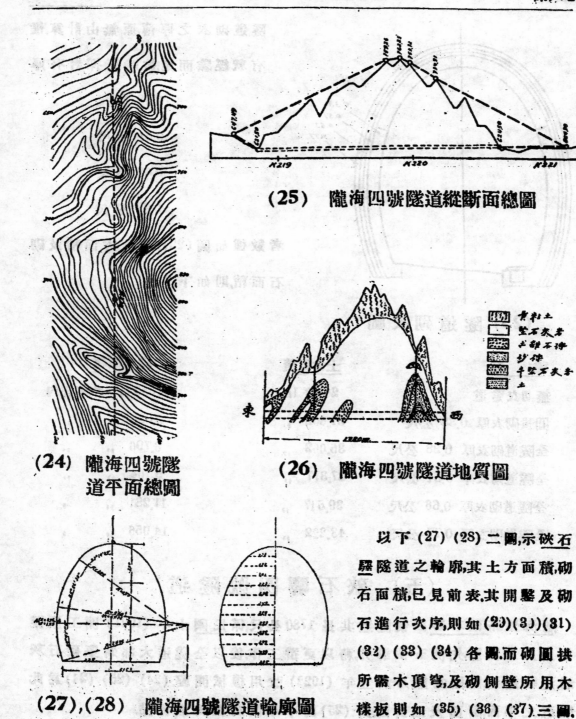

(25) 隴海四號隧道縱斷面總圖

(24) 隴海四號隧道平面總圖

(26) 隴海四號隧道地質圖

(27),(28) 隴海四號隧道輪廓圖

以下(27)(28)二圖,示硤石壘隧道之輪廓,其土方面積,砌石面積,巳見前表,其開鑿及砌石進行次序,則如(29)(30)(31)(32)(33)(34)各圖.而砌圓拱所需木頂穹,及砌側壁所用木樣板則如(35)(36)(37)三圖.

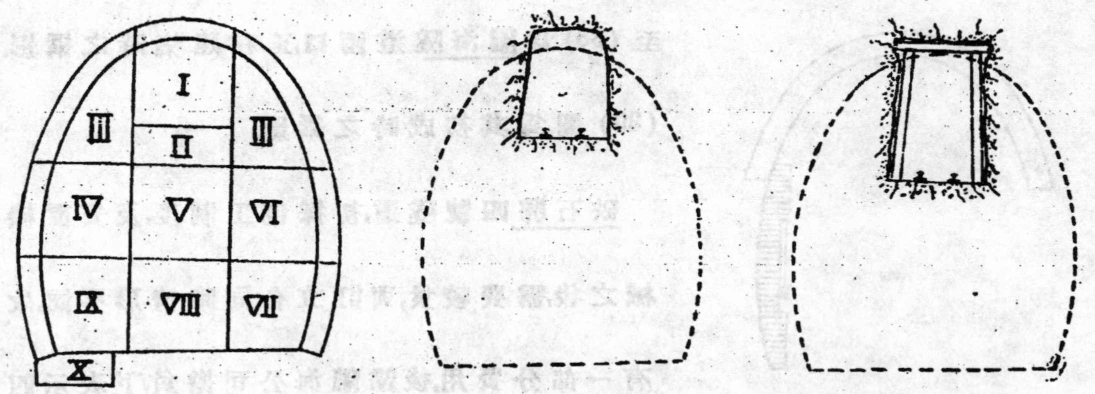

(29),(30),(31) 隴海四號隧道進行次序圖

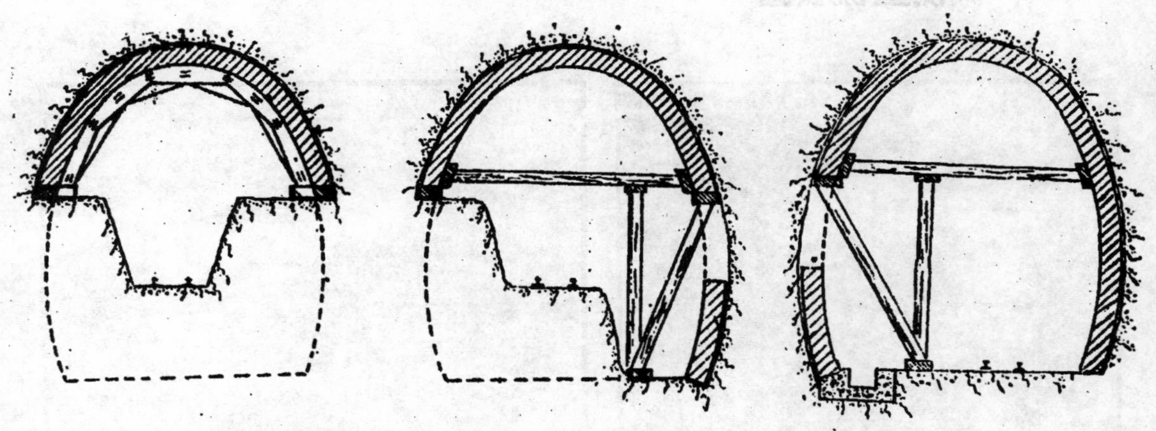

(32),(33),(34) 隴海四號隧道進行次序圖

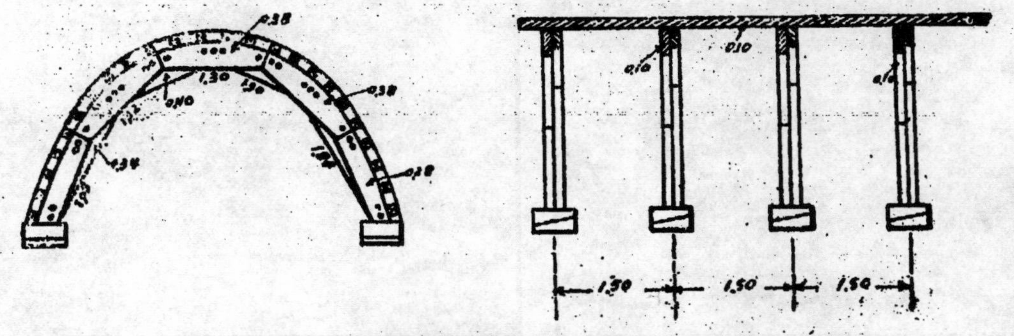

(35),(36) 隴海四號隧道砌圓拱木頂穹圖

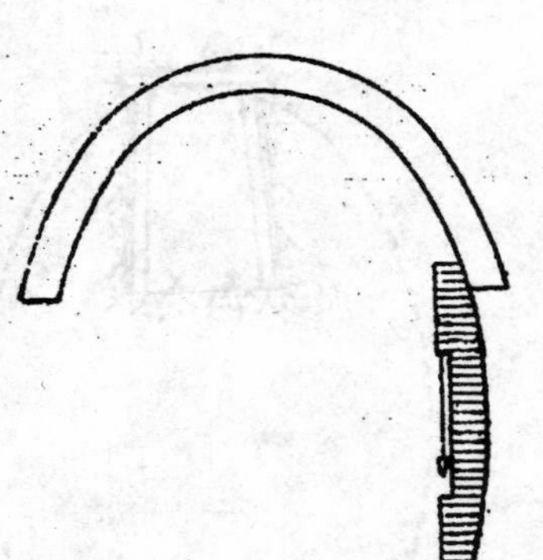

(37) 隴海四號隧道砌側壁樣板圖

至(38)爲隴海隧道西口,正在建築時之攝影.(39)圖爲其初成時之攝影.

硤石驛四號隧道,初採包工制度,及安置機械之後,需費較貴,與訂立合同時情形不同,故有一部分費用,後歸隴海公司擔負,下表示四號隧道開鑿費:

(38),(39) 隴海四號隧道西口圖

(E)四號隧道開鑿費用(隧道長1,780公尺)

1. 鐵路公司用費

類別	單位	數量	價格	總價
軟石	m^3	11,804	3.35元	39,543.40元
硬石(人工開鑿)	m^3	42,873	4.20	180,066.60
硬石(機械開鑿)	m^3	7,303	8.45	61,710.35
由井上出土加費	m^3	3,203	0.80	2,561.60
			共	283,881.95元

$$每公尺=\frac{283,881.95}{1,780}=159\tfrac{1}{2}元$$

2. 包工公司用費

類別	單位	數量	價格	總價
本地木	m^3	921	25.00元	23,025.00元
黑火藥	斤	182,356	0.15	27,353.00
黃炸藥	磅	44,700	1.20	53,640.00
鐵鑛及零件	磅	27,497	0.18	4,949.00
器具	——	——	——	11,200.00
燈油費	——	——	——	1,400.00
人工	日	425,080	0.33	140,276.00
包工公司開消	10%	——	——	26,184.00
遇險損失	——	——	——	4,000.00
營業損失	——	——	——	(—) 8,145.00
			共	283,882.00元

開鑿堅硬岩石導坑,四號隧道為一好例.茲略言其開鑿方法.其進行次序已見(29)(30)(31)(32)(33)(34)各圖,導坑面積平均為 $6\,m^2$.今分鑿眼,用藥,裝藥,三項言之.(1)鑿眼如(40)圖,導坑平均共置炮眼二十個,分為三組:第一組四眼,各長1.10 m;第二組四眼,各長1.00 m;第三組十二眼,各長0.90 m.此項炮眼全徑各為4公分(cm).平均機械鑿眼二十個需五小時,即第一組第一時

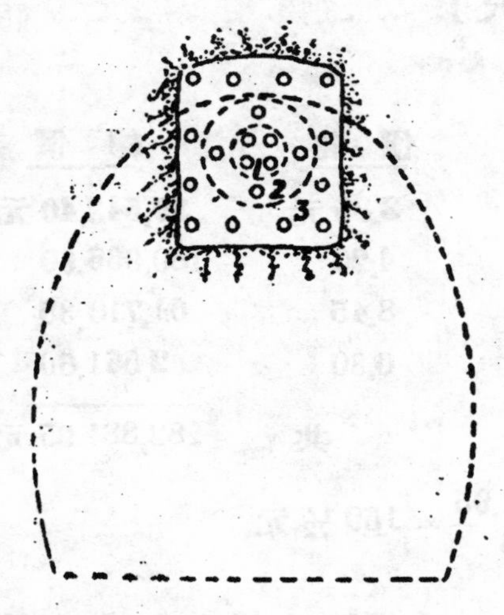

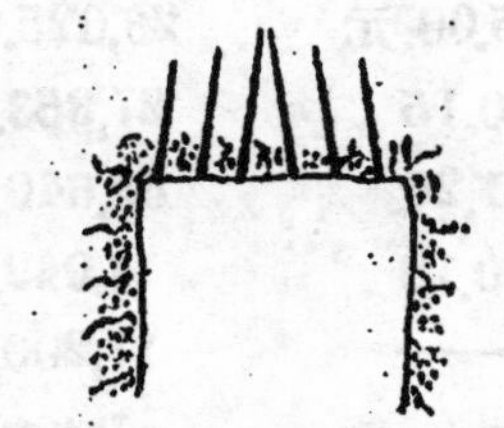

(40) 隴海四號隧道鑿眼圖

半,第二組一小時,第三組二小時也.此外裝炮,點炮一小時,出石等一小時.(2)用藥,視炮眼之深淺大小,及石質之硬軟而定.平均每班約用小黃炸藥250根至280根,約重11¼公斤(kgs)至12¾公斤.而每深一公尺之炮眼,可納藥重680克蘭姆(即公分).毎轟裂一公立方尺,約需黃藥2½公斤(kgs),或5.55磅(lbs).(3)裝藥,其法先裝黃藥,次納藥線,次封以黃土,以防洩氣.裝置旣畢,以第一組四眼之藥線爲最短,第二組視第一組長一公寸(10 cms),第三組又視第二組長一公寸.如此同時燃燒藥線,大約二分鐘後可轟第一組,三分鐘後轟第二組,四分鐘可轟第三組矣.藥線以購現成者爲便.因其入水不濕,且其燃燒速度,至爲均勻也.卽黃藥亦應善爲選擇.例如75%黃藥含硝酸虞利斯林(Nitroglycerine)過多,空氣混濁至速,工人工作,時多悶絕,不如60%黃藥爲便.以上爲導坑用機械及黃炸藥開鑿情形.

導坑以外多用人工開鑿.此項開鑿,兩礦工,一人提鑽,一人擧槌,每八時可打炮眼二公尺,大約此處炮眼每個長八公寸,約需黑藥375克蘭姆.每一公立方尺(m^3)需黑藥1½公斤,惟亦有兼有黃炸藥者,則視地質情形而定也.

隴海四號隧道如(41)(42)所示,以機械開鑿,導坑最多每日亦有進行六公尺者,但視美國Rogers Pass隧道之每月進行932英尺者,尙有遜色.

(43)圖擧美國Rogers Pass隧道用機械開鑿情形以見一般.至硤石驛四

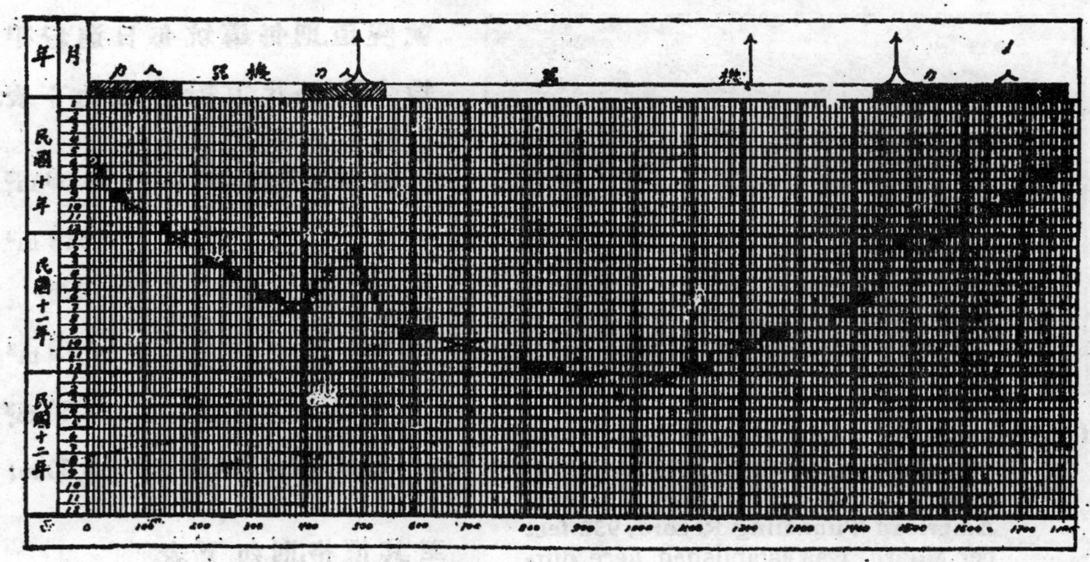

(41)　隴海四號隧道導坑進行圖

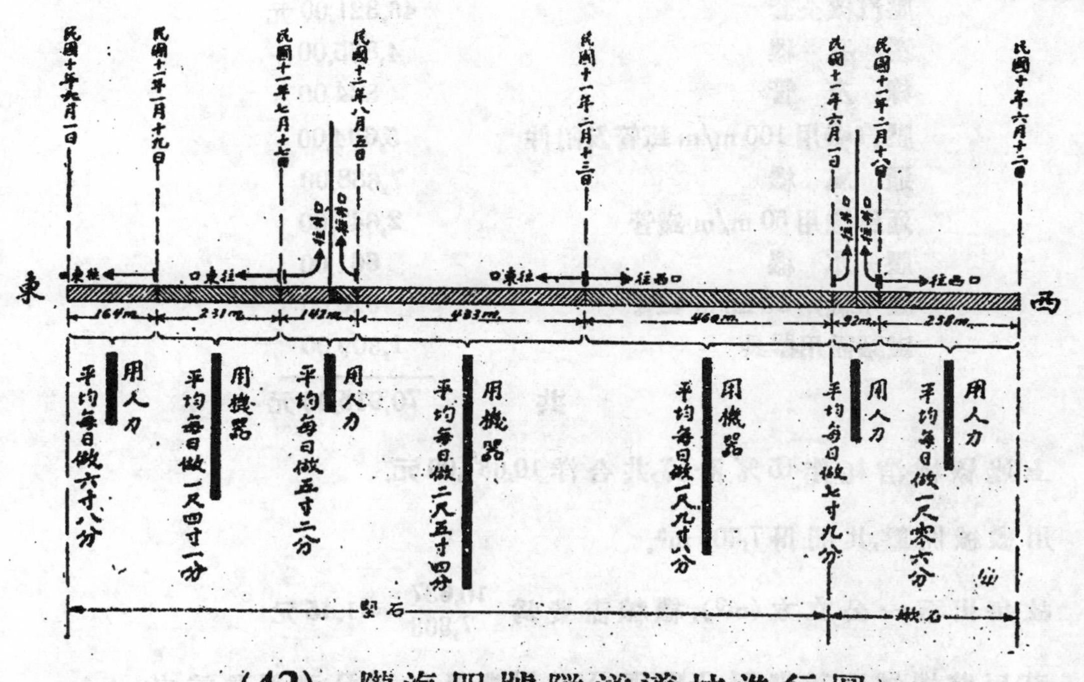

(42)　隴海四號隧道導坑進行圖

(43) Leyner-Ingersoll Drills in Rogers Pass Tunnel of Canadian Pacific R. R., at Glacier, British Columbia. A new American Tunneling Record, 932 feet per Month, was established here during January, 1915.

號隧道,則每導坑每日進行平均 2.50 公尺,其計算法如下表:

由機械開鑿;	東口	7,303 m^3
	西口	5,813 m^3
日數		226
每日		25.72 m^3
每導坑每日出石		12.86 m^3
每導坑每日進行		2.50 m.

至其價格,則如下表.

建設機械費用

壓汽機全套	46,321.00元
鑽孔機	4,855.00
橡友管	824.00
壓汽機用 100 m/m 鐵管及附件	5,604.00
通風機	7,688.00
通風機用 50 m/m 鐵管	2,647.00
壓水機	661.00
壓水機用 50 m/m 鐵管	916.00
鐵廠需用器具	1,309.00
共	70,915.00元

上述機械消耗作 15% 計算,共合洋 10,637.00 元.

用機械開鑿,共開得 7,303 m^3.

故每出石一公立方(m^3),機械需費爲 $\frac{10,637}{7,303}=1.45$ 元

茲再將機械開鑿費用,分隴海公司用費及包工公司用費言之:

(1) 隴海公司用費(每出石一公立方 m^3 計算)

機械需費		1.45 元
水(0.64),煤(1.39),石油(0.27),脂油(0.24),		
雜件(0.34),修理費(0.52)	共計	3.40 元
錙鑛之消耗		0.07 元
人　工 { 薪水		2.20 元
人　工 { 獎金		0.42 元
	共計	7.54 元

(2) 包工公司用費

		(每人工資)	總　計	
人　工……鑛工頭一人		1.25 元	1.25 元	
每導坑……工頭三人		0.60 元	1.80 元	
鑛工十八人		0.45 元	8.10 元	
小工二十六人		0.30 元	10.80 元	
共開 12.86 m^3 計………………			21.95 元	
每日開 1 m^3 計 $\frac{21.95}{12.86}$				1.70 元

炸　藥 { 60 %	4.75 磅		
炸　藥 { 75 %	6.38 磅		
	11.13 磅	平均 $\frac{1}{2} \times 11.13 = 5.56$	
	計	5.56×1.20	6.66 元
油燈費			0.05 元

轉運;	租七個小鐵車	14.00 元	
	及 1,200 m 小鐵道	4.00 元	
	每月需	18.00 元	
	共開 12.86 m^3 每日需	0.60 元	
	每日開 1 m^3 需	0.60 元	
		12.86 元	0.04 元
		共計	8.45 元

(1),(2) 兩項隴海公司用費及包工公司用費共計 (7.54+8.45)　15.99 元

由上表研究之結果,鐵路公司規定付於包工公司之硬石,用機械開鑿價格,每立公方爲8.45元.至其總價,每公尺開鑿費前已說及,計159.5元.就中建設機械費用,尙未計及.其砌石費用,亦與第一,二,三號相類,每公尺爲155元.

(六) 隴海靈潼段隧道

隴海靈潼段工程,係民國十三年(1924)開工.數年來時局多變,迄今尙未完成.此線沿黃河而行,該處地質,係黃土而雜入多量砂礫者.且黃河河流,南北頻年移動,河岸時見崩塌.而函谷關附近,南倚高原,北臨黃河,建築上尤感困難.前此本計畫於靈寳高原上鑿一長六公里隧道,後以需費太大,變更計劃,於黃河岸建一臨時路線,以代此六公里隧道.此線計有六號至十二號七座,共長二千九百公尺.惟因太近黃河,民國十五年(1926)夏間河水高漲之際,河流南岸衝成一凹灣,十號十一號隧道間,河岸發生崩塌.隧道隨而動搖,此項臨時路線計畫,無由實現.今擬另行計畫十號A,十號B兩隧道,選擇其一,以備應用.在潼關附近,城下鑿一隧道,今亦在計畫中.其他各隧道工程,已十九完成.玆將各隧道情形,列表於次頁.

以下各圖示隴海靈潼段隧道進行次序及其完成前後情形:

(44) 第十六號隧道東口圓拱初成之圖

(45) 第十號隧道土工進行之圖

號數	位置(公里)	長度	坡度	灣度	距路基高度	總值	每公尺價	百分數
	b　　　　b	m		m	m	元	元	
6	286+547,00——286+635,00	88,50	0,008	R=600	32,77	27,469,09	294	100%
7	286+549,80——287+171,00	621,20	0,008	600	42,25	138,970,55	224	100%
8	287+270,00——287+360,30	90,30	0,008	—	28,49	31,512,60	349	100%
9	287+418,50——287+523,90	105,40	0,008	—	34,47	35,655,77	338	100%
10	287+851,34——288+639,46	788,12	0,008 0,004	800 350	52,38	135,575,11	—	約 57%
11	288+793,00——289+385,00	592,00	0,002	350	64,60	52,291,03	—	約 30%
10 A	287+828,00——289+100,00	1,272,00	0,004	350	79,55	未詳	未詳	計劃中
10 B	287+828,00——290+560,00	2,732,0	0,004 0,006	400 500	210,00	未詳	未詳	計劃中
12	290+765,00——291+387,60	622,60	0,007	350 600	50,16	176,714,35	284	100%
13	323+630,00——324+325,00	695,00	0,010 0,009	1,000 500	51,90	約240,000	345	約 61%
14	329+450,00——330+　80,00	630,00	0,062	—	69,11	未詳	未詳	100%
15	339+399,20——339+794,20	95,00	0,010	1,000	36,42	95,320,15	241	100%
16	351+190,80——352+　64,90	874,10	0,010	1,000	62,16	約262,000	300	約 54%
17	352+680,00——353+760,00	1,080,00	0,005	1,400 1,500	76,15	未詳	未詳	計劃中

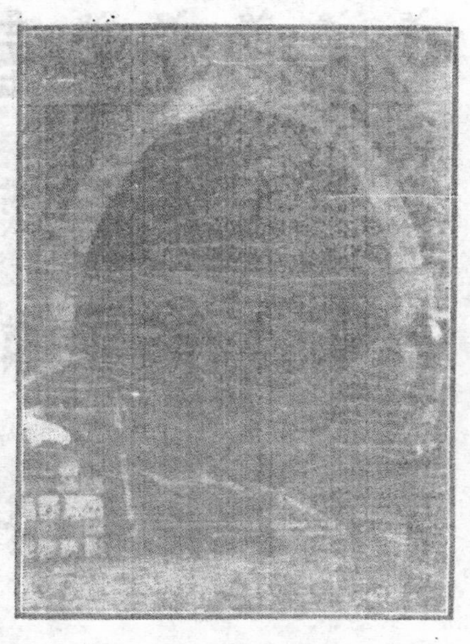

(46) 第十六號隧道西口砌石初成之圖

(47) 隴海六號隧道東口，及函谷關圖

(48) 隴海七號隧道東口圖

(49)　隴海九號隧道西口圖

以前開鑿四號石質隧道,用工人425,030人,每公尺約需233人,就中12%用於導坑,餘用於他處.而工人中52%爲普通工人,46%爲礦工,2%爲雜工也.砌工共用工人60,000人,每公尺約需33人,就中60%用於圓拱,40%用於側壁.而工人中54%爲普通工人,42%爲瓦匠,4%爲雜工.就中用木,每公尺凡1立方公尺.

今六號以後土質隧道,則每公尺約需30人,就中20%用於導坑,80%用於他處.砌工每公尺約需33人,就中50%用於圓拱,50%用於側壁.就中用木,每公尺凡1立方公尺.而每公尺隧道土方,包工公司用費計:

類別	單位	數量	價格	總價
本地木	m^3	1½	25元	37.50元
人工	日	30	0.30	9.00
器具	—	—	—	2.50
公司開消	10%	—	—	4.90
			共	53.90元

土質隧道厚0.51 m者,面積37.370 m^2

故 $\frac{53.900}{37.370}=1.445$ 元爲每立方公尺用費.

陜潼段因石料運輸不便,常就地製西門土磚.此項西門土磚厚 0.38 公尺,備一鏇圓拱長千公尺之用,共需385塊,每塊平均體質爲0.0297 m^3.由D號西門土漿製成.每立方公尺由 $1/1.2$ m^3 石子,$1/2$ m^3 砂,及170 kgs 或0.170 噸(T) 配成.圓拱厚 0.38 公尺者,面積爲 3.038 m^2.今將此項西門土磚每立方公尺造價列表計算如下:而(50)圖則爲圓拱厚 0.51 公尺,西門土磚分配之圖.

西門土磚造價表:

	項	目	單位價	總 計
材 料:	石 子	$385\times0.0297\times1/1.2=9.528$ m^3	0.75元	7.146元
	砂	$385\times0.0297\times1/2=5.717$ m^3	0.45	2.573
	西門土	$385\times0.0297\times0.170=1.944$ T	40.00	77.760
製磚工人:	工 頭	2 人	1.00	2.000
	小 工	60 人	0.30	18.000
	木 匠	26 人	0.50	13.000
				120.479 (A)
砌石用灰漿		10 % (A)	——	12.048
砌圓拱木頂穹		0.35 m^3	25	8.750
砌石用襯板		1.00 m^3	25	25.000
砌石用工人:	小 工	72	0.30	21.600
	瓦 匠	18	0.40	7.200
轉運用工人:	運 石	46 人	0.30	13.800
	運 水	4 人	0.30	1.200
				210.077 (B)
公司開消		10 % (B)	——	21.008
		共計		210.077+21.008=231.085元

故每立方公尺西門土磚造價 $=\frac{231.085}{4\times3.038}=19.00$ 元

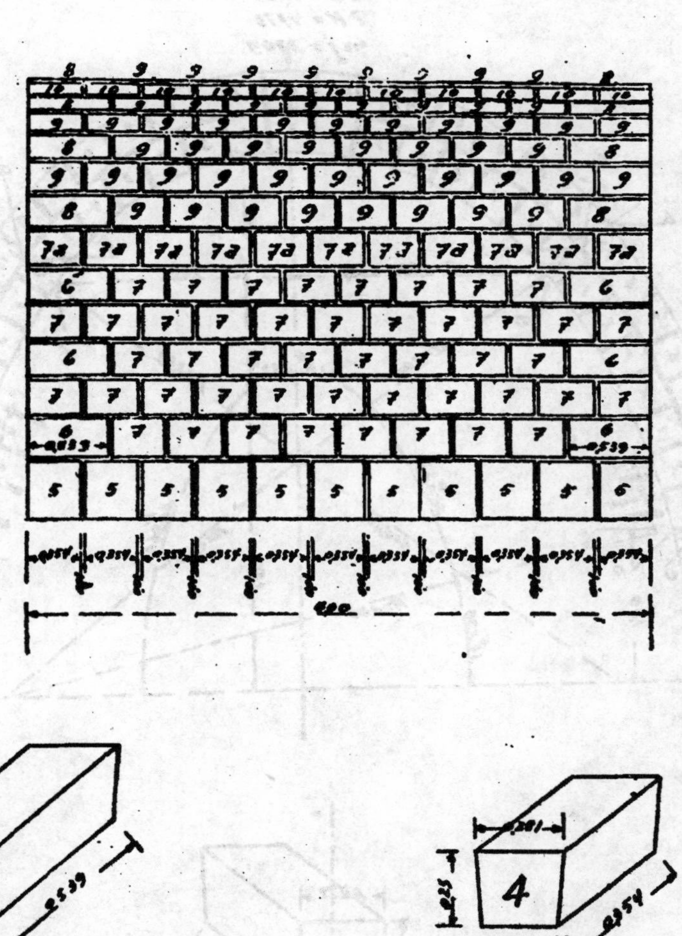

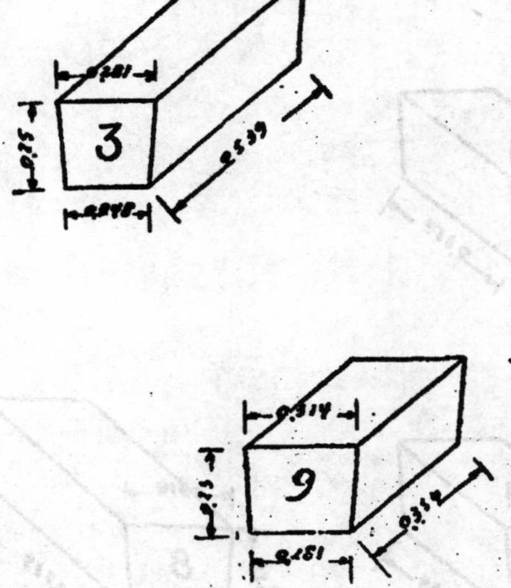

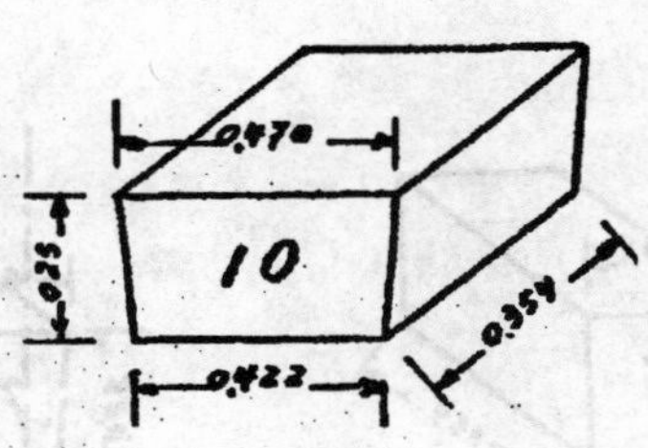

(05) 圓拱厚0.51公尺西門土磚分配圖

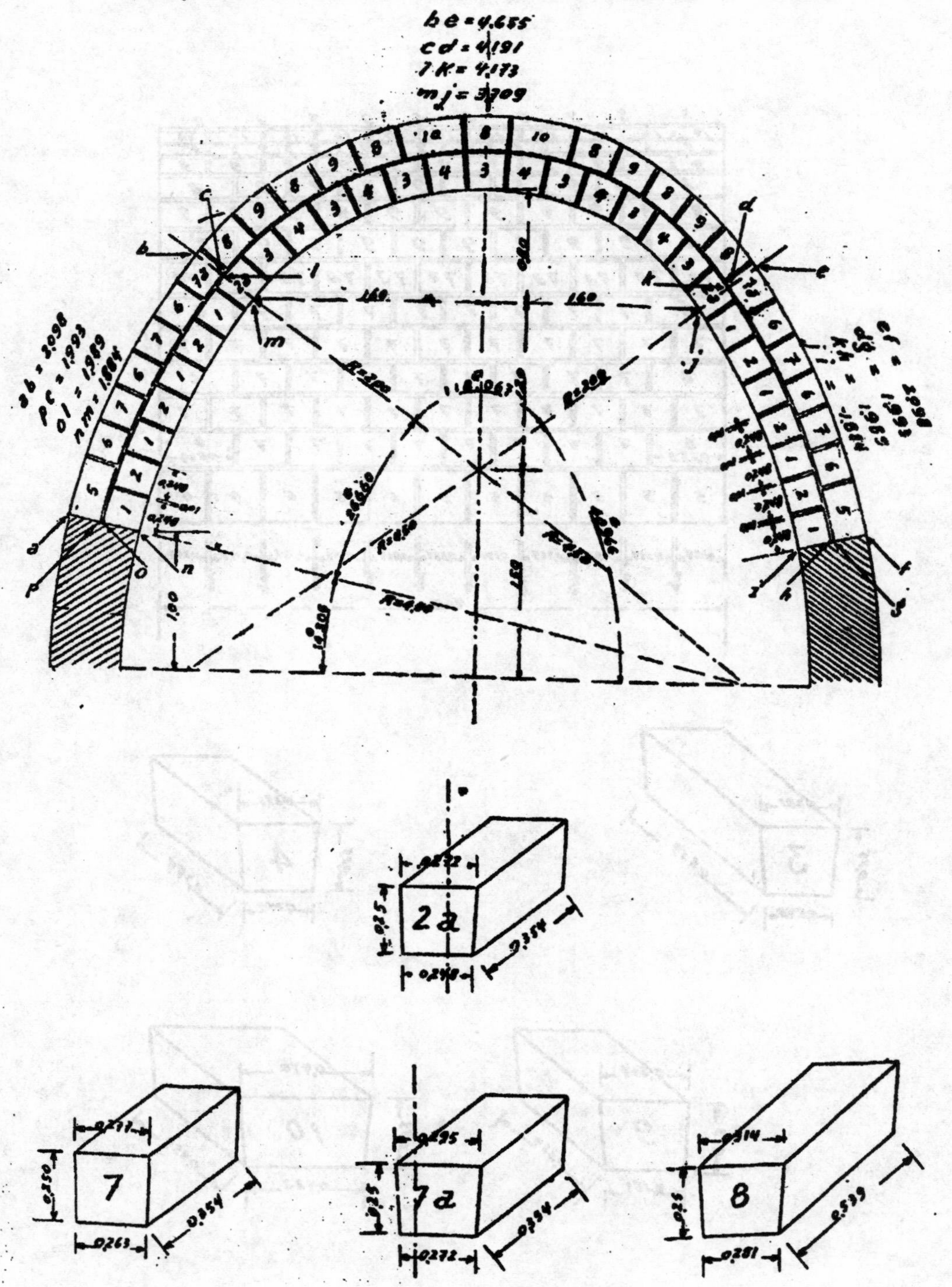

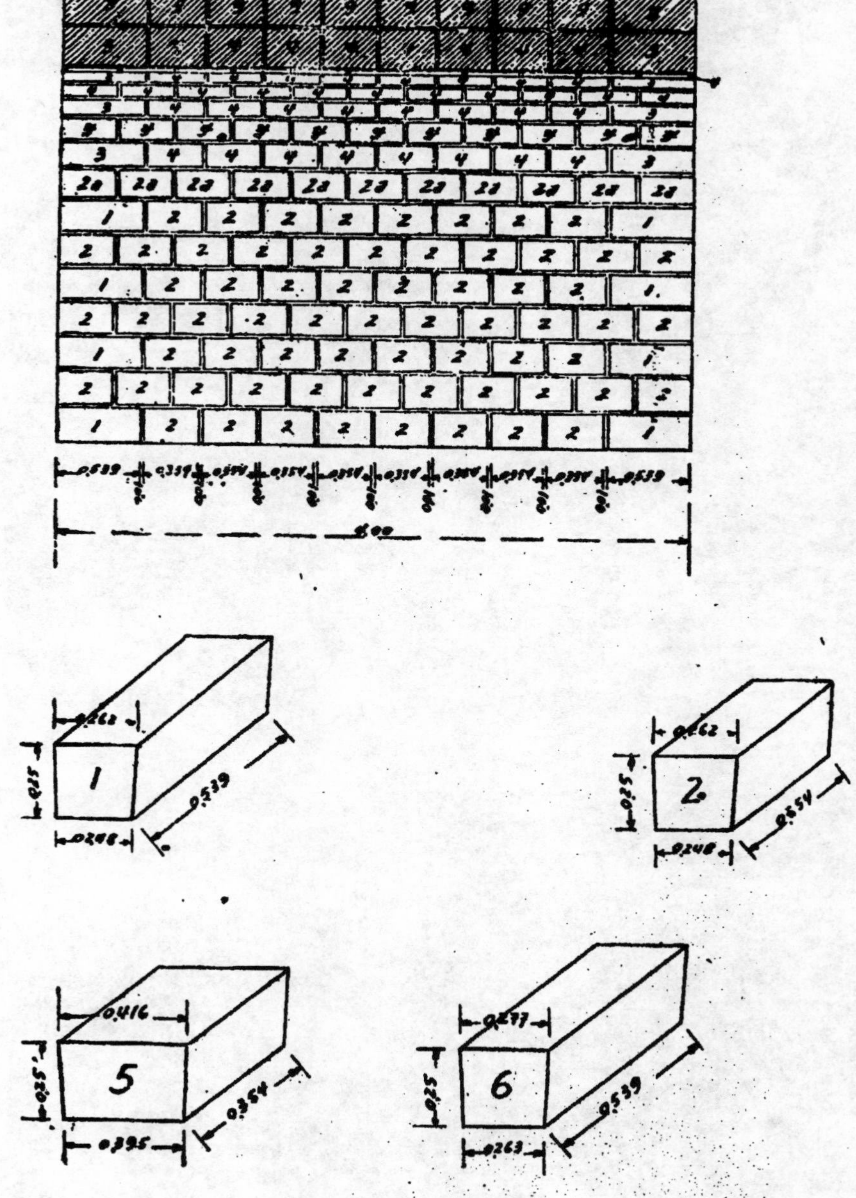
0.262
0.25
1
0.539
0.248
0.262
0.25
2
0.554
0.248
0.416
5
0.354
0.395
0.277
0.25
6
0.539
0.263

INVESTIGATION ON GRADE IMPROVEMNET TO PEIPING-MUKDEN RAILWAY

BY EVEREST T. F. LIU. (劉峻峯)

I. General Observations.

In Chinese Government Railways, Peiping-Mukden Line is classified as the first class line regarding her operating revenues as well as her equipments. Soon after the Tahushan-Tungliaohsien Branch Line (大通支路) was opened to traffic and had the newly built lines in the Manchuria as connecting lines in transportation, this line has increased its power of transporting goods and passengers from both South and North Manchuria. Consequently the volume of business in the line has increased from a total annual operating revenue of under $30,000,000 to $37,000,000 last year (1929).

Furthermore, the construction of Hulutao Harbor (葫蘆島港) which is going to start soon, the rendition and improvement to Ching Wang Tao Harbor (秦皇島港) and the project of constructing the Harbour of North China (北方大港) and the four more railway lines there-from (1 from Harbor joining P.M.R. at Chang Li (昌黎) for Manchuria; 1 from Harbor crossing P.M.R. at Tangshan (唐山) for the Mongolia. 1 from Harbor going south crossing P.M.R. at Tientsin joining C.T. railway to Sian joining L.H. railway; 1 from Harbor along the coast passing Chishan (岐山), Yenshan (鹽山) joining P.H. railway at Konkow (港口) to Hankow) all have additional increments to emphasize the importance of this line. In order to render this line to meet the changed condition in foresight, let me say something from the engineering point of view for her improvement.

When the volume of traffic on one line increases we have to get means to cope with it accordingly by either increasing the number of trains run, or the load on each train, or both. The number of trains, can be increased by sufficient rolling stock, adequent signaling, and double tracks. The way for increasing train load, is to use heavier locomotive engines, and secure better alignment of track, that is, betterment to alignment, curvature and gradient. Of course the maintenance of truck in good condition must be also counted for, but it is of only minor importance as compared with the others. All of these have their own independent effects but they are also closely interrelated. Doubtlessly the one showing prominent effect and most economical in the direction of operating expenses should be taken up first. We shall first take up the improvement to alignment.

Peiping-Mukden railway main line is 523.89 miles with the sharpest curve of 6° (2 more sharp curves of 5°-10′ & 4°-5′) and maximum grade of 1% (1 in 100) but the ruling grade is the $2\frac{1}{16}$ mile long haul of 0.7575% (1 in 132). Fortunately the foresaid 3 sharp curves are all located near the stopping stations and not on steep grades, while all the other curves in this line are flatter than 3° (2000′ radius). So curvature is not to be worried. As to the distance I have not much to say since neither the general plan at hand nor betterment survey is available for study. I am only possible to discuss the improvement to grade.

II. Effect of Existing Grade to Train Load and Speed.

Soon as we get sight of the present working profile of track of the whole line we are brought to understand that the gradient between C.Y.M. and T.N. is rather satisfactory; then the train goes through a rise and fall of about 147 ft. to L.S.Y. with the worst part at L.H. No sooner the train gets out of the Great Wall at S.H.K. than the undulating grade plays its part. The rise and fall from K.L.C. to C.S. of 120 ft. within $4\frac{1}{8}$ miles, from H.C. to H.C.K. of also about 120 ft. within $8\frac{15}{16}$ miles and from H.C.K. to L.S. of 90 ft. within $2\frac{5}{16}$ miles are the governing points for train load.

Now turn to the average speed diagram which was constructed on the same profile sheet, based upon the present working time schedude. Except for the limited speeds in running through the suburbs of Peiping, Tientsin and Mukden, the passenger speeds in passing L.H. K.L.C. and H.C.K. all are much affected by the heavy grades there, from the average through speed 31.20 m.p.h. to 21.04, 19.84 & 16.70 m.p.h. respectively. Same low speeds for freight train are clearly shown in the same spots.

Again, from the present working train loading table (see the Appendix) we further appreciate the effect of heavy grade in the S.H.K.—C.H. section on the reduced train load. Passenger train are running through the whole line, so it is not practicable to vary the train load by sections according to the grade conditions; therefore, the engine takes the same load all through, expect that in passing easy grade the surplus effort can be spared for accelerating the train. For freight trains, take G class engine for example, hauling 154 axles (1386 tons for the engine divisions C.Y.M.—T.N. T.N.—S.H.K.

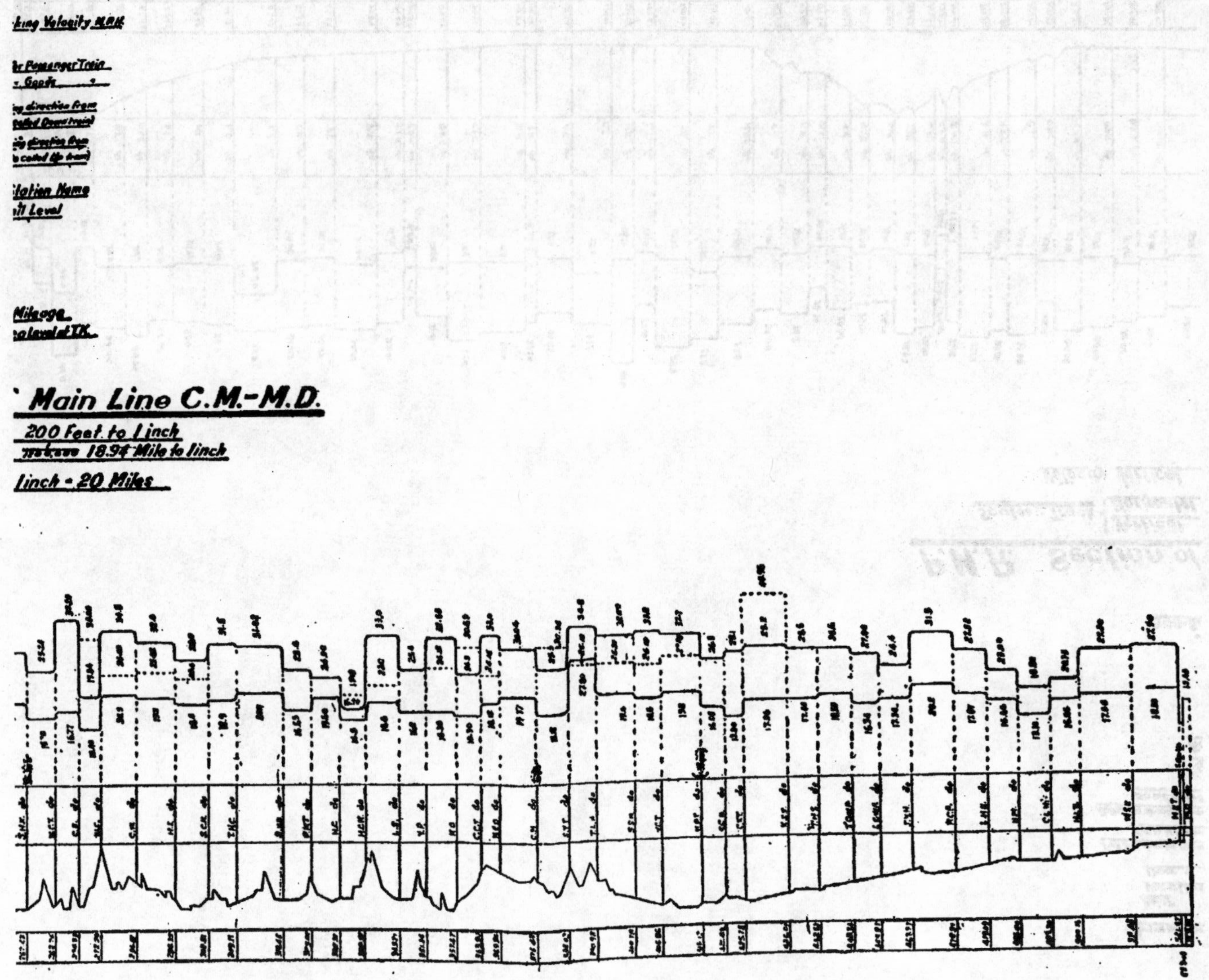
Station Name
Mileage
Main Line C.M.-M.D.
200 Feet to 1 inch
18.94 Mile to 1 inch
1 inch = 20 Miles

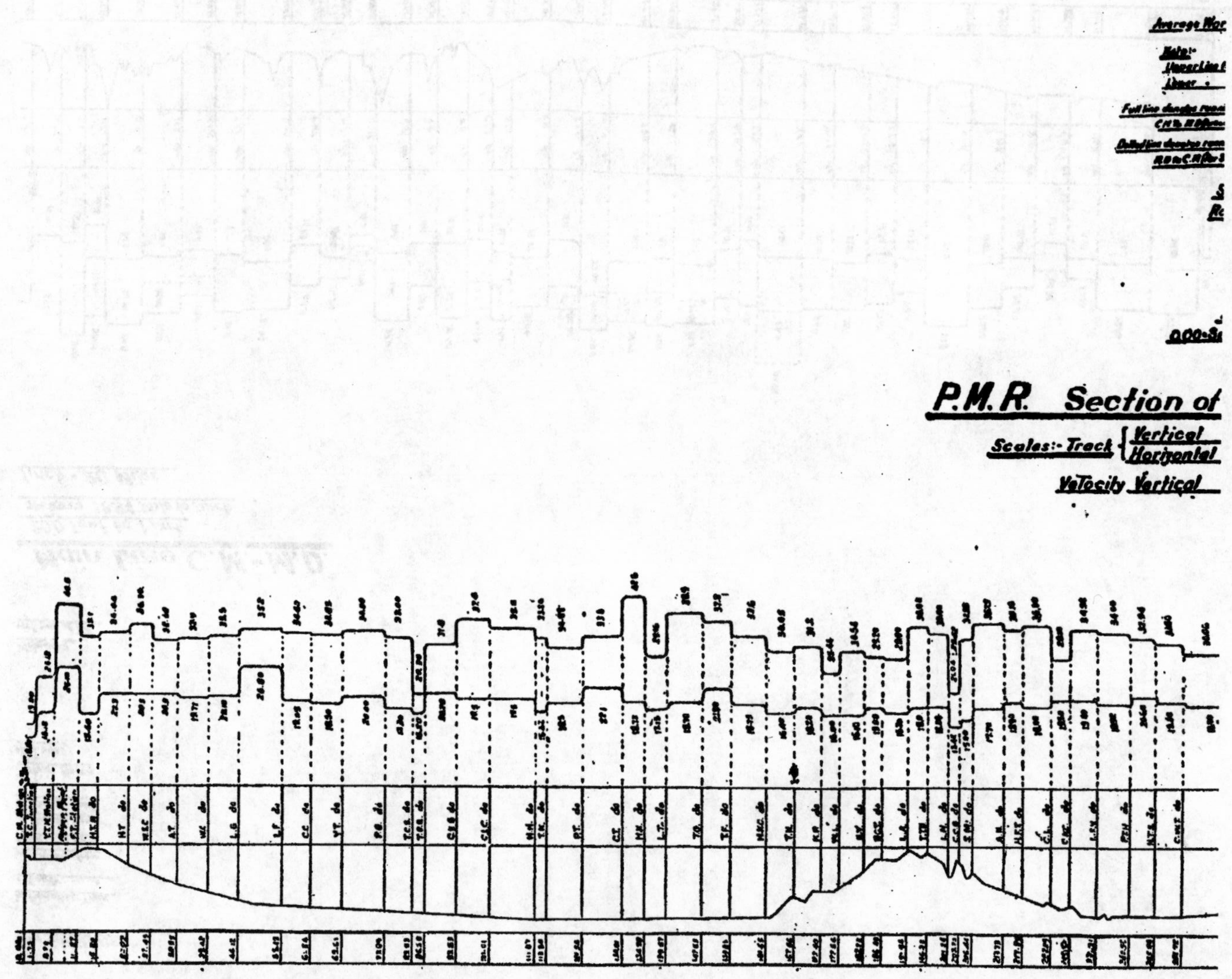
P.M.R. Section of
Scales:- Track { Vertical
Horizontal
Velocity Vertical

K.P.T.—M.D. but only 104 axles (936 tons) for S.H.K.—C.H. and 124 axles (1116 tons) for C.H.—K.P.T. We will then notice the extent to which the train load was checked by the said grades in S.H.K. and K.P.T. section. If the grade thereof is improved to get near the same equivalent gradient throughout, 40 more axles can be added to each through train. That is to say: with equal number of trains 38% more business may be done; or when goods and traffic remain unchanged, 1/3 of the present number of trains may be reduced for this section. Recently the Head Office in Tientsin has organized a committee for buying more engines to cope with the increased volume of traffic. The importance of increasing train load by improving the grade is made more evident here. Now let me go into the detail to see in what way the greatest economy may be secured.

III. Characteristics of Locomotives.

Before taking the consideration of grade improvement, we must know clearly about the working force of locomotives. In P.M.R. present equipment (see the appendix) there are 39 shunting engines, 20 class A, 45 class B, 3 Class C, 48 class D, 10 class E, 3 class F, 39 class G, 10 class H, and 24 class J engines, giving a total of 241 engines. Among these only A, D & E classes of engines are allowed to run on the branch lines (301.86 miles or 500.48 km. in total) laid with 60 lbs rails. Statistics shows that Chinese Government railways have on the average per 100 km. of line 15.4 locomotives, so that all the said branch lines only will require $15.4 \times \frac{500.48}{100} = 77$ pieces. The A, D and E, classes of engines (78 pieces) are therefore, not to be counted as the main line power. The C and E classes are small in number, so their service is to be neglected. Here the engines left for consideration are B class for mixed trains, H class fast passenger, and G and J classes freight trains. 17 out of the G class engines are specially assigned for K.M.A. coal use. From the sake of the larger numbers in use and the general tendency of applying heavier engines I select the typic engines of class H for passenger and J for freight, for a detailed study as follows:—

A. *Principal Dimensions*:—

Class	H	J
Whyte symol	4-6-2	2-8-2
Kind of service	Passenger	Freight

Date of service	1920	1918
Manufacturers	Baldwin	Baldwin
Rigid wheel base	11′-8″	15′-0″
Weight on drivers	175,224 lbs.	181,384 lbs.
Total engine and tender	295,878 lbs.	302,038 lbs.
Heating surface	2,362 sq. ft.	2,365 sq. ft.
Super-heating surface	508 sq. ft.	508 sq. ft.
Working pressure	180 lbs. per sq. in.	180 lbs. per sq. in.
Cylinders	20″ dia. × 26″ stroke	21″ dia. × 28″ stroke
Diameter of drivers	66″	54″

B. *Tractive efforts:—*

(1) Tractive efforts of adhesion:

H. $Ta = fw = \frac{1}{4} \times 175,224 = 43,806$ say 43,800 lbs. in summer

$Ta = fw = \frac{1}{5} \times 175,224 = 35,044.9$ say 35,000 lbs. in winter

J. $Ta = fw = \frac{1}{4} \times 181,384 = 45,346$ say 45,340 lbs. in summer

$Ta = fw = \frac{1}{5} \times 181,384 = 36,277$ say 36,280 lbs. in winter

(2) Cylinder tractive effort:

H. $$Tc = \frac{Pd^2l}{D} = \frac{180 \times .85 \times (20)^2 \times 26}{66} = 24,100 \text{ lbs.}$$

J. $$Tc = \frac{Pd^2l}{D} = \frac{180 \times .85 \times (21)^2 \times 28}{54} = 35,000 \text{ lbs.}$$

(3) (a) Critical speed:

H. $$Tbc = 24,100 = \frac{80\,H}{Sc}$$

$$Sc = \frac{80 \times 2,362 \times 1.6}{24,100} = 12.54 \text{ miles per hour.}$$

J. $$Tbc = 35,000 = \frac{80\,H}{Sc}$$

$$Sc = \frac{80 \times 2,365 \times 1.6}{3,5000} = 8.65 + \text{ miles per hour.}$$

(b) Normal speed:

H. $Sn = 12.54 \times 2 = 25.08$ m.p.h.

J. $Sn = 8.65 \times 2 = 17.30$ m.p.h.

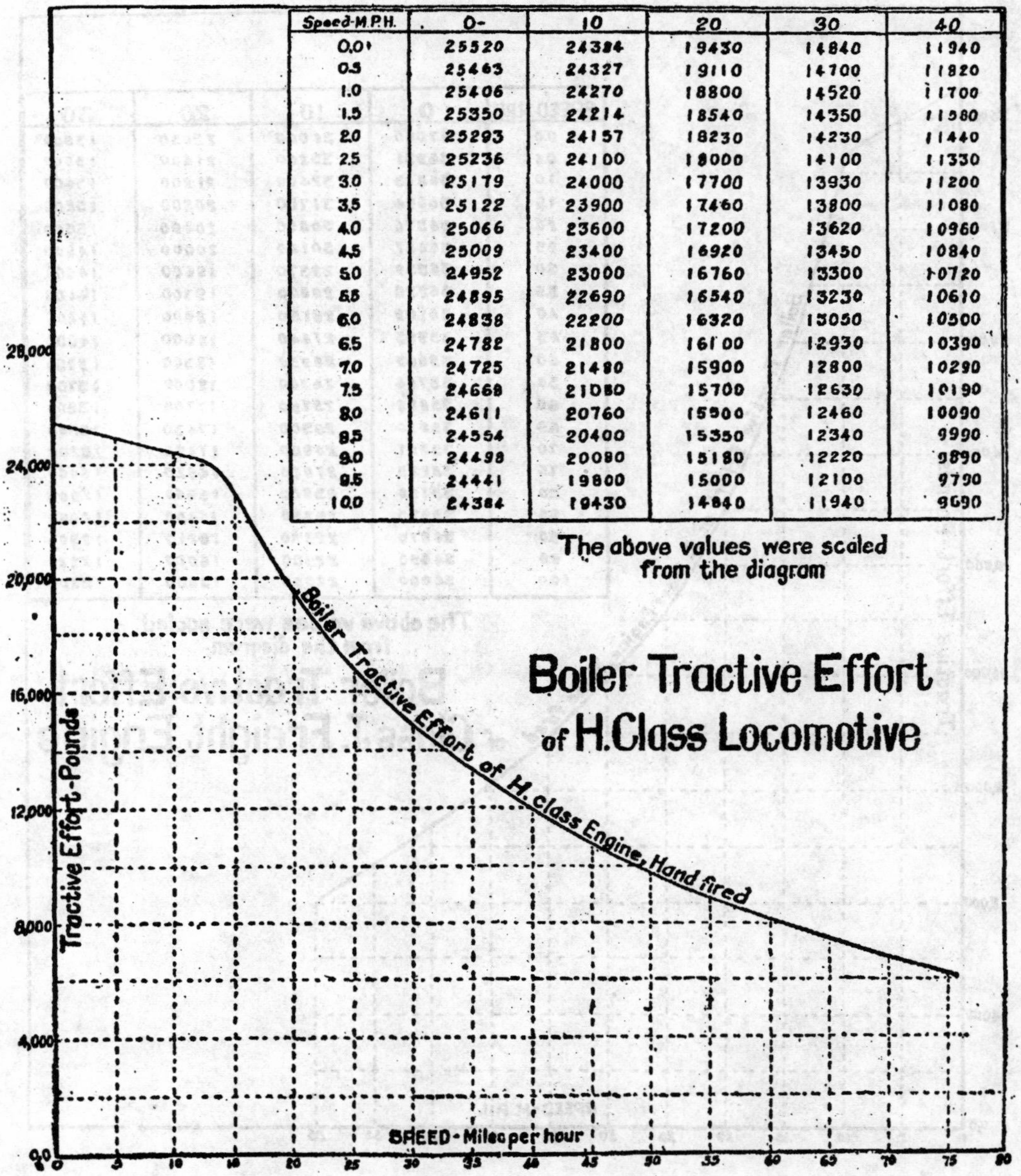

Speed-M.P.H.	0-	10	20	30	40
0.0	25520	24384	19430	14840	11940
0.5	25463	24327	19110	14700	11820
1.0	25406	24270	18800	14520	11700
1.5	25350	24214	18540	14350	11580
2.0	25293	24157	18230	14230	11440
2.5	25236	24100	18000	14100	11330
3.0	25179	24000	17700	13930	11200
3.5	25122	23900	17460	13800	11080
4.0	25066	23600	17200	13620	10960
4.5	25009	23400	16920	13450	10840
5.0	24952	23000	16760	13300	10720
5.5	24895	22690	16540	13230	10610
6.0	24838	22200	16320	13050	10500
6.5	24782	21800	16100	12930	10390
7.0	24725	21480	15900	12800	10290
7.5	24668	21080	15700	12630	10190
8.0	24611	20760	15900	12460	10090
8.5	24554	20400	15350	12340	9990
9.0	24498	20080	15180	12220	9890
9.5	24441	19800	15000	12100	9790
10.0	24384	19430	14840	11940	9690

The above values were scaled from the diagram

Boiler Tractive Effort of H.Class Locomotive

Boiler Tractive Effort of H.class Engine, Hand fired

Tractive Effort-Pounds

28,000
24,000
20,000
16,000
12,000
8,000
4,000
0.0

SPEED-Miles per hour

0 5 10 15 20 25 30 35 40 45 50 55 60 65 70 75 80

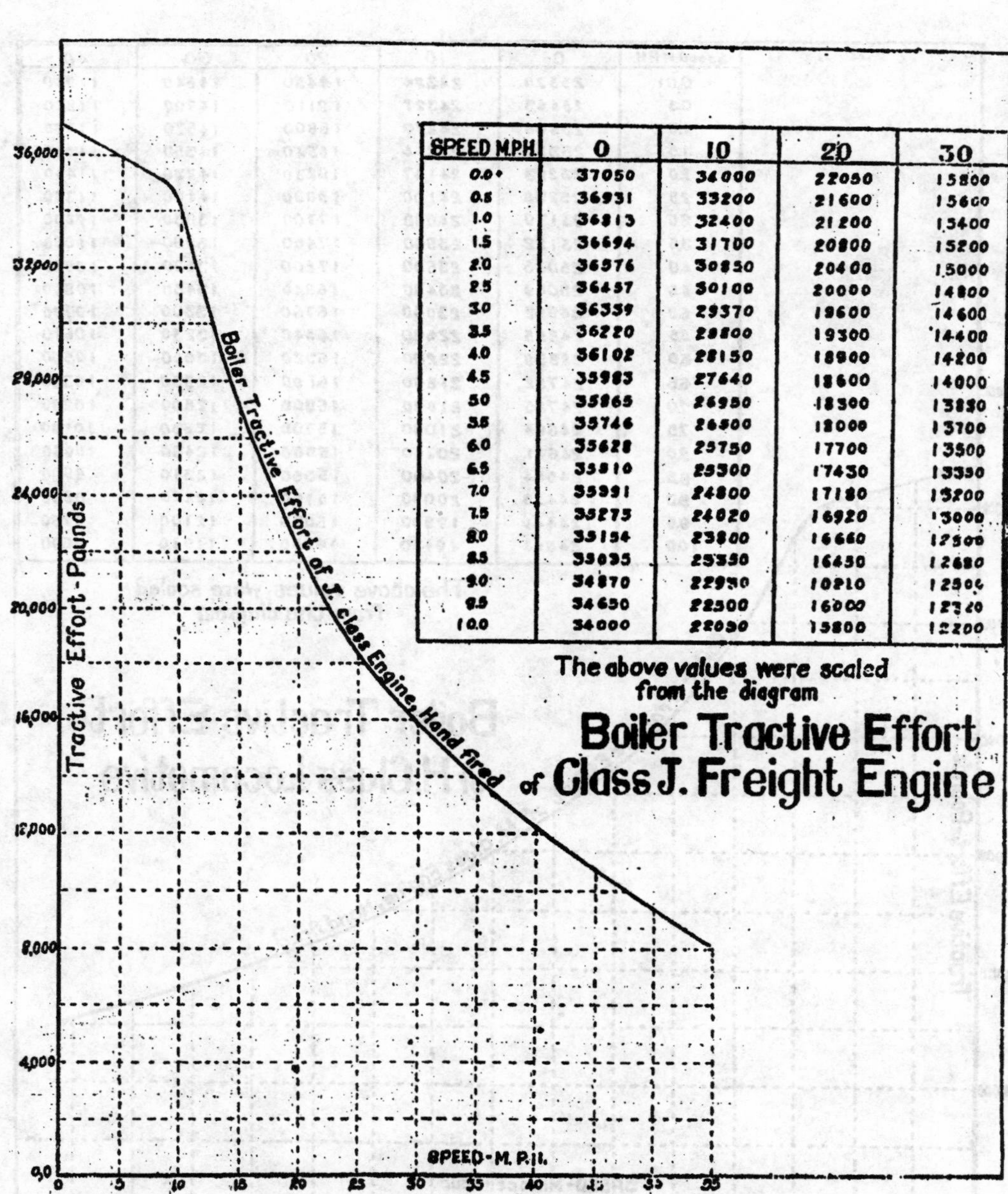

SPEED M.P.H.	0	10	20	30
0.0+	37050	34000	22050	15800
0.5	36931	33200	21600	15600
1.0	36813	32400	21200	15400
1.5	36694	31700	20800	15200
2.0	36576	30850	20400	15000
2.5	36457	30100	20000	14800
3.0	36339	29370	19600	14600
3.5	36220	28800	19300	14400
4.0	36102	28150	18900	14200
4.5	35983	27640	18600	14000
5.0	35865	26950	18300	13850
5.5	35746	26500	18000	13700
6.0	35628	25750	17700	13500
6.5	35510	25300	17430	13350
7.0	35391	24800	17180	13200
7.5	35273	24020	16920	13000
8.0	35154	23800	16660	12500
8.5	35035	25350	16450	12680
9.0	34870	22950	10210	12500
9.5	34650	22500	16000	12340
10.0	34000	22050	15800	12200

The above values were scaled from the diagram

(c) Boiler tractive effort at normal speed:

$$\text{H.} \quad Tbn = \frac{111H}{Sn} = \frac{111 \times 3,779}{25.08} = 16,721 \text{ lbs.}$$

$$\text{J.} \quad Tbn = \frac{111H}{Sn} = \frac{111 \times 3,784}{17.30} = 24,280 \text{ lbs.}$$

(d) Boiler tractive effort at twice normal speed:

$$\text{H.} \quad Tb^2n = \frac{128H}{2Sn} = \frac{128 \times 3779}{2 \times 25.08} = 9,643 \text{ lbs.}$$

$$\text{J.} \quad Tb^2n = \frac{128 \times 3,784}{34.60} = 14,000 \text{ lbs.}$$

(e) At starting or 0 + m.p.h the mean effective pressure may be 90%.

$$\text{H.} \quad Tc = \frac{180 \times .90 \times (20)^2 \times 26}{66} = 25,520 \text{ lbs.}$$

$$\text{J.} \quad Tc = \frac{180 \times .90 \times (21)^2 \times 28}{54} = 37,050 \text{ lbs.}$$

From the above datas boiler tractive effort diagrams were constructed and the values for each increment of ½ miles speed were scaled therefrom and tabulated.

C. *Maximum Loads the above engines can haul on the given grade at given speed:—*

TAKE SOME OF THE WORST GRADES AS SHOWN IN TABLE 1

TABLE I

Locality	Chianages	Grade			Sharpest curve on the grades			Remarks.
		%	length in feet	Ref. No.	Degree	Radius in feet	length. in feet	
K.L.C.	852187-862705	0.7519 $(\frac{1}{133})$	10,518	Z	2° 52′	2,000	1235 app.	
K.L.C.	864900-875850	0.7575 $(\frac{1}{132})$	10,750	Y	1° 55′	3,000	1340 app.	
H.C.K.	1177600-1189719	0.7575 $(\frac{1}{132})$	12,119	X	1° 9′	5,000	1645 app.	
H.C.K.	1162500-1163600	1.0 $(\frac{1}{100})$	1,100	W	1° 9′	5,000	678 app.	
Y. P.	1230200-1235800	0.8333 $(\frac{1}{120})$	5,600	V	1° 9′	5,000	663 app.	

In climbing over the ruling grades, the critical speed at which full cut-off of the engine can be maintained i.c.

8.7 miles per hour for freight train and

12.5 miles per hour for passenger train may be used for calculations here, but it is perhaps wise to use the greater speed to give a safety factor, so take

10 m.p.h. for freight trains and
15 m.p.h. for passenger trains.

Furthermore, the curve resistances on the previous grades were not properly compensated during construction so they will further increase the grades by the equivalent amount:

0.04% for 1—degree curve
0.06% for 2—degree curve
0.09% for 3—degree curve

The curve length for W. and V. grades are not long enough to hold the whole train as the ordinary freight trains are all more than 700 ft. long, but taking only 3/4 of them into consideration. Hence we have the new cquivalent grades.

Z	Y	X	W	V
0.842%	0.818%	0.798%	1.03%	0.863%

Hence

J. Re (Resistance due to Engine) $=6.0+0.0035\ (S-10)^2 = 6.0+$ lbs per ton.

Rt (Resistance due to Train) $=3.5+0.00555\ S^2+\frac{16}{(s-1)}=4.2$ lbs. per ton

Rg (Resistance due to Grade) $=20\times .842=16.84$ lbs. per ton for Z
$=20\times .818=16.36$ lbs. per ton for Y
$=20\times .798=15.96$ lbs. per ton for X
$=20\times 1.03=20.6$ lbs. per ton for W
$=20\times .863=17.26$ lbs. per ton for V

Net loads behind the tender are

$$\frac{34{,}000-(6+16.84)\times 151}{4.2+16.84}=\frac{30{,}551}{21.04}=1{,}450 \text{ tons for Z}$$

$$\frac{34{,}000-(6+16.36)\times 151}{4.2+16.36}=\frac{30{,}624}{20.56}=1{,}490 \text{ tons for Y}$$

$$\frac{34{,}000-(6+15.96)\times 151}{4.2+15.96}=\frac{30{,}684}{20.10}=1{,}525 \text{ tons for X}$$

$$\frac{34{,}000-(6+20.6)\times 151}{4.2+20.6}=\frac{29{,}983}{24.8}=1{,}210 \text{ tons for W}$$

$$\frac{34{,}000-(6+20.6)\times 151}{4.2+17.26}=\frac{30{,}487}{21.46}=1{,}420 \text{ tons for V}$$

Similarly we obtained the values for H class engine and tabulated as in Table II

TABLE II

Grade / Engine		Z	Y	X	W	V	Remarks
J		1,450	1,490	1,525	1,210	1,420	Short ton
		1,320	1,355	1,386	1,100	1,290	metric ton
H		906	930	950	750	885	short ton
		824	845	863	680	804	metric ton

From the above data we will note how our loading table in use was fixed. Between the S.H.K.—C.H. engine division the grade W (1%) gives the restriction for heavy loads. For the fast passenger trains a larger factor of safety is to be provided and about one-twelveth of the gross load should be reduced for winter condition. Hence 750 short ton or 680 metric tons should give a working figure of only 680-1/12×680=623 tons (metric), giving a difference=623-595 (the last figure from the working loading table)=28 tons, while for goods trains the difference is 1116 (from the working loading table)-1100=16 tons. The small discrepencies here are due to the different lowest speeds assumed in the two calculations.

D. *The ruling grades by practical experiences:—*

Regarding the loading the maximum grade 1/100 gives the restriction as shown in the last article, which is theoretically correct. From the practical experience of the engine drivers, however, the ruling grades for working trains are not the W grade 1% but as follows:—

(1) 1/150 grade at L.H. The down train running all down grade from T.T.U. with full momentum can stop at L.H. station by only applying brake with great care, other-wise it will surely over run the station.

(2) 1/150 grade between S.H.K. & W.C.T. In climbing the grade goods trains in some cases had to be broken into two portions; in running down, brakes must always be applied.

(3) 1/125 grade near C.S. station. Too great resistance for starting up trains, and equally too bad for down trains likely to over run the station. The 1/124 grade there are on the opposite side is of momentum grade, which is not very serious.

(4) C.S. station. This station is in a sag, having the heavy grades at both sides. Slight carelessness of the engine drivers of the up train will let the train pass over station. It is very easy to cause accidents.

(5) 1/133 & 1/132 grades at K.L.C. At these two grades, both up and down trains are very often, expecially in the winter season, brought to a stand still on the half way up, and the train had to be broken into two portions. It is a bad practice.

(6) 1/121 grade between T.H.C. and S.H.S. The grade sometimes causes trouble also but not a serious one.

(7) 1/132 grades near H.C.K. This is a rather long and heavy grade in P.N.R. Up trains are very likely brought to a full stop at the spot of the revered curves thereon. (although there is about 200 ft. tangent inserted into the revered curves) Train load is not uncommon to be cut into two portions if full axles on.

(8) 1/110 & 1/120 grades near Y.P. They hindered the train also very much, but the grades being not long, are not serious.

(9) 1/200 grade, near T.L.H. The grades are too near the T.L.H. station, so that sometimes down trains were very easy to pass over station when brakes were not applied in time.

On all the above grades, sand is always used on climbing up if weather is unfavorable and brakes applied when coming down.

But among the above 9 places we don't find the W grade 1/100 plays any hindrance although our train load is based upon that point. For reasons see the following details.

E. *Varied character of the engine in practice:—*

The maximum tractive effort of the engine boiler was calculated on the basis of good working conditions, that is the working pressure to be 180 lbs. per sq. in. and not less. Less pressure, less tractive effort. The usual working gressure is only 175 lbs. and never reaches 180 lbs. per sq. in. This alone will reduce 1/20 of the tractive efforts.

Again, the pressure depends upon the kind of fuel and the rate at which it is burned. When the train is climbing up the long grade the tractive effort is always at its maximum, that means to say the heating elements should be always at their greatest efficiency. A little under fired, the pressure will drop and the tractive effort will be reduced, especially when the engine has rendered enough engine mileage (about 1500 miles for S.H.K.—C.H. section at present conditions) and her boiler will undergo washing out. As the train on up grade is about to stop, one can see the exhaust steam out of the smoke stack not dry vapors, but small water drops, which proves that the steam is not superheated high enough, and that is because the heating efficiency is poor.

Furthermore, the average life of locomotives is about 23 years if properly maintained. Most of our locos are more than 20 years and only a few have been 10 years services. The abuse during military transportations, in addition, shortens their life. Most of the locomotive should have had been out of service long ago.

Finally, the efficiency of coal combustion is the greater soon after the ashes are cleaned out. So if the grade comes near the water station where the ashes are being cleaned, it can be climbed up easier. The 1/100 grade therefore does not give too much troıble as the long 1/132 grade in the opposite side at H.C.K. since the down train has her ashes cleaned at H. C. water station and the grade is only 1,100ft. long.

Weather is also a very important factor, dew, small rainning, and frost on rails will decrease the tractive effort of adhesion very much.

To verify the foregoing, please see the following actual records (Table III):—

Table III. *Train Accident Records for Engine Divisions T.N.-S.H.K. and S.H.K.-C.H. Taken from L.I., S.H.K.'s Office.*

Date	Weather	Locality		Loco No.	Trains	Caused delay to train	Nature of accident	Cause of accident	Remarks given by L. I., S.H.K. Mr. Y.K. Chu.
		m.p. East of T. K.	Between Stations						
5/5/29	stormy	123.25 118.75	Y.P. and L.S. L.S. and H.C.K.	218 (J)	Upfreight	1.54 hr.	Train load divided for 2 portions	Slippy rail on Grade	This train of 115 loaded axles.
14/5/29	fine	46.25	L.H. and C.C.G.	152 (B)	Dn „	5.19 „		Train over loaded	S. M./L. H. to be responsible for this.
17/5/29	„	122.75 86.00	L.S. and Y.P. C.S. and K.L.C.	129 (B)	„ „	1.43 „		Loco under steamed	The steam pressure is only 120 lbs./sq. in. although 180 lbs./sq. in. is to be maintained.
2/6/29	„	86.00	C.S. and K.L.C.	232 (B)	„ „	2.15 „	do	Heavy loading	I consider too much load on train.
16/6/29	„	83.00	C.S. and W.C.T.	206 (J)	Up „	0.57 „	do	Heavy train and rails slippy.	
19/6/29	„	123.00	L.S. and Y.P.	P. S335	Dn „	1.33 „	do	Heavy train and grade	
21/7/29	Rainy	123.25 118.50 87.50 83.00	Y.P. and L.S. L.S. and H.C.K. C.W. and K.L.C. C.S. and W.C.T.	218 (J)	Up „	7.00 „	do	do	The train is fully loaded with provisions and is running on slippy and gradient rails.
1/8/29	Under dark	117.75 122.75	H.C. and H.C.K. L.S. and Y.P.	152 (B)	Dn „	5.00 „	do	do	40 tonners loaded with cement are heavy for Loco 152 to climb up over the gradient section.
17/9/29	„	125.75	Y.P. and K.O.	167 (G)	Dn spl.	1.30 „	do	Loco under steamed	The gradient between Y. P. and K. O. is not so heavy as any other sections between S. H. K. and C. H. So consider driver failed to make full steam during the journey.

TABLE III. *(Coutinued)*

Date	Weather	Locality		Loco No.	Trains	Caused delay to train	Nature of accident	Caused of accident	Remarks given by L.I., S.H.K. Mr. Y.K. Chu.
		m.p. East of T. K.	Between Stations						
25/10/29	Under dark	85.75	C.S. and K.L.C.	203 (J)	Dn M spl.	3.10 hr.	Train load divided for 2 portions	coupler bolt broken	Driver concerned is not to be blamed as train was too heavy and bad weather.
10/11/29	fine	86.75	K.L.C. and C.W.	130 (B)	Upfreight	4 00 „	do	do	Driver is not to be blamed as train is too heavy.
12/11/29	Under dark	118.00	L S. and H.C.K.	129 (B)	„ „	0 50 „	do	heavy load	
14/11/29	fine	86.75 83.00	C.W. and K.L C. C.S. and W.C.T.	130 (B)	„ „	2 20 „	do	do	
14/11/29	Under dark	83.00	C.S. and W.C.T.	167 (G)	„ „	0 30 „	do	do	
22/11/29	„	118.00	L.S. to H.C.K.	152 (B)	„ „	0.50 „	do	do	
29/12/29	„	118.00	L.S. to H.C.K.	176	„ spl.	2.00 „	do	Loco under steamed	Engine under steamed due to the leakage of the fire box.

N O T E : The above was compiled from Locomotive Inspector Mr. Y. K. Chu's file; before his coming to this job, there was no such data available. His kindness for his contribution is much appreciated by the writer.

R E M A R K S : The above should be counted only as half the chance of occurrances for the said sections, while the other half probabilities are to be found from the L. I.'s files at T. N. and C. H. Offices since the engine drivers report to their own superiors only, although they have equal chances to run on the T. N.-S. H. K. and S. H. K.-C. H. divisions respectively.

The data in the records shows that the Max. number of trains broken loose in one day is 2, and the longest delay to one train is 7 hours. Besides the accident at L.H. due to the fault of S.M., which should be excluded from our discussion, all the rest should be counted against the grade problem and partly the engine side.

After multiplying the above frequency of breaking loose the train load on passing the heavy grades mentioned therein by 2 as remarked (L.H. is excluded), we obtain the following summary with the good probability.

8 times between C.S. & G.W.T. at ch. 83.00 with 1/124 grade.
12 „ at K.L.C. between chs. 85.75-87.50 with 1/133 & 1/132 grade.
12 „ at H.C.K. between chs. 117.75-118.00 with 1/110 and 1/132 grade.
10 „ Near Y.P. at ch. 123 with 1/110 grade.
2 „ between Y.P. & K.O. at ch. 125.75 with 1/130 grade.
44 times in total for a period of 8 months, and the time delayed to trains 2 × 35 hr. 32m. =71 hr. 4m.

In the average the train load had to be broken loose once in every 5.57 days and delayed 1 hr. 40 m. each time. From these we are brought to understand without any further explaination how the heavy grades affect the train load, traffic and the running expenses. Of course, a lot of other kind of accidents which are also most likely to have on the heavy grades than any other part of line adds more to the operating expenditures.

In conclusion, the grades should be improved without hesitation, although the fore-said more concerns the engine problem than the grade.

IV. Limit for Grade Improvement.

Our principal purpose and greatest economical consideration for the grade improvement is not only such as to eliminate the present difficulty from train running as stated in the last chapter but also to have the hindrance which reduces the train load for the inside wall sections so as to have the same load thoughout the line. In such a case, our freight train loading must not be less than 154 axles, that is 1,386 metric tons. Based upon this load, we shall see to what extent the poor grades should be reduced.

Net load 1,386 tons (metric) or 1,525 tons (short)

Taking 1/12 of the load as the factor of safety for winter we have the summer loading 1,525+127=1,652 tons.

Therefore

$$\frac{34,000-(6+G)\times 151}{4.2+G}=1652$$ Where G is resistance force due to grade.

$$34,000-906-151G=4.2\times 1,652+1,652\ G$$

$$1,803G=26,156$$

or $G=26,156/1803=14.5$ lbs. per ton of load.

The maximum grade should be $\frac{14.5}{20}=0.725\%$ when the curves properly compensated. For our case, we take up the 3-degree curve compensation and the maximum grade must be below

$$0.725-0.09=0.635\%\ \text{or}\ 1/158\%\ \text{say}\ 1/160\ \text{or}\ 0.625\%$$

For this grade, we have the net load behind tender for H class engine, passenger train,

$$\frac{23,000-(6.1+12.5+0.9)\times 148}{4.81+(12.5+0.9)}=\frac{20,114}{18.21}=1,100\ \text{short tons or}\ 1,000\ \text{metric tons.}$$

Taking 1/12 load for the safety factor for winter load we get the final passenger train load, =917 tons (metric).

This limited grade is only correct by theory. By practice, however, we should take at least the 1/200 or 0.5% as the ruling grade. But due to the better rolling stock and the immensely increased volume of traffic it will be deemed wise and economical to have the ruling grade further cut down to 1/250 or 0.40% as the topography in the fore-said spots are not inherently difficult. Considering the P.N.R. as a whole I grouped the poor grades to be improved as Table IV;—

TABLE IV

Grade		Length		% to whole line of P. N. R. (524 miles)	Remarks.
1 in	%	in ft.	in miles		
100-109		1100			
110-119		7861			
120-129		28096			
130-139		49588			
140-149		7197			
150-159		22360			
Total		116202	22.00	4.20%	for 1/160 ruling grade.
160-169		12131			
170-179		12501			
180-189		7800			
190-199		19055			
Total....		51577	9.77	1.86%	for 1/200 ruling grade.
200-209		68348			
210-219		10300			
220-229		0			
230-239		14817			
240-249		3464			
Total.....		96129	18.25	3.48%	for 1/250 ruling grade.
Grand total			50.02	9.54%	

From the table IV we get,

1st. scheme with ruling grade reducted 1/250 or 0.4% with 9.54% or 50.02 miles of the track to be improved.

2nd. scheme with ruling grade reduced to 1/200 or 0.5% with 6.06% or 31.77 miles of the track to be improved, and

3rd. scheme with ruling grade reduced to 1/160 or .625% with 4.20% or 22.00 miles of the track to be improved,

V. Estimated Cost of the Improvement

(1) *Existing Cut*:—A glanee over the existing track profile shows that only a small part of the line was in cutting while most of the line is in filling. The result was poor grade between S.H.K. and T.L.H. The total length of all cuttings between S.H.K. and T.L.H. (678,816 ft. apart) the most hilly counting in P.N.R, amounts to only 69300 ft. (10.2%) with a progressive average height of cut 7.4 ft. although once a maximum height of cut of 39.36 ft. was disposed of at the mile post 123 near Y.P., while at K.L.C. a maximum cut of 23.97 ft. and at H.C.K. of 25.67 ft. but all of only very short lengths. The general details of cuts as Table V:—

TABLE V

Depth of cut	length of cut
0 ft. to 5.0 ft. inclusive	29,200 ft.
5.01 ft. to 10.0 ft. inclusive	23,900 ft.
10.01 to 14.0 ft. inclusive	13,400 ft.
20 ft. sharp	2,800 ft.
Total	69,300 ft.

This further proves that the topography of this country is not inherenty difficult for grade improvement.

(2) *The New Cut*:—From he existing working profile, a rough sketch was drawn, giving the following table VI for the height of new cut., the length of track affected thence and the corresponding rise and fall reduced for the 3 schemes of different ruling grades:—

The lengths affected in the above table for grade improvements were a little different from those in the last chapter and is due to the fact that some of the grade too difficult for improvement was omitted and some other parts affects a longer length for the sake of smoother grade.

(3) *Earthwork*:—1/160 ruling grade, assuming 35% to be in cut to the existing cut; 65% to be in cut to the existing filling.

TABLE VI

Locality near the mile post east of T. K.	1/160 grade			1/200 grade			1/250 grade		
	Max. depth of cut in ft.	length of track effected in ft.	Rise & fall rebured in ft.	Max. Cut.	Length	Rise and Fall	Max. Cut.	Length	Rise and Fall
79	3.01	6,300	0.0	7 2	7,640	7.2c	17.12	18,770	17.12c
81	4.32	3,829	1.47b	9 87	4,829	5.72b	13.22	8,282	9.32b
83	12.20	12,749	12.00c	20.27	12,749	19.00c	26.92	12,749	25.50c
84	2.2	5,760	0.0	9.0	6,800	3.61b	14.4	8,750	3.61b
86-87	13.05	23,463	11.70c	26.45	23,463	25.00c	37.30	23,463	35.50c
89				3.38	5,760	0.0	7.68	7,800	4.00b
91				3.98	8,094	3.98	8.11	8,094	8.11b
94				3.75	8,200	3.75b	7.55	8,200	7.55b
101							1.4	2,750	0.0
105	9.15	5,200	3.00c						
	4.25	8,000	0.0	14.90	14,600	18.40c	19 5	14,600	19.50c
110				6.82	10,950	6.82b	10.00	12,500	10.00b
116-118	1.00	4,600	0.0	6 86	5,976	0.0	1 .46	7,465	0.0
	12.05	11,589	12.05c	21.70	11,589	21.70c	29.50	11,589	29.50c
	14.73	15,630	14.73c	29.93	15,630	29.93c	42.03	15,630	42.03c
123	14.25	13,695	13.00c	24.33	15,644	22.30c	32.40	15,644	32.00c
126	5.95	8,467	5.95b	11.66	8,467	11.60b	15.14	8,467	15.14b
144							10.4	20,200	10.4b
145LH	2.00F	9,400	2.0b	7.0F	10,650	7.0b	11.2F	12,700	11.2b
Total	98.19	128,682 24.4 m	75.90	207.07	171,041 32.4 m	181.01	315,33	217 653 41.1 m	230.48
	Progressive average height for the length 12,868 ft. is 6.34 ft.			Progressive average height for the length 171,041 ft. is 8.00 ft.			Progressive average Height for the length 217,653 ft. is 9.95 say 10.00 ft.		

We have cut in cut $128,682 \times .35 \times \frac{h}{2}\ (6H + 3h + 50)$

$= 128,682 \times .35 \times \frac{6.34}{2}\ (6 \times 7.4 + 3 \times 6.34 + 50)$

$= 45,036 \times 3.17 \times 113.42 = 16,200,000$ cu. ft. $= 147,500$ fg. and cut in fill

$128,682 \times .65 \times \frac{3h + 40}{2}\ h = 128,682 \times .65\ (\frac{3 \times 6.34 + 40}{2} \times 6.34)$

$= 83,646 \times 29.51 \times 6.34 = 15,700,000$ cu. ft. $= 142,600$ fg.

Total E. W. $= 290,100$ fg.

The cost for E. W. $= 290,100$ @$0.50 = $145,000 1/200 ruling grade assuming 45% to be in cut to the existing cut; and 55% to be in cut to the existing filling.

We have cut in cut $\frac{171,041 \times .45 \times 8}{2} \times (6 \times 7.4 + 3 \times 8 + 50)$

$= 76,968 \times \quad \times 118.4 = 36,400,000$ cu. ft. $= 351,000$ fg. and cut in fill

$171,041 \quad .55 \times \frac{3 \times 8 + 40}{2} \times 8$

$= 94,073 \times 32 \times 8 = 24,080,000$ cu. ft. $= 219,000$ fg.

Total E. W. $= 550,000$ fg.

The cost for E. W. $= 550,000$ @$0.50 = $275,000 1/250 ruling grade assuming 65% for cut in cut.

and 35% for cut in fill.

We have cut in cut $217,653 \times .65 \times \frac{10}{2} (6 \times 7.4 + 3 \times 10 + 50)$

$= 141,474 \times 5 \times 124.4 = 91,000,000$ cu. ft. $= 826,000$ fg.

cut in fill $217,653 \times 0.35 \times \frac{3 \times 10 + 40}{2} \times 10$

$= 76,179 \times 350 = 26,630,000$ cu. ft. $= 243,000$ fg.

Uotal E. W. $= 1,069,000$ fg.

The cost 1,069,000 fg. @0.50 = $534,500

(4) *Bridges Affected*:—Due to the fact of raising or lowering tracks, some of the bridges are affected, requiring to be raised, lowered, or even reconstructed. Some parts, however, cannot be altered without giving due considerations to the clearance of waterway. From drawing I put down the following as a reference:—

1/160 ruling grade

1,122 ft. of girder bridge openings effected. The cost for same say 33.3% of the general average cost $150.00 per ft. run=$50 per ft. run.

240 ft. of covered drains effected with an average cost of $135.00 per ft. run for the re-construction expenses since the existing ones can no longer be serviceable soon as the track lowered below or too near the top of barrel.

Therefore the total cost=1,122 × 50+240 × 135=$88,500 1/200 ruling grade.

4,141 ft. of girder Br. (incl. 2060 ft. of L.H. Br.) and 299 ft. of covered drains effected.

Similarly we got the total cost=4,141 × 50+299 × 135=247,415

1/250 ruling grade.

4,583 ft. of girder bridges (incl. 2,060 ft. of L.H. Br.) and 318 ft. of covered drains effected.

The total Cost=4,583 × 50+318 × 135=$272,080.

(5) *Stations Effected*:—When track lowered or raised, of course the station in question must be changed also to meet the new positions. It would be so much the better if station site can be removed to a better position instead of sticking to the original places. Among the stations below L.H. is of the 1st. consideration as there are more business, and the double track construction of Lan Ho Br. there. It is only necessary to alter the old site to meet the requirements after grade improvement. So just put down the following rough figures for the expenditures:—

1/160 ruling grade.

Station	Change	Cost
K. L. C.	12 ft. lowered	say $ 40,000
H. C. K.	12 ft. lowered	

1/200 ruling grade.

Station	Change	Cost
L. H.	7 ft. raised	say $100,000
K. L. C.	25.5 ft. lowered	
P. M. T.	6.8 ft. lowered	
H. C. K	21.7 ft. lowered	

1/250 ruling grade.

Station	Change	Cost
L. H.	11.2 ft. raised	say $250,000
K. L. C.	36.4 ft. lowered	
P. M. T.	10 ft. lowered	
H. C. K.	29.5 ft. lowered	

The **TOTAL COST** for the different schemes was tabulated as shown in Table VII:—

TABLE VII

	1st scheme 1/250 grade.	2nd. scheme 1/200 grade	3rd. scheme 1/160 grade.
E. W.	$ 584,500	$275,000	$145,000
Bridge.	272,080	247,415	88,000
Station	250,000	100,000	40,000
Total . .	$1,056,580	622,415	273,000
Plus 25% for the mis work	264,145	155,604	68,250
Grand Total.....	$1,320,725	$778,019	$341,250

VI. Saving after Grade Improvement.

(1) *Saving due to the reduction of ruling grade*:—

The existing ruling grade is 1/100 or 1% for the division from S.H.K. to T.L.H. of 128.56 miles long; or 207. km. using H class engines for passenger & J for goods service as stated in chapter III. Let the new ruling grades to be 1/160 (0.625%), 1/200 (0.50%) & 1/250 (0.40%)

Again let the min. speeds on the ruling grades 1/100, 1/160 & so on to be

10 miles per hr. & 15 miles per hr.

For goods and passenger train respectively we have the loadings behind the tender and excluding the brake van 25 tons as Table VIII:—

TABLE VIII

Grade	Passenger	Goods	Rank
1/100	570	1,091	Metric tons
1/160	892	1,359	„ „
1/200	983	1,595	„ „
1/250	1158	1,716	„ „

One train-kilometer cost in this line by the data of 1923 is $2.86 since no other reliable data are obtainable during these later 6 years and the value is used for the following estimate.

The present daily trains between S.H.K. & T.L.H. are 4 fast passengers, 2 mixed trains and 4 goods trains. We put 1 mixed trains into passenger and 1 in goods.

So the present equivalent No. of daily trains, is 5 passenger and 5 goods. Therefore there may be only

$\frac{570}{893}\times 5=3.2$ daily passenger trains and

$\frac{1,001}{1,359}\times 5=4.02$ daily goods trains on 1/160 grade;

$\frac{570}{983}\times 5=2.90$ daily passenger trains and

$\frac{1,091}{1,595}\times 5=3.42$ daily goods trains on 1/200 grade;

$\frac{570}{1,159}\times 5=2.50$ daily passenger trains and

$\frac{1,001}{1,7.6}\times 5=3\ 20$ daily goods train on 1/250 grads

The saving is

10—(4.02+ 3/2)=2.78 daily trains on 1/160 grade
10—(2.50+3.42)=3.68 „ „ „ 1/200 „
10—(2.50+3.20)=4.30 „ „ „ 1/250 „

Or the daily saving in train kilometers are

207.00 × 2.78 × 2 = 1,150 on 1/160 grade,
207.00 × 3.68 × 2 = 1,522 on 1/200 „ ,
207.00 × 4.30 × 2= 1,780 on 1/250 „ .

The running expenses, however, due to the less number of daily trains will not be affected as a whole since the number of traffic staffs must be kept the same; the track maintenance will be only little saved due to the decreased density of traffic, etc. It is estimated that only 43% of the above estimated cost of these train-kilometer can be saved, or 0.43 × 2.86=$1.23 per train kilemeter.

The annual saving may therefore appear to be

1,150 × 365 × $1.23 = $516,000 on 1/160 grade,
1,522 × 365 × $1.23 = $685,000 on 1/200 „ ,
1,780 × 365 × $1.23 = $800,000 on 1/250 „ .

But on account of heavier train run, there will be additional loss due to the mechanical wear to rolling stocks and rails. More labour will be required to maintain the track once worn down. The average higher cost of per train kilometer and portion of time lost on the revised lines and etc. may to taken as 30%. Then we have the final annual saving due to reducing the ruling grade as follows:

$510,000 × (1-30%) = $361,200 on 1/160 grade
$685,000 × 70% = $479,500 on 1/200 „
$800,000 × 70% = $460,000 on 1/250 „

(2) *Saving due to reduced rise and fall:*—

The rise & fall in vertical feet are affected per table of last chapter as follows:

TABLE IX

Grade	Class A.	Class B.	Class C.
1/160	0.0	9.42	66.48
1/600	0.0	38.80	142.14
1/250	0.0	75.72	204.76

Take the cost per year per daily train (round trip) of a foot of rise and fall estimated by Dr. Wellington on minor gradients to be

$0.84 × 2.86 = $2.40 for B. class,
$2.67 × 2.86 = $7.64 for C. class.

Therefore annual saving due to reduced rise and fall for 10 daily trains is

1/160 10×(9.42 × 2.40 + 66.48 × $7.64) = $ 5,300
1/200 10×(38.87 × 2.49 + 142.14 × $7.64) = $11,780
1/250 10×(75.72 × 2.40 + 204.76 × $7.64) = $17,440

Therefore the total annual saving due to this improvement is

$366,500 for 1/160 ruling grade,
$491,280 for 1/200 „ „
$577,440 for 1/250 „ „

This saving is on present traffic only; when the number of daily trains is doubled, the amount of annual saving will be probably doubled also. How much of the total saving on the running expenses in the long run can not be accurately estimated.

VII. Procedure of Carrying out the Improvement and some Improtant Points for Construction.

From the last two chapters on the first cost for improvement and annual saving there from, it is evident that the first scheme to have the present ruling grade reduced to 1/250 is the most economical although the first cost requirements are the greatest. We shall now discuss the procedure in detail.

(1) *Betterment survey*:—What I have discused so long was based upon the profile only. In actual work of course, some difficulty may be met with. A betterment survey is absolutely necessary so as to detect the difficult points and better deviation line for the grade.

(2) *Carefully study and design for the grade line*:—The minor details for the grade improvement, taking the line as a whole, will reguire a careful study and layout so as to bring out once for all the economical and most easy riding track such as to eliminate the sharp curves, heavy grades, dangerous sags, inserting the long easy vertical curves, considerations to those spots where accidents have occured the most, etc.

(3) *Co-operation with the traffic and locomotive departments*:—For this kind of work it has a very close connection with the other two departments. Although engineering department constructs the track, the traffic and engine departments really profits by or suffers from the actual conditions of good or poor track. We should have good connections with them so as to get the necessary informations for those they suffered so long. Practical experience is a very good guide.

(4) *Double tracking*:—For the grade improvement, tracks should not be disturbed too much. When the work is going on, the track can be

shifted side to side as the writer has done for the grade improvement at Ch. 4,700 on T.T.B.L. But this may not be perfectly safe with the main line of dense traffic. Our District Engineer C.Y. Liang well said that since sooner or later the double track to the out-side wall section will be necessary, viewing from the increased volume of traffic, it is wise for grade improvement construction work to-day to construct the double track first so as not to disturb the main track at all. This is a very good suggestion, as the work is rendered easier and whatever money spent this time will be utilized later.

(5) *Widening the Cuttings*:—Cutting is a source of trouble in track maintenance. So the common engineering practice is to avoid it as much as possible. The snow and sand find their favorable recess in the cuttings. The last winter snow fall drifted to the cuttings, blocked up all trains for almost two days on the main line. The administration is just trying to make snow fence and apply snow plow for its removal. As the cuttings are much increased in depth, some more trouble will happen. This may be a strong reason for not improving work started. This, however, need not be wooried, as can be avoided by windening the cuttings. The sand trouble in T.T.B.L. is notorious at the spring season, but it is now much reduced and scarcely so serious as to block trains, although trains usually get to stop for one or two days before last year (1929). The cuttings there were widened to 50ft. to 65ft. from center of track in the windward side of the general strong wind direction while 15ft. to 25ft. in the lee. The last snow storm in T.T.B.L. was more severe than that at main line, but did not block up the windened cutting in the former, which proves the above contention.

VIII. Conclusion

Engineers who originally projected the line tried hard to save the first cost, so they avoided to cut the hilly country as little as possible, although the countries passed through were not inherently difficult. They also were impressed too much by the trouble brought by the cuttings that in some parts the high fillings would be required to meet the grade of shallow cuttings.

Railways in foreign countries have had the tendency to make certain improvements in track lines by reducing the ruling grades to 1/500 or 0.2%. Our first scheme with ruling grade reduced to 1/250 or 0.4% is evidently not too much. Besides, in the long run it is the most economical scheme.

Mr. Mentel Adviser to the Ministry of Railways, remarked that all the improvement to P.M.R. should have had been taken up 20 years ago. How can it be put off any further now? Furthermore, the work of Hulutao Harbour is going to be started and soon the volume of business will be surely immensely increased after the completion of harbour. We should lose no time to effect the improvement.

The first scheme with the ruling grade 1/250 requires a first cost of $1,325,725 and with an annual saving of $577,440 if number of daily train are cut down to meet the present traffic. But this is not the usual case by practice to reduce the number of daily passenger trains too much so as to affect the convenience of the public althouth the train load may be increased. Therefore, if the load on passenger trains is not increased so much to her full capacity, but say to the load of the 1/160 grade, we then have the saving at least $366,500 annually, and the speed of trains is consequently multiplied due to less load. The economy in time is also another gain.

We should not try to save a little first cost again to adopt the 2nd (1/200) or the 3rd. (1/160) scheme, for either of the latter is not so economical as the first in the long run. We must do it for permanent good; otherwise the line will and must be done over again to its final improvement in the furture. If the finance is in doubtful question, the whole piece of work may be divided into 3 or 5 years but any part we do must be the final and best thing for that part.

On the safe side, taking the construction cost of the 1st. scheme $1,320,725 and the annually saving of the 3rd. $366,500 we note that the cost can be saved back 3 years and 7 months after the completion of the work. If the fund is to be raised at 10% interest it would be worth while to use the justifiable expenditure of $366,500/10.—$3,665,000 to better this line.

After this improvement, if the annual saving put into bank at 8% compound interest year by year we will get a total sum of capital and interest of

$366,500 × 14.65=$ 5,369,255 after 10 years.
$366,500 × 48.43=$17,747,595 „ 20 „
$366,500 × 111.33=$40,801,545 „ 30 „
$266,500 × 368,72=$98,485,760 „ 40 „

It means that 40 years after the savings will be nearly equal to the whole capital of the P.M.R. today. This is however, only taking the present number of daily trains. If the number of trains increased the amount of saving will be increased in proportion; 30 years will be enough to round out the last figure. If my proposed scheme is accepted and put into force at once, I'll have the pleasure to see the resulted saving years after.

APPENDIX

PEIPING-MUKDEN RAILWAY

Chinese Government Railway.

LOAD TABLE-AXLES-AND CLASSIFICATION OF ENGINES

(I) *Classification of Engines*

Class A—13. 14. 22. 31 33. 35. 36. 38. 58 to 69 inclusive.
Class B—40 to 42. 126 to 153. 225 to 238. inclusive.
Class C—44. 45. 46.
Class D—15. 16. 18. 19. 25. 27. 29. 30. 73 to 106 120 to 125 inclusive.
Class E—107 to 116 inclusive.
Class F—117. 118. 119.
Class G—156 to 176. 260 to 277 inclusive K.M.A.
Class H—190 to 199 inclusive.
Class J—200 to 223 inclusive.

Class	Sections on which allowed to run.
Class A. D. E.	No restrictions.
Class B.C.F.H.J. & G. class	Not allowed on branch lines laid with 60 lb rails

(II) *Load Table and Axles.*

Section	Class of loco.	Passenger Train				Mixed Trains				Goods Trains				Empties (Axles L)	
		Summer Axles	Summer Tons	Winter Axles	Winter Tons	Summer Axles	Summer Tons	Winter Axles	Winter Tons	Summer Axles	Summer Tons	Winter Axles	Winter Tons	Summer Axles	Winter Axles
CYM-TN	A	...	...	...	...	69	621	59	531	84	756	64	576	94	39
	B	70	490	55	385	104	936	94	846	134	1,206	124	1,116	144	144
	C	52	364	44	308	84	756	74	666	94	846	84	756	104	104
	D	60	420	50	350	84	756	74	666	114	1,026	104	936	134	124
	E	60	420	50	350	84	756	74	666	114	1,026	104	936	134	124
	F	70	490	55	385	104	936	94	846	114	1,026	104	926	134	124
	G	85	595	85	595	124	1,116	114	1,026	154	1,386	154	1,386	204	204
	H	85	595	85	595	114	1,026	114	1,026	114	1,026	104	936	134	124
	J	85	595	85	595	134	1,206	134	1,206	154	1,386	154	1,386	204	204

LOAD TABLE AND AXLES. *(Continued)*

Section	Class of Loco.	Passenger Train				Mixed Trains				Goods Trains				Empties (Axles L)	
		Summer		Winter		Summer		Winter		Summer		Winter		Summer	Winter
		Axles	Tons	Axles	Tons	Axles	Tons	Axles	Tons	Axles	Tons	Axles	Tons	Axles	Axles
KPT-MD	A	—	—	—	—	69	621	59	531	84	756	64	576	94	89
	B	70	490	55	385	104	936	94	846	134	1,206	124	1,116	144	144
	C	52	364	44	308	84	756	74	666	94	846	84	756	104	104
	D	60	420	50	350	84	756	74	666	114	1,026	104	936	134	124
	E	60	420	50	350	84	756	74	666	114	1,026	104	936	134	124
	F	70	490	55	385	104	936	94	846	114	1,026	104	936	134	124
	G	85	595	85	595	124	1,116	114	1,026	154	1,386	154	1,386	204	184
	H	85	595	85	595	114	1,026	114	1,026	114	1,026	104	936	134	124
	J	85	595	85	595	134	1,206	134	1,206	154	1,386	154	1,386	204	204

Note :— (1) Total loads in axles not to exceed empty axle load.

(2) In case o goods trains loads include one goods brake van.

(3) When two engines are on train single engines load may be increased 30% of length if station loop permits.

(4) Rating of cars :-

Goods	No. of axles
BV	2
BV	3
BV	4
20 ton	4
30 „	5
40 „	6
50 „ well	8
Refrigerator car	6

Passenger	No. of axles
55 S. P. T. P., Private and service	4
cars 65 and 68 F. S. C, P. L. C.,	6
F. P., F. D. R. B. V. S. Postal, Guard	6
and Baggage T. P. L. Blue cars	8

Load Table and Axles. *(Continued)*

Section	Class of Loco.	Passenger Train Summer Axles	Passenger Train Summer Tons	Passenger Train Winter Axles	Passenger Train Winter Tons	Mixed Trains Summer Axles	Mixed Trains Summer Tons	Mixed Trains Winter Axles	Mixed Trains Winter Tons	Goods Trains Summer Axles	Goods Trains Summer Tons	Goods Trains Winter Axles	Goods Trains Winter Tons	Empti-s (Axles L) Summer Axles	Empti-s (Axles L) Winter Axles
TN-SHK	A	...	...	...	...	69	621	59	531	84	756	64	576	94	89
	B	70	490	55	385	104	936	94	846	124	1,116	114	1,026	144	144
	C	52	364	44	308	84	756	74	666	94	846	84	756	104	104
	D	60	420	50	350	84	756	74	666	114	1,026	104	936	134	124
	E	61	420	50	350	84	756	74	666	114	1,026	104	936	134	124
	F	70	490	55	395	104	936	94	846	114	1,026	104	936	134	124
	G	85	595	85	595	124	1,116	114	1,026	154	1,386	154	1,386	204	204
	H	85	595	85	595	114	1,026	114	1,026	114	1,026	104	936	134	124
		85	595	85	595	134	1,206	134	1,206	154	1,386	154	1,386	204	204
SHK-CH	A	...	¾	...	...	60	540	54	486	66	594	52	468	94	64
	B	70	490	55	385	64	846	84	756	64	864	84	756	134	124
	C	...	...	...	...	...	...	...	...	...	...	...	...	...	...
	D	55	385	50	350	80	720	70	630	80	720	70	630	104	94
	E	55	385	50	350	80	720	70	630	80	720	70	630	104	94
	F	55	385	50	350	80	720	70	630	80	720	70	630	104	94
	G	85	595	85	595	114	1,026	104	936	114	1,026	104	936	154	144
	H	85	595	85	595	114	1,026	114	1,026	114	1,026	104	956	134	124
	J	85	595	85	595	114	1,116	114	1,026	124	1,116	114	1,026	164	154
CH-KPT	A	...	¾	...	...	69	621	59	521	74	663	59	531	94	89
	B	70	490	55	385	104	936	94	846	114	1,026	104	936	144	134
	C	60	420	50	350	84	756	74	636	94	846	84	756	124	114
	D	...	...	...	...	...	...	...	...	...	...	...	...	...	...
	E	60	420	50	350	84	756	74	666	89	801	80	720	124	114
	F	...	...	...	...	...	...	...	...	...	...	...	...	...	...
	G	85	595	85	595	134	1,026	124	1,116	134	1,206	124	1,116	154	144
	H	85	595	85	595	114	1,026	104	936	114	1,026	104	926	134	124
	J	85	595	85	595	130	1,170	130	1,170	144	1,296	134	1,206	174	164

(5) Summer loads from 1st. April to Nov. 30th but dependent on weather.

Provision of W. H. brake fitted wagons coupled up to Loco on goods trains.

All Class Engines:

Locos working mixed and goods trains on C. Y. and T.T. branch lines must be coupled up with two W. H. brake fitted wagon irrespective of load because there are heavy gradients along these branch lines.

Locos working goods trains between M. D. and S. H. K. and on Y. K. branch line if take more than 100 axles (loaded or empty, must be coupled up with two W. H. brake fitted wagons).

156, 161 & 200 (G. & J.) class engines.

154	empty	axles	and	over	required	four	W.H.	brake	fitted wagons.
164	”	”	”	”	”	six	”	”	” ”
174	”	”	”	”	”	eight	”	”	” ”
184	”	”	”	”	”	ten	”	”	” ”
194	”	”	”	”	”	twelve	”	”	” ”
204	”	”	”	”	”	fourteen	”	”	” ”

154 loaded axles require four W. H. brake fitted wagons.

劉氏檢積籌說明書

著者：劉增冕

第一節　梅氏倍數籌之改良

梅定九氏之橫式倍數籌,如

等,十籌

經增冕改爲豎式囘碼,以便使用,計十種,如

0		1		2		3		4		5		6		7		8		9	
			1		2		3		4		5		6		7		8		9
			2		4		6		8	1		1	2	1	4	1	6	1	8
			3		6		9	1	2	1	5	1	8	2	1	2	4	2	7
			4		8	1	2	1	6	2		2	4	2	8	3	2	3	6
			5	1		1	5	2		2	5	3		3	5	4		4	5
			6	1	2	1	8	2	4	3		3	6	4	2	4	8	5	4
			7	1	4	2	1	2	8	3	5	4	2	4	9	5	6	6	3
			8	1	6	2	4	3	2	4		4	8	5	6	6	4	7	2
			9	1	8	2	7	3	6	4	5	5	4	6	3	7	2	8	1

用法舉例一,　求32之各倍數

檢 3 籌, 2 籌,對列之,即可讀得各倍數之積.

3		2	
	3		2
	6		4
	9		6
1	2		8
1	5	1	
1	8	1	2
2	1	1	4
2	4	1	6
2	7	1	8

倍	積
1	32
2	64
3	96
4	128
5	160
6	192
7	224
8	256
9	288

┆至┆爲位界,在位界┆┆之內,如第五横列之5, 1,應讀其和數爲6.第六横列之8, 1,應讀其和數爲9.此等┆┆位界,仍以( )括弧式爲佳,作者爲便於繕寫,故用┆┆線爲位界.

用法舉例二. 求7 8 9之各倍數.

7		8		9	
	7		8		9
1	4	1	6	1	8
2	1	2	4	2	7
2	8	3	2	3	6
3	5	4		4	5
4	2	4	8	5	4
4	9	5	6	6	3
5	6	6	4	7	2
6	3	7	2	8	1

檢7籌,8籌,9籌對列之.

其7倍之積爲5523,讀出較難,至4倍,8倍,9倍,等亦均不易讀出,遂有劉氏檢積籌之創製,可以一目卽得

第二節 劉氏檢積籌之創製

因倍數籌讀積較難,增晃遂創爲檢積籌一組,以便讀積,卽將本位之積的個位與次位之積的十位,業已計和,製就此籌,凡百種,爲一組.

籌頂標字 本位用大字體,次位用小字體.

茲將檢積籌式樣百種,列陳於次.

0_1	0_6	1_1	1_6	2_1	2_6	3_1	3_6	4_1	4_6	5_1	5_6	6_1	6_6	7_1	7_6	8_1	8_6	9_1	9_6
		1	1	2	2	3	3	4	4	5	5	6	6	7	7	8	8	9	9
	1	2	3	4	5	6	7	8	9		1	2	3	4	5	6	7	8	9
	1	3	4	6	7	9	1	2	3	5	6	8	9	1	2	4	5	7	8
	2	4	6	8	1	2	4	6	8		2	4	6	8	1	2	4	6	8
	3	5	8		3	5	8		3	5	8		3	5	8		3	5	8
	3	6	9	2	5	8	1 1	4	7		3	6	9	2	5	8	1 1	4	7
	4	7	1 1	4	8	1	5	8	1 2	5	9	2	6	9	1 3	6	1	3	7
	4	8	1 2	6	1	4	8	2	6		4	9	1 2	6	1	4	8	2	6
	5	9	1 4	8	1 3	7	1 2	6	1 1	5		4	9	3	8	2	7	1	6
0_2	0_7	1_2	1_7	2_2	2_7	3_2	3_7	4_2	4_7	5_2	5_7	6_2	6_7	7_2	7_7	8_2	8_7	9_2	9_7
		1	1	2	2	3	3	4	4	5	5	6	6	7	7	8	8	9	9
	1	2	3	4	5	6	7	8	9		1	2	3	4	5	6	7	8	9
	2	3	5	6	8	9	1 1	2	4	5	7	8	1	1	3	4	6	7	9
	2	4	6	8	1	2	4	6	8		2	4	6	8	1	2	4	6	8
1	3	6	8	1	3	6	8	1	3	6	8	1	3	6	8	1	3	6	8
1	4	7	1	3	6	9	1 2	5	8	1	4	7	1	3	6	9	1 2	5	8
1	4	8	1 1	5	8	2	5	9	1 2	6	9	3	6	1	1 3	7	1	4	7
1	5	9	1 3	7	1 1	5	9	3	7	1	5	9	1 3	7	1 1	5	9	3	7
1	6	1	1 5	9	1 4	8	1 3	7	1 2	6	1	5	1	4	9	3	8	2	7
0_3	0_8	1_3	1_8	2_3	2_8	3_3	3_8	4_3	4_8	5_3	5_8	6_3	6_8	7_3	7_8	8_3	8_8	9_3	9_8
		1	1	2	2	3	3	4	4	5	5	6	6	7	7	8	8	9	9
	1	2	3	4	5	6	7	8	9		1	2	3	4	5	6	7	8	9
	2	3	5	6	8	9	1 1	2	4	5	7	8	1	1	3	4	6	7	9
1	3	5	7	9	1 1	3	5	7	9	1	3	5	7	9	1 1	3	5	7	9
1	4	6	9	1	4	6	9	1	4	6	9	1	4	6	9	1	4	6	9
1	4	7	1	3	6	9	1 2	5	8	1	4	7	1	3	6	9	1 2	5	8
2	5	9	1 2	6	9	3	6	1	1 3	7	1	4	7	1 1	1 4	8	1 1	5	8
2	6	1	1 4	8	1 2	6	1	4	8	2	6	1	1 4	8	1 2	6	1	4	8
2	7	1 1	1 6	1	1 5	9	1 4	8	1 3	7	1 2	6	1 1	5	1	4	9	3	8
0_4	0_9	1_4	1_9	2_4	2_9	3_4	3_9	4_4	4_9	5_4	5_9	6_4	6_9	7_4	7_9	8_4	8_9	9_4	9_9
		1	1	2	2	3	3	4	4	5	5	6	6	7	7	8	8	9	9
	1	2	3	4	5	6	7	8	2	1	1	2	3	4	5	6	7	8	9
1	2	4	5	7	8	1	1 1	3	4	6	7	9	1	2	3	5	6	8	9
1	3	5	7	9	1 1	3	5	7	9	1	3	5	7	9	1 1	3	5	7	9
2	4	7	9	2	4	7	9	2	4	7	9	2	4	7	9	2	4	7	9
2	5	8	1 1	4	7	1	1 3	6	9	2	5	8	1 1	4	7	1	1 3	6	9
2	6	9	1 3	6	1	3	7	1	1 4	7	1 1	4	8	1 1	1 5	8	1 2	5	9
3	7	1 1	1 6	9	1 3	7	1 1	5	9	3	7	1 1	1 5	9	1 3	7	1 1	5	9
3	8	1 2	1 7	1 1	1 6	1	1 5	9	1 4	8	1 3	7	1 2	6	1 1	5	1	4	9
0_5	0_0	1_5	1_0	2_5	2_0	3_5	3_0	4_5	4_0	5_5	5_0	6_5	6_0	7_5	7_0	8_5	8_0	9_5	9_0
		1	1	2	2	3	3	4	4	5	5	6	6	7	7	8	8	9	9
		3	2	5	4	7	6	9	8	1		3	2	5	4	7	6	9	8
1		4	3	7	6	1	9	3	2	6	5	9	8	2	1	5	4	8	7
2		6	4	1	8	4	2	8	6	2		6	4	1	8	4	2	8	6
2		7	5	2		7	5	2		7	5	2		7	5	2		7	5
3		9	6	5	2	1 1	8	7	4	3		9	6	5	2	1 1	8	7	4
3		1	7	7	4	4	1	1 1	8	8	5	5	2	1 2	9	9	6	6	3
4		1 2	8	1	6	8	4	6	2	4		1 2	8	1	6	8	4	6	2
4		1 3	9	1 2	8	1 1	7	1	6	9	5	8	4	7	3	6	2	5	1

第三節 檢積籌之用法

(甲) 整數

求 789 之各倍積數

應檢出以下各籌 注意 {檢籌時,以籌頂大字爲本位,小字爲次位,依次啣接,至本數首位籌取(首位數 0),尾位籌取(尾位數取 0).

0_7		7_8		8_9		9_0	
·			7		8		9
	1		5		7		8
	2		3		6		7
	2	1	1		5		6
	3		9		4		5
	4		6	1	3		4
	4	1	4	1	2		3
	5	1	2	1	1		2
	6	1	1	1			1

倍	積
1	789
2	1578
3	2367
4	3156
5	3945
6	4734
7	5523
8	6312
9	7201

檢各倍積數,一目卽可讀出,較之倍數籌容易,毋庸再事計和進位手續矣.

(乙) 遇有零位數,如求 3 03 之各倍數.

應檢 (0_3) (3_2) (2_0) (0_8) (8_0) 五籌,列之以檢積.

(丙) 多位數乘多位數,將其各積定位加之,卽得總積.

(丁) 多位除法,列除數籌,檢積由實數減之,遞次得商.

(戊) 遇有重複數,應多造檢積籌數組,而每籌可作正背兩面用.

第四節 檢積籌正背並用

玆將籌之正背面籌頂標字,列表如下.

籌之正背面標頂字本次位互換,爲易檢也.

正	背	正	背	正	背	正	背	正	背	正	背	正	背	正	背	正	背	正	背
0_1	1_0	1_1		2_1	1_2	3_1	1_3	4_1	1_4	5_1	1_5	6_1	1_6	7_1	1_7	8_1	1_8	9_1	1_9
0_2	2_0	1_2	2_1	2_2		3_2	2_3	4_2	2_4	5_2	2_5	6_2	2_6	7_2	2_7	8_2	2_8	9_2	2_9
0_3	3_0	1_3	3_1	2_3	3_2	3_3		4_3	3_4	5_3	3_5	6_3	3_6	7_3	3_7	8_3	3_8	9_3	3_9
0_4	4_0	1_4	4_1	2_4	4_2	3_4	4_3	4_4		5_4	4_5	6_4	4_6	7_4	4_7	8_4	4_8	9_4	4_9
0_5	5_0	1_5	5_1	2_5	5_2	3_5	5_3	4_5	5_4	5_5		6_5	5_6	7_5	5_7	8_5	5_8	9_5	5_9
0_6	6_0	1_6	6_1	2_6	6_2	3_6	6_3	4_6	6_4	5_6	6_5	6_6		7_6	6_7	8_6	6_8	9_6	6_9
0_7	7_0	1_7	7_1	2_7	7_2	3_7	7_3	4_7	7_4	5_7	7_5	6_7	7_6	7_7		8_7	7_8	9_7	7_9
0_8	8_0	1_8	8_1	2_8	8_2	3_8	8_3	4_8	8_4	5_8	8_5	6_8	8_6	7_8	8_7	8_8		9_8	8_9
0_9	9_0	1_9	9_1	2_9	9_2	3_9	9_3	4_9	9_4	5_9	9_5	6_9	9_6	7_9	9_7	8_9	9_8	9_9	
0		1_0	0_1	2_0	0_2	3_0	0_3	4_0	0_4	5_0	0_5	6_0	0_6	7_0	0_7	8_0	0_8	9_0	0_9

造200枚籌已足重複四字數之用,外加製本位次位同數者如 3_3 4_4 ……等十籌.

第五節　置籌法

如紙製可置分室袋,如竹木籌可置百孔箱,隨手抽用,若特製帶彈性之運送械,或纏捲現數,而欲避免搬運送還之頻煩手續,乃係機械設計範圍,用者可意近構造,毋庸鄙人再贅.

本刊啟事

本刊每年四期,每逢一,四,七,十各月出版.凡會員諸君及海內外工程人士,如有鴻論鉅著,務望隨時隨地,不拘篇幅,賜寄本會,刊登本刊爲幸,此啓.

總務 支秉淵 啟

揚子江之概要與其性質

著者：宋希尙

江河淮海,天之奧府,具備四德,敷潤萬物,尤以揚子江橫貫東西,關繫綦鉅.其發源於岷山,亙綿九千餘里,棟飛珠濺,高屋建瓴,經過流域,均屬重鎭,土地之肥沃,戶籍之興盛,物產之饒富,宜如何研求江流性質,賴以發揚交通運輸便利,則功用深宏,裕國惠民,如操左劵.惟是溝洫有志,河渠成書,禹貢爲治水之鼻祖,桑經係窮源之專編,而世之言水利者,類皆視爲常談.卽淮南子所載窩窴洄流,蘆灰止水,渼水重安而宜竹,雒水輕利而宜禾,其防澗災淫,辨別水性,雖爲時代變遷古今易勢,法久斯弊,用非所宜,然苟能本其意,深思遠慮,棄粕存菁,則治水之學識,殊不讓西方爲專備也.揚子江自實行測量以來,於以知河床之改度,寬深之異態,流緩沙停者,則積而爲航道之梗,激湍氾濫者,則轉而爲農田之害人民損失,固難綜計,工商梗阻,竟滯其機.若長此以往,不籌導浚,以洩水勢,不謀築護,以防氾濫,是揚子江流,匪獨失厥敷潤萬物之功,而妨害農商,其禍將與黃淮相等矣!揚子江水道整理委員會,負有整理全江之職責,將江分作三段,以吳淞至漢口,規定爲第一大段,年來因財力人力之關係,不得不專注力於第一段之水道測量.舉凡淤淺之處,如崇文洲,太子磯,姚家洲,張家洲,江家洲,糧洲,湖廣沙,戴家洲,蘿蔔洲.鴨蛋洲,及得勝洲等,其水深斷面,及水面流向等項,因測量而獲其概要.而全江之性質,若面積之計算,距離之遠近.坡度之變遷,水位之高低,流速之緩急,流量之容積,雨量之記載,泥沙之增減,考諸測量所得,均足爲設計之依據.原治水原則,必自下始,則整理揚子江第一大段之計畫,已具有相當之資料矣.余忝長揚子江工務,因就所知,將江之概要與性質,附以簡表,詮次說明,聊供注意揚子江水利者之參考云爾.

第一節　流域及湖泊面積

揚子江流域,及洞庭鄱陽兩湖,蓄水面積,歷來均無確切之數.茲就前英國海軍測量圖,並參考其他之圖,擬定面積之數如下:

(甲) 流域全面積　計一,九五九,三三三平方公里

(乙) 洞庭湖面積　計三,四四三平方公里

(丙) 鄱陽湖面積　計二,八三〇平方公里

第二節　揚子江之相間距離

揚子江全長,計爲三千二百海里,合約五千九百公里.其每處相間距離,自上海起,至宜昌止,業經實測分列五表如下:(附後頁)

第三節　揚子江之坡度

揚子江自上海至宜昌間距離,由精確水平隊實測外,其自宜昌,萬縣,重慶間距離水平,則由川漢鐵路所測查考得之.更於重慶,萬縣,宜昌,岳州,漢口,九江,蕪湖,南京,鎮江,吳淞等處沿江重要都市,與海關合作,設立水尺,按時記錄,水位坡度,藉以推算洪水位坡度.至江水流速,亦可從水位坡度,及河床濕半徑,用季然式Chezy公式$V=C\sqrt{rs}$求之.茲計算揚子江重慶吳淞間之坡度如下:

重慶至萬縣	0.000180
萬縣至宜昌	0.000210
宜昌至岳州	0.000043
岳州至漢口	0.000023
漢口至九江	0.000025
九江至南京	0.000025
南京至吳淞	0.000012

觀此,知坡度最大,江流最急之處,係在萬縣,宜昌之間.

第四節　揚子江之水位

水位與流速及河床剖面積,俱有連帶關係,用爲推求流量之要素.蓋在某段河床剖面積,乘其水流速度所得之果,即爲某段揚子江之流量.流速每隨水漲落而變.水位漲,則流速增.水位落,則流速減.如於各種水位高度時,測驗流速,積多時間之繼續觀測,以其平均數值,編製流率曲線,(Rating Curve)則以後如知水位高度若干,即可由流率曲線上,推求而得流量.至若每年水位漲落情形,及河床變遷程度,足以表示此處航道狀况,以爲航業之指南,關係更覺重要.自光緒二十六年至民國十五年間,揚子江各處水位漲落情形,列表如下:

（第六表）　揚子江各站水位漲落表

地名	最高水位	最低水位	備攷
吳淞	一七•一二英尺	○•二英尺	
鎮江	一九•○○英尺	○•八英尺	
南京	二四•三○英尺	○•○八英尺	
蕪湖	二九•七○英尺	○•六英尺	
九江	四四•九○英尺	負一•○英尺	
漢口	五○•五○英尺	負二•○英尺	
岳州	四八•四○英尺	負二•○英尺	
宜昌	五一•二○英尺	負一•八英尺	
重慶	一一八•○○英尺	負二•○英尺	

第五節　揚子江之流速率

揚子江洪水時,水面坡度,水位漲落情形,既已明瞭.玆自民國十一年至十六年間,實測各站流速,列表如後:

（第 一 表）

上海至南京距離計長二三一海里
（合四八二公里七四三華里）

上海							
15.0	吳淞						
81.0	66.0	南通					
117.5	102.5	36.5	江陰				
139.5	124.5	58.5	22.0	定興洲			
154.5	139.5	73.5	37.0	15.0	口岸		
183.0	168.0	102.0	65.5	43.5	28.5	鎮江	
231.0	216.0	150.0	113.5	91.5	76.5	48.0	南京

東流					
29.0	小孤山				
39.0	10.0	流龍灣			
48.0	19.0	9.0	八里江口		
50.5	21.5	11.5	2.5	扁担洲	
66.0	37.0	27.0	18.0	15.5	九江

猴子磯		
10.0	陽邏	
26.0	16.0	漢口

（第 二 表）

南京至九江距離計長二五〇海里
（合四六三公里八〇四華里）

南京										
43.0	東梁山									
53.0	10.0	蕪湖								
75.5	32.5	22.5	黑沙洲							
82.0	39.0	29.0	6.5	板石磯						
115.0	72.0	62.0	39.5	33.0	大通					
127.0	84.0	74.0	51.5	45.0	12.0	崇文洲				
144.0	101.0	91.0	68.5	62.0	29.0	17.0	太子磯			
152.0	109.0	99.0	76.5	70.0	37.0	25.0	8.0	新洲		
162.0	119.0	109.0	86.5	80.0	47.0	35.0	18.0	10.0	安慶	
173.0	130.0	120.0	97.5	91.0	58.0	46.0	29.0	21.0	11.0	姚家洲
184.0	141.0	131.0	108.5	102.0	69.0	57.0	40.0	32.0	22.0	11.0
213.0	170.0	160.0	1[illegible]7.5	131.0	98.0	86.0	69.0	61.0	51.0	40.0
223.0	180.0	170.0	147.5	141.0	108.0	96.0	79.0	71.0	61.0	50.0
232.0	189.0	179.0	156.5	150.0	117.0	105.0	88.0	80.0	70.0	59.0
234.5	191.5	181.5	159.0	152.5	119.5	107.5	90.5	82.5	72.5	61.5
250.0	207.0	197.0	174.5	168.0	135.0	123.0	106.0	98.0	88.0	77.0

（第 三 表）

九江至漢口距離計長一四五海里
（合二六九公里四六七華里）

九江										
16.0	江家洲									
27.0	11.0	武穴								
36.0	20.0	9.0	田家鎮							
49.0	33.0	22.0	13.0	蘄州						
64.0	48.0	37.0	28.0	15.0	道士洑					
70.0	54.0	43.0	34.0	21.0	6.00	黃石港				
73.0	57.0	46.0	37.0	24.0	9.0	3.0	戴家洲			
84.0	68.0	57.0	48.0	35.0	20.0	14.0	11.0	羊[illegible]磯		
92.0	76.0	65.0	56.0	43.0	28.0	22.0	19.0	8.0	黃州	
103.0	87.0	76.0	67.0	54.0	39.0	33.0	30.0	19.0	11.0	鴨蛋洲
119.0	103.0	92.0	83.0	70.0	55.0	49.0	46.0	35.0	27.0	16.0
129.0	113.0	102.0	93.0	80.0	65.0	59.0	56.0	45.0	37.0	26.0
145.0	129.0	118.0	109.0	96.0	81.0	75.0	72.0	61.0	53.0	42.0

(第　四　表)

漢口至城陵磯距離計長一一三·五海里

(合二一〇公里三六四·五華里)

漢口								
16.5	金口							
24.5	8.0	大嘴鎮						
44.0	27.5	19.5	簰洲					
55.0	38.5	30.5	11.0	燕子窩				
75.0	58.5	50.5	31.0	20.0	寶塔洲			
88.0	71.5	63.5	44.0	33.0	13.0	新堤		
108.5	92.0	84.0	61.5	53.5	33.5	20.5	白螺磯	
113.5	97.0	89.0	69.5	58.5	38.5	25.5	5.0	城陵磯

郝穴											
13.0	突起洲										
26.0	13.0	沙市									
33.0	20.0	7.0	太平口								
45.0	32.0	19.0	12.0	大復街							
51.0	38.0	25.0	18.0	6.0	江口渓						
59.0	46.0	33.0	26.0	14.0	8.0	松滋					
70.0	57.0	44.0	37.0	25.0	19.0	11.0	石岡子				
77.0	61.0	51.0	41.0	32.0	26.0	18.0	7.0	枝江			
88.0	75.0	62.0	55.0	43.0	37.0	29.0	18.0	11.0	宜都		
100.0	87.0	74.0	67.0	55.0	49.0	41.0	30.0	23.0	12.0	虎牙灘	
110.0	97.0	81.0	77.0	65.0	59.0	51.0	40.0	33.0	22.0	10.0	宜昌

（第 五 表）

城陵磯至宜昌距離計長二四三海里

（合四五〇公里七八一華里）

城陵磯										
14.0	尺八口									
19.0	5.0	上反嘴								
28.0	14.0	9.0	左家灘							
39.0	25.0	20.0	11.0	上車灣						
53.0	44.0	39.0	30.0	19.0	監利					
65.0	51.0	46.0	37.0	26.0	7.0	新河口				
74.0	60.0	55.0	46.0	35.0	16.0	9.0	劉家灘			
85.0	71.0	66.0	57.0	46.0	27.0	20.0	11.0	調絃口		
98.0	81.0	79.0	70.0	59.0	40.0	33.0	24.0	13.0	新碼頭	
119.0	105.0	100.0	91.0	80.0	61.0	54.0	45.0	34.0	21.0	天星洲
133.0	119.0	114.0	105.0	94.0	75.0	68.0	59.0	48.0	35.0	14.0
146.0	132.0	127.0	118.0	107.0	88.0	81.0	72.0	61.0	48.0	27.0
159.0	145.0	140.0	131.0	120.0	101.0	94.0	85.0	74.0	61.0	40.0
166.0	152.0	147.0	138.0	127.0	108.0	101.0	92.0	81.0	68.0	47.0
178.0	164.0	159.0	150.0	139.0	120.0	113.0	104.0	93.0	80.0	59.0
184.0	170.0	165.0	156.0	145.0	126.0	119.0	110.0	99.0	86.0	65.0
192.0	178.0	173.0	164.0	153.0	134.0	127.0	118.0	107.0	94.0	73.0
203.0	189.0	184.0	175.0	164.0	145.0	138.0	129.0	118.0	105.0	84.0
210.0	196.0	191.0	182.0	171.0	152.0	145.0	136.0	125.0	112.0	91.0
221.0	207.0	202.0	193.0	182.0	163.0	156.0	147.0	136.0	123.0	102.0
233.0	219.0	214.0	205.0	194.0	175.0	168.0	159.0	148.0	135.0	114.0
243.0	229.0	224.0	215.0	204.0	185.0	178.0	169.0	158.0	145.0	124.0

（第七表）揚子江各站流速表

地名	最大流速	最小流速	備攷
大通	每秒一·三一〇公尺	每秒〇·五二〇公尺	
湖口	每秒一·九四五公尺	每秒〇·四二〇公尺	
九江	每秒一·七四五公尺	每秒〇·五二五公尺	
武昌	每秒一·六九五公尺	每秒〇·六〇〇公尺	
漢口	每秒一·八一〇公尺	每秒〇·四七〇公尺	
城陵磯	每秒一·六六五公尺	每秒〇·九〇五公尺	
尺八口	每秒二·四六五公尺	每秒〇·七五〇公尺	
枝江	每秒一·七七〇公尺	每秒〇·五八五公尺	

第六節　揚子江之河床剖面

揚子江各測站河之最大寬度,最大深度,最小深度,列表如次:

（第八表）揚子江流量測站河床剖面尺度表

流量測站	河寬		河深		備攷
	最大寬度	最小寬度	最大深度	最小深度	
大通	二〇八〇公尺	一九〇〇公尺	二八·〇公尺	一二·〇公尺	
湖口	一五〇〇公尺	九六〇公尺	四二·〇公尺	二六·〇公尺	
九江	一九一〇公尺	一〇一〇公尺	二九·二公尺	八·〇公尺	
漢口	一六四〇公尺	一三八〇公尺	二六·〇公尺	七·五公尺	
城陵磯	一九八〇公尺	一六二〇公尺	一六·五公尺	九·〇公尺	
尺八口	八〇〇公尺	七六〇公尺	一四·〇公尺	七·〇公尺	
枝江	一二六〇公尺	八四〇公尺	二三·〇公尺	七·〇公尺	

第七節　揚子江之流量測驗

揚子江所測最大最小流量,列表如次:

(第九表) 揚子江各站流量表

測站地名	最大流量	最小流量	備攷
大通	每秒五六,九〇〇立方公尺	每秒七,七二一立方公尺	
湖口	每秒六五,八八〇立方公尺	每秒五,五九五立方公尺	
九江	每秒六三,九七〇立方公尺	每秒四,八一八立方公尺	
漢口	每秒六〇,七五〇立方公尺	每秒五,二〇八立方公尺	
城陵磯	每秒三七,九六〇立方公尺	每秒六,五三〇立方公尺	
尺八口	每秒二一,〇〇〇立方公尺	每秒四,八二〇立方公尺	
枝江	每秒四九,三七〇立方公尺	每秒四,一九〇立方公尺	

參觀上表,知揚子江洪水時,最大流量約每秒六萬立方公尺,枯水時最小流量,約每秒五千立方公尺,相差如此.而河床深度,在洪水時,湖口已深四二公尺,下游當不至此.枯水時,漢口,九江,俱僅深八公尺左右,而在沙洲淤澱之處,河淺更可想見.無怪每年冬春之季,航運至感困難,今年春間,尤覺更甚!

第八節 揚子江之沙泥數量

查揚子江水中,所挾沙泥量,雖不若永定河,黃河之甚,然統計每年所挾數量之鉅,亦足驚人.據海關及揚子江水道整理委員會測驗所知,每年隨江流入海者,當在五萬噸以上.如以一年中所挾之量堆積上海全埠面積,則可增高八公尺之多.至於中游下游,因沙而成之沙洲,星羅棋布,尤足以表現泥沙淤積之多.揚子江整理委員會,為欲研究揚子江水流挾沙能力,及上游來水挾沙數量,以便設計相當之河床,維持相當之流速,使水中挾沙,不致中途停滯,航道不致日益淤淺起見,已於沿江各測站,繼續測驗其挾沙量,平均統計之,藉作他日規劃整理之依據.茲將各站挾沙量列如下表:

（第十表）揚子江各站挾沙量表

地名	挾沙數量	平均次數	備考
大通	百萬分之三五四·五	測驗七次之平均數	
湖口	百萬分之三九八·五	測驗八次之平均數	
九江	百萬分之三六七·五	測驗八次之平均數	
漢口	百萬分之三三六·三五	測驗五一次之平均數	
城陵磯	百萬分之五五二·七	測驗一一次之平均數	
尺八口	百萬分之八五三·九	測驗一一次之平均數	
枝江	百萬分之七〇九·九五	測驗二〇次之平均數	

第九節　揚子江之雨量記錄

揚子江流域內雨量,關於揚子江本身至重且切.除揚子江測量隊記錄外,其餘各處,均委託各地天主教堂,代爲就近記載.其雨量器則由揚會置備,分發各地應用.每月報告,由各天主教堂彙由上海徐家匯天主教堂,寄交揚會,歷年成績極佳.玆自民國十三年起至十七年止,每月各站雨量,彙刊如下表.

（附後頁）

第十節　揚子江淤淺各處之狀況

揚子江自吳淞至漢口,計水程爲一千一百三十公里（約合六百十一海里）.平時航運,素稱便利.惟每屆冬春之季,水落江枯,重載之輪,凡吃水至十五英尺者,即感困難.然考其發生困難之處,當以崇文洲,太子磯,姚家洲,張家洲,江家洲,糧洲,湖廣沙,戴家洲,羅菊洲,鴨蛋洲,及得勝洲,等十一處爲最甚.其他若黑沙洲,新洲,東流水道,葉家洲,及馬當等處,亦有日見淤淺之趨勢.現揚子江水道整理委員會,爲整理第一大段航道起見,着意測量,經營年餘,玆將最感困難之十一處淤淺狀況,及其發現年月,彙列一表如下:

（第十三表） 吳淞漢口間揚子江淤淺各處之狀況

地 點	日 期	淺灘之長度以公尺計	淺灘之高度以公尺計
崇文洲	民國十七年正月	一,五〇〇	一•二
太子磯	民國十七年二月	三,〇〇〇	一•六
姚家洲	民國十四年七月	五,八〇〇	二•三
馬當	民國十七年正月	一,一〇〇	〇•六
張家洲北港下游	民國十五年正月	五〇〇	〇•五
張家洲北港上游	民國十五年正月	三,〇〇〇	一•五
（張家洲南港下游）	（民國十五年三月）	（五,〇〇〇）	（二•二）
（張家洲南港上游）	（民國十五年三月）	（二,二〇〇）	（一•七）
江家洲	民國十七年五月	七,〇〇〇	一•七
戴家洲	民國十四年二月	五,〇〇〇	一•四
得勝洲	民國十三年二月	一,八〇〇	一•〇
羅蔔洲下游	民國十二年二月	一,〇〇〇	〇•九
羅蔔洲上游	民國十六年二月	一,二〇〇	一•三
湖廣沙	民國十七年三月	三,五〇〇	一•三
漢口沙洲	民國十八年四月		
總計		三五,四〇〇	

第十一節 揚子江第一大段內航行困難時日之統計

前表所列,爲十一處淤淺狀況,茲將歷年第一大段內各淤淺之處,發生航行阻礙之年月日,自民國十二年起至十八年止,作一統計,吃水分十二呎,十五呎,兩種,俾知現在揚子江自吳淞至漢口間航行困難狀況,及其影響於工商業之程度

(第十一表)民國十三年至十七年揚子江流域各站雨量表

Tables Of Monthly Rainfall In MM. At Various Y. R. C. Stations In Yangtse Basin For The Years 1924-28

站名 Station	年 Year \ 月 Month	一月 Jan.	二月 Feb.	三月 Mar.	四月 Apr.	五月 May	六月 June	七月 July	八月 Aug.	九月 Sept.	十月 Oct.	十一月 Nov.	十二月 Dec.	每年總數 Yearly Total	每月均數 Mean Monthly Rainfall
巴塘 四川	1924						162.2	213.9	187.7	136.0	11.5	0.1	0		
Batang Szechuan	1925	0	0	0.3	1.2	1.2	52.9	73.0	83.2	187.7	44.0				
	1926	0	4.0	2.0	15.0	85.0	109.0	130.0	113.0	59.0	19.0	0	0	536.0	44.7
	1927	0	0	5.0	20.0	18.0	71.0	157.0	101.0	110.0	44.0	0	0	526.0	43.8
	1928	0	0	0	0	60.0	85.0	61.0	120.0	71.0	118.0	0	0	515.0	42.9
打箭爐	1924	0.2	4.2	26.5	118.0	437.0	237.0	151.6	145.0	147.5	50.9	46.0	5.4	1,369.3	114.1
Tatsienlu ,,	1925	30.5	12.2	17.1	57.7	117.6	112.0	42.5	75.5	169.0	83.5	7.0	1.5	726.1	60.5
	1926	0	20.0	37.5	43.5	77.5	216.0	151.0	157.1	86.0	45.5	4.0	0	838.1	69.8
	1927	4.0	19.5	34.3	51.5	104.0	256.0	98.5	139.5	179.5	69.0	28.5	2.0	986.2	82.2
	1928	0	11.5	33.0	25.5	218.5	205.7	66.5	168.4	127.9	121.0	23.0			91.0
寧遠	1924			4.0	14.1	94.0	353.1	417.6	206.0	147.8	69.6	8.7	4.7		
Ningyuanfu ,,	1925	22.3	7.3	57.2	58.5	79.5	128.0	164.3	268.3	179.8	149.5	23.5	3.2	1,141.4	95.1
	1926	2.9	32.6	35.1	15.4	70.0	410.1	260.3	246.4	135.8	94.7	23.8	0	1,327.1	110.6
	1927	0	31.0	20.1	63.0	72.8	179.4	115.8	80.0	126.0	41.0	34.7	24.1	787.4	65.6
	1928	0.5	12.1	0.6	13.5	127.6	287.9	86.8	234.1	228.4	129.9	36.3	0	1,157.7	96.5
成都	1924	22.0	7.0	13.0	49.0	63.0	99.0	86.0	195.0	28.0	27.0	4.0	0	593.0	49.4
Chengtu ,,	1925	7.5	19.5	24.0	34.0	49.5	99.0	365.0	175.7	51.5	53.5	25.0	7.0	911.2	75.9
	1926	1.5	12.5	23.5	39.5	127.0	102.0	82.0	179.0	118.0	31.0	7.0	9.0	732.0	61.0
	1927	16.0	15.5	12.0	120.0		118.0	39.0	341.0	77.0	24.0	9.0	0		70.5
	1928	0	13.0		16.0	93.0	55.0	70.0	68.0	81.0	74.0	8.0	1.0		39.9

敘府 Suifu	1924	5.8	22.9	29.7	62.9	51.0	196.3	206.4	85.1	68.2	63.0	22.0	16.9	830.2	69.2
,,	1925	15.7	38.1	37.3	54.1	94.1	252.1	126.5	116.9	219.0	195.8	34.1	16.8	1,200.5	100.0
	1926	6.9	21.7	74.2	101.3	165.9	206.0	206.2	230.2	113.5	104.2	58.4	21.8	1,310.3	109.2
	1927	17.8	29.1	41.1	121.3	94.2	184.7	159.8	198.3	167.4	63.2	33.0	14.9	1,124.8	93.8
	1928	10.6	24.0	85.8	46.4	91.9	115.1	347.6	389.9	100.1	105.4	11.5	36.7	1,365.0	113.8
綏定 Suiningfu	1924		57.0	8.6	79.8	67.9	236.5	97.6	145.1	160.3	82.9	37.3			
,,	1925	25.3	21.0	13.7	81.4	94.0	137.2	69.2	185.7	70.3	16.8	56.0	22.2	792.5	66.0
	1926	45.5	27.5	27.0	36.0	144.0	63.2								
忠州 Chungchow	1924		15.0	54.4	256.8	100.1	112.2	302.8	112.9	175.9	105.5	103.7	21.5		
,,	1925	37.0	19.0	43.0	144.9	174.0	125.0	112.0	383.0	273.0	166.0	126.0	47.0	1,640.9	137.5
	1926	36.0	57.0	63.0	163.0	195.0	119.0	415.0	158.0	98.5	154.0	91.0	45.0	1,594.5	132.9
	1927	37.0	24.0	60.0	164.0	110.0	172.0	218.0	116.0	214.0	100.0	57.0	32.0	1,304.0	108.7
	1928	26.0	20.0	74.0	131.0	191.0	209.0	150.9	151.5	153.0	70.0	19.0	12.0	1,207.4	100.6
東川 雲南 Tungchwan Yunan	1924	2.8	1.0	1.2	16.7	85.2	181.0	320.6	217.5	71.2	30.0	10.5	4.4	942.1	78.5
	1925	69.6	20.9	15.1	20.3	111.5	91.6	156.8	221.4	97.9	65.3	25.0	5.8	907.2	75.6
	1926	16.3	49.1	72.7	11.1	56.9	178.2	241.1	143.7	178.9	76.3	31.4	0	1,055.7	88.0
	1927	1.0	2.8	21.3	63.0	71.4	76.3	111.0	92.0	112.3	63.7	64.8	2.2	681.8	56.8
	1928	0	25.0	0	43.1	103.0	106.7	288.0	58.5	246.5	61.0	41.0	0	972.8	81.1
貴陽 貴州 Kweiyang Kweichow	1204	18.7	23.7	11.1	160.9	152.9	183.8	301.2	82.0	97.2	136.6	5.5	1.3	1,174.9	97.9
	1925	41.6	14.9	27.1	38.7	144.3	135.9	120.5	124.0	262.7	116.9	54.0	3.5	1,084.1	90.3
	1926	23.1	28.9	45.3	56.9	200.1	173.8	322.0	56.8	87.6	124.8	54.5	29.2	1,028.0	100.7
	1927	5.1	27.3	25.2	36.0	162.5	213.6	303.4	53.2	74.9	76.5	82.7	5.3	1,065.7	88.8
	1928	11.9	42.1	13.0	58.4	233.3	256.7	174.3	106.4	81.4	49.6	33.5	28.8	1,089.4	90.8
城固 陜西 Chengku Shensi	1924	4.5	3.3	11.1	7.7	69.6	60.8	44.2	259.0	143.7	100.6	24.2	15.6	744.3	62.0
	1925	1.7	5.0	5.9	94.1	132.1	215.2	152.2	131.8	104.4	23.8	20.8	4.0	891.0	74.3
	1926	3.0	10.0	23.1	20.4	91.4	18.5	171.0	298.6	92.2	49.6	43.9	1.2	822.9	68.6
	1927	45.0	14.2	38.8	40.5	14.7	36.2	142.2	75.1	62.1	67.5	104.0	11.5	651.8	54.3
	1928		4.0	19.6	31.3	8.4	64.4	98.3	98.5	98.6	19.0				

興安 Hingan	1924	2.5	4.2	7.0	16.5	94.0	22.5	98.8	97.2	176.8	60.7	32.9	12.4	625.5	52.1
	1925	3.6	0.8	10.9	83.1	130.0	68.4	108.4	221.8	91.3	19.0	32.0			
	1926								51.0	184.1	66.2	36.5	3.2		
	1927	32.9	9.8	27.3	77.1	17.1	90.6	48.4	134.5	66.6	92.1	8.0	3.1	607.5	50.6
	1928	9.0	3.0	8.0	50.4	9.1	55.1			38.2	11.3	8.0	3.0		19.5
永州 湖南 Yungchowfu Hunan	1924								33.8	16.7	126.5	35.5	1.9		
	1925	97.6	102.8	272.5	238.0	325.5	50.5	22.0	49.5	67.5	30.5	64.0	19.0	1,339.4	111.6
	1926	57.5	158.2	142.0	106.0	174.2	470.5	136.0	76.5	33.0	129.5	28.1	113.7	1,634.2	136.2
老河口 湖北 Laohokow Hupeh	1924								74.0	72.0	18.5	63.0	4.0		
	1925	9.0	5.0	62.0	65.0	152.0	125.0	112.0	286.0	72.0	58.0	40.0	15.0	1,010.0	84.2
	1926	14.0	10.0	99.0	38.0	1 6.0	57.0	407.0	222.0	157.0	169.0	40.0	37.0	360.0	113.3
	1928	44.0	13.0	38.0	109.0	22.0	76.3	50.1	45.0	66.0	3.8	31.7	34.9	533.8	44.5
牯嶺 江西 Kuling Kiangsi	1924	34.3	74.3	162.7	121.0	230.3	140.9	285.0	21.0	80.0	88.0				
	1925	1.0	49.0	67.0	112.0	163.0	113.0	177.0	280.0	40.0	2.0	92.0	0	1,126.0	93.8
	1926	16.0	9.0	31.0	153.0	239.0	274.0	75.0	41.0	63.0	133.0	96.0	47.0	1,177.0	98.1
贛州 Kanchow ,,	1924	58.3	18.9	59.0	281.0	236.0	382.0	170.0	212.0	201.0	45.0	1.0	5.0	1,678.2	139.9
	1925	102.0	104.0	217.0	274.0	281.0	140.0								
	1926	34.0	173.0	135.9	102.0	138.8									
建昌 Kienchangfu ,,	1924	76.2	169.9	163.5	165.0	358.6	402.9	16.7	46.3	72.4	39.5	2.6	1.7	1,515.3	126.3
	1925	113.7	94.1	225.8	135.9	177.9	105.4	81.7	145.9	32.4	4.3	71.6	4.7	1,193.4	99.5
	1926	18.2	220.6	190.2	150.5	97.4	447.2	60.5	111.0	154.7					
禹觀山 湖北 Yukwanshan Hupeh	1926		10.5	100.4	72.6	124.7	61.6	222.7	40.9	91.0	45.8	42.5	37.5	1,037.0	86.4
	1927	62.0	34.0	61.0	105.0	33.0	143.0	323.8	119.2	99.8	25.0	13.5	18.0		
	1928	51.0	34.0	62.0	160.0	75.0	182.0	74.0	67.0	103.0	6.0	140.0	49.0	1,033.0	86.1
廟嘴 Miaochui ,,	1926									60.0	92.0	43.0	49.0		
	1927	45.0	31.0	44.0	136.0	30.0	125.0	116.0	96.0	43.0	19.0	8.0	11.0	704.0	58.7
	1928	37.0	9.0	66.0	149.0	13.0	53.0	13.0	28.0	45.0	3.0	109.0	32.0	557.0	46.4

(第十二表) 緒光六年至民國十七年漢口雨量表

Table Of Monthly Rainfall In MM. At Hankow For The Years 1880-1928

年 Year \ 月 Month	一月 Jan.	二月 Feb.	三月 Mar.	四月 Apr.	五月 May	六月 June	七月 July	八月 Aug.	九月 Spet.	十月 Oct.	十一月 Nov.	十二月 Dec.	每年總數 Yearly Total	每月均數 Mean Monthly Rainfall
1880			19.0	106.4	266.7	122.4	151.6	106.9	45.0	0.0	7.6	63.2	(888 8)	(88.88)
1881		71.1		83.1	246.5	206.2	130.8	101.6	82.0	5.7	80.5	25.9	(1,133.4)	(113.34)
1882	31.2	149.9	34.3	164.3	243.3	113.0	140.5	86.4	67.1	207.5	74.7	0.0	1,312.2	109.35
1883	11.4	56.4	42.7	128.5	197.4	274.8	231.1	21.6	56.9	68.1	15.2		(1,104.1)	100.37)
1884														
1885	63.0	4.6	17.0	156.0	261.6	236.0	95.0	183.4	31.0	15.5	17.0	48.5	1,128.6	94.05
1886		40.6	101.6	202.4	259.8	264.2	38.9	37.3	29.2	278.9	9.4	0.0	(1,262.3)	(114.75)
1887	172.2	37.1	86.6	29.2	325.6	819.9	95.5	83.1	12.2	58.5	8.9	1.5	1,730.2	144.18
1888	89.9	40.4	142.5	191.5	91.9	22.9	169.4	231.4	36.8	49.3	19.8	7.6	1,093.4	91.12
1889	37.8	77.7	124.0	178.1	232.9	597.2	169.4	81.3	314.7	229.6	61.5	1.3	2,105.5	175.45
1890	46.2	40.9	111.5	220.0	100.3	205.5	115.8	92.5	0.0	5.1	34.3	50.3	1,022.4	85.20
1891	27.2	62.7	62.7	146.8	86.1	98.8	262.4	91.9	5.8	137.4	23.6	39.6	1,045.0	87.08
1892	21.3	63.0	58.4	281.5	231.9	204.0	150.9	144.3	44.2	9.9	51.1	33.0	1,293.5	107.79
1893	72.9	47.8	69.6	125.3	255.5	400.3	220.1	20.1	88.6	92.5	13.7	9.7	1,415.1	117.92
1894	47.2	66.8	155.4	118.1	178.3	198.1	126.0	32.0	240.0	69.6	38.9	41.4	1,311.8	109.31
1895	13.5	74.2	173.0	137.9	123.2	72.1	158.5	19.8	15.2	68.6	2.8	61.7	920.5	76.71
1896	31.7	14.7	88.4	296.9	176.5	323.8	316.0	100.6	40.9	87.6	56.1	51.6	1,584.8	132.06
1897	59.9	5.6	246.4	126.0	234.4	205.2	137.2	158.5	146.6	59.4	115.3	4.6	1,499.1	124.92
1898	58.4	98.5	95.8	80.0	392.2	251.2	6.3	97.3	30.7	15.7	2.8	2.5	1,131.4	94.28
1899	28.4	89.2	96.3	170.7	67.8	283.0	97.8	139.2	127.3	37.6	73.4	131.8	1,342.5	111.88
1900	9.4	3.1	23.9	173.2	119.6	108.5	305.3	1.3	41.9	111.8	35.8	27.9	961.7	80.14
1901	136.4	3.8	37.6	95.2	149.6	231.9	533.9	23.6	38.1	108.2	0.8	2.3	1,361.4	113.45

1902	20.3	1.3	59.4	160.0	125.5	5.1	86.4	5.8	4.1	54.1	11.2	42.7	575.9	47.99
1903	18.8	81.3	193.0	121.7	260.6	114.0	263.4	213.4	66.5	34.5	20.3	3.0	1,390.5	115.87
1904	30.5	43.4	54.1	319.3	118.6	67.8	19.0	242.6	115.3	215.1	10.2	2.8	1,238.7	103.22
1905	39.4	24.1	97.0	107.2	101.1	60.2	68.1	122.9	91.4	59.4	29.2	61.2	861.2	71.76
1906	23.9	66.3	77.5	47.0	99.3	288.8	83.3	456.4	20.8	1.8	21.1	7.1	1,193.3	99.44
1907	66.5	5.6	12.2	3.6	80.0	90.9	337.6	19.6	94.0	188.0	44.2	0.0	942.2	78.51
1908	73.7	33.0	45.7	131.1	152.4	38.5	224.8	22.9	100.6	136.9	93.5	5.1	1,401.2	116.76
1909	41.1	20.3	123.4	62.7	67.3	502.9	531.4	72.4	32.5	140.8	58.7	18.5	1,672.0	139.33
1910	89.7	15.7	94.2	68.3	130.8	186.9	318.5	65.8	29.2	47.2	48.0	0.0	1,094.3	91.19
1911	60.2	84.3	165.6	150.1	244.1	361.7	225.0	138.2	78.7	31.0	149.9	25.4	1,714.2	142.85
1912	32.0	60.7	216.9	142.0	153.2	337.6	342.4	8.4	28.4	47.8	154.7	47.5	1,571.6	130.96
1913	17.0	73.0	62.5	494.5	158.2	216.2	116.1	3.6	14.2	9.4	82.0	27.2	1,274.8	106.23
1914	21.6	109.7	61.5	227.3	161.0	118.4	5.6	88.1	52.1	133.6	78.5	17.3	1,074.7	89.59
1915	20.3	25.1	72.4	149.1	119.1	508.8	176.0	118.1	414.8	188.2	117.3	2.5	1,911.7	159.31
1916	28.2	56.1	31.2	218.7	134.4	264.7	233.7	167.1	38.9	110.2	61.0	20.3	1,364.5	113.71
1917	14.7	35.1	52.6	68.6	85.3	299.0	243.8	65.5	16.8	54.4	30.5	2.5	968.8	80.73
1918	0.0	26.6	120.6	185.7	89.2	166.4	120.6	272.0	29.5	17.3	226.1	128.0	1,378.0	114.83
1919	44.7	7.9	183.4	55.6	228.2	457.7	220.5	89.4	149.9	30.2	41.1	19.0	1,527.6	127.30
1920	91.7	100.3	251.7	72.1	166.1	84.8	59.4	47.8	75.2	107.7	40.4	72.6	1,169.8	97.48
1921	20.8	27.2	104.1	356.1	189.2	425.2	54.1	146.6	196.6	110.5	1.5	21.1	1,653.0	137.75
1922	64.5	83.8	86.4	127.0	142.5	229.4	88.6	20.6	10.4	44.7	2.0	2.5	897.4	74.78
1923	14.5	36.6	77.2	114.3	254.0	172.5	204.7	19.0	18.8	13.5	41.7	22.1	988.9	82.41
1924	40.4	52.3	87.4	104.1	200.9	104.9	293.9	20.3	9.7	26.7	5.3	8.1	954.0	79.50
1925	77.1	21.4	84.1	37.6	350.6	39.7	63.9	91.5	66.0	37.8	85.3	32.2	987.4	82.25
1926	45.5	50.8	71.4	83.3	125.0	114.9	314.8	38.1	95.7	53.8	46.4	38.5	1,078.2	89.85
1927	61.2	35.4	55.1	142.8	47.5	196.8	319.6	178.8	128.5	51.6	7.1	16.5	1,240.9	103.44
1928	41.1	24.3	91.9	152.4	118.9	308.0	73.4	35.4	90.4	16.4	140.9	48.1	1,141.2	95.10

（第十四表） 揚子江第一大段內航行困難時日之統計表 民國十二年冬至十八年春

年份	限度	湖廣沙	戴家洲西港
十二年春至十三年冬	十英尺	自一月廿一日至二月五日　計十五日	
	十二英尺	自一月三日至二月十九日　計四十七日	自一月十七日至廿二日　又自一月廿六日至二月六日　計十六日
	十五英尺	自十二月廿三日至三月十九日　計八十六日	自十二月廿九日至二月十一日　又自三月七日至十日計四十七日
十三年冬至十四年春	十英尺	自十二月十五日至一月十一日　計二十七日	自一月十一日至二月二日　計二十二日
	十二英尺	自十二月九日至二月一日計五十四日	自一月一日至二月七日　計三十七日
	十五英尺	自十一月卅日至三月五日　計九十五日	自十一月十九日至十二月十日　又自十二月廿五日至二月十五日　計七十三日
十四年冬至十五年春	十英尺	自二月一日至十三日　計十二日	自十二月廿四日至二月十七日　計五十五日
	十二英尺	自一月十六日至二月十八日　計三十三日	自十二月廿日至二月十八日　計六十日
	十五英尺	自十二月三十日至三月廿七日　計五十九日	自十二月十三日至二月廿五日　又自四月七日至十五日　計八十二日
十五年冬至十六年春	十英尺	自二月十五日至十七日　計二日	
	十二英尺	自二月十四日至廿一日　計七日	
	十五英尺	自一月廿七日至二月廿六日　計三十日	
十六年冬至十七年春	十英尺	自一月二日至三月三日　計六十日	
	十二英尺	自十二月廿七日至三月十二日　又自四月五日至十五日　計八十五日	
	十五英尺	自十二月十一日至三月十五日　又自三月廿六日至四月十五日　計一百十四日	自一月五日至十五日　又自一月廿八日至三月十日　計五十一日
十七年冬至十八年春	十英尺	自十二月十二日至四月十五日　計一百廿四日	
	十二英尺	自十二月五日至四月十五日　計一百卅一日	自十二月十九日至廿三日　又自十二月卅日至一月三日　又自一月十三日至二月一日　又自三月十六日至廿日　又自三月廿六日至四月二日　計卅八日
	十五英尺	自十一月卅日至四月十五日　計一百卅六日	自十二月十日至三月八日　又自三月九日至四月十五日　計一百廿五日

戴家洲冬港	羅蕩洲及鴨蛋洲
自十二月十七日至廿日　又自十二月廿七日至二月四日　計四十二日	自一月廿六日至二月四日　計九日
自一月三十一日至二月四日　計四日	
自十二月廿五日至一月廿日　又自一月廿四日至二月八日　計四十一日	
自十一月廿四日至十二月六日　又自十二月十三日至二月十一日　計七十二日	
自一月十九日至廿三日　又自一月廿六日至卅一日　計九日	自二月十二日至十九日計七日
自二月十四日至廿二日　計八日	
自一月三日至十五日　又自一月廿三日至三月九日計五十七日	自一月十一日至十三日　計二日
自十二月十九日至廿四日　又自十二月卅一日至一月二日　又自一月十七日至廿三日　又自一月廿七日至二月廿日　又自三月廿日至卅一日　計四十八日	
自十二月十日至四月四日　計一百十五日	

江家洲北港

十二月卅一日至二月四日　計卅五日

自一月三日至十九日　計十六日

自十二月十二日至二月十一日　計六十一日

自十二月廿七日至一月六日　又自二月三日至十一日　計十八日

自十二月十六日至二月十九日　計六十五日

自二月二日至十二日　計十日

自二月七日至廿一日　計十四日

自十二月廿七日至一月十一日　又自一月廿三日至三月十三日　計六十四日

自十一月廿四日至三月廿一日　又自三月廿七日至四月七日　計一百廿八日

自一月三日至廿九日　又自二月一日至十九日又自四月十三日至十五日　計四七十日

自十二月九日至三月二日　又自三月十三日至四月十五日　計八十三日

自十一月廿九日至四月十五日　計一百三十七日

江家洲南港

自十二月十八日至三月四日　計七十六日

自十二月十三日至三月廿八日　計一百零五日

自十二月六日至四月九日　計一百廿四日

自二月廿八日至四月二日　計卅三日

自一月卅一日至四月十五日　計七十四日

自十二月廿日至四月十五日　計一百十六日

張家洲北港

自一月十六日至廿八日　計十二日

自十二月十二日至一月廿三日　計四十二日

自一月十日至十七日　計七日

自十二月十三日至　二月十日計五十九日

自一月十日至二月十八日　計卅九日

自十二月廿日至一月十五日　又自二月九日至三月十日　計五十五日

自一月二日至七日　計五日

自十二月十四日至二月十九日　計六十七日

張家洲南港

自十二月八日至三月八日　計九十日

自十二月六日至三月十四日　計九十八日

自十一月廿九日至三月廿二日　計一百十三日

年份	限度＼地名	太子磯	崇文洲南港
十二年冬至十三年春	十英尺		
	十二英尺		
	十五英尺	自一月十日至二月二日　計廿三日	自十二月廿五日至二月五日　計四十二日
十三年冬至十四年春	十英尺		
	十二英尺	自一月五日至廿日　計十五日	自十二月廿八日至一月廿日　計廿三日
	十五英尺	自十二月十五日至二月廿三日　計七十日	自十二月三日至二月廿二日　計八十一日
十四年冬至十五年春	十英尺		
	十二英尺		自十二月廿九日至二月十一日　計四十四日
	十五英尺	自十二月卅日至二月十一日　計四十三日	自十二月十七日至三月五日　計七十八日
十五年冬至十六年春	十英尺		
	十二英尺	自二月九日至十九日　計十日	自二月七日至十八日　計十一日
	十五英尺	自一月廿六日至二月廿二日　計廿七日	自一月廿七日至二月廿三日　計廿七日
十六年冬至十七年春	十英尺		自十二月廿九日至一月九日　又自二月十日至廿一日　計廿二日
	十二英尺	自十二月卅一日至二月廿日　又自二月廿五日至三月六日　計六十日	自十二月十八日至三月九日　計八十一日
	十五英尺	自十二月十一日至三月十四日　計九十三日	自十二月三日至三月十六日　又自三月卅日至四月七日　計一百十一日
十七年冬至十八年春	十英尺		自十二月廿九日至二月十九日　又自三月十四日至四月十五日　計八十四日
	十二英尺		自十二月八日至二月廿八日　又自三月五日至四月十五日　計一百廿三日
	十五英尺	自十二月廿六日至二月廿日　又自三月十三日至四月十五日　計八十九日	自十二月一日至四月十五日計一百卅五日

崇文洲北港

自十二月廿三日至廿五日　計二日

自十二月廿三日至一月一日　又自一月七日至十八日、又自三月十六日至四月四日　計卅九日

得勝洲

自一月六日至二月六日　計卅一日

自三月十二日至十四日　計二日

姚家洲	馬當上游
自十二月十一日至二月廿六日　計七十七日	
自十二月八日至四月一日　計一百十四日	
自十一月卅日至四月十五日　計一百卅六日	
自一月十日至二月五日　計廿六日	
自十二月廿六日至二月十五日　計五十一日	
自十二月二日至三月二日　計九十日	自二月三日至廿一日　計十八日
	自十二月六日至三月十五日　計九十九日
	自三月廿日至廿六日　計六日
	自十二月五日至三月一日，又自三月十日四至四月十五日　計一百廿二日

漢口沙洲	漢口水尺最低位
	一月廿六日廿七日及廿八日爲四英尺一英寸
	一月八日爲三英尺三英寸
	二月七日及八日爲二英尺九英寸
自十二月十七日至一月二日　又自一月廿二日至二月八日　計卅二日	二月十七日爲五英寸
	二月十二日十三日及十四爲七英寸
自一月十日至十五日　又自二月四日至十八日　計十九日	
自十二月廿五日至三月十日　計七十五日	三月廿八日爲零度下一英寸
自一月十三日至廿四日　又自三月廿一日至廿八日　計十八日	
自十二月十四日至四月十四日　計一百廿一日	

結　論

綜覽上列各表,揚子江之概要與其性質,已可略見一斑,嘗讀　總理實業計畫,對於揚子江之改良及整理方法,鴻謨遠慮,籌畫周詳,治本治標,灼然燭照.際此物質建設時期,允宜奉行謹恪,以人力之支配,增進天然之利益,國計民生,日臻強富,豈惟饒灌溉,便舟楫,惠工商,利農田,免旱潦而已哉.況整理揚子江之動機,出於外人,因航業關係,欲求疏浚.而利交通,豈非越俎之謀,實遺借箸之誚.現在測量,先後已將八載,年製圖表,亦已成帙.惟是測量爲設計之準備,而疏浚爲整理之結果,徒事測量而不施工,雖可得久長可恃之資料,究非實行疏浚之目的.際此勵行建設,深望當局者,對於揚子江之整理,必能使之實現也.

隴海鐵路工程局招投土方工程

隴海西路工程已由靈寶起繼續向西展築,所有潼關城東有七公里之道基及車站土方工程,約十萬立方公尺,將於四月十五日在鄭州工程局開標,已登京滬津各報招投,並在鄭州局內及上海九江路六號比公司,備有章程圖樣標單等件,可以價領取云.

三相交流電標準制論

著者：周琦

曩昔吾國用電工程各種制度，上無政府機關之規劃及監督，下無學術團體之研究及倡議，因循將事，漸至城鎮各自爲政，鄰廠不相爲謀，雖電報，電話，屬於國營，亦復章程分歧，至電燈電力（電車等）多屬民營，尤爲混雜，遍查全國各小發電廠，迄今尚有用直流一二百伏脫之低壓者甚多，實爲不統一及不能擴大之主要原因，比種不統一之損害頗多，舉其犖犖大者如下：

（一）設備不經濟　向來國內創辦發電廠者，其發起人因無一種工程制度之依據，頗少通盤籌劃，貿然向各機器洋行，查存報價，其有現貨及價廉者，往往得標，不問其直流與交流，電壓之高低，及週波之大小，第求其能裝用發電，不知直流及低壓，須受傳電之損失，週波不符，須受鄰廠聯絡之影響，且初無鄰廠聯絡之預備，則其發電機容量，必失之太大或太小，如其太大，負荷率必低，卽以多量之燃料，僅獲少許之能量，如其太小，負荷率過高，機體壽命必減短，而電力不足，用戶反對，營業難於發達，凡此種種，皆以多量之金錢獲少量之效果，卽所謂不經濟也。

（二）供電不連續　發電廠圖設備之經濟，必由設備之切當，尋常供電，必無餘量以濟意外之急，然電燈關於一地之治安，電力關於各廠之秩序，皆所謂不可須臾離者，供電旣一刻不可間斷，偶遇原動機發生障故時，全恃鄰廠之聯絡，今因電流之異類，電壓之異度，週波之異數，竟致不便聯絡，或聯絡而所費不貲，有所不願，則供電有時必須間斷，卽所謂不連續也。

（三）用電不簡便　各地發電制旣不劃一，同一電器適於甲地者不適於乙地，或必須另加機具以改用，皆足以阻用戶之樂購，卽電機製造廠，每一出品，亦必常備合於各種電制之多件，以應銷售，然製造方面驟增繁複，成本售價,亦無不增加.此皆直接影響於用戶之心理,間接影響於電業之擴大也.

自國民政府定都南京,趨重建設.各省各市政府,多設公用局,監督公用事業,如水電兩項.同時建設委員會直營電業,或監督民營電業.於是政府機關之制裁,始粗具規模.然對於劃一電制標準,尙未釐訂,總未能充量施其效用.至於學術團體,對於此既重且要之工程法規,亦未聞有所擬議.本會顧名思義,尤當其衝,允宜急起設定標準制,以迅速時機.鄙人不敏,敢貢其議於後.

吾國發電機應定三相交流五十週波之電爲標準,而以單相交流五十週波之電爲附麗,其理由甚長,容於異日另著專篇,綜其大要,則因三相交流五十週波之電爲(一)發傳最經濟,(二)應用最廣,(三)電燈電力可同機供用.

至於各種電壓分發電,傳電,及用電之三類.應須有相當之聯絡標準及說明.今詳諸附圖.

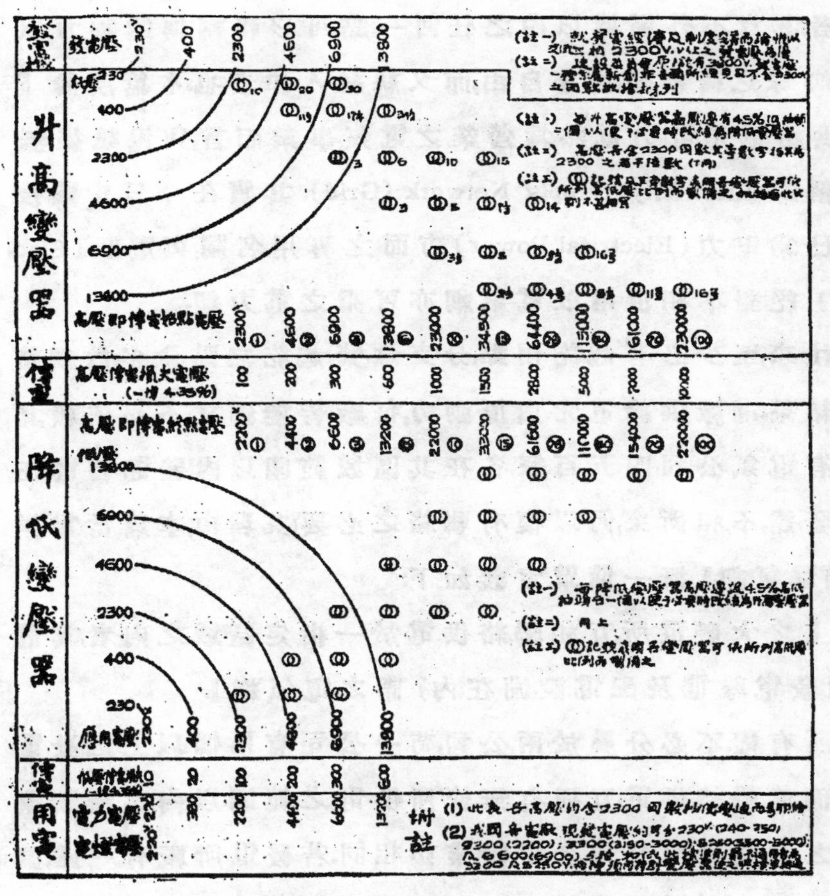

標準三相電壓發傳及用處聯絡圖

電 氣 網

著者: 惲震

(一) 何謂電氣網

電報電話之以多處相互接通者,曰電訊網.鐵路之縱橫貫穿,相互通達者,曰鐵路網.所謂網者,其意不外區域以內之任何一點,可以輾轉傳遞物力以達其地各點,一如蜘蛛之營網絡,往來自由,而又絲絲入扣者也.本篇所論『電氣網』,專指以供給電光電力電熱爲營業之電氣事業而言,與電訊絕無關係,在英文中可稱爲 Electricity Supply Network (Grid). 其實在今日中國技術界中,電氣二字已爲電力 (Electrical Power) 方面之專用名詞,與電訊 (Electrical Communication) 絕對不相混淆,故電氣網亦可謂之電力網.

電訊收發,火車往來,至少必有起訖兩點,分支愈繁,起訖之點愈多,故網絡之形成,乃當然之結果.電廠供給電光與原動力,每廠皆能獨立,不必依賴其他電廠.例如中國有電氣公司四五百家,各在其區域範圍以內發電售電,在表面上觀之,各個發達,不相衝突,何以復有聯絡之必要?其理由,本篇當爲詳細闡明之.茲先爲『電氣網』擬一簡單定義如下:

『凡以兩個以上之大發電所,互相聯絡供電於一指定區域之內者,其電氣制度(包括發電輸電及配電設備在內)謂之電氣網』.

上述發電所之所有權,不必分屬於兩公司,倘一公司有兩個以上之發電所,或水力,或熱力,而能聯絡供電,互通有無,使所供電之範圍以內,常無斷電之虞,則其電氣網之功能,與兩公司合作營業亦相同.若發電所祇有一處,其發電容量無論如何巨大,輸電線路無論如何遠達,仍不宜稱之爲電氣網.其故蓋以萬一發電所機器停止,此外別無電源,用戶之機器電具,皆將暫時停網,其範圍愈廣,斷電之危險亦將愈大也.

(二) 電氣網之重要

每一電廠既可獨立生存,則其聯絡供電,互通有無,似爲一種發達後之附帶現象,而非電氣事業之根本條件.今日中國之少數民營電廠,日夜惟恐爲大力者所覬覦併呑,見電氣網三字,尤畏之如洪水猛獸,充耳不欲有聞.此其不明大勢,不識大體,實以囿於見聞知識之故,無足深怪.吾國生產落後,農事衰落,工商凋敝,憂國之士,無不焦心極慮,欲爲全國人民謀一出路.然則發展工商增加農產,開採礦利,苟無『取之不盡』『隨處可得』之廉價原動力,則任何創業,必仍感困難,進行亦受限制,成本既不經濟,建設終歸失敗.此巨大之原動力,决非直接燒煤運汽所可勝任.若用電力,亦非電廠各自爲謀不相聞問所可成就.欲求一切農工鑛業之電氣化,惟有建設電氣網一法.故電氣網之重要,專從電氣事業主觀方面視察,决難得其究竟.必從全國整個經濟建設立場着想,始能窺其全豹.此固非少數人之利害,而實爲大多數國民福利之樞紐也.茲試爲條舉其理由如下:

(1) 供電普及與穩定　一電廠之供電,線路至長不過百餘里,範圍至大不過二三城市.供電區域,電氣事業,必擇其人口密而市面盛者.鄉野之區,用戶少而線路長,在廠方極不經濟,故延線入鄉,必不得已而後爲之.結果城市愈便利,鄉村愈艱苦.農田工作,必以重價,自購機器以資運轉.鄉野山地,開設工廠,亦祇能自備動力.工商農鑛,卽有進展,亦爲畸形的,而非系統的,普及的.今若以高壓電線,聯絡各電廠,使輸電配電各線路,縱橫貫連,達於四野,城市依然保持其用電重心,而鄉僻之區,曩之以去市太遠不易得電者,今則俯拾卽是.製造工廠,曩之以電廠力量太小而不能供電者,今則數廠相連,供過於求,購電遂無問題,且電源衆多,斷電之機會減少至最小限度,此電氣網之大利一也.

(2) 成本減輕　假定某省區中,有電廠四十所,其發電容量在一千基羅瓦特以上者有十所,其餘皆爲百餘基羅瓦特之小廠,或用燃煤,或用柴油,皆不甚經濟.電氣網設立後,將十廠容量擴大,以高壓電線相連絡,復於適當地點,加設兩大發電所,以爲基本電廠,其餘小廠停止運用.原有之各電氣事業人,均得向電氣網躉購電流,照舊販售.昔有之四十廠,平均每度電假定耗煤六磅,現有之十二廠,平均假定二磅,則在燃料上已可節省三倍費用.此即集中發電,常用效率較高之發電所之結果.燃料既省,成本自可減輕,用電之需要自可激增,電氣事業之獲利愈大,此電氣網之大利二也.

(3) 電多而價廉　發電成本既輕,其躉售之價必廉.而零售之電亦可由政府法令規定,不得故意非法抬高價格.如是則不僅大工廠或電力用戶可得甚廉之電價,即家庭用電,烹飪工作,皆可儘量採用.用途愈多,供給愈便,若復開發水力,擴大供給,則電價將愈趨愈廉,以達全國電氣化之境界,如挪威,意大利皆可爲佳證,此電氣網之大利三也.

(4) 節省全國燃料　電氣網設立後,不僅發電廠之燃料可以節省,其他一切工廠,經過電氣化後,其本廠原用之燃料,全可保留不用.又如鐵路之用煤,爲量甚鉅,若改用電氣機車,取給於中央電廠,則所省之煤,爲數亦必可觀.據Hugh Qu'gley 估計,英國1922年用煤160,000,000噸,若能儘量電化,則每年可省60,000,000噸燃煤,此項省下之煤以之提煉煤油,可得600,000,000加侖.此就英國一國而論,在他國當然亦有同樣情形.吾人知燃料之鑛藏,有其窮時,吾人若不及早求使用之經濟,後人將必感受莫大之痛苦.電氣網可以節省全國燃料,此電氣網之大利四也.

(5) 利用水力　水力發電廠建築費雖較鉅,經常費用極省,故欲得廉價電力,必設法利用水力.例如挪威有發電總量1,580,000基羅瓦特全部皆爲水力,其人民每人每年用電達1,5880KWH,爲全世界首屈一指,較之英國每人每年用電246 KWH 尚超過七倍半.但水力所在地常與用電中心點相去甚遠,又水流量在四季中往往不能平均,水少之時必賴熱力電廠爲之補充.故若以高壓線路,組成電氣網,則水力之發展,可以無慮險阻,千里相聯,此外復有適當地位之熱力廠,互通有無,相與調節,使燃料盡量節省,水力盡量利用,各得其所此電氣網之大利五也.

(6) 減少備用機量　每一電廠,爲預防機器損壞停駛起見,必有備用機量,少者爲常用機量之半數多則或爲常用機量之一倍.以全國統計,此項不生產之機器投資,爲數殊屬可驚.以增加機器資產之故,發電之固定費用(Fixed costs)亦隨之增加,電價成本遂多一非必要之負擔.若用電氣網將各廠聯絡,則此項備用機量,可以減低至最小限度.例如甲乙丙三廠,各有一萬基羅瓦特常用機量,如無電網聯絡,則每廠或須各置備用機量一萬基羅瓦特,三廠總數卽爲六萬千瓦,其中半數爲不生產的.若三廠早已組成一網則機器決不致同時損壞,備用機量祗須於一廠中設置一萬千瓦已甚安全夠用,可以省去二萬千瓦.據英國 J. M. Kennedy 之報告,其全國備用機量,在推行電氣網之數年內,已自70%減至15%,結果得異常滿意之經濟.此電氣網之大利六也.

(7) 調劑負荷分配　甲地與乙地之最高負荷,未必在同一時間發生,例如甲乙二廠,常用機量及最高負荷各皆爲五萬千瓦,甲廠負荷最重時在每日正午十二時,乙廠在下午八時,其他時間,各僅需二萬千瓦.若兩廠相聯,則每廠當最高負荷時,可得他廠之電力補助,不必由一廠單獨負担.如是則兩廠之常用機量總數至少可以減小四分之一,甲廠置四萬基瓦,乙廠置三萬基瓦,已足應付甲乙兩地聯合之需要,機器投資既可減輕,發電成本又可減少,此電氣網之大利七也.

(8) 分散人口與振興農村　近世人口多集中都市,勞力者漸自鄉村向城市移殖,都市人口旣繁,公用之供給,衛生之維持,彌感困難,而罪惡之產生,亦日見其多.電氣網若經過鄉野,便於住家反可設工廠範圍驟增數培,一部份人民自然復由都市分散而囘入於鄉村.小鎮市之繁榮,易如反掌.如此則社會學家之一大問題,不解决而自解决,此電氣網之大利八也.

根據以上八大理由,電氣網之宜於積極推行,殆無疑義.若美國之以二千七百萬基羅瓦特領袖全世界,其各地電氣網皆已陸續完成,未藉政府提倡之力,而其各個網上之發電量尙有加無已,繼長增高,以應各業之需要.英文謂偉大之電廠爲 Super-power station 謂電氣網爲 Super-power zone. Super-power station 集中電量於一處,其極也或反得不經濟之結果,故論者公認 Super-power zone 之重要當在 Super-power station 之上.德國分全國爲五區,法國分爲七區,其高壓聯接皆在十萬伏而脫以上,成效甚爲顯著.英國由政府主持,分全國爲十五區,其推行較遲,而近數年進步則極速.俄國新經濟政策,以電氣網爲主要動力,近年建設,孟晉可觀其他電氣發達之國家,如加拿大,意大利,瑞士,挪威,瑞典,日本,無不採用電氣網制,蓋電氣網在今日已成爲發展工業及提倡電化之必要條件矣.

(三) 電氣網之困難

電氣網非萬能也,亦非有求必應之靈方,苟不問其當否而盲目推行,其結果亦必陷入於不經濟.高壓線路,目前所最通用者爲六萬至二十萬伏而脫,其建設費用至鉅,維持亦非容易.雷擊之危險,暴風之襲擊,如在中國則更有盜竊之患,在在皆足以使高壓線路增加困難.又如人口太稀,負荷不重,而欲建若干高壓線路,超越千里,徒事發展,不問用途,亦非有經濟常識者所許.又如初次擬用高壓線路,其選用之電壓,每以途程較短,不必甚高,日後線路展長,超出從前預計,改造線路,其絕電物油開關變壓器等,皆須更換,所費之大,往往使人躊躇難決.故在設立電氣網之前,第一必須有經濟比較之考查,第二必須有遠大之設計程序.

電氣網之建設,或由民營電廠自動組合以經營之,如美國之例是也.或由政府出資建造,強迫聯接,如英國之例是也.然無論原動力之屬於何方面,凡屬電氣事業人,皆須澈底明瞭電氣網之根本利益,政府尤宜廣爲宣傳,俾来喩解,無所疑惑.否則即使輸電線網造成,各電廠不肯儘量利用,或設法使彼此交互供給之電度數量相抵銷,惟恐利益爲對方所侵佔,其愚雖不可及,要亦宣傳未盡得力之失也.

今日中國之發展電氣網,宜先自工商業較繁盛之區域試辦,高壓線路,暫以十萬伏爲限.一省試辦有效,他省自易倣行.茲將技術組織及行政組織分述如下,以爲研究者之一助.

(四) 電氣網之技術組織

電氣網最重要之條件有三:其一,各廠置機,均須依照標準電壓及週率之規定;其二,效率較低之發電所,必須停止使用;其三,技術上之指揮,必須統一.在電氣網中,各發電所皆失去其特殊之地位,即使其所有權分屬於各公司,各電氣事業人亦不得對於其發電多少,或使用時間長短,有所爭議.蓋一切電力既已集中,各電氣事業人祇須向電氣網購用最經濟之電流,而逐度記錄之,不必問電自何處來也.例如甲公司有新機五萬基羅瓦特,未加入電氣網之前,每月發電15,000,000度售諸各用戶;加入之後,以其電廠效率較高之故,使用時間增多,每月發電25,000,000度,按照成本售諸電氣網,其自需之15,000,000度,再向電氣網購來,分售各用戶.假定從前甲公司每度成本爲二分五厘,加入電氣網之後,其成本減至二分,電氣網若照每度二分二厘售與甲公司,則甲公司因加入電氣網所得之利益當爲$\$0.003\times15,000,000=\$45,000$即每月四萬五千元也.

每一電氣網之區域,不宜過大,過大則線路太複雜,控制不易.如在中國,一區不宜大過於一省,然亦不宜以省界爲限,有時天然發電及用電之配合,或須關聯三四省區,亦未可定,在電氣網初具雛形之時,二三公司試行聯合,不妨暫從簡略,不設中央控制室(Load Dispatcher's Office),亦不將各發電所所發之電統歸電氣網支配,祇以互通有無爲原則,每月結算,僅憑電表上彼此流動之過剩數,如此則簡單易行,且無行政上之問題.如發電所逐漸增多,或配電線網逐漸複雜,則中央控制室之設置,實爲事實上所需要,不復能如向之簡易矣.

所謂中央控制室者,卽全區總工程師之駐在地,與全區各發電所及變壓所必須用電話接通,其各路高壓線之使用,及分段油開關之啓閉,亦必須用適當方法傳達至中央控制室,使全網電氣狀況,管理員得以一目瞭然.何廠應多受負荷,何廠應暫時停駛,一一皆須聽命於管理員,以求全網使用上之最大經濟.各發電所變壓所之工程師,祇須受命於中樞,執行其命令,隨時將實況報告,自可得最良之結果,無所謂偏枯偏榮也.

電氣網之基本電廠,其地位之選擇最關重要.水力發電所祇能就水力所在地而建廠.熱力發電所則有運煤取水二大問題.以定取舍.運煤便利則燃料成本低廉,此點當比較運煤與運電孰爲經濟而定.發電中心,能與用電中心愈近愈妙,然有時爲就燃料便利之故,祇能放棄此種利益.水量之多少,足以測定熱力發電所之最大擴充容量,故電廠宜設於大河或巨湖之沿岸,以便取用多量之凝汽冷水(Circulating water).若迫於環境,不能取得活流之水,則可設大規模之冷水塔(cooling tower)以爲替代,此法在德國用者甚多.

電氣網造成後,在技術上有數點必須特別注意.第一,線路上之『力來』保護(Relay Protection)應愼重設計,使油開關之因變故而啓斷,及重行閉合,有一定之次序,不致妨害全網之運用.第二,各線路之重要油開關,其斷路容量(Rupturing capacity)必須有充分之能力,使在任何變故情形之下,能按照預計時間啓斷.

假如在已有電氣事業中,其發電所各用不同之週率及輸電電壓,則一旦若欲實行電氣網制,必將感受異常之困難.若另購新機,則成本太鉅;若用變週率機及加多一重變壓,則不但費用浩大,在運用上亦極笨拙不便.日本此類困難甚多,至今尚受其累.吾國電氣事業正在萌芽,然對於此點,大衆皆負有責任.政府所規定之週率與電壓標準,任何電廠皆應遵守,蓋此實有關於全國電氣界之合作也.

每一電氣網之區域,既可自成單位,其與鄰近之電氣網,自不宜在多處聯接,以增加其組織上之繁複.但於區域邊界上,宜預設適當之配電所,使兩個不同之電氣網,亦可交換電流之供給.在平時則各不相關,在特別變故情形下,則彼此皆多一重保障,有恃而無恐.如此甲乙相聯,復更迭與丙丁聯接,雖有數十網,亦可於必要時合而爲一,全國貫通,呵成一氣.此事在今日雖若渺茫難期,然逐步做去,實現亦非難事也.

(五)電氣網之行政組織

吾國將來之電氣網,在行政組織上將採取何種方式,誠爲今日亟應討論而極有興趣之問題.世界各國之電氣網,多由政府與私家公司合力經營,政府負提倡督促及籌集巨欵之責,各公司努力合作,皆無異議,蓋以公利之所在,原無所用其私鬥.惟美國代表資本主義,俄國代表社會主義,其組織方式,皆各趨極端.在美國之各電氣事業及電氣網其發電量總數百分之八十三操於二十大公司之手,政府完全不加干涉,在名義上各州雖有 Public Utility Commission,司節制業務電價之責,然在積極發展及行政管理上,政府皆不過問.俄國新設大電廠若干所,並已設立電氣網多處.完全由政府經營,私人無加入之權利.此二者吾國皆不宜仿行.英國電氣發展較遲,近十年來勵行電氣網制,頗足供吾人取法.茲試於本節略述之.

英國之電氣事業人,或爲私家公司,或爲地方政府,各有其註冊營業區域,每一電氣網區域,有一類似參事會組織之聯合電氣委員會(Joint Electricity Authority)由各電氣事業人,用戶及工人之代表,及區域內有關係之官吏組織之.各區域之業務管理,由中央電氣局(Central Electricity Board)負責,至行政計劃及仲裁之權,則歸政府所指派之電氣委員八人處理之.區域內之電廠,均須由中央電氣局指揮支配,所發之電,由局按照實在成本全部收買,作爲電氣局所有之電.各電氣事業人所需若干電度,即向該局躉購.(其實全根據電表上紀錄計算並無買賣手續)電氣局並得要求電氣事業人增加設備,改換週率,以便電氣網之聯絡,必要時須以無利息之借款貸與電氣事業人.如被選之電廠,不願爲電氣局所指授之擴充,得請中央主管部長派員公斷之.如無結果,電氣局得呈准部長收買,交由另一電氣事業人或本局經營之.電氣局又得發行電氣公債,建造高壓輸電桿線,惟重要設施,須得電氣委員之同意.

由上所述,可知英國政府對於電氣事業人完全取干涉之態度,不但干涉也,且代爲經營其全部發電輸電,祗餘配電及零售不加干涉.各電氣事業人對於其一切資產,仍保持其所有權,絕不疑政府之恃勢侵凌,此種合作態度,吾人至當效法,然其詳細組織,則以國情互有異同,不宜亦步亦趨.吾國各公司之營業區域,政府自當予以保障,電氣網之管理權,各區亦宜設局以主持之,至發電輸電之應否全部由局辦理,抑仍由各廠分別自辦,而由局監督,則應依各省情形各爲規定,不宜强納全國於一種方式.總之電氣事業以國營爲主幹,以民營爲分支,本合作之精神,圖全體之利益,中國雖大電氣化不難推行盡利也.

(六) 結語

電氣網不僅爲電氣事業一方面之問題,而爲全國共同之問題,已於上文中詳言之.全國經濟改造與實業建設,實賴電氣網爲之樞紐. 總理於其遺教中(民族主義第六講,民生主義第三講,及建國方略各章)已再三諄示,吾人殊不必更事懷疑.今後問題,惟在如何逐步施行.在政府方面,在民電公司方面,在各業用戶方面,在技術人員方面,除通力合作外,殆無其他出路.作者倉卒草此短文,對此光明之前途,實抱無窮希望,幸吾電氣界同人注意及之.

國防與工業

著者:馮朱棣

近日列強於縮短軍備條約聲中,某國以軍艦下水聞,某國以潛水艇製造聞.盟誓具在,不崇朝而自渝,掩耳盜鈴,孰有甚於此者.近更有聯世界各國,締息爭之盟,會簽凱樂Kellogg公約者.奈弭兵之說,終古無成,利害之爭,豈息爭二字得以限制之乎.正恐弱小者無志遠圖,因非戰而偷安苟且,更弛軍事.不逞者密布軍備,以逞雄於異日.則強弱之勢更殊,弱肉強食之勢,稱有更甚於今日者.然則軍備之施設,容可戢乎.近日各國學者,知軍備之不能或戢也,故於研兵器之計畫.不遺餘力.回視吾國,則兵器之制式,恆步武他國,終落人後.良以兵器一項,研究乏人,以致工藝兵器,未能融通貫一.甚者且以兵器爲殺人利器,仁者不爲.然人方磨礪以須,伺吾以隙,己不殺人,人且殺我.蹩躠爲仁,踶跂爲義,國亡無日矣.天下興亡,匹夫有責.說文:國,邦也,从□,从或.又曰:或,邦也,从口,戈以守其一.一,地也.余則謂國从□,从或,或从一,口从戈,□國境也,一口者一人也.字於六書爲會意,則執干戈以衛社稷,人人之責也.造字之初,固具此理矣.吾儕雖不能投筆從戎,亦當擘畫利器,爲士卒用,即所以報國也.工業與國防,關係絕鉅.以言乎設計,則槍砲一內燃機也.火藥之爆發,氣壓作用也.是以槍砲之動作,槍筒砲筒中之氣壓,彈丸之速率及加速,火藥燃燒氣體之溫度,彈丸之形狀,空氣之阻力,射程若何,侵徹力若何,舉凡內彈道學 innere Ballistik 及外彈道學 aussere Ballistik 所研究者,靡不以力學 Mechanik 熱力學 Thermodynamik 及熱化學 Thermochemie 爲本.吾儕機械工程師對此,自不難一舉三反,求得線索.然吾國向日鄙視工業,兵器一項尤視爲殘忍,誰肯研究.是以習用他國成法,鮮有剏造能力,依式製造,未肯探其原理.索其究竟.拾人唾餘.未能出人頭地.然吾國民族,未可云弱.當夫漢唐盛隆,四裔震驚蓄短兵相接,固可出奇制勝.殆夫火器盛行,拙象始現.是以有清一代,所遇被挫.雖因將卒老朽,然兵器不良,亦必敗之道也.今日科學昌明,新兵器之剏造,日新月異,是正工程師効命之秋也.德諺云:『戰爭爲萬事之父』.Der Krieg ist der Vater aller Dinge. 余謂工業爲戰爭之父.非有工業,不足以造兵器,儆兵器,則不足以言戰.火藥剏自吾國,而國人不知用,外人襲其成法,幾經改良,於是易黑色藥爲無煙藥矣.槍之初造,口徑至鉅.拿破崙以一七·五公釐鉛彈,橫行歐洲.後世欲增高彈道能力,ballistische Leistung. 縮小口徑.近減至八公釐以下,而初速則自三百公尺,增至八百公尺以上.砲之初造,口徑至小,增高戰力,幾經放大,於是歐戰時四十二公分口徑開於世.近更有擬造四十二公分以上者,更以槍彈未能及遠,砲位運施不便,欲求戰力活動,飛機是尙.於是飛機投擲彈,Abwurf geschoss 應時而生.他如毒瓦斯之應用,防禦器之完備,舉凡種種,靡不由工程師之悉心規畫.國家之盛衰,亦卽與工業之興替爲向.試觀工業昌明之國,海陸空軍之戰鬥力必強.以其工業發達,兵器精良也.國防之繫於工業,毋待言矣.更言乎製造.製造不難,而難在原料之完備.吾國兵器原料,在在均缺,事事外求.主要原料爲鋼鐵,而國內煉鐵爐寥寥可數.漢冶萍之不振,龍煙之荒閉,鐵且不給,遑足論鋼.他如銅鋅鎳鉛,以及炸藥雷管毒氣等等之化學原料,無莫而非外貨.主要原料固缺,甚且輔助工作之品,國貨亦絕少.往往以一物之微,國產無可代用,而影響工作者有之.若是而欲固國防,不亦難歟.萬一國際違言,疆埸起釁,敵人祇須封鎖海口,杜絕原料.則不損一兵,不費一彈,巳足致我死命.現雖民智漸開,民氣漸旺.奈赤手空拳.爲圖何濟.國勢之不競,工業不振有以致之也.是以事事受壓迫,處處不平等.誠欲挽回頹氣,非生產原料,製造利器,不克有濟,士卒爲國防前驅,而爲之後盾者,舍工程師其誰屬.本會同人,各科咸備,深願各竭所長,生產原料,製爲利器,以固國防.旣不負所學,亦且効忠於祖國,願本會同人勉諸.

長途電話橋接法

著者:陳紹琳

長途電話兩局之間,吾人有時欲另設一通話之處,如添設一局,或加放專線,則經濟上容有所未許.若將普通之電話機,橋接於兩線之上,則一處搖鈴,他處皆響,雖可用搖鈴之次數及長短,藉資識別,然終覺令人討厭,且易發生錯誤,决非妥善之辦法也.玆所述說之橋接法,能將上述之困難,悉行除去.兩局之間,可橋接一具或兩具電話.除此兩具電話,不能直接通話外,其餘兩局與二電話之間,皆能互通信號打接電話,而不驚動其他二處.若兩橋接之電話,欲彼此通話,則須先行通知任一局之接線生,然後再由接線生將被叫者喚出,始能達到通話目的.此爲長途電話所必須經過之手續,蓋非如此,兩橋接電話,可自由通話,局中無所記錄,不能照收話費,殊非所宜也.

橋接電話分一橋單接,一橋雙接,及二橋單接,與二橋雙接四種,以適應各種不同之需要.試分述之於下:

(一) 一橋單接法

於兩局間,橋接一具電話,但祇能與一局直接通話.其接線法示於第一圖內.甲乙爲兩局,丙爲橋接於兩線上之電話.甲乙間打電話之時,可照普通之方法將塞子插進接線孔 (Jack) 2 或 4,搖動磁石發電機 (Magneto-Generator) (圖上未表) 而使對方之號牌 (Drop) 3 或 1 跌落.如甲欲打電話至丙,須將塞子 (plug) 插進接地接線孔 7,而搖動磁石發電機,則電流必通過塞流線圈 (Retardation Coil) 5,及兩線而至丙,再經過塞流線圈 8 與電鈴 9,然後入地而回至甲.於是 9 響,丙出而提起聽筒,卽能與甲通話.然此時甲須將另一塞子插進接線孔 2,蓋 7 祇供搖鈴而不能談話也.橋接電話機之構造,除電

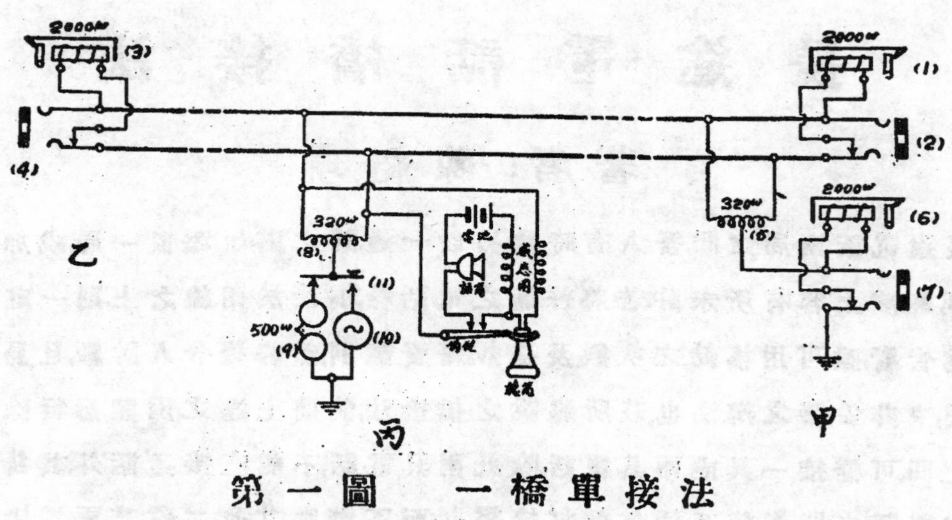

第一圖　一橋單接法

鈴與發電機之接線法,稍有更改外,餘均與常用者相同.當電流通過塞流線圈時,兩半圈之電流相等而方向則相反,故不發生阻塞作用.如丙欲與甲通話,須將電紐(Button) 11 下押,而搖動其磁石發電機 10 使電流通過 8 與兩線而至 5,再經過號牌 6 入地而回至丙.於是 6 跌落,甲可將塞子插入 2,而與丙談話.當甲丙互通信號之時,乙可毫不知情.然任何兩方談話之時,第三者皆得而聞之,故各方於搖動發電機之前,須先聽對方是否正在談話也.

(二) 一橋雙接法

此法能使橋接電話與兩局間,皆可直接通話.其接線法如第二圖所示,三處各裝一斷續直流發電機 (pulsating D. C. Generator) 2, 及一塞流線圈.甲乙二處,各裝一定向電鈴 (Biased Ringer) 3, 對於一定方向之斷續直流,始能發生鈴聲.丙處之 6, 係一繼電器 (Relay), 7 係一振動電鈴 (Vibrating Bell), 繼電器動作一次, 7 亦敲響一次,如有斷續直流通過 6, 則 7 之鈴聲可不斷而大作矣.甲與乙之通話手續,與上節所述者相同.甲與丙通話之前,須先將電紐 4 下押,而搖動發電機 2,則斷續直流由地傳至丙,過繼電器 6, 與塞流線圈 5

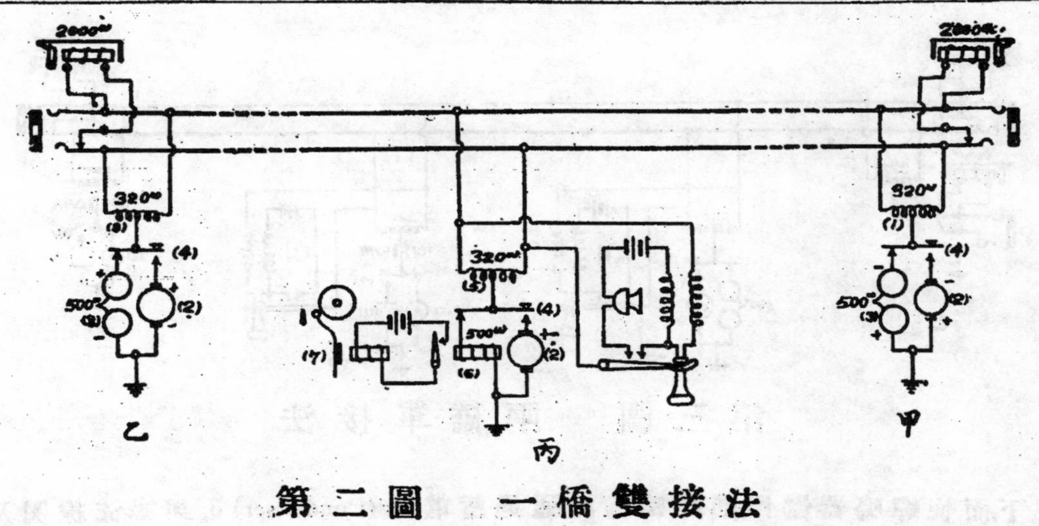

第二圖　一橋雙接法

及線路而回至甲.於是丙之鈴聲大作,可出而接話.當甲搖鈴之時,有一部分電流通過乙之電鈴,但因此種自地而出之電流,不能使乙之電鈴體作,故可默然無聲.於乙搖鈴至丙之時,亦同樣可以不驚動甲.至於丙搖鈴至甲時,可先由兩極雙投開關(D. P. D. T. Switch),將發電機之陽極接地,陰極接於電紐之下端,而押下電紐 4,然後搖動發電機 2,則斷續直流,自地至甲,通過定向電鈴 3,與塞流線圈 1,及線路而回至丙.於是甲之定向電鈴發聲,可將塞子插進接線孔而與丙對談.此時亦有一部分電流通過乙之定向電鈴,但不發聲耳.若丙由兩極雙投開關,將陰極接地而接陽極於電紐上,則可搖響乙之定向電鈴,而遂與乙直接通話之目的.如此三局間可彼此直接通話,而不假手於他人.但於搖鈴之前,仍須先聽線路是否空閒也.

(三) 兩橋單接法

甲乙兩局間,橋接丙丁兩具電話,但丙祇能與甲,丁祗能與乙,直接通話耳.第三圖內之 1,係一蟬鳴器(Buzzer),2 爲通常用之聽筒.當蟬鳴器內發生之電流,通過聽筒時,能發生宏大之蟬鳴聲.甲丙間互通信號之時,可將電紐 4

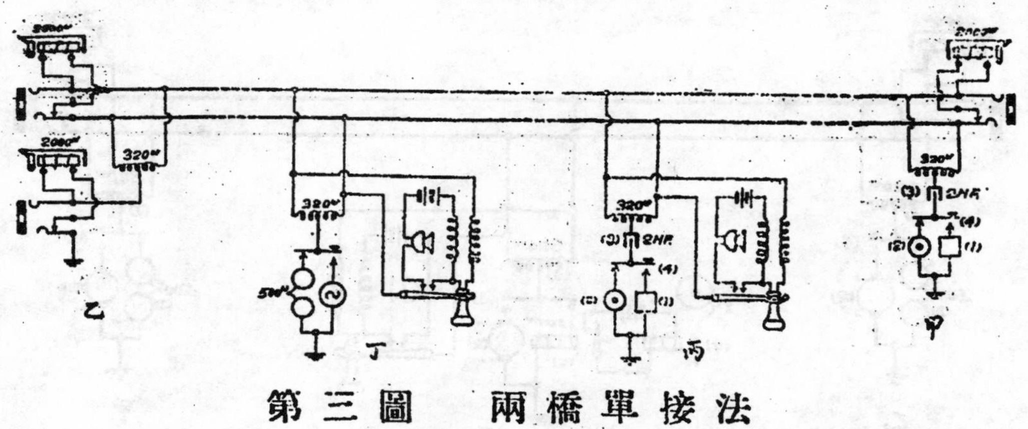

第三圖　兩橋單接法

押下,而使蟬鳴器働作,則蟬鳴電流,通過蓄電器(Condenser) 3,與塞流線圈及線路,而至對方,再經過塞流線圈與蓄電器及聽筒而入地,於是對方之聽筒內,即發出蟬鳴聲,而促其前來接話.至於乙丁之通話手續,則與第一節所述者相同,不復贅.

(四) 兩橋雙接法

甲乙兩局間,橋接丙丁兩具電話,除丙與丁以外,其餘皆能彼此直接通話,各不相擾.此種接線法,示之於第四圖內.4 爲單極單投開關 (S. P. S. T. Switch),於打電話時須打開,而不打電話時則須關閉.甲乙丁三處之通話手

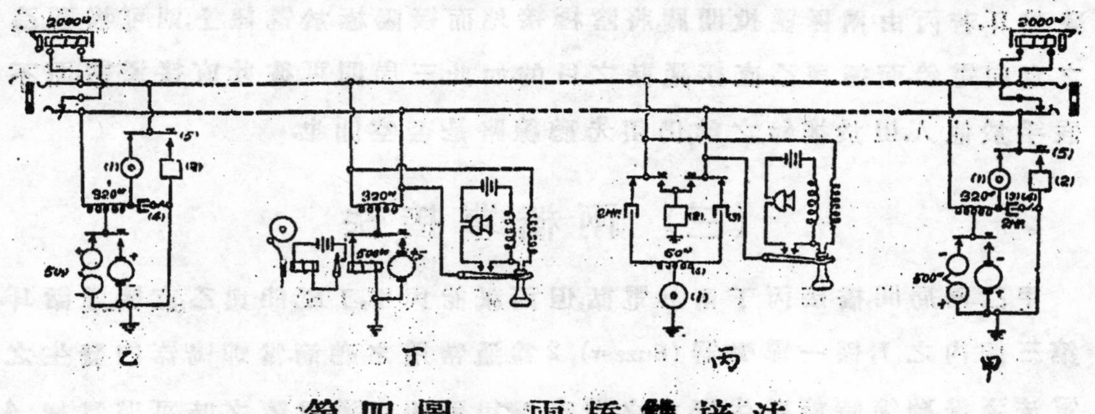

第四圖　兩橋雙接法

續,與第二節所述者相同.今若甲欲與丙通話,可先將 4 打開,而押下電鈕 5,使蟬鳴器動作,則丙之聽筒 1 內,必發生蟬鳴聲.蓋自蟬鳴器發出之電流,由電線傳至丙,通過蓄電器 3 與塞流線圈 6 之一半,及聽筒 1 而入地故也.乙與丙之通話手續亦如之.次設丙欲與甲通話,則祇須將右方之電鈕下押,接線路於蟬鳴器上,而使之働作,則電流自電線傳至甲,通過聽話筒 1,開關 4,與蓄電器 3 而入地.故甲之聽筒發聲,可將塞子插入接線孔,而與丙對談.丙與乙通話時,則將左方之電鈕下押即得.當一處送蟬鳴電流至第二處時,亦有一部分電流通過其他二處,但不發聲耳.

附註一　一橋單接法之信號,亦可應用蟬鳴裝置,其接線法如第三圖之甲丙.又一橋雙接法之信號,亦可應用蟬鳴裝置,其接線法如第四圖之甲乙丙,且可拆去甲乙二局之對於丁處信號裝置.

附註二　本法所使用之斷續直流發電機,可由普通電話機上之磁石發電機改造之.即在軸上附一絕緣環,而於其外周之適當處,所裝一小銅片,連此銅片於發電子線圈 (Armatvre Winding) 之一端,而押一彈簧於絕緣環上,使絕緣環旋轉一周,彈簧與銅片可接觸一次.再由發電子線圈之另一端,接出一線.則此線與彈簧之間,當發電子旋轉之際,可發生數十伏而次之斷續直流電壓.其斷續之次數,適與發電子線圈旋轉之次數相同,一秒間約可達十餘次也.

附註三　蟬鳴器之構造甚爲簡單,以圖示之如下:

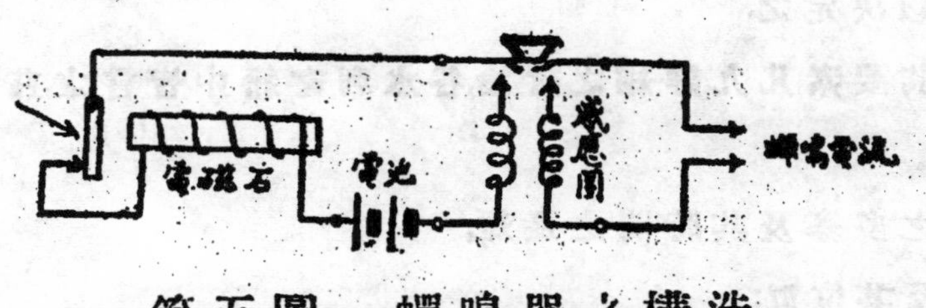

第五圖　蟬鳴器之構造

附註四　本法所使用之機件,均係電話上所常用者,雖有一二處須稍加以改造,亦無多大困難之事,故普通之機匠皆能裝配裕如也.

水道横切面大小之討論

著者：張含英

篇端語

本篇的範圍，約可分爲三部.第一部爲搜求關於水道横切面大小(Area of Waterway)之公式凡三十九;第二部將各家意見,按諸流量之統計,作一比較之研究;第三部就作者之意見,擬定水道横切面大小之公式.然見聞淺陋,結論容有錯誤,參考缺乏,張本或有未遇.故極欲於中國工程學會第十三次年會之機會,公佈而討論之,尚希有以教正!

(一) 總論

昔日水道横切面大小之規定,全由工程師根據該河域之大小,最高水位之訪問,利用個人之經驗,而判斷之.一則昔日之測量不備,再則橋空建築多由木料爲之,其費甚省,故所關不甚重要.近數十年來鐵路,道路橋樑之建築,既多改爲永性者,如鋼鐵及洋灰等,而水道之治理,及河患之防禦,又日見切要,於是乎水道横切面之大小問題,不得不詳加討論.而求得精密之解决矣.

求水道横切面大小之方法可分爲兩種:(一)利用經驗公式(Empirical Formula),或(二)由觀察以决定之.

影響於流量之大小者,其要素凡九.詳細之討論,各水利書籍中皆言之,玆特路舉如次:

(一)雨量之大小,次數之多寡及其時間之長短;

(二)流域之大小形狀及其位置;

(三)流域地勢之情形;

（四）地質之狀況；

（五）地文之情形，如溫度，蒸發量，氣壓之變化；

（六）地面植物之狀況；

（七）地下水；

（八）湖泊及其他蓄水之情形；

（九）河水之利用及河道治理之情形．

若根據以上九項要素，而推演一合理之計算流量公式，乃不可能之事實．蓋以各項皆複雜，其影響於流量之準確關係，尚難明瞭，不能推演一合理之公式也．所幸者十分精確之結果，非今日所需要者，亦非能由已有之張本所能推求者，故根據經驗而擬定之公式尚焉．

（二）水道橫切面已有之公式

欲作各公式之研究，不可不先述其擬定之經過，今爲統一符號起見，除另有聲明外，本篇皆採取以下之符號，茲分述之如下：

a ＝所求之水道橫切面，以平方英尺計；

A ＝流域之大小，以英畝（Acre）計；

C ＝係數；

M ＝流域之大小，以平方英里計；

q ＝最大流量，以每平方英里每秒若干立方英尺計；

Q ＝河流之總量以每秒若干立方英尺計；

R ＝水力徑（Hydraulic Radius）以英尺計；

S ＝河面之坡度；

V ＝平均流速，以每秒英尺計．

（1）馬堯（Myer）氏公式：此公式 1879 年馬堯（E. T. D. Myers）所擬定，美國大西洋沿岸各州多用之，

$$a=C\sqrt{A}$$

若為平坦之區,C之最小數為1,若為高下不平之區,C為1.6,若為高山及石地之區,C之最大數為4.0.

(2) 塔寶悌(Talbot)氏公式: 此公式乃1887年塔寶悌(A. N. Talbot)教授所擬定,為一般工程師所通用者.

$$a=CA^{\frac{3}{4}}$$

若在山地及多石之區,C=⅔至1,若在較有高低之農產區域,當雪溶之時,亦常影響於洪流,且流域之長較其寬約大三四倍時,C=⅓,若在不受雪溶影響之區域,其流域之長數倍於其寬時C=⅕或⅙.

(3) 范明(Fanning)氏公式: $Q=200M^{5/6}$

若速率為每秒8英尺時,則$Q=25M^{5/6}$

(4) 溫德華(Wentworth)氏公式: $a=A^{\frac{1}{2}}$

此公式之擬定乃專為瑙威鐵路(Norfolk and Western Railway),故在該線附近頗為適用.普通言之,在美國東南各州適用之.若在雨量較少之區或流域地勢平坦,該公式所得之數較大.溫氏觀察之結果,在此等情形時,可用該數百分之六十.

(5) 皮克(Peek)氏公式: $a=\frac{A}{C}$

C之變化為4至6,按地勢之情形而變.

(6) 哲費斯(Jarvis)氏公式: 經哲費斯(C.S. Jarvis)對於流量加一詳盡之研究後,該氏對於馬堯之公式略加變更,而成為流量之基本公式

$$Q=CM^{\frac{n-1}{n}}$$

$$\text{或}\ \frac{Q}{10}=a=C'M^{\frac{n-1}{n}}$$

其中C, C′及n皆為常數.

其對馬堯氏公式之變更如下:

$$a=C\times25.3\sqrt{A/640}=25.3C\sqrt{M}$$

此與哲費斯之基本公式頗合(C′=25.3C, n=2).

或 Q=aV,若 V=10 英尺/秒,則

Q=253CM,

或 Q=RM,在此公式R爲每平方英里每秒若干立方英尺之流量,其變化約爲 100 至 10,000.

(7) 台得瀉太鐵路(Tidewater Railway)公式:

$$a=0.62A^{\frac{7}{10}}$$

(8) 弗利才耳(Frizell)氏公式: 弗利才耳根據美國麻省康乃悌求悌河(Connecticut River)在豪姚(Holyoke)壩頂流量五十年測量之結果而擬定一公式,則在同樣氣候之區域其流域爲 M 時,則洪量爲

$$q=17.35\sqrt{8006/M}$$

(9) 笛根(Dicken)氏公式: 在印度關於流量之研究,其發表之公式如下:

$$Q=27CM^{\frac{3}{4}}$$

C 爲一常數,笛根氏對東印度用 8.25.

(10) 古雷(Cooley)氏公式: 古雷(L.E. Cooley)在美國米梭里(Missouri)州測量各河之結果,河之流域自10平方英里至2,500平方英里,擬定公式如下:

(1) $Q=200\ M^{\frac{2}{3}}$　或 $a=20\ M^{\frac{2}{3}}$

(2) $Q=180\ M^{\frac{2}{3}}$　或 $a=18\ M^{\frac{2}{3}}$

(11) 歐康乃爾(O'Connel)氏公式: 歐康乃爾提議用拋物線公式,以代表流量及流域之關係,設 X爲弧上一點之X坐標表,示流域爲若干平方英里;y 爲同點之 y 坐標,表示該河每秒若干立方英尺之流量,按照普通拋物線公式,$y=k\sqrt{X}$;k爲一常數因各流域而不同.

在印度各河,k 為自 40 至 302.7. 在北美洲各河,k 為自 13.06 至 57. 在英國各河,k 為 0.13 至 37.

(12) 弗來(Fuller)氏公式;

$$Q_{平均}=C_1 M^{0.8}$$

$$Q=Q_{平均}(1+0.8\log T)$$

$$Q_{最大}=Q(1+2M^{-0.3})$$

C_1=係數因各河而變 T=欲估計之年代(例如欲計算十年或二十年之最大流量之類).

(13) 葛雷(Gray)氏公式:

$$Q=5.89\times 640M^{\frac{5}{6}}$$

(14) 工程新聞雜誌(Engineering News)公式:

$$a=\sqrt[3]{8A}$$

(15) 穆飛(Murphy)氏公式: 穆飛研究美國東北各州流量之結果,擬定公式:

$$q=\frac{46.790}{M+320}+15 \quad \cdots\cdots(1)$$

計算河流流量所常用之公式為:

$$Q=aV=aC\sqrt{RS} \quad \cdots\cdots(2)$$

$$\text{其中,}\quad C=\frac{a+\frac{b}{n}+\frac{c}{s}}{1+\left(a+\frac{c}{s}\right)\frac{n}{\sqrt{R}}}$$

a, b 及 C 為常數, n 為河底之粗糙係數.自第(2)公式得

$$a=Q/V=Q/C\sqrt{RS} \quad \cdots\cdots(3)$$

$$\text{水力徑}\ R=A/p=\frac{(I+2d)\ x}{(I+2d)\sqrt{I+x^2}}\times d$$

其中 I 為梯形河底之寬, d 為深, x 為邊之坦坡.

(16) 芝伯困鐵路(Chicago, Burlington & Quincy Ry. Co.)公式:

$$Q=\frac{3,000M}{3+2\sqrt{M}}$$

$$a=\frac{300M}{3+2\sqrt{M}}$$

(17) 可太(Kutter)氏公式:

$$Q=\frac{1421M}{0.311+\sqrt{M}}$$

此公式多用瑞士河流.

(18) 意大利(Italian)公式:

$$Q=\frac{1819M}{0.311+\sqrt{M}}\cdots\cdots\cdots\cdots(1)$$

$$Q=\frac{2,600M}{0.311+\sqrt{M}}\cdots\cdots\cdots\cdots(2)$$

第一公式適用於意大利北部,第二公式適用同一區域之小河流.

(19) 尼靈(Knichling)氏公式:

$$q=\frac{44,000}{M+170}+20\cdots\cdots\cdots\cdots(1)$$

$$q=\frac{127,000}{M+370}+7.4\cdots\cdots\cdots\cdots(2)$$

此公式之擬定,乃根據歐美各洪流之結果.第一公式爲常有之洪水流量,第二公式爲罕有之洪水流量.

(20) 易柳悌(Elliot)氏公式:

$$q=\frac{20}{M}+3.63\cdots\cdots\cdots\cdots(1)$$

$$q=\frac{24}{M}+6\cdots\cdots\cdots\cdots(2)$$

第一公式適用於密西西比河上游,第二公式適用於亞利塔薩新(Arkansas)東北部.然皆爲牴濕之區.

(21) 馬可勞力(McCrory)氏公式:

$$q=\frac{35}{\sqrt[6]{M}}$$

此公式適用於亞利堪薩斯之西蒲利斯(Cypress)河.

(22) 毛根工程公司(Morgan Eng. Co.)公式:

$$q=\frac{28.0}{\sqrt{M}}+7.2 \quad\ldots\ldots\ldots\ldots(1)$$

$$q=\frac{38.0}{\sqrt{M}}+8.0 \quad\ldots\ldots\ldots\ldots(2)$$

第一公式適用於亞利堪薩斯之密西西比流域,第二公式美國凱赤(Cache)河流域.此二公式皆適用於低濕之區.

(23) 鄧(Dun)氏張本: 第一表爲據美國米蘇里(Missouri),堪薩斯(Kansas),印地安(Indian)及台可塞斯(Texas)各州之觀察而適用於山達費(Santa Fe)鐵路者.

(24) 馬可馬斯(McMath)氏公式: 馬可馬斯(R. E. McMath)研究美國聖路易(St. Louis)城情形之結果,而擬定以下之公式:

$$Q=0.75\times2.75\sqrt[5]{15A^4}$$

若改公式中之文字爲符號時,則變爲

$$Q=Cr\sqrt[5]{SA^4}=ACr\sqrt[5]{S/A}$$

其中C=雨水能流至河中之比例數,r=當急雨之時,落於每英畝上每秒若干立方英尺之雨量,此數幾與每點落雨若干英寸相等.

(25) 蟄支(Dredge)氏公式: 蟄支在印度專門報告中,擬定下列之公式,[亦名寶支(Burge)氏公式]:

$$Q=1{,}300M/L^{\frac{2}{3}}$$

其中L爲流域之長,以英里計.

(26) 亞當(Adam)氏公式:

$$Q=ACi\sqrt[6]{Si/Ai}$$

其中 C=1.837 　　i=1.0

S=平均河坡每1000英尺之尺數.

此公式多適用於城市之區,用以設計下水道者.

第一表　鄧氏張本

水道横切面大小及流域表

用於亞赤森透皮加及山達費鉄路

(ATCHISON, TOPEKA & SANTA FE RAILWAY)

流域平方英里	水道切面 A MISSOURI & KANSAS	CAST PIPE	BOX & ARCH CULVERTS	ILLINOIS	IND. TER.	TEXAS	NEW MEXICO	SANTA FE PACIFIC		流域平方英里	水道切面 A MISSOURI & KANSAS			ILLINOIS	IND. TER.	TEXAS	NEW MEXICO	SANTA FE PACIFIC	
.01	2.0	1-24"	2X1 B							24	1060					110	94		
.02	4.0	1-24"	2x2 "							26	1100					"	92		
.03	6.0	1-30"	2x3 "							28	1140					"	"		
.04	7.5	1-36"	2½x3"							30	1180					"	"		
.05	9.0	1-42"	3x3 "							32	1220					"	"		
.06	105	1-42"	3½x3 "							34	1255					"	"		
.07	120	1-48"	9X4 "							36	1290					"	91		
.08	135	2-36"	D2½x3 "							38	1320					"	"		
.09	15	2-36"	" " " " "							40	1350					"	"		
.10	16	2-36"	"3x3 "							45	1435					"	"		
.15	25	2-48"	"3x4 "							50	1510					"	89½		
.20	32	3-42"	6X4 A							55	1580					115	"		
.25	38	3-48"	6x5 "							60	1650					"	"		
.30	44		6x5½ "							65	1720					"	88		
.35	51		8x4½ "							70	1780					"	"		
.40	56		8x5 "							75	1840					"	"		
.45	62		8x6 "							80	1000					"	86½		
.50	66		" " " "							85	1960					"	"		
.55	70		8x6½ "							90	2015					"	"		
.60	74		10x4½ "							95	2065					"	"		
.65	78		10x5 "							100	2120					120	85		
.70	81		10x5½ "							110	2220					"	"		
.75	85		10x6 "							120	2315					"	"		
.80	88		10x6½ "							130	2405					125	83½		
.85	91		" " " "							140	2500					"	"		
.90	84		" " " "							150	2580					130	82		
.95	97		12x5 "							160	2665					"	"		
1.0	100		" " " "				98½			170	2745					"	80½		
1.1	110		12x6 "			105	"			180	2820					"	"		
1.2	120		12x7 "			"	"			190	2900					"	79		
1.3	130		12x8 "			"	"			200	2870					"	"		
1.4	140		14x6½ "			"	"			220	3115					"	77½		
1.5	150		14x7 "			"	"			240	3245					"	"		
1.6	160		16x6½ "			"	"			260	3370					"	76		
1.7	170		16x7 "			"	"			280	3495					"	"		
1.8	180		16x7½ "			"	"			300	3615					"	74½		
1.9	190		16x8 "			"	"			325	3770					"	"		
2.0	200		18x7 "			"	"			350	3300					"	73		
2.2	220		18x8 "			"	"			375	4035					"	"		
2.4	240		18x9½ "			"	"			400	4165					"	71½		
2.6	260		20x8 "			"	"			450	4385					"	70		
2.8	280		20x9 "			"	"			500	4610					"	68½		
3.0	300		20x9½ "			"	"			550	4825					"	67		
3.2	321		22x8½ "			"	"			600	5030					"	65½		
3.4	340		22x9 "			"	"			650	5230					"	64		
3.6	357		24x8½ "			"	"			700	5420					"	62½		
3.8	373		24x9 "			"	"			750	5610					"	61		
4.0	388		26x7 "			"	97			800	5800					"	59½		
4.2	403		26x7½ "			"	"			850	5990					"	58		
4.4	417		26x8 "			"	"			900	6080					"	56½		
4.6	430		26x8½ "			"	"			950	6230					"			
4.8	443		26x9 "			"	"			1000	6380					"			
5.0	456		26x9½ "			"	"			1100	6705					"			
5.5	483		26x10 "			"	"			1200	6960					"			
6.0	509		32x7½ "			"	"			1300	7230					"			
6.5	533		32x8 "			"	"			1400	7480					"			
7.0	556		32x9 "			"	"			1500	7725					"			
7.5	579		32x10 "			"	"			1600	7960					"			
8.0	601		32x11 "			"	"			1700	8195					"			
8.5	622		32x11½ "			"	"			1800	8390					"			
9.0	641		32x12 "			"	93½			1900	8625					"			
9.5	660		32x12½ "			"	"			2000	8820					"			
10	679		32x13 "			"	"			2200	9240					"			
11	710					"	"			2400	9606					"			
12	740					"	"			2600	9970					"			
13	775					"	"			2800	10320					"			
14	805					"	"			3000	10840					"			
15	835					"	"			3500	11445					"			
16	865					"	94			4000	12160					"			
17	890					"	"			4500	12825					"			
18	920					"	"			5000	13500					"			
19	945					"	"			5500	14080					"			
20	970					"	"			6000	14520					"			
22	1015					"	"			6500	15140					"			

- BOX & ARCH CULVERTS (rows 11–22): BRIDGES DESIGNED TO PROVIDE AREA ACCORDING TO CIRCUMSTANCES
- ILLINOIS: EAST OF STREATOR USE 50% / WEST OF STREATOR USE 80%
- IND. TER.: SOUTH OF PURCELL USE TEXAS COLUMN / NORTH OF PURCELL USE COLUMN "A"
- TEXAS: USE COLUMN "A" ↓
- NEW MEXICO: USE COLUMN "A" ↓

- BRIDGES TO PROVIDE AREA ACCORDING TO CIRCUMSTANCES
- ILLINOIS: EAST OF STREATOR USE 60% / WEST OF STREATOR USE 60%
- IND. TER.: SOUTH OF PURCELL USE TEXAS COLUMN / NORTH OF PURCELL USE COLUMN "A"

(27) 巴里木雷(Paremley)氏公式:

$$Q=ACi\sqrt[6]{S^{3/2}/A}$$

C=0至1.0　　　　i=4.0

此公式亦適用於城市之區.

(28) 葛里高里(Gregory)氏公式:

$$Q=ACi\times\frac{S^{0.186}}{A^{0.14}}$$

Ci=2.8(若地面滲透性小時用此式)

(29) 脊木(Chamier)氏公式:

$$Q=640RCM^{\frac{2}{3}}$$

其中Q爲於出口處最大之流量;R爲最急雨時每點所落之平均英寸數,平均之時期以自最上游流至出口處所用之時間計之;C爲雨量及流入河中之比例數.

(30) 何乃(Horner)氏公式: 根據美國聖路易城之張本,得以下之公式:

$$Q=320\left(\frac{A^2}{L}\right)^{4/7}S^{1/5}\ldots\ldots\ldots\ldots\ldots\ldots\ldots(1)$$

$$a=4.5\left(\frac{A^2}{L}\right)^{3/7}\left(\frac{1}{S}\right)^{1/5}\ldots\ldots\ldots\ldots\ldots(2)$$

其中L=流域之長以英尺計,

S=河底之平均坡度.

何乃將該公式變爲 $a=CA^{\frac{6}{7}}$. C之數値,因河底坡度及流域長寬比例之不同,而列表如下:

長寬比＼坡度	0.001	0.005	0.01	0.015	0.02	0.04
L=W	1.76	1.32	1.15	1.06	1.00	0.87
L=2W	1.52	1.14	0.99	0.91	0.86	0.75
L=3W	1.39	1.04	0.91	0.83	0.79	0.69
L=5W	1.24	0.93	0.81	0.75	0.71	0.62
L=10W	1.06	0.80	0.70	0.64	0.61	0.53

(31) 可來蠻(Cramer)氏公式:

$$Q=\frac{C_3R_3mMS_2^{\frac{1}{3}}}{9+(0.0658mR_3M)^{\frac{1}{2}}}$$

其中　Q＝每秒若干立方英尺流量,

R_3＝流域之平均雨量,以英寸計,

S_2＝流域之平均坡度,

C_3＝係數,

＝186,若流域高低不平時用之,

＝697,若流域平坦且不滲透如城市時用之.

m因全流域面積M,易受水漲湮沒之地F,及每年平均雨量R_3而變,其關係如下:

$$m=1-\sin\left(-\tan^{-1}\frac{709F}{MR_3}\right)$$

此公式應用於若沿河皆屬汎濫者.如汎濫之區祇集中於下游,則

$$m=1-\sin\left(-\tan^{-1}\frac{1418F}{M\ R_3}\right)$$

(32) 可里支(Craig)氏公式: 可里支(James Craig)擬定以下之公式:

$$Q=400\ Bc\ Vi\times hyp.\ \log\frac{4L}{B}\left(L+\sqrt{L^2+\frac{B^2}{16}}\right)$$

其中　Q＝流量,以每秒立方英尺計,

B＝流域之平均寬,以英里計,

C＝流量係數,

V＝平均速率,以每秒英尺計,

i＝雨量以英寸計,

L＝流域之長以英里計,

$\frac{B^2}{16}$項可以減去(CVi)項可以N_{n}代之因地文及地勢之情形而變.如此則公式可書之如下:

$$Q+440\ BN\times hyp.\ \log 8\ L^2/B$$

(33) 包生悌(Possenti)氏公式:

$$Q=C_2\left[\frac{R}{L_1}\right]\left[M_2+\frac{M_1}{3}\right]$$

其中 C_2 = 係數,普通用1010,

R = 雨量,以二十四點鐘若干英寸計,

M_2 = 流域有山部分之面積,以方英里計,

M_1 = 流域平坦部分之面積,以方英里計,

L_1 = 自河之發源至觀察點之長,以英里計,

此公式適用於亞盆乃尼斯(Appenines)之山地.

(34) 伯可里及蔡哥拉(Burkli-Ziegler)氏公式:

$$q=Cr\sqrt[4]{S/A}$$

其中 q = 流量,以每秒立方英尺計,

r = 最大雨量,以全流域每點落之英寸計,

S = 流域之普通坡度,以1000英尺若干英寸計,

C = 0.75,應用於洋灰鋪地之城市,

= 0.625,應用於普通之城市,

= 0.30,應用鄉村有草地及石子鋪地者,

= 0.25,應用於普通農田.

此公式觀察之根據坡度至大爲每1000英尺10英尺,流域不過50英畝者.故適宜於城市下水道之設計.

(35) 寇蘭(Coghlan)氏公式:

$$P=tCR^{1.9}$$

P = 河流排去之雨量,以英寸計,

t = 時間之係數,因氾濫時間之長短而變,

C = 雨量之係數,因雨量之大小而變,

R = 氾濫時每日之雨量,以英尺計,

此公式適用於新南威爾斯(New South Wales)較小之流域.

t之數如下:

氾濫期,以日計	350	300	250	200	150	100	50	25
t,以英寸計	2.6	2.5	2.4	2.2	2.1	2.0	1.9	1.8

C之數值如下:

雨量,以英寸計	0.40	0.35	0.30	0.25	0.20	0.15	0.10
C,以英寸計	0.95	0.90	0.85	0.80	0.75	07.0	0.65

(36) 哈鑑(Herirg)氏公式: 設以 t 表示落雨時間之長短, V 表示流至涵洞(Culvert)前水之平均速率, d 表示每單位寬之一帶地每秒能將雨量流入於河中之數,則流量 Q 爲

$$Q=tdV \quad \cdots\cdots(1)$$

若設 t_1 爲流域邊界所落之雨,以速率 V 流至涵洞所用之時間, ι 爲水流之平均距離,則

$$\iota=vt_1 \text{ 或 } v=\iota/t_1 \quad \cdots\cdots(2)$$

代入第一式,則

$$Q=td\,\iota/t_1 \quad \cdots\cdots(3)$$

若應用於全流域面積,爲 M, td 雨量流去之部分,若以 D 代之,則

$$Q=\frac{DM}{t_1} \quad \cdots\cdots(4)$$

可知流量因雨量流去之部分,流域之大小,及邊界之雨流至涵洞之時間而變。

第四公式仍可改爲

$$Q=DMv/\iota \quad \cdots\cdots(5)$$

於1889年哈鑑又推演以下之公式:

$$Q=CiA^{0.85}S^{0.27}$$

$$\text{或 } Q=CiA^{0.833}S^{0.27}$$

其中 i = 最大雨量率,以每點英寸計,

S = 地面之坡度,以每1000之英尺計,

Ci之變化爲自1.02至1.64,此二公式之差,約爲百分之15。

(37) 勞得浦(Lauterberg)氏公式:

$$Q=0.03171ChF$$

其中 Q = 每秒之流量,

F = 流域面積,以平方公尺計,

h = 全年之雨量,以英寸計,

3945

$C=0.20$ 應用於鬆壤之地,
$=0.25$ 應用於平原之地,
$=0.30$ 應用於不甚平坦之地,
$=0.35$ 應用於小山之地,
$=0.45$ 應用於有山之地,
$=0.55$ 應用於森林之地,
$=0.70$ 應用於高山峻嶺之地,

勞得浦之張本,多採自瑞士.

(38) 豪塘(Harton)氏公式:

$$Q=4021.5\frac{\sqrt[4]{T}}{M}$$

其中T表示欲計算最大流量之年限.

此公式多用於尼設木內(Neshoming),桃吉康(Tochickon)及百吉門(Perkiomen)等地方.

(39) 愛斯考斯吉(Iszckowski)氏公式: 此公式爲奧國工務部總工程師愛斯考斯吉(R. Iszckowski)所擬定,可名之曰『根據流域之觀察歸納而擬定計算普通流量及洪水流量之公式』.

$$Q_{最大}=(0.022C_1+m\,C_2)\,R\,M,$$

其中 $Q_{最大}$ = 最大可能之最大流量,以每秒立方公尺計,
R = 全年之平均雨量,以公尺計,
M = 流域面積,以平方公里計,

C_1 因流域地勢及坡度而變,平坦或低濕之地,約爲0.20,山地約爲0.65.

C_2 因地面土壤情形及其滲透性而變,對於地面有植物生長及可滲透之地,可用0.035,對不能滲透之山地或凍地可用0.70.

$$m=\frac{0.59(11{,}050+M)}{818+M}$$

尼鑒用 $C_1=0.385$ 及 $C_2=0.40$, 以便適用於普通情形,例如有小山之區,土壤略可滲透且有植物生長之地,將此數目代入以上公式,並將流量之單位

改爲每方英里每秒立方英尺,面積爲平方英里,每年平均雨量爲英寸,則

$$q_{最大}=\frac{0.568\ R(4,129.5+M)}{315.8+M}\text{......................}(1)$$

若在山嶺之區,其地多石或凍 $C_1=0.50$ 及 $C_2=0.60$,則

$$q_{最大}=\frac{0.848\ R(4,147.4+M)}{315.8+M}\text{......................}(2)$$

以上各公式多爲自經驗結果而擬定之公式.大多數公式中祇有一個或兩要素.最要者似爲流域之面積.各公式之用途似皆按地域之不同,而稍有異者.

(三) 各公式之研究

今就上述公式中之較常用者,按流量之統計,比較研究之.流量統計乃採自哲費斯(C. S. Jarirs)發表於美國土木工程師學會會刊 (Proceeding of A.S. C.E., Dec., 1924) 之『洪流之特性』(Flood flow Characteristics) 一文中.其張本之採集,頗稱完備,最多者爲美國,各國亦皆有之,故用之以作比較研究之張本.

欲作比較,必先將各公式變爲同樣之項數,以 q 表流量,以每平方英里每秒之立方英尺計之,M表流域面積,以平方英里計之,今設平均之流速爲每秒 1.0 英尺,則

馬堯氏公式: $q=253\frac{C}{\sqrt{M}}$,

賓堵悌氏公式: $q=1270\frac{C}{\sqrt[4]{M}}$,

溫德華氏公式: $q=742\frac{1}{\sqrt[3]{M}}$,

范明氏公式: $q=200\frac{1}{\sqrt[6]{M}}$,

哲費斯氏公式: $q=\frac{C}{\sqrt{M}}$,

弗利才耳氏公式: $q=1550\frac{1}{\sqrt{M}}$.

葛雷氏公式: $q=3760\frac{1}{\sqrt[4]{M}}$,

馬可馬斯氏公式: $q=1980\frac{1}{\sqrt[5]{M}}$,

工程新聞雜誌: $q=172\frac{1}{\sqrt[3]{M^2}}$,

以上各公式中q之數值於第二表中列之.

以下各公式中,q之數值於第三表中列之.

第二表 由各公式所計算之流量

以每方英里每秒立方英尺計

流域 平方英里	馬克			塔爾梯			范明	溫德華	哲費斯		費利米耳	葛雷	馬可馬斯	工程新聞雜誌
C	1	1.6	4.0	2/3	1/3	1/6			100	10,000				
1	25.3	405	1010	845	422	211	200	742	100	10,000	1550	3760	1980	172
100	25.3	40.5	101	267	133	66	93.5	160	10	1,000	155	1155	792	7.9
10,000	2.5	4.0	10	84.5	42.2	21	43	34	1	100	15	376	316	.4

第三表 自各公式所計算之流量

以每方英里每秒立方英尺計

流域 平方英里	穆飛	芝伯困	可太	意大利		尼靈		易柳悌	馬可勞力	鞏文	包生悌	伯可里	弗來		愛斯考斯吉斯	
				(1)	(2)	(1)	(2)	(2)					qav	q	(1)	(2)
1	161	600	1080	1980	1390	277	349	30	35	1300	10,100	12,050				
10	156	328	410	747	526	264	341	13.6	23.8	800	3200	6806	63	164	260	390
50	141	177	193	352	246	220	309	9.4	18.2	352	1430	4540	45	120		
100	120	131	138	252	171	183	277	8.4	16.2	280	1010	3520	30.8	104	206	312
500	72	62.5	62.5	114	80.2	85.7	154	7.0	12.4	162	450	2540	29	75.3		
1000	50	45.5	44.5	81.4	57.0	57.5	100	6.7	11.1	130	326	2140	25.2	65	80	119
5000	23.8	20.8	20.0	36.6	25.6	28.5	31	6.3	8.5	76	143	1440	18	47.3		
10,000	19	14.8	14.1	25.8	17.7	24.3	19.7	6.2	7.6	60	101	1205	16	41.4	28	42
50,000	15.9	6.5	6.3	11.6	8.1	20.9	9.9	6.0	5.8	35	45	804	11.5	30		
100,000	15.	4.7	4.5	8.24	5.6	20.0	8.7	6.0	4.0	28	32	680	10	26		

穆飛氏公式, 芝伯困鐵路公式,

可太氏公式, 意大利公式,

一 圖

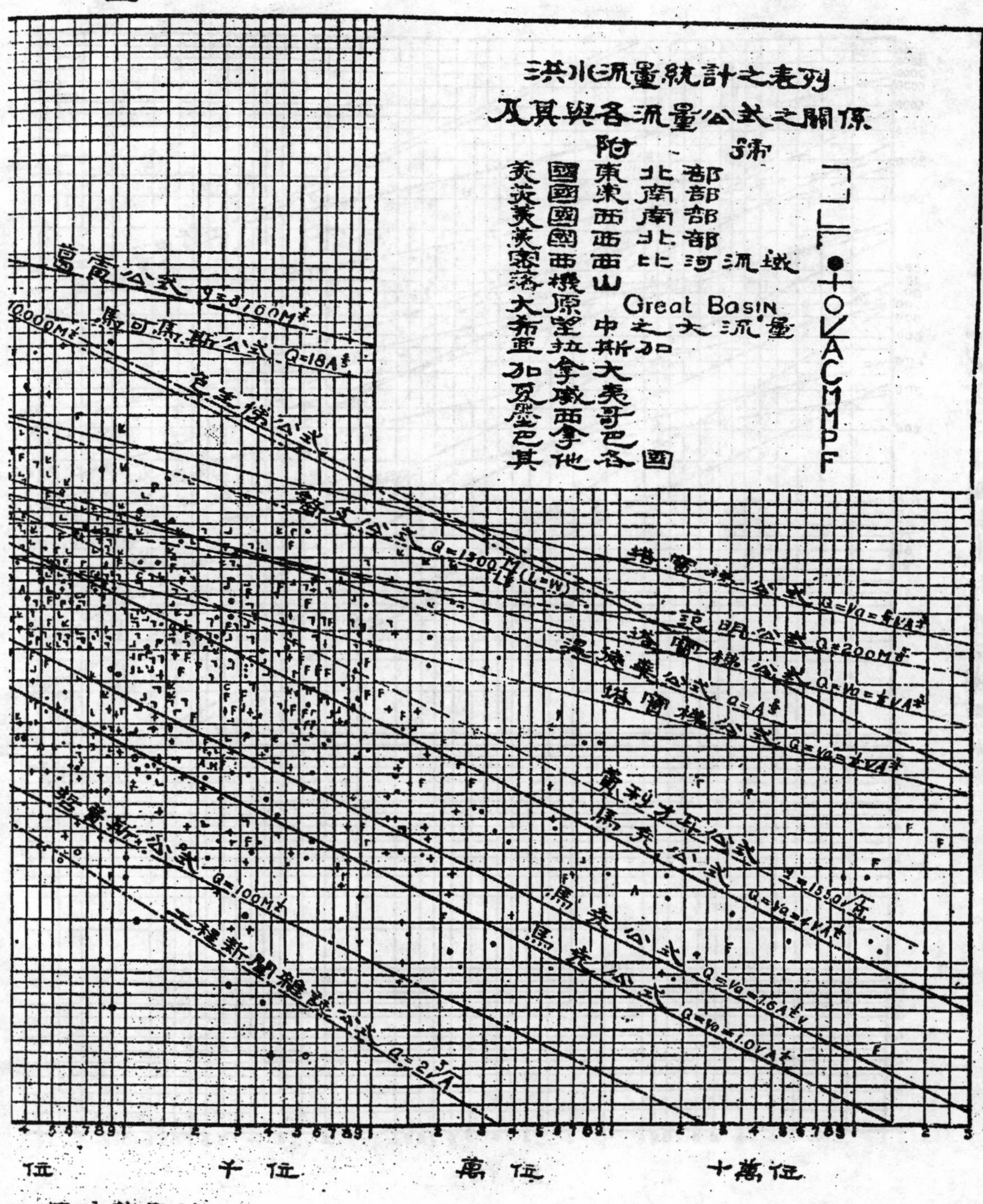
洪水流量統計之表列
及其與各流量公式之關係
附　號
美國東北部
美國東南部
美國西南部
美國西北部
密西西比河流域
落機山
大原 Great Basin
希望中之大流量
亞拉斯加
加拿大
夏威夷
墨西哥
巴拿巴
其他各國
高肯公式
馬可馬斯公式 Q=18A
苞生塔公式
第公式
塔窩楼公式
范明公式
塔窩楼公式
泥樊葉公式
塔窩楼公式
黃利才耳公式
馬先公式
馬先公式
馬先公式
哲奧斯公式 Q=100M
工程新聞雜誌公式
位　千位　萬位　十萬位

以平方英里計

第

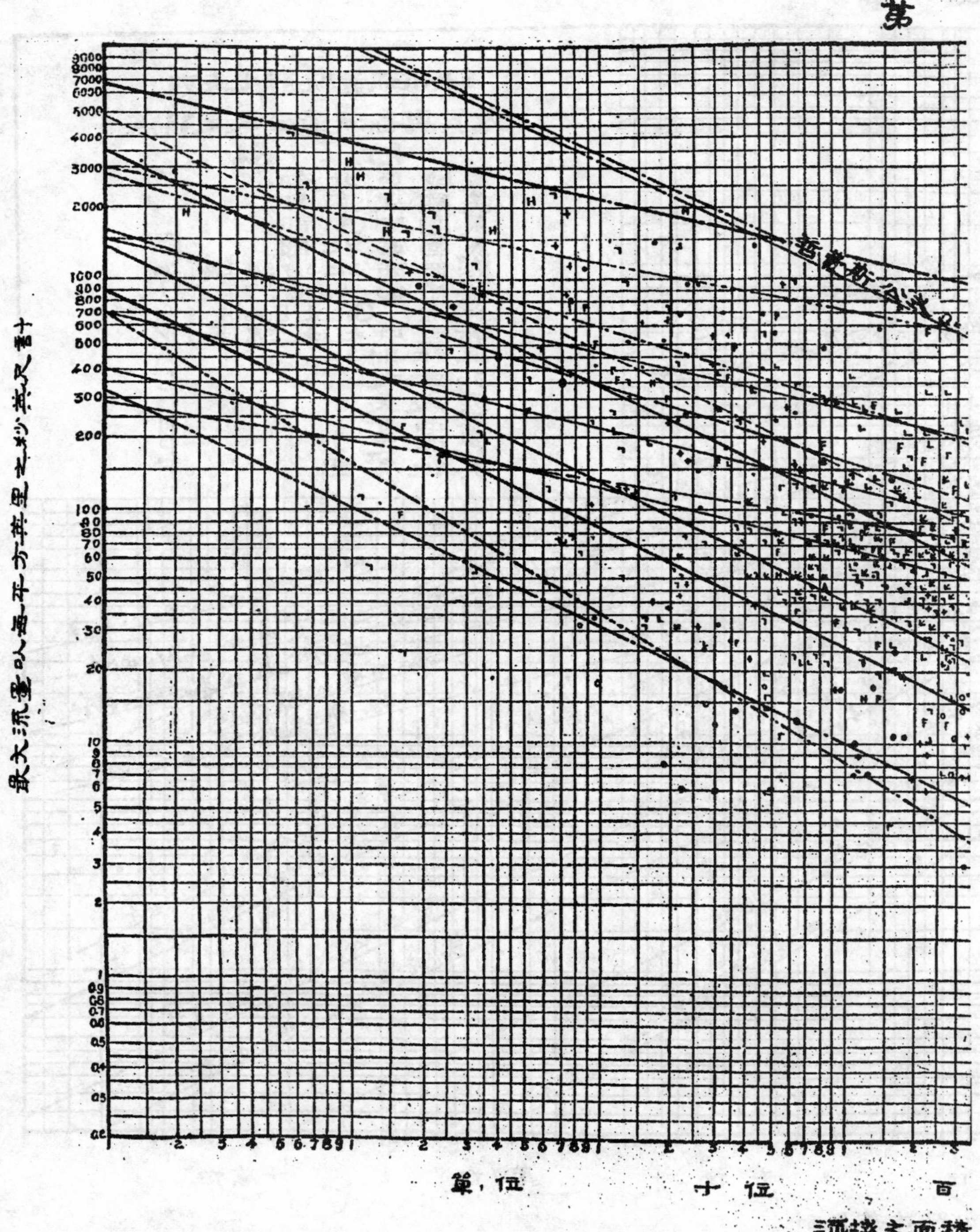
最大流量以每平方英里之秒英尺計
9000
8000
7000
6000
5000
4000
3000
2000
1000
900
800
700
600
500
400
300
200
100
90
80
70
60
50
40
30
20
10
9
8
7
6
5
4
3
2
1
單位
十位
百
流域之面積

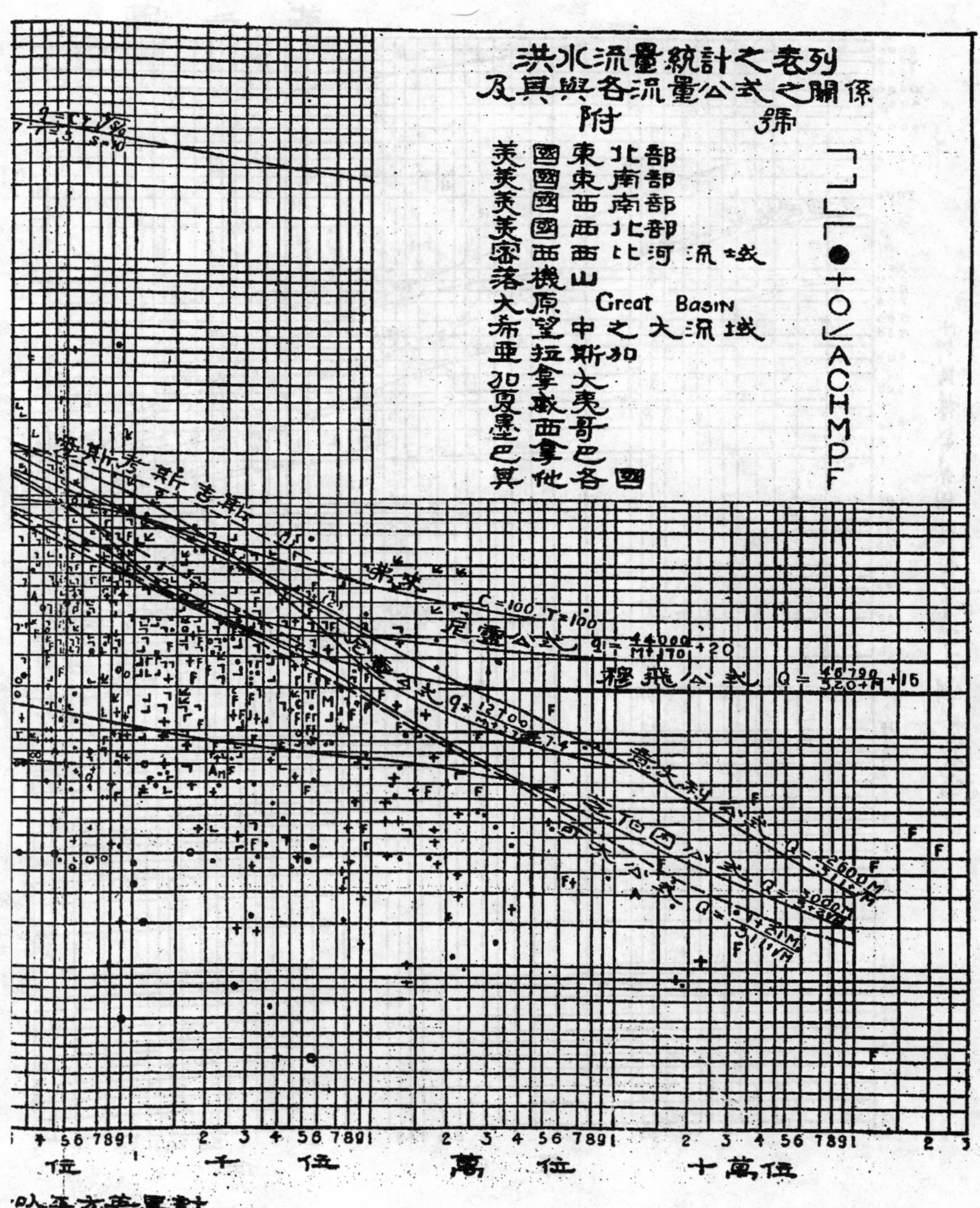

洪水流量統計之表列
及其與各流量公式之關係
附 號
美國東北部
美國東南部
美國西南部
美國西北部
密西西比河流域
落機山
大原 Great Basin
希望中之大流域
亞拉斯加
加拿大
夏威夷
墨西哥
巴拿巴
其他各國
C=100, T=100
定靈公式 $q=\frac{44000}{M+170}+20$
穆飛公式 $Q=\frac{46790}{320+M}+15$
千位
萬位
十萬位
以平方英里計

第二圖

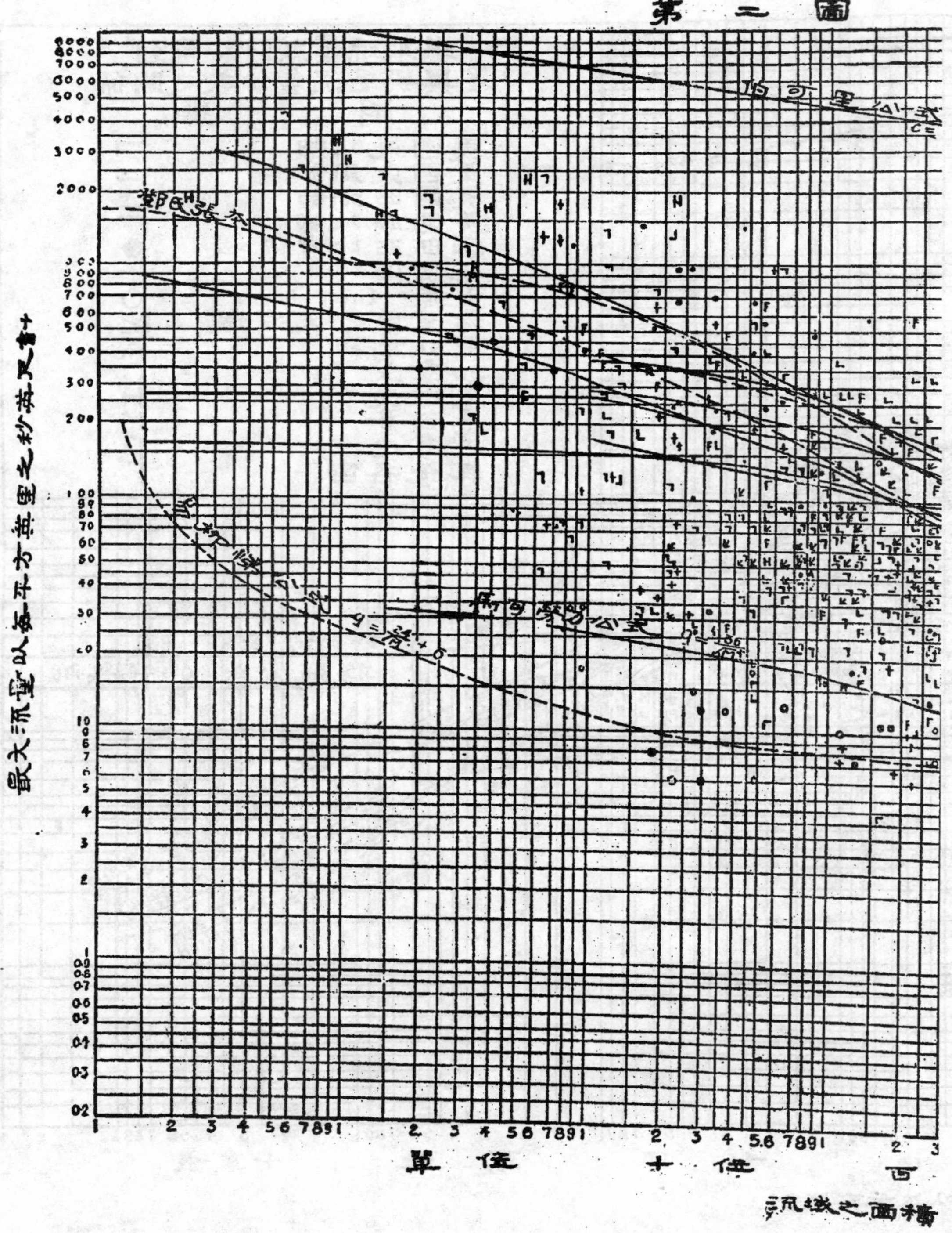

尼鑑氏公式，　易柳悌氏第二公式，

馬可勞力氏公式，

以下各公式中 q 之數值亦列表第三表中，惟於公式各有其假定之數。

墊支氏公式：　$q=1300\frac{1}{\sqrt[3]{L^2}}$，　設 $L=\sqrt{M}$

包生悌氏公式：$q=10100\frac{1}{\sqrt{M}}$，　設 $R=10$，

$C_2=1010$，$L=\sqrt{M}$　且在山地，

伯可里及蔡哥拉氏公式：　$q=Cr^4\sqrt{S/A}$

設 $r=3$，$S=10$，$C=0.70$，$q=12050\frac{1}{\sqrt[4]{M}}$

弗來氏公式：　$q_{平均}=C\frac{M^{0.8}}{M}$，

$q=q_{平均}(1-0.8\log T)$，

設 $C=100$ 及 $T=100$，

各表之數值皆繪於第一及第二圖上，以茲比較。

馬堯氏公式，自第一圖中觀察之，與統計極為符合，惟若在最大洪流，$C=4$，似屬較小。其弧線頗能與統計張本之方向相合，惟應用時於 C 之數值必須特別注意。

塔竇悌氏公式以可應用流域在 10 至 1000 平方英里之面積，若流域較 10 平方英里為小時，自該公式所得之值似太小，流域較 1000 平方英里大時，所得之值似太大。係數 C 在 ⅙ 至 ⅔ 之間，似較適宜。

范明氏公式，q 對於 M 之減低比例似較小，計算較大洪水位，流域在 100 至 1000 平方英里間可以用之。

温德華氏公式除流域在 40 方英里以下所得之數較小外，尚合統計之張本。

哲費斯氏公式之 q 及 M 之關係與馬堯者同，惟係數有變化耳。

弗利才耳氏公式在普通情形尚可應用。q 及 M 之關係與馬堯者同

葛雷氏公式在流域較大時,所得之值較大.對於最大洪流可應用至 100 平方英里之流域.對馬可馬斯氏公式,可用同樣之批評.

包生悌氏公式按所假定之常數,與流量統計頗相符合,惟對常數之選定,較爲困難耳.

墊支氏公式在普通情形似可應用至流域 10,000 平方英里.

工程新聞雜誌公式之數值太低.

皮克之假設不正確,蓋以流量與流域之面積成正比例也,故其結果不甚當.

古雷氏公式之 q 及 M 之關係與溫德華氏相同,惟前者較後者之係數稍低.

歐康乃爾之弧形與馬堯者之方向頗相同.因歐康乃爾氏對其常數並無指定之數值,故難加詳判.

穆飛氏公式所得之值,在流域小時爲小.在流域大時爲大.流域在 100 至 0,000 平方英里時頗可用之.尼靈氏公式之結果頗與之相似.

芝伯困鐵路公式,可太氏公式,及意大利公式所得之結果頗相似.鄧氏張本亦如之.其與流量之統計頗爲密合,故可用爲設計水道切面大小之用.惟流域在 1,000 平方英里以內,流量較最大洪量爲小,故可用之以計算普通洪量.

伯可雷及蔡哥拉氏公式所得之值太高,蓋以此公式乃用於下水道設計者.易柳悌及馬可勞力二氏之公式所得之值太低,蓋以此二公式乃用於低濕之地者.

其他含有多項之公式,其數值因假定而變,除確知各項之數值,難以校對該公式之精確程度.故關於此類之公式,祇舉出墊支氏,包生悌氏伯可里及蔡哥拉氏公式討論之.

弗來及蒙塘爾氏公式,在普通情形不能應用,蓋以若將公式變爲 $Q=4021.5^4$ 斤,全河之流量,不因流域之大小而變,似不合理.

愛斯考斯吉氏公式在流域波小時所得之値較小惟彼解決此問題之方法,極為聰明,可以深加研究.

玆就以上討論之結果,馮堯氏公式較為精確;芝伯因鐵路公式,可太氏公式,意大利公式及鄭氏張本炎之塔實梯氏公式用於流域小於1,000平方英尺者為適用.若流域小於100平方英尺者爲 Ⅱ公式得最高之値,塔實梯氏公式所得之値似較適中.

(四) 作者擬定之公式

叅照第三圖,自第一弧,可得以下之關係:

$$q \propto \frac{1}{M^{0.65}},$$

$$\text{或 } q \propto \frac{1}{\sqrt[3]{M^2}},$$

$$\text{或 } q = C\frac{1}{\sqrt[3]{M^2}},$$

此公式表示每平方英里之流量,與流域大小之關係.換言之,流量之總數與流域之三次方根成比例,或

$$Q = CM\frac{1}{\sqrt[3]{M^2}},$$

$$\text{則 } Q = C\sqrt[3]{M}$$

此弧雖能表示各點大概之變化,然在較大或較小流域之地,似有不符之處,故不得不加以變通,以求密切之符合.

今姑採公式之形式如下:

$$q = \frac{x}{y + M^{\frac{1}{2}}} + z$$

此式中未知數凡三,今自流量之統計,以求此三數.在罕有之最大流量,自統計圖得,若 M=1, q=4,200; 若 M=10, q=2,800; 若 M=100,000, q=20. 代入前

式,則

$$4200=\frac{x}{y+1}+z\cdots\cdots\cdots\cdots(1)$$

$$2800=\frac{x}{y+4.56}+z\cdots\cdots\cdots\cdots(2)$$

$$20=\frac{x}{y+2160}+z\cdots\cdots\cdots\cdots(3)$$

∴ 自(1)及(2)

$$1400=\frac{x}{y+1}=\frac{x}{y+4.65}$$

或 $$1400=\left(\frac{1}{y}-\frac{1}{y+4.65}\right)x$$

$$\therefore\ x=\frac{1400}{\frac{1}{y+1}-\frac{1}{y+4.65}}\cdots\cdots(4)$$

自(1)及(3)

$$4180=\left(\frac{1}{y+1}-\frac{1}{y+2160}\right)x$$

$$\therefore\ x=\frac{4180}{\frac{1}{y+1}-\frac{1}{y+2160}}\cdots\cdots(5)$$

自(4)及(5)

$$\frac{1400}{\frac{1}{y+1}-\frac{1}{y+4.65}}=\frac{4180}{\frac{1}{y+1}-\frac{1}{y+2160}}$$

$$\frac{1}{\frac{1}{y+1}-\frac{1}{y+4.65}}=\frac{2.99}{\frac{1}{y+1}-\frac{1}{y+2160}}$$

$$\therefore\ \frac{1}{y+1}-\frac{1}{y+2160}=\frac{2.99}{y+1}-\frac{2.99}{y+4.65}$$

$$\therefore\ \frac{1.99}{y+1}-\frac{2.99}{y+4.65}+\frac{1}{y+2160}=0$$

$$1.99-\frac{2.99y+2.99}{y+4.65}+\frac{y+1}{y+2160}=0\ldots(5)$$

若y之值爲甚小,則 $\frac{y+1}{y+2160}$ 亦必甚小,故可省去.

$$\therefore\ 1.99-\frac{2.99y+2.99}{y+4.65}=0$$

$$1.99\,y+9.25=2.99\,y+2.99$$

$$\therefore\ y=9.25-2.99=6.26=6.3$$

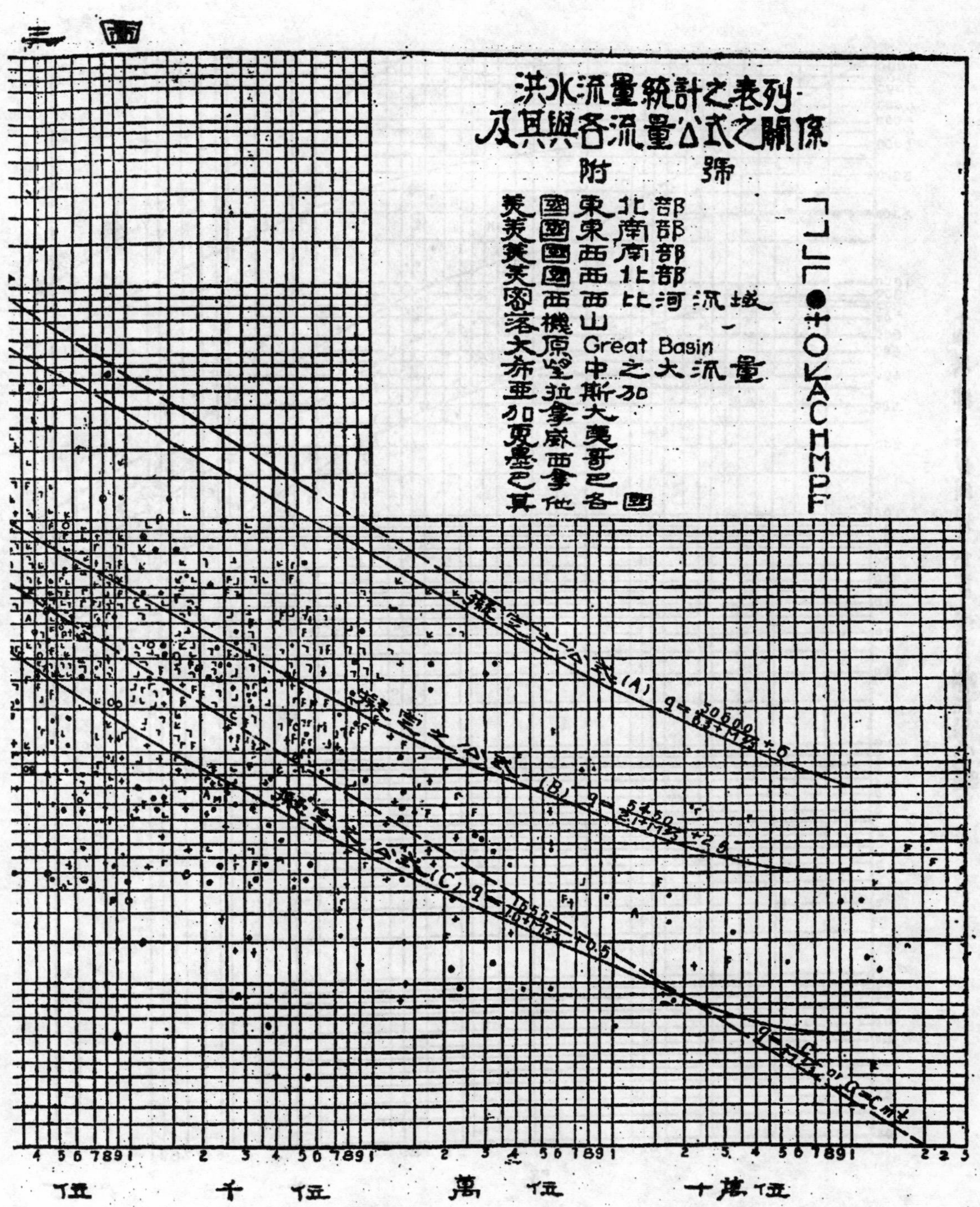
三圖
洪水流量統計之表列
及其與各流量公式之關係
附 號
美國東北部
美國東南部
美國西南部
美國西北部
密西西比河流域
落機山
大原 Great Basin
大沙漠中之大流量
亞拉斯加
加拿大
夏威夷
墨西哥
巴拿巴
其他各國
(A)
(B)
(C)
4 5 6 7 8 9 1 2 3 4 5 6 7 8 9 1 2 3 4 5 6 7 8 9 1 2 3 4 5 6 7 8 9 1 2 3
位 千位 萬位 十萬位

積以平方英里計

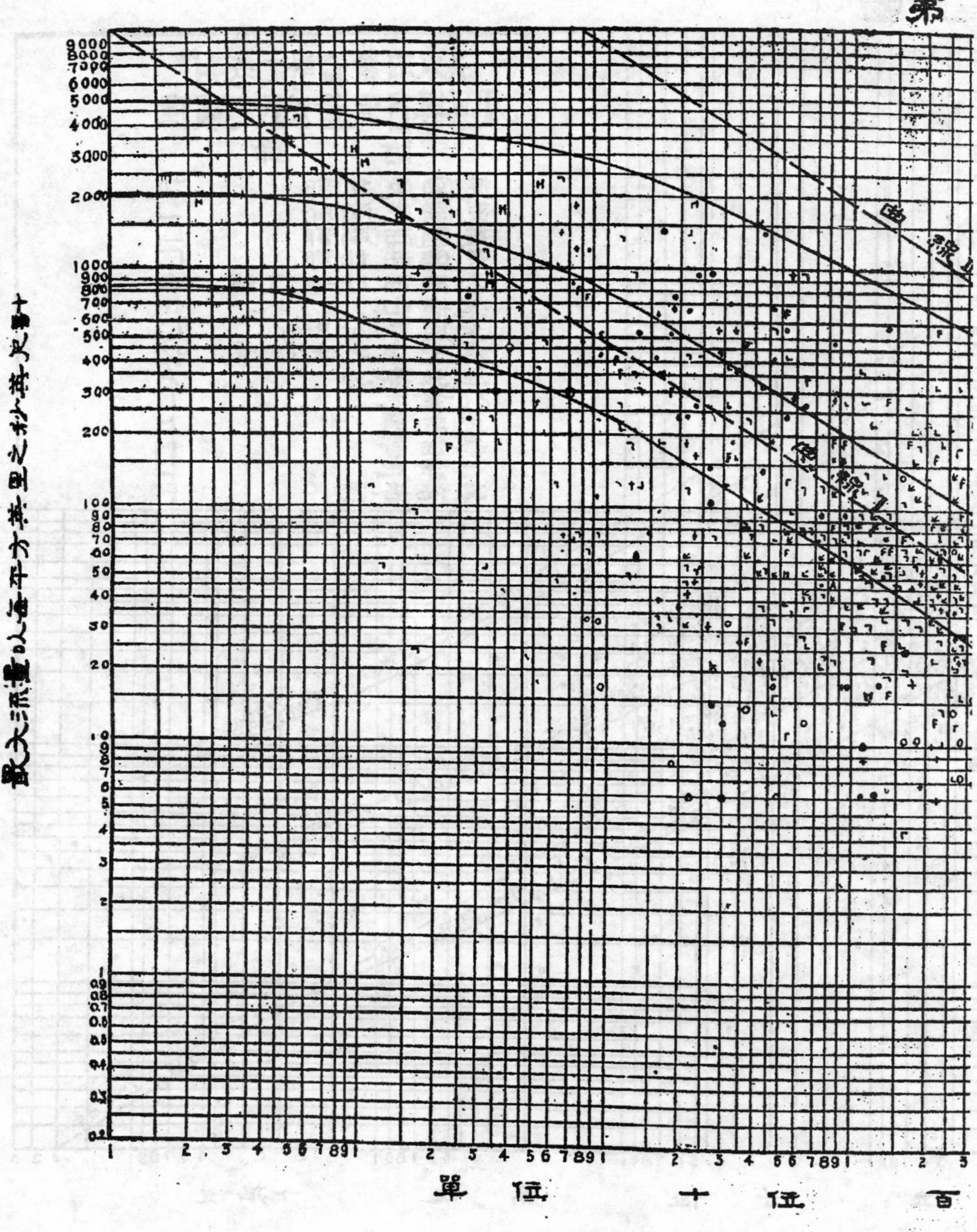
第 圖
最大流量以每平方英里之秒英尺計
9000
8000
7000
6000
5000
4000
3000
2000
1000
900
800
700
600
500
400
300
200
100
90
80
70
60
50
40
30
20
10
9
8
7
6
5
4
3
2
1
0.9
0.8
0.7
0.6
0.5
0.4
0.3
0.2
曲線II
曲線I
1 2 3 4 5 6 7 8 9 1 2 3 4 5 6 7 8 9 1 2 3 4 5 6 7 8 9 1 2 3
單位 十位 百位
流域之面:

$$\left(\frac{y+1}{y+2160}=0.00033\right.$$ 其數至小,故省去無礙.)

將 y 之值代入 (5) 式

$$x=\frac{4180}{\frac{1}{6.3+1}-\frac{1}{6.3+2,160}}=30,600$$

將 X 及 y 之值代入 (3) 式

$$z=20-\frac{30,600}{6.3+2,160}=20-14=6$$

$$q=\frac{30,600}{6.3+M^{\frac{2}{3}}}+6 \ldots\ldots\ldots\ldots (A)$$

用同法可以求得以下二式:

$$q=\frac{5450}{2.1+M^{\frac{2}{3}}}+2.5 \ldots\ldots\ldots\ldots (B)$$

$$q=\frac{1555}{1.6+M^{\frac{2}{3}}}+0.5 \ldots\ldots\ldots\ldots (C)$$

(A) 公式應用於罕有之洪流量,

(B) 公式應用於最大之洪流量,

(C) 公式應用於普通之洪流量,

推演此等公式,乃假設世界各河之流量而爲一河之流量.究竟此假設合理否,或距事業之遠近如何,尙希有以討論而指正之也.此三公式之弧.如第三圖.

附錄權度比較表:

1 英尺 = 0.3048 公尺 = 0.9525 營造尺

1 英寸 = 0.0254 公尺 = 0.0791 營造尺

1 英里 = 5280 英尺 = 1.8519 公里 = 3.215 華里

1 英畝 = 0.4047 公畝 = 6.5867 華畝

1 英畝 = 4,3560 平方英尺

1 公畝 = 100 平方公尺

1 華畝 = 6000 平方尺

DRY FORDS PROVIDE ECONOMICAL AND ADEQUATE BRIDGE FACILITIES

BY T. K. KOO

In many parts of the world, ares of alternate aridity and heavy rainfall develop peculiar conditions for the roadway engineer to face when adequate and economical structures across drainage beds must be provided.

These areas of extremes as a rule develop wide, shallow drainage beds which lie almost dry during the major part of the year and then for a limited time during the rainy season become raging torrents. Perhaps the high water stage lasts but a few hours or it may last for several days, but rarely longer than that. It usually subsides as quickly as it rises.

Very frequently it is necessary to carry a road over such a stream bed and during the dry season this is a simple matter as no bridge is needed, there being only a negligible amount of water in the stream, which is easily forded.

However, during the rainy season this method of crossing is out of the question as the stream reaches the proportions of a wide, swiftly flowing river and if traffic is to be kept moving during this period, an adequate means of crossing must be provided.

In locations where the amount of traffic warrants an all-year passable structure, but road building and maintenance funds are meager, a difficult situation is faced by the engineer.

The cost of building a bridge long enough to extend entirely across the wide stream bed and strong enough to endure the ferocity of the flood stage of the stream, would be prohibitive.

It is under circumstances such as this that the "dry ford" or submersible causeway finds favor. It is a drainage structure inexpensive to construct; 100 per cent efficient in operation except for a very limited time during the extremely high water; and is permanent, needing no upkeep.

The "dry ford" or submersible causeway usually consists of a battery of Armco Nestable corrugated culverts with a concrete fill; headwalls, built on a slant to lessen resistance to extreme high water; and a concrete surfaced road passing over the structure.

1—A submersible causeway on the Ellichpur-Anjangaon-Akot Road, Berar, India. This causeway is deigned to meet the demand for an economical and adequate bridge structure in an a:ea of alternate aridity and heavy rainfall. Installed 1923.

These fords are designed with a capacity to carry the average rainfall runoff of the district through the culverts and to allow extremely high water to flow over the road without damage to the structure.

Even while the water is flowing ovor the pavod con:rete surface these structures are passable the greater part of the time. Markers showing the height of the flood waters and defiring the edges of the pavement are erected to guide the traveler. Of course, during extremely high water no attempt is made to cross the stream but the construction is such that except in very extreme cases and for short periods the causeway is passable.

2—This is another submersible causeway, or "dry ford" on the Ellichpur-Anjangaon-Akot Road in Berar. The Armco Nestable culveris care for the usual drainage and during rainy season the water flows over the concertepavement.

Two submersible causeways or "dry fords" which have been giving excellent service in Berar, India, since their construction in 1923, are shown in the accompanying photographs.

Photograph number one shows a submersible causeway on the Ellichpur-Anjangaon-Akot Road about six miles from Ellichpur at what is known as the Chandabargah Crossing. There are twenty 30" diameter Armco Nestable culverts in the structure, each 24' long, all of which are in excellent condition after seven years service.

This causeway is submerged part of the year and during extremely heavy rains in the monsoon season it is not fordable for short periods. These periods are rare however, as the ford is designed to carry most of the flow through the culverts.

Photo number two shows another submersible causeway at the Sharmur River Crossing at Anjangaon on the Ellichpur-Anjangaon-Akot Road. There are ten 30" diameter and four 24" diameter Armco Nestable culverts, each 22' long, in the structure. This "dry ford" is also submerged during the extremely heavy rainy season and for brief periods is not fordable.

Structures such as these are giving very satisfactory service in many parts of the world and are providing the solution for the problem of constructing economical and adequate bridge facilities.

通信欄

隴海西路山洞工程不日招標

隴海路靈寶至潼關一段,七十二公里建築工程,前因戰事停頓,茲已於上年十二月起完全復工,所有山洞橋梁涵洞工程,皆由原承包人繼續承辦.惟在靈寶附近函谷關之第十號新山洞,長一千七百餘公尺,及潼關之第十七號山洞,長一千零八十公尺,尚未開工,茲聞鄭州隴海工程局,已將此兩洞圖樣計畫,送請鐵道部審核,一俟奉准,卽行招標開工,又該路文底鎮,至潼關,尙有一小段土方未動工,亦擬不日招標云.

又該路靈寶裝配十二孔大橋工程,前登報招工承投,嗣因標價太貴,改變方針,自行雇工裝配,二月底開工,兩個月可完工.

湯姆森電機廠之最近成績

英國湯姆森電機廠(The British Thomson-Houston Co. Ltd.),開設於英國之魯格比(Rugby),新近倫敦市政府向該廠定造透平發電機一架,共有二萬基羅瓦特之多,並備有加水抽水機,發熱器,蒸發器,以及凝結器等.此項定貨,將裝置於格林威士電站(The Greenwich Power Station)供給倫敦市電車電力之用.

在此之後,該廠又有七萬五千基羅瓦特之透平之定貨,以裝設於巴金電站(Barking Station),故該廠在過去十二個月內,以倫敦一市而論,對於電機之供給,及尙在製造中者在內,已達有二十五萬基羅瓦特之巨.

客歲,蘭開夏電氣公司之伯啟漢電站(The Padiham Power Station of dhe Lancashire Electric Power Co.)內,有三萬基羅瓦特之透平機,亦爲湯姆森廠出品,在一千九百二十九年中,以全英國電站中認爲效率最高見著.

現在蘭開夏電氣公司之寇司來電站(Kearsley Power Station),有湯姆森廠製造之三萬二千基羅瓦特透平發電機二架,亦以熱力效率高至百分之二四・三六著稱,因之不論大小與產量各方面,在一千九百三十年中均推爲全英效率最高之電站.

SIR CHARLES PARSONS

News reached Shanghai some days ago of the death of Sir Charles A. Parsons, the inventor of the modern turbine, and the Head of the famous firm that bears his name.

The deceased gentleman was a great scientist and engineer,—a man who ranks with the eminent engineers of the Ages, with watt, Stevenson, Kelvin. His great invention has had an immense social influence. Nearly every large electricity station is equipped with steam turbines, and it is these steam turbines which have made it possible to generate with such economy as to make the use of electricity as widespread as it is to-day. In marine work,—in the propulsion of ships,—the introduction of the turbine has had still more phenomenal results. The adoption of the turbine within recent years for the propulsion of nearly all important vessels has become so much a matter of course that the public mind has almost lost sight of the debt it owes to the man who first invented the turbine and then first applied it to driving ships and driving generators.

Early in life Sir Charles Parsons revealed his extraordinary genius in scientific and mechanical matters, and when at Cambridge, where he graduated in 1877, he made models of an epicycloidal engine. This engine differed from all others of the period, the cylinders revolving around the crankshaft and making half as many revolutions. A good many engines of this type were made, and it was from his early invention that Sir Charles came to attack the problem of the steam turbine.

Before he produced his first successful turbine in 1884 a good deal of hardship and labour were experienced,—the usual concomitants of any great work. The first machine was of 6 H.P., and others followed in quick succession till by 1889 there were about 300 turbo-generators ranging in capacity up to 75 K.W. From this stage onwards progress became continuous and rapid. The size of machines grew larger in size and more economical in operation. In 1892 a machine of 100 K.W. was produced. In 1900 it was possible to make one of 1000 K.W., in 1907 a machine of 5000 K.W. was produced.

The turbine was first applied to ship propulsion by Sir Charles Parsons in 1897 when the now famous "Turbinia" created history by outstripping the fastest destroyers of the day at the Naval Review at Spithead. Two years later the British Admiralty ordered Sir Charles Parsons to fit a turbine in the destroyer "Viper," and this was so successful that other orders rapidly followed. France was the first of the Foreign Powers to fit the turbine to a warship; that was in 1903, but now nearly all British and Foreign warships are fitted with Parsons turbines.

Thus from the small 6 H.P. turbine made in 1884 we come to the giant machines of 200,000 H.P. installed in our modern electricity works and driving our modern liners.

Early in the history of the Parsons firm it was decided to grant licenses to Makers in other countries who were interested in developing turbines. Many manufacturing Firms availed themselves of this arrangement, and thus it came about that Parsons tuhbines were made under license in practically all important manufacturing countries in the World, including the United States, Switzerland, France, Germany and Japan.

It seldom happens that an inventor lives to witness the remarkable developments of his genius, but that has been the happy lot of Sir Charles. Throughout these years he has remained the presiding genius of the famous Works at Newcastle, being actively employed in directing its affairs right up to the end.

In manner he was quietly composed, conveying a sense of quiet victory and much hard work well done. None of the eccentricities which the World likes to associate with genius were traceable in his collected and simple manner, which at once commanded the respect and the admiration of all who came in contact with him. He has contributed greatly to the World, and his memory will be reverenced by all engineers and scientists and by those who love a disinterested Worker.

中國工程學會職業介紹委員會

介紹職業簡章

(一)宗旨　本委員會以介紹相當技術人材發展工程事業爲宗旨

(二)範圍　本委員會介紹人材以曾經專門技術訓練或具有相當經驗者爲限

(三)手續　凡招聘或待聘者均須先向上海寧波路四十七號領取委託書或志願書塡明寄交本委員會指定之審查委員詳細審查分別登記後介紹相當人材或位置

(四)用費　凡委託本委員會代聘技術人員者概不取費惟經委託者之同意登載廣告或發送電報等費用須由委託人負擔之其有經本委員會介紹而得有位置者永久會員得自由捐助普通會員應捐第一月薪十分之一非會員十分之三以資彌補

(五)責任　本委員會關於招聘者與應聘担保事項概不負責應由雙方自行辦理之

(六)證書　應聘者如有證書或像片等物經本委員索取或自行寄交本委員會者請附帶寄回所需郵費否則本委員會不負寄回之責

(七)附則　本簡章得隨時修改之

民國二十年七月

工程

第六卷 第三號

中國工程學會會刊

THE JOURNAL OF THE CHINESE ENGINEERING SOCIETY

VOL. VI, NO. 3 JULY 1931

中國工程學會發行 總會會所：上海寧波路四十七號 電話：一四五四五
每册三角預定全年四册一元每册郵費本埠二分外埠五分國外三角六分

3967

中國工程學會職員錄

（會址上海寧波路四十七號）

歷任會長

陳體誠（1918—20） 吳承洛（1920—23） 周明衡（1923—24） 徐佩璜（1924—26）
李屋身（1926—27） 徐佩璜（1927—29） 胡庶華（1929—1930）

民國十九年至二十年職員錄

董事部

凌鴻勛 鄭州隴海鐵路工程局　　陳立夫 南京中央執行委員會秘書處
李屋身 上海四川路112號大興建築事務所　　吳承洛 南京實業部
徐佩璜 上海市教育局　　薛次莘 上海南市毛家弄工務局

執行部

（會長）胡庶華 吳淞同濟大學　　（副會長）徐佩璜 上海市教育局
（書記）朱有騫 上海新西區楓林路公用局　　（會　計）朱樹怡 上海四川路215號亞洲機器公司
（總務）支秉淵 上海江西路378號新中公司

基金監

惲　震 南京建設委員會　　裘燮鈞 杭州市工務局

工程

中國工程學會會刊

季刊第六卷第三號目錄 ★ 民國二十年七月發行

總編輯 周厚坤 總務 支秉淵

插圖

正文：

中國工程學會發行

3969

中國工程學會會章摘要

第二章　宗旨　本會以聯絡工程界同志研究應用學術協力發展國內工程事業爲宗旨

第三章　會員

(一)會員　凡具下列資格之一由會員二人以上之介紹再由董事部審查合格者得爲本會會員

(甲)經部認可之國內外大學及相當程度學校之工程科畢業生并確有二年以上之工程研究或經驗者

(乙)曾受中等工程教育并有六年以上之工程經驗者

(二)仲會員　凡具下列資格之一由會員或仲會員二人之介紹並經董事部審查合格者得爲本會仲會員

(甲)經部認可之國內外大學及相當程度學校之工程畢業生

(乙)曾受中等工程教育并有四年以上之工程經驗者

(三)學生會員　經部認可之國內外大學及相當程度學校之工程科學生在二年級以上者由會員或仲會員二人之介紹經董事部審查合格者得爲本會學生會員

(四)永久會員　凡會員一次繳足會費一百元或先繳五十元餘數於五年內分期繳淸者爲本會永久會員

(五)機關會員　凡具下列資格之一由會員或其他機關會員二會員之介紹並經董事部審查合格者得爲本會機關會員

(甲)經部認可之國內工科大學或工業專門學校或設有工科之大學

(乙)國內實業機關或團體對於工程事業確有貢獻者

(八)仲會員及學生會員之升格　凡仲會員或學生會員具有會員或仲會員資格時可加繳入會費正式請求升格由董事部審査核准之

第四章　組織　本會組織分爲三部(甲)執行部(乙)董事部(丙)分會(本總會事務所設於上海)

(一)執行部　由會長一人副會長一人書記一人會計一人及總務一人組織之

(三)董事部　由會長及全體會員舉出之董事六人組織之

(七)基金監　基金監二人任期二年每年改選一人

(八)委員會　由會長指派之人數無定額

(九)分　會　凡會員十人以上同處一地者得呈請董事部認可組織分會其章程得另訂之但以不與本會章程衝突者爲限

第六章　會費

(一)會員會費每年國幣六元入會費十元　(二)仲會員會費每年國幣三元入會費六元

(三)學生會員會費每年國幣一元　(四)機關會員會費每年國幣十元入會費二十元

隴海鐵路灞橋風景

橋在西安東郊約十二華里隴海路綫現由橋北約二百公尺處經行其地風景甚佳

第二次世界動力會議總報告

報告者本會出席代表黃伯樵

(一)組織上之成效

參加世界動力會議之人數,共有三千九百人,在開會兩月以前,即將三百八十種報告,照來稿原文刊印,分佈於與會人員,嗣復歸納爲總報告三十四章,用三種通用語言,(德文英文及法文),於開會前印就現已分編三册出售.每一總報告中含有:

(一)各項報告內容摘要;

(二)關於發展趨向之綱領;

(三)預備討論之問題.

在此總報告內,巳含有此次會議之大部分成績.

以總報告爲基礎之三十四類專門會議,聽衆總數逾一萬人.專門會議時,聽衆最多時,超過一千二百人.各專門會議中之論文平均逾十二篇.

一切總報告及討論資料,與口頭演講,皆用三種通用語言之一種並以傳話機連同其他二種語言,傳播於聽衆.此項試驗,成績良好.凡在會議以前交到會中之文稿,因繙譯者之訓練有方,及晝夜工作逐譯皆極正確;即完全自由之討論,其同時迻譯之成績,亦出意料之外:蓋無論如何,總勝於無傳話機之繙譯.但在演講者利用文稿,繙譯者事前未能預備時,當然不免失敗.總之爲節省時間,舍用傳話機之方法,無由使四百三十參加討論者,皆得演講機會.傳話機經此實地試驗,獲得極爲可感之改良意見七種.普通重大演講,足以表示新穎學理.此種極有價值之貢獻,能使動力會議之精神,灌注於向不注意此項問題之一般人士各種專門會議,得有多數重要議決案,國際委員總會,已授權於委員會,將此項議案,遞送各國分會研究,俟下次卽一九三一年會議時,提出報告,在此議會中,再詳細討論,見諸實行.

(二) 國際委員總會議事紀要

國際委員總會根據建議,特組一委員會,其任務爲徵詢各國分會意見後,研究一種極有効力而合理的工作方法,以備提出於下次卽一九三一年之會議;尤應努力者,爲使範圍廣泛之報告,在將來可以避免,并使與各國分會,得以不時交換動力統計材料,及指示重要公佈文件.

國際委員總會認可第二次世界動力會議德國籌備委員會之議決案.將會議報告分類訂成廿一册發表.此項報告亦可零購,第廿一册中,載有悉依當時語言之七篇.普通大演講刻已出版.

國際委員總會與其他國際機關之友誼關係,有極可欣慰之進步,關於彼此工作範圍之界限,已與國際谷壩委員會及國際高壓會議成立一種諒解.世界動力會議於將來會期,可得此類國際機關之特別報告,一如此次國際谷壩委員會之協助.國際委員總會因此採納建議:與國際電話線聯會,及國際地下溝渠聯會,隨時交換學識.自從國際標準委員會,國際電工委員會.及新國際檢驗材料聯合會.各將其組織宗旨及工作情形,向國際委員總會爲友誼之報告後,國際委員總會決意將所有之標準工作,付託於以上三團體,而聽其爲適當之工作分配.

國際委員總會收到關於創立世界工程師聯會或世界工程師永久聯會之建議,認爲此類新機關,足以聯貫原有舊團體而成國際高級機關.

一九二八年,世界動力會議在倫敦舉行燃料會議時之四項議決案,有三項正在實行.其關於「國際調節熱量問題」之一項,在第二次世界動力會議六月二十日之特別會議,得有最後之解決:當一致決議,以世界動力會議之全副力量,促成將來燃料熱量之顯明記載.或爲毛熱量.或爲淨熱量.爲達此目的,第一須將兩種熱量,用各種語言正式記載;第二須將兩種熱量,註以國際統一之指數.此種記載與指數,本年內當可發表.至於毛淨熱量,如何予以國際統一之定義,已請上述之標準委員會辦理.

國際委員總會已决議應瑞典及斯干的那維亞半島各國之邀請,於一九三三年在彼處開下次世界動力會議局部會議,此項局部會議以供給動力於大規模工業爲討論資料.

第三次動力會議當於一九三六年在美國舉行.

國際動力委員會應聲明者:即希望各方對於出售電力,以啓羅華特年,或啓羅華特時,或二者合用爲標準,於其利弊,加以研究與討論,據悉國際工業及電力分配聯合會,已將此問題列於其下次議程中.

國際委員總會最後向各種科學家及團體表示感謝,以答其供給委員會及各國分會多數有工業與法律價值之書報.

(三) 總結果

第二次世界動力會議,各種專門會議中,數量豐富而重要之各個結果,欲於短時間內一一詳述,勢非可能,唯有於會議報告册中求之,刻正由總報告編纂者編成撮要,數星期內,可分致各國分會,并在報章發表.

茲篇所舉,特普通結果中犖犖數端,臚列如次,會議之最大結果,在使全世界多數饒有經驗之專家,獲得私人間之聯絡.因此不但有種種利益,更可增進民族間之親睦.

其次在會議期內,各地新聞界充分合作,將本會議工作與目的盡量宣傳於全世界各專家及一般人士,尤爲本會所感謝.

最後以整個的第二次世界動力會議,向各國領袖及全世界對於其所發之重要問題,作下列之答復:

世界動力會議,認減低動力價格之途徑,爲(一)用在本會議討論之許多技術上與組織上的方法,以減輕動力製造與分配設備之價格;(二)提高動力製造與分配設備的使用時間,本會議各種報告及論文,幾無不以此二者爲目的.故可謂爲本次動力會議之總題.

第二次世界動力會議又證明以機力光熱供給世界,足以減少失業程度.因有動力之供給製造物品之方法,年多一年,每一新方法,發生一新工作,又發生一新出品.同時又推廣出品之銷路.據近一百五十年之經驗,因此而增加之工作,實超過因機力替代人力而減少之工作.非然者,各文明國家之人口,在此時期,決無如許之增加.第二次世界動力會議又爲科學界技術界經濟界社會界開一新局面.各項普通演講,對此尤爲努力.此項演講,能道破在今日人類複雜思想與行爲間,所認爲性質不同的各界之相互關係,而詔示吾人,已進入一經過極端專門化而趨向一致精神文明的新時代.

鎮句路第一段告成

鎮江至句容公路,爲京杭線通江蘇省會之幹線,共長四十公里.第一段自鎮江起至迴峯庵止,約七公里,業已興築工竣,試行通車.該路全身均用最新方法築成,路面寬約四公尺,建築頗爲堅固.其工程約分六部:(一)開路床,(二)舖大石塊,(三)舖小石塊,(四)灌泥漿,(五)舖瓜子石片,(六)舖細沙.該路路面工程之精,爲省城中山路所不及,五六年可以無須修理.且於轉灣處仿照鐵路築法,一邊高起,使乘者不覺轉灣,雖開快車,亦無妨礙.路之兩旁,均栽樹木,多半爲法國梧桐.迴峯庵適當五州山與十里長山之交界處,峯迴路轉,車行其中,雙峯夾道,兩山之麓,小松葱蔚,澗水滲滲,風景之佳,勝過招隱.汽車速度,可開四五十碼.現建廳擬於迴峯庵後,建造茅屋三楹,以作避暑療病之用.沿路一帶,不久定可繁榮.至自迴峯庵起至句容各段,現亦積極籌備.興工建築,完成之期,當不遠也.

機車之選擇

著者：程孝剛

(一) 機車類別

機車式別,通常以輪組表示之.首字爲導輪數,次數爲動輪數,末數爲隨輪數.其常用之各式,有下列之數種:—

(1) 2—6—0 摩古式　　(2) 2—6—2 平原式

(3) 2—8—0 凝結式　　(4) 2—8—2 天皇式

(5) 4—6—0 十輪式　　(6) 4—6—2 太平式

(7) 2—10—0 千足式　　(8) 2—10—2 山達費式

(9) 4—8—2 山地式　　(10) 2—8—4

(11) 4—6—4　　(12) 4—8—4

(註) 調車用機車及聯組式機車未列入

就上列多種之式樣中,欲選定一二式,爲客貨運輸之用,其意義甚爲重要.蓋不獨機車價值甚鉅,投資須謹慎經濟,而且影響於將來之運輸,尤爲鉅大.業務能否盡量發展,或處處受限制,胥視機車之式樣而定.

茲爲愼重研究起見,特詳細討論,初釋各種輪組之意義,次述各方之關係,而殿以審擇標準.

(二) 輪組之意義

(1) 動輪　動輪上之重量,爲發生牽引力之源泉,輪軌間之澀率 (Coefficient of Friction) 通常爲百分之廿五.因天氣及軌道之情形,而略爲高低.換言之,即粘着率 (Factor of Adhesion) 昇降於四之上下.故初步約計法,以四除動輪上之重量,卽可得牽引力之大概.

因橋梁軌道之載重,有一定之限度,故每軸之負重,不能逾越規定之數.在此範圍內,欲求增加鉅量之牽引力,惟有增加動輪之軸數.惟此項增加,仍須受橋梁集中負荷量,及軌線曲度之限制.

若動輪軸已增加至最多數,則增加牽引力之方法,惟有利用更多數汽缸,或爲三汽缸,或爲交複式四汽缸,以減低粘着率,或爲四汽缸六汽缸之聯組式,以增加動輪之數.

故就上述情形而論,在一定之軌道情形之下,動輪數卽可表示牽引力之大小.

鍋爐之供汽量,與汽缸之徑,如無問題,則增加動輪之徑,可得較高之速度,而牽引力不至受任何影響.惟輪徑旣大,則輪盤之距離長,而所受之曲度限制亦愈甚,雖中間各輪間,可取消輪緣,稍資補救,然要不可視爲常法也.

(2) 導輪 力學定例,凡動力非受他力阻撓,其方向不變.故機車當轉入曲線軌道時須有導力以變其方向,輪緣與外軌道之內側相切,雖亦能發生此項導力,而在高速度時則易發生出軌之危險.故須用轉向架以資導引,而導輪卽支持此轉向架所負重量之輪軸也.

轉向架之構造,有懸桿或倒心式之支鈑,機車前部之重量,卽置於懸桿或支鈑之上.當經過曲道時,機車之中線,與轉向架之中線分離,藉懸桿或支鈑之功用,有將所荷重量向上抬高之勢,於是藉重量下墜之力,使兩中線仍歸密合,而機車卽逐漸變易其方向.故導輪所負荷之重量,雖於運轉上爲無用,而轉向之際則深恃此力之作用.負荷愈重,轉向愈靈.

機車前部,有龐大而重之汽缸,且汽鍋重量之一部分,亦支配於汽缸之上,故其負荷亦有不得不分配於導輪之勢.

導輪軸之負荷量,亦同樣受橋梁軌道之限制,故採用四輪轉向架,雖大多爲轉向之便利,而有時亦爲載重所限,不得不然.

（3）隨輪　燃燒之效率,係於火箱之構造.爐條之面積欲其大,以便煤之散佈.火箱之容積欲其深,以便得完全之燃化.顧機車之構造,爲淨空規所限,難於向上伸高.且軌道之寬度有定,重心過高,亦所不宜.故欲得適宜之火箱,非向旁向下展拓不可.若將火箱置於動輪之上,則因動輪徑大,其上地位有限,勢不能深.若將火箱夾於動輪之中,則爐條之面積,又勢不能大.故改良之法,惟有將火箱置於隨輪之上,則輪徑既小,兩旁及下方,均有展拓之餘地.

最近機車構造之進展多注意於充分之供汽量,及完全之燃燒.前者以求牽引力及速度之增大,後者以求煤量之節省.於是火箱之中,如缸磚拱,(Brick Arch)如熱脊,(Thermic Syphon)如自運供煤機(Mechanical Stoker)均次第裝設.所需於火箱面積體積者愈大,所增之重量愈多,而隨輪亦有由兩個而增爲四個之傾向.

煤之品質,常爲計劃火箱之要件.就世界之趨勢觀之,上等之煤日少而日貴,於是劣等之煤,逐漸變爲機車之主要燃料.顧欲利用劣等之煤,則火箱構造,尤非增大其面積體積不可,隨輪之數,常以此爲決定之主要條件.

綜上觀之,可知輪組之構成,爲機車全部之關鍵,以此爲機車式別之分野,誠屬扼要也.

(三) 機車與各方之關係

（1）軌道　在普通情形之下,每三百分之一之坡度,即須將列車重量,減少一半.坡度愈陡,則照單純之正比例而遞減,惟坡道短者,可藉列車之動能量(Momentum)衝上坡道,犧牲速度,以作重量之代價.反之,如坡道適位置於列車停止之後,則列車所需之加速度,有更行減低重量之必要,故於約計牽引力時,上項情形,必須詳晰考慮.

曲度之在平道者,其所施於機力之限制,常不如坡道之甚.其在坡道上者,又須化入坡度計算.故曲度之影響於機車式別者,多在輪盤之長度,若影響於牽引力者,惟在特別情形之下,方始發生.

軌道鋪置於軌枕之上,其兩枕之間,恰如一連桁式小橋.機車車輪之載重,因平衡重(Counter Balance)之作用,在高速度時,幾倍於靜止狀態.故軌枕之距離,與鋼軌之重量,亦與機車式別頗有關係.蓋三汽缸機車,固可減少平衡重之衝擊,而特別之往復配件,亦可得同等之效力.何去何從,宜事先抉擇也.

(2) 橋梁　橋梁之關係機車者,為淨空,為載重.載重雖有規定之每軸最大限,而因車輪平衡重之設計不一,橋梁所受衝擊,亦不一致.如認為太高時,則須採三個至四個汽缸式,或用特別輕固之往復配件,以資解決.至平衡重之量,則不宜過輕否則往復機件前後之衝擊不但機車受損,且將生搖頭(Nosing)之現象,而使橋梁發生過度之平面震動.

(3) 運輸狀況　機車之目的,即在因應運輸之需要.故此方面關係,最為繁複,茲分述之:—

(甲) 方向　鐵路業務,除客運外,其在貨運方面,普通分進口出口及聯運經過三種.其中以出口之礦產佔最大量.而農產次之.其進口貨量,能相抗衡者,實不多睹.故大部之運輸,多為單向,回程時不過空車行程而已.故貨運噸量同,而移動之方向不同,則所需之機力總量亦異,此應注意者一.坡度之方向,如與貨運流動之方向有利,則所需之機力少,反之,如與貨運流動之方向有損,則所需之機力多.此應注意者二.

(乙) 風軔　列車當下坡時,行動固不成問題.而欲其停止如意,以免危險,則有賴於風軔故車輛風軔不完全時,列車重量,往往受其限制,雖有牽引力鉅大之機車,無所用之.

(附註) 命機車煤水車之總重量 = W

動輪上之重量 = w

牽引力 = T

則機車軔力 = .75 W (國有路標準)

= .75 × 2 w (約計)

$$= .75\times2\times4\text{ T}\ (\text{約計})$$
$$= 6\text{T}\quad(\text{約計})$$

命列車車輛總重　$= \text{Wc lb.}$

皮　重　$= .4\text{ Wc}$ (約計)

總軔力　$= .7\times.4\text{ Wc}$ (國有路標準)

設每噸之阻力　$= 20\text{ lbs,}$

則列車全阻力　$= \text{T} = \frac{\text{Wc}}{2,000}\times20$

代入上式,總軔力 $= .7\times.4\text{ Wc}$

$$= .7\times.4\times\frac{2,000\text{ T}}{20}$$

$$= 28\text{ T.}\ (\text{約計})$$

普通行車規章,每三輛載車最少一輛應有風軔.

其軔力　$= 9.3\text{ T}$ (約計)

故機車軔力,不過為列車軔力之20%以至43%.

(丙)貨車載重量　機車之牽引力.係對於列車全重量所發生之阻力而定.而列車全重,係由各個車輛組合而成.因其大小不同,不但皮重與載重之比例不同,即所發生之阻力亦異.通常之十噸十五噸之車輛,皮重約為載重之50%.標準四十噸車,約為43%.百噸至120噸之車,則約為35%.故每千噸之列車,其載重量用十五噸車時,僅為660噸.用四十噸車,可增為700噸.用百噸車,則可增為740噸.

次之,十五噸車之內部濕阻力,約為四十噸車之一倍.若假定濕阻力,為列車全阻力之半數,則四十噸車可牽引千噸時,若用十五噸車,僅得六百六十噸.

復次,十五噸車之長,約當四十噸之三分之二.而重量則不及一半,故錯車環道之長度,往往於用十五噸車時,感覺不足,轉運空車時為尤甚.機車之牽引量,常因受此限制,而不能掛足軸數.

(丁)站間之列車密度　鐵路貨運,以礦產及農產為大宗.農產固隨季候為轉移.即礦產如煤類亦因冬夏季需要不同,而生差異.通常均以秋冬為最忙之時.故單線運轉之列車密度,如在某時期內,已達最高點.則除改敷雙軌外應設法增加列車重量,以減少列車數次.惟當增加機車牽引力,以運轉最過之列車時,須注意於上舉之他項限制.

增加列車速度,不能發生減少列車密度之效力,蓋增加速度,卽需減少重量,所得或不償所失也.

改良列車號誌,可以減少在站之停留時間,爲補救列車密度之良法,費輕而易舉.且於一般之運轉,均有裨益.故在可能範圍內,脩改良號誌而置大型機車爲緩圖,以其牽涉甚多,費用亦鉅也.

(戊) 客貨通用機車　客運機車之要點爲速力,貨運機車之要點爲牽引力.顧國有路之客車,通常以三十至四十英里爲標準速度.與美國之習慣,迥不相侔.若以六十餘吋動輪之天皇式機車運轉之,斷能勝任愉快.故間有提議採用此式,作爲通用機車,於運輸上可得不少之便利.頗有考慮之價值.惟軍事時期,雖有利益,而鐵路入於正當營業時,則仍須另行改正.否則殊非改良進步之道耳.

(4) 燃料　機車之燃料,有煤,木柴,油三種,木柴與油二種,近於暴殄天物,自將日就淘汰之列.煤則將繼續爲機車之惟一燃料.

就煤之品質分析,大概有無煙煤(硬煤),半無煙煤,煙煤,半煙煤,草煤五種.鐵路每取其價目廉供給便利者而利用之,國有各路所用之燃料,多屬半無煙煤,煙煤,及半煙煤三種.

因上項各種煤炭之熱量不同,燃燒狀況亦異,機車之鍋爐計畫,隨之不同.若漫不分別,則機車之效率,勢必減低.

就煤之形態而分.大概爲塊煤,末煤,粉煤,及煤磚數種.塊煤指一吋以上者而言,末煤則爲大小不等之細末,粉煤則爲特別研磨而成之勻淨細末,煤磚則爲由末煤加瀝青後壓成之磚塊.其中以塊煤及磚煤最易應用.末煤則需較高之技藝及細號爐格或添煤機,粉煤則需研磨廠,存煤庫,裝卸機等設備,機車火箱之構造,亦異於其他添煤需用吹風管,而尤以進風之路,須有嚴密裝置,以便精密之管理.

考查中國煤礦之現狀,其煤質良者不多見.塊煤價格,亦日趨高貴.將來之趨勢,大約非注重煤磚及粉煤二者不可.如採用粉煤,則今日機車之火箱構造,似尙多改良餘地也.

(5) 給水　水經煮沸,則所溶之雜質,沈澱鍋底,或結成硬塊,(俗稱水銹)而附着於鍋鈑,爲熱之不良傳導體.故通常須將水質由硬性變成軟性,然後注入鍋內.惟市上所售各種軟水化學品,多含腐蝕性,雖免水銹,而鍋鈑仍受損.歐洲各國多喜用紫銅火箱鈑,蓋不但不虞腐蝕,且水銹亦易於洗落也.

(附註)煤炭如富硫質,則硫經氧化後,溶解於水.侵蝕銅鈑,而成硫化銅.故銅鈑不僅價貴,環境亦宜斟酌.

水站之距離過遠,或中間有長距之坡道,則機車給水,須特別充分,煤水車之設計,應以此作爲根據.

(6) 工廠設備及轉盤　工廠機器太小,或轉盤太短,有時影響於機車之大小及長度.惟運輸狀况,如有必要時,則寧增加設備,而不限制機車,較爲得計.

爲修理之便利,機車之零件,以採用標準式爲宜.蓋不獨木模鋼模,均可公用,修理時工速而費省.即材料及技巧方面,亦可節省不少.惟含有改良之性質者,則以斷行新式爲宜.否則積重之勢,愈久愈難改革,必將永無改良之望.

(7) 人事　機車行於鋼軌之上,所受之震動甚強,熱度之伸縮甚劇.雨雪塵土之侵襲甚烈,其所以歷久不敝,日行千里者,全持檢查之勤恪,修理之敏捷而已.機件之構造過繁,則檢查難週.裝置過於嚴密,則則修理不便.人情大低畏難苟安,而小患遂釀成大病矣.交互式四汽缸與三汽缸機車,其設計甚精巧,其行動極平穩,其費用甚經濟,然而不能競爭優勝者,則繁複與簡單之分而已.

顧機車之勝於人力者有三,曰匀,曰速,曰鉅.苟此三者,有改良之必要,則機件雖繁複,有所不避,要當斟酌於利害之間,而擇其中耳.

茲爲明晰起見,特將上述各項關係,列爲一表,以便檢查:—

項目	關係事項		加於機車之限制	應付方法
1.	軌道	坡度	牽引力	減低列車速度
		曲度	輪盤長度	去中間動輪之邊緣
2.	橋梁		軸之載重	
			平衡重	採用三個或四個並列汽缸,或用特別輕固之往復配件
3.	運輸狀況	方向	機力總量	
		風軔	不能載滿軸	添裝客貨車之風軔
		貨車載重量	牽引之有用重量	用較大之貨車
			列車長度受限制不能滿軸	延長道岔,或改用較大之貨車
		列車密度	牽引力不足	用大型機車,或設雙軌,或改良號誌
		通用機車	貨物列車之牽引力	
			旅客列車之速度	斟酌環境
4.	燃料		火箱之構造及隨輪之軸數	機車構造應適合於燃料
5.	給水		火箱之構造及水車之容量	機車及煤水車之構造應適合水質及水量
9.	工廠設備		機車之大小及標準	增加設備
7.	人事		複雜之構造及附件	力求簡單,或訓練員役

(四) 審擇之標準

機車之式別,及各方之關係既明,乃可進述審擇之標準.分別言之,則有四端:—

(1) 辨中外之異同 吾國鐵路,方在幼稚時代,舉措設備,自不能不參酌先進國之規模.顧中外情勢,非能盡同,則生吞活剝,亦非得計.英日二國,地狹民稠,與吾國大異者無論矣.卽歐洲大陸,此疆彼界,制度各殊,實與島國無大異.至於各殖民地,則鐵路政策,胥為殖民政策所支配,且亦不足稱為先進.其疆域遼廓,人煙稠密,物產豐富,俱與我國相似者,惟北美合衆國而已.第美國農工業之發達,為全球之冠,吾國鐵路之力量及其需要,有未能一蹴幾及者,師其意則智,師其迹則泥矣,故人工有貴賤,則機器效用之範圍不同而添煤機進退機(Power Reverse)之採用,須考慮矣.資金之大小難易不同則運用之緩急輕重,必須斟酌矣.材料之補充有難易,則新奇之品,不能貿然採用者矣.土產之物質不同,則粗者有所不避,精者有所不取矣.國有各路,往往受條約之束縛,主持者又或輕信洋顧問或商家之言論,於以上諸端殊不可不深長思也.

(2)察運輸之趨勢　機車之購置,影響於運輸者,至深且鉅.購買之初,最少須為十年乃至二十年之地步.在此長期間中,運輸之發展,將至若何程度,不可不預為熟計.所重者在客運,抑在貨運.客運之趨勢,偏於重量,抑為快捷.貨運之趨勢,為長途或短途,或在平道之段,或坡道之段,抑延長於全線.且現在之缺點安在,何賓以為補救.將來之業務發展,留餘地以資改良.審擇之際,必須研究既熟,胸有成算,方不至貽淺見之失.

(3)審經濟之價值　經濟價值者,非指機車之價格之高下而言.乃計算所投資本,能得若干益處.其益處或為增加收入,或為減少支出,鐵路既屬營業性質,自不能不詳加考慮.

機車之新式零件或附件,其構造品質功用,確有經濟價值者,均能用試驗及成本會計證明之.採用與否,嚴行考查後,即可積極決定.

至關於式別及重量之選定,則其考查之範圍較繁而又難確定.如由古拔標準之E35改為E50,則所有橋梁軌道轉盤等,均受影響.又如機車速度,若由三十英里增至五十英里之標準,則風軔信號道岔等,均受影響.又如採用三或四個汽缸,則工人之特殊訓練,及檢查修理方法,均受影響.故當決定機車標準時,其改良範圍內之支出,往往不僅限於機車一端.故計算其經濟之價値,非有嚴密之調查研究不可,否則機車之功用,不能發揮,資本等於虛擲矣.

(4) 與技術之規範 上述三項,均就機車之功用,加以審擇.但尙一主要條件,卽機車之構造,必須精良,然後能發揮其功用至預期之程度.故除重要之尺寸及材料,均宜照普通手續,詳細規定外,卽製造方法,亦宜擇要規定.如鍋鈑之眼,或壓或鑽,整軸之鈑,或鑄或模打(Drop Forge)等等.次之,則通用性(Inter Changeability),應設法保存,以免式樣過繁,存料過多之弊.又次,則機車之構造,必須與所有工廠之設備,及工人之技術相應,否則將來運用時,檢查修理,必至難期完善.雖有精良之機車,不旋踵已歸破敗矣.

(五) 結論

綜上述之各點觀之,可得一結論.卽選擇機車,雖係鐵路機務人員之專責,而考慮範圍之所及,則與運輸工務財政均牽連而不可分,要當以整個鐵路爲單位,而研究其得失,則庶幾矣.

路市展覽會之籌備

中華全國道路建設協會,自民十創立迄今已歷十載.愛會中當局有十週紀念籌備路市展覽大會之組織,原定本年四月舉行,嗣因邊遠省區及國外各地,製造寄遞陳列物品,頗費時日,故經十六次執行委員會議決延期.定八月至九月十二日以前正式開幕.俾各地得以從容籌備,俾路展開幕,得有完全品物以壯觀瞻.已由協會呈請行政院通令所屬,積極趕製應徵物品.一面並函請鐵道,內政,交通,實業各部查照,並函請外交部電令駐外各使領,就近向各駐在國路市機關與所分各廠,敦請派員參加,並將出品運華以廣見聞.其徵品方法,則以路市交通範圍內已成之工作,與現在進行之工作,均拍十二寸以上之精美畫片,如以古蹟名勝地段內,新修之道路橋梁公園等.大小新建築及新發明築路養路與運輸機器車輛油輪種切,均須詳註中西文說明,書,黏列照片與各模型.及各車機展品之上端,以便中外代表,及中外人士之記載或攝影,帶回各該國與各省市陳列,永留紀念.一面由協會編印展覽特刊.以廣宣傳而資借鏡.是以道路協會本屆之路市紀念展,集中外路市交通成績於一爐,實空前未有之國際路市展覽大會也.

眞茹國際無線電台之國貨變壓器

著者：周琦

眞茹國際無線電台，系國民政府建設委員會所創辦，後由交通部完成之，於民國十九年十二月六日開幕發報。發電處設眞茹，收電處設瀏河，控制機關，營業，及總務歸滬上沙遜大廈所設辦事處掌理之。該台之發電裝置，依無線電發播方向，可分爲舊金山（美）柏林（德）及巴黎（法）三分台，共需電量約400 KVA。照建設委員會原來設計，備有內燃引擎拖動交流發電機二部自行發電。迨閘北剪松橋新電廠告成，路線密邇眞茹，乃決計改用閘北來電，而以原有內燃引擎等，列爲副機。閘北電廠來電，系6,400伏脫高壓，接到電台後，經過變壓器，降低至380/220伏脫，以應用於無線電台。供電於舊金山及柏林分台者，爲安利洋行經售之325 KVA變壓器；供電於巴黎分台者，爲國人自辦之益中機器公司所製之225 KVA變壓器。（參考第一圖）均系三相，6,400至380/220伏脫，50周波油冷式且均設兩部，一用一備，庶無停頓之虞。

第一圖 變壓器外觀

著者在益中機器公司曾躬自監同製造及試驗225 KVA變壓器，頗屬精審。玆詳述各項於後，以貢同人之參考。

（甲）製造特殊情形

油冷式變壓器製造方法，種類雖繁，然其主要部份不外乎鋼片，線圈，及桶蓋三部，現就三者構造之特點集要如下.

(一)鋼片

切面　此 225 KVA 變壓器鋼片內心,用圓狀鐵心式. Cruciform Core 其切面如第二圖,即以寬狹不同二種以上之鋼片搭配而成.四圍留有充分油孔,利用油力以絕緣及冷却線圈,經濟耐用,相得益彰.

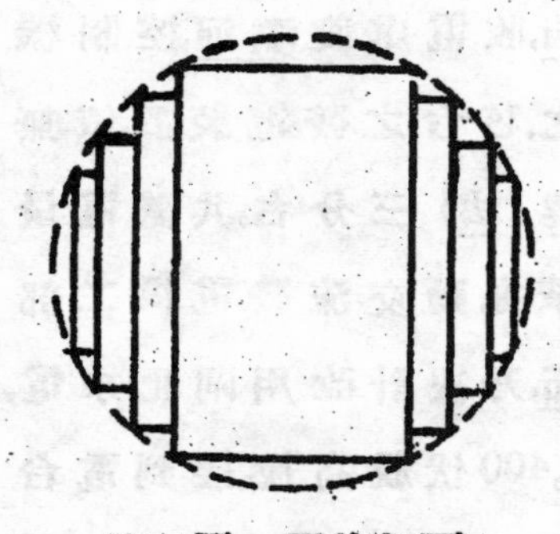

第二圖　圓片切面

漆乾　鋼片均系#29號 B. & S. (0.014 吋厚)着名紗質鋼皮剪衝而成.此鋼皮一面塗以特種絕緣漆之薄膜.此層薄膜不得超過 0.005 吋之厚度,故加漆之鋼片,其厚度約爲 0.0145 吋,換言之,每八張半鋼片之厚等於英吋之⅛.每一內心叠起所須張數,即依此算定,裝壓,毫無遺漏,務使鐵耗(Core Loss)低至最小極限,較之普通兩面塗漆或一面貼紙之絕緣鋼片,優勝多矣.

鋼皮漆後即盛入電烘箱中,通電烘熱至法氏 200度,歷六小時之久,始取出衝剪,則鋼片上之薄膜,永無溶解於高熱變壓器油之可能.

壓縛　鋼片心柱搭配齊全,先經過五噸重或五噸以上之螺絲壓床壓至所需時度.如係 200 KVA 以下用包帶縛緊,以上則用毛釘毛牢.鋼片上下橫擋處,預衝圓孔,裝配後以闊厚之水流鐵 (Channel Iron) 夾緊,再以粗壯之對梢螺絲旋牢.(參考第三圖)總之鋼片上各種配件,務使 (1)應用時鋼片之震響減至最低極限;(2)油路暢達毫無阻礙(3)機心穩持於桶內,不致因運轉而移動.

(二)線圈

式樣　每相有內低壓1圈,外高壓8小圈.低壓圈係螺旋式 Spiral Coil 即以扁銅帶 Copper Ribbon 六根併繞一層,形似螺旋.高壓圈係對稱式 Cross Over Coil 即以薄扁銅帶單根繞若干奇數層,線頭線尾對面引出,(參考第

第　三　圖

三圖）線圈輪廓均系八角形,(或圓形)如第四圖所示.線圈既不厚,四角內外均通油道縱線圈中層最熱處 Hottest Spot, 之溫度亦極安全.

接線　每相低壓依 Y 式結成,隔相電壓

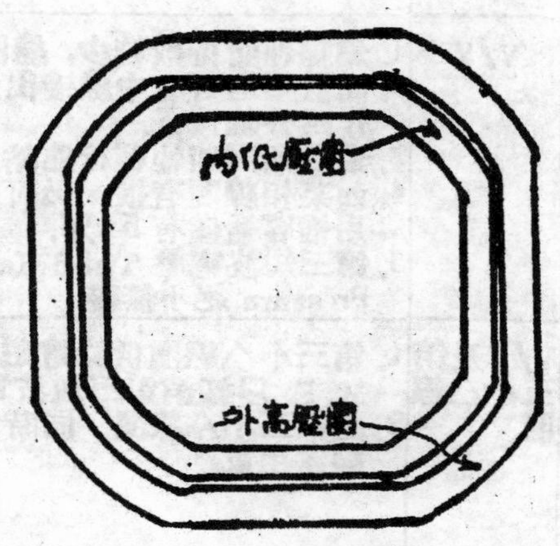

第四圖　線圈平面

為 380 伏脫,相與中線間電壓為 220 伏脫.每相高壓,則依△式結成,隔相電壓為 6,400 伏脫,并備有 6,600 及 6,200 伏脫兩抽頭,其簡明連結法,如第五圖所示.

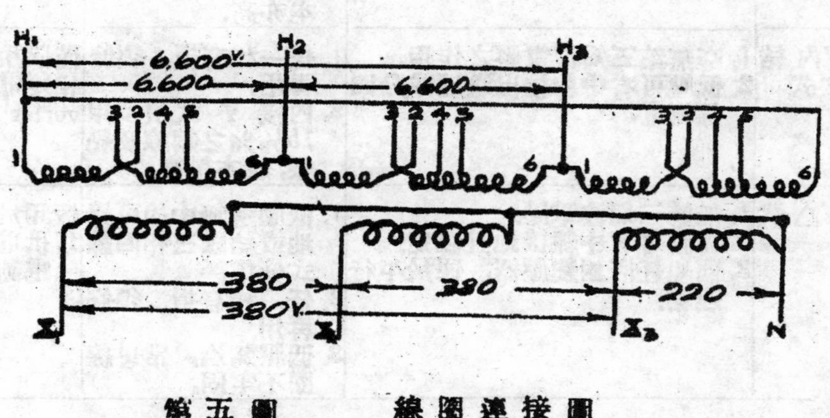

第五圖　　線圈連接圖

此種連結法,至少有兩點足以說明其長處.

(1)△—Y式之適用: 三相變壓器高低壓之連結方式,(專指三相鐵心式而言凡三相鐵殼式或三隻單相結成三相者不在此限)共計八種.又依兩兩並用之可能限度,分爲兩組.今就每組各種,略舉其優劣,列表如後:

組別	連結方式	優點	劣點	應用處
第一組	Y/Y	1. 銅線粗而捲數較少,線圈牢固. 2. 高低壓均可有中線接出通地或作四分線傳電. 3. 最易與同組他種行連結. 4. 如某相線圈有損,仍可供單相用惟容量僅有58%. 5. 第三短波電壓 Third Harmonic Pressure 絕小無礙.	1. 中線必須接地,否則不穩定. 2. 某一相有損,即不能再供三相用. 3. 銅線過粗,則線圈構造困難而費多.	僅限用於供給三相小量電力用之電.
	Y/Y及第三小△線圈	1. 第三小△線圈供給電磁力 M. M. F. 以打消第三短波電壓. 2. 此第三小△線圈,同時何供給線路用電.	1. 增加製造成本. 2. 第三小△線圈如有損則高低壓兩方亦易受損而不能用.	僅限於與Y/Y, Y/結Y結及內結Y/Y各三相變壓器平行聯結以打消第三短波電壓之影響.
	△/△式	1. 最易與同組他種行連結. 2. 如某一相有損,其餘兩相仍可照開口△或V式,以供58%之三相電量 3. 毫無第三短波電壓之作用. 4. 可用於三相不平均之儎重,僅使電壓稍不平均	1. 銅線較小,捲數絕緣均增,線圈不牢固. 2. 高低式均無中線. 3. 不能依三相四線傳電. 4. 線圈構造較艱,成本亦高.	用處極少.
	△/內結Y式	1. 毫無第三短波電壓之作用. 2. 低壓可有中線結出通地或分四線傳電.	1. 任一相有損,須修再用. 2. 內接Y式須多用15½%之銅線及絕緣,成本較高.	僅限用傳電至三相變電機. Converter
第二組	Y/△式	1. 無第三短波電壓. 2. 高壓可有中線接地且穩定. 3. 同組任何兩變壓器,便於平行連結.	1. 低壓邊無中線可接地或結成三相四線式傳電. 2. 任一相有損,須修再用. 3. 低壓爲△,常患線圈不牢固.	恆用於降低電壓以供電力用途.(電動機大宗)

第二組	△/Y 式	1. 無第三短波電壓. 2. 低壓有中線接地或依三相四線制傳電. 3. 無論三相平均或不平均之電量均可由低壓供給，且電壓不甚受影響.	1. 任一相有損，須修再用.	最適用于降低電壓以同時供電于電燈及電力，又適用于升高電壓因高壓邊既無第三短波電壓且線圈牢固而有中線保護.
	內接Y/Y式	1. 高低壓均有中線接地，或依四四線制供不平均之電量. 2. 高壓邊無第三短波電壓. 3. 高低壓線圈均極牢固.	1. 高壓須多用15½%之銅線及絕緣，成本較高. 2. 因構造關係內結Y式宜於低壓而不宜於高壓，故僅合於升高電壓之用. 3. 低壓邊仍有小量之第三短波電壓.	爲代用 Y/△ 或 △/Y 式變壓器之用，惟價較昂.
	Y/內接Y	1. 3. 同　上 2. 低壓邊無第三短波電壓. 4. 合於降低電壓之用.	1. 低壓須多用15½%銅線及絕緣，成本較高.	爲代用 △/Y 式變壓器而降低電壓，惟價較昂，又不合升高電壓之用.

注意　上表均假定高壓邊爲進電邊 Primary 低壓邊爲送電邊 Secondary 讀者留意.

細玩上表，當以△/Y連結方式之變壓器，實爲最適用於一般之狀況，理由甚明.

（2）高壓線圈中段抽頭之安全：變壓器非僅用於降低或升高一定量電壓，且可預防線路來電壓之變動，仍維持應用之電壓於不變，或於固定之來電壓時，增減其應用電壓之量.凡此非於線圈中抽出線頭不可.譬如此225 KVA 降低變壓器，明定來電壓系 6,400 伏脫，用電壓系 380/220 伏脫.今必欲如下兩表，以對付電壓之變遷.（參考第五圖）

來電壓	抽線頭結法	用電壓
6,600	2 結 3	380/220
6,400	2 ,, 4	380/220
6,200	2 ,, 5	380/220

來電壓	抽線頭結法	用電壓
6,600	2 結 5	404/234
	2 „ 4	392/227
6,400	2 „ 5	392/227
	2 „ 3	368/213
6,200	2 „ 4	368/213
	2 „ 3	357/207

不論任何線圈,均可抽此種電壓調整頭.惟以高壓邊抽出爲便,因高壓可用較小銅線及地位以引長至桶口結線處,且高壓恆繞在外圈,抽線處易於保護也.普過高壓抽線頭頗多於頂線圈,卽近 H_1, H_2, H_3 出線處抽出者,殊不穩妥,因高壓頂線圈設計時,銅線捲數特少,絕緣加厚,末尾數捲 End Turns 尤須加重絕緣,所以防接電一瞬時間末尾捲數電壓之獨高.(此因高低壓線圈及鐵桶於接電時有凝電器 Condenser Action 之作用)今如於近末尾捲數 H_1, H_2, H_3 處抽出線頭,則各線頭處時或等於末捲處,勢非隨處加厚絕緣不可,構造旣繁,絕緣費鉅且不穩妥,今特於高壓第三小圈抽出,一祛其弊.(參考第六圖)

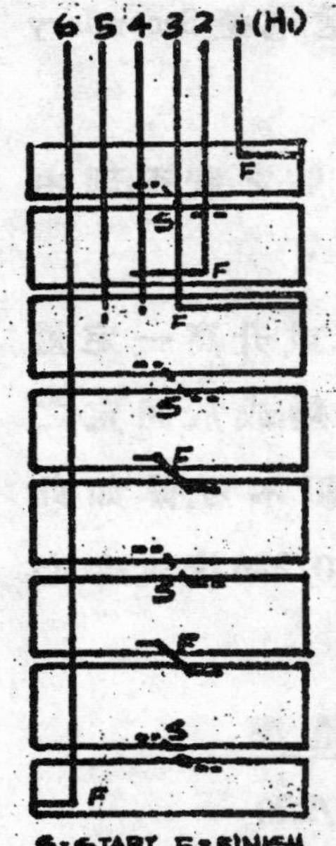

第六圖
線圈抽圖

絕緣 變壓器通電部份,互相絕緣.其絕緣方法之優劣,實與變壓器之溫度及壽命最有關係,蓋無論何種絕緣物質,皆有一定保安溫度之極限,過此極限,則或枯焦,或化散,不更保持其絕緣性.此際變壓器銅線互觸,發生短電;各圈相接,發生跳電;又或銅鐵兩部溝通,發生碰地.種種收象,卽其一端,足以毀壞全器.凡變壓器製造廠常分變壓器絕緣爲二部,卽圈內絕緣 Winding Insulation 及裝配絕緣 Assembly Insulation, 二類.前者注重在防短電及跳電,限制線圈隔層電壓不得逾 250 伏脫,每圈電壓不得逾 1000 伏脫.凡線圈繞成烘乾後,必加其本身二倍電壓之試驗.後者注重在防跳電及碰地,卽高壓與低壓圈間及高壓圈與銅片間,使其絕緣足勝二倍高壓更加 1000 伏脫之電壓試驗.又低壓圈與銅片間足勝約 3 倍低壓之電壓試驗,此種試驗歷一分鐘.內須不發生變動.(此名 Puncture Test 詳後).

變壓器絕緣物質最大之敵,莫如濕氣與高溫,此濕氣 Moisture 即空氣中之水份,或繞線時之手汗,或由其他媒介,傳入絕緣物中大足減少其絕緣本能,譬如最佳之變壓器油,僅加入 0.04% 之水份,即減少其絕緣量 Dielectric Strength 至半數.其餘絕緣物,如棉紗布,膠紙,膠布,木料,漆膠及瓷料等,如含濕氣,其絕緣力皆較完全乾燥時減少頗鉅.

溫度過高足使絕緣物質枯焦或化散,已如前述.惟各物質各有其一定保安溫度之極限,設計時,儘可避免其危害而提高安全因子.故變壓器造成後,除偶因特種情形外,(如儎量過高,油量淺薄或久用沉澱流動甚緩各情)溫度決不致達於極限,而損壞絕緣質.惟濕氣之減除,不能於設計時限制之不能於製造時防範之,且濕氣之於線圈內成爲氣囊,既難傳熱,又致虛弱,銅線熱度不易發散則溫度必高.是高溫之患,亦由於濕氣,雖謂濕氣爲絕緣質唯一之敵,未嘗不可.製造廠當盡力排除之.

排除絕緣質之濕氣方法,視質體而異.其對於線圈內部,如棉紗帶,膠紙,膠布,木料等固體恆用烘漆法 Impregnation, 其對於線圈外部,如變壓器油,恆用濾清法 Filtering & Drying 今分述之於後.

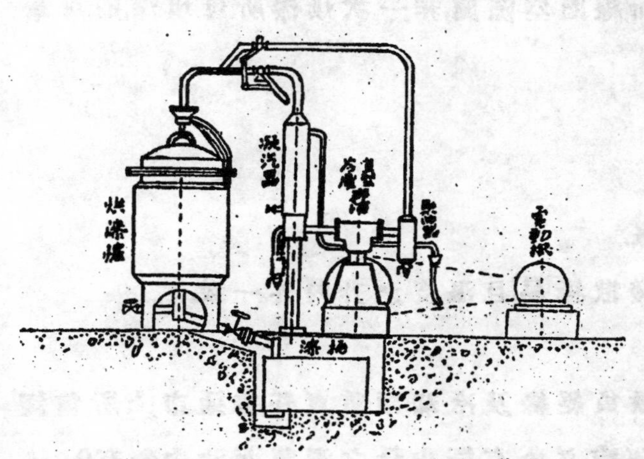

第七圖(甲)　益中公司烘漆機器裝置正面圖

(1) 烘漆　已繞成之線圈或已裝配變壓器之內部,於舉行烘漆法時,須先烘熱於電熱之烘房內,以排除表

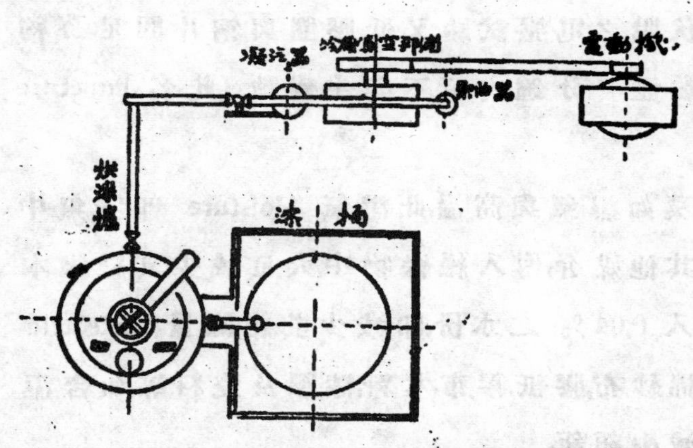

第七圖(乙) 益中公司烘漆機器裝置平面圖

面大部份之濕氣,再放入烘漆爐中.(參考第七圖益中公司裝置現狀)將蓋嚴閉,并用多數螺絲釘旋牢,此爐周圍隔層通以蒸氣,以保持其溫度.爐內照以電燈,爐蓋有玻璃小孔,可以窺見內部烘漆狀况.次以電動機開動眞空邦浦,抽淨爐內空氣,約至水銀柱27½吋之眞空.此際殘餘線圈內之濕氣於低氣壓下,極易蒸發,排除盡淨.次開放漆桶之活門,一種特製漆膠即升入爐內,淹沒線圈,或內部全體.斯時眞空邦浦依槓桿之調整,變爲冷氣邦浦.打入氣壓於爐內約至每平方吋八十磅左右.此壓力足使漆膠充溢線圈內部,淪肌浹髓,無微不至.約數十分鐘後,仍收回漆膠於漆桶,停止氣壓,吊出線圈或內部,使其自冷.按漆膠系膠質與溶解劑之混合物,此種烘漆法前部份注重在溶解劑之蒸發,後部份注重在膠質之養化.必須蒸發完全,始行養化,恰到好處.如操之過急,則不得其益.故運用之妙,貴乎經驗.如線圈空隙處,非一次烘漆所可填滿,則可舉行二次或三次以填滿之.

烘漆之優點凡四:—

A. 絕緣量充分增加.

B. 烘漆後不易吸收濕氣.

C. 線圈成爲實體,傳熱發散極易,且溫度之升內外一致.

D. 線圈強固,不易受損.

(2)濾清: 變壓器所用油兼負絕緣及冷却兩重責任.普通市上所售變壓器油無論如何愼審裝桶及運輸,必含有極少量之濕氣,而油中含有0.04%

之水份,已足減其絕緣能量至半數.故新購之油,必須濾淨始能應用.又變壓器油應用略久,常留沉澱,Sludge而於高溫度時所遺沉澱尤多.此時沉澱足以塞沒線圈油孔,弛緩油之周流,使線圈愈熱,沉澱愈多,循環不已,終至溫度過高,線圈燬壞.故常時滿儲之變壓器油,必須濾清始能持久.

變壓器油於舉行濾清時,先以油桶傾油入濾油機.(參考第八圖)此機裝有多數之濾板,(第八圖之左下邊)及濾框.(第八圖之右下邊)濾框兩面,均襯以1/32″厚吸水紙五張,機下電動機轉動打油邦浦,壓油於濾框內約每方吋至八磅,油須透過各吸水紙以達他端之出路.油中水份漸由吸水紙吸收,照此重複過濾,油中水份可排除盡淨.普通現購之油,如濾過十次左右,可得絕對乾燥之狀況,惟吸水紙必須烘乾,應用電烘箱爲之.(參看第九圖)油之乾燥狀況,可由試油機定之,此機(參看第十圖)爲一高壓絕緣試驗變壓器.其進電低壓邊直接於一多頭搖柄抵抗器,并附裝刀開關,鉛絲自動切斷器,警燈及電壓表等件.如將油置於試驗缸中連結於高壓邊,并將缸中兩極即1″圓塊之距離定爲0.1″時,漸轉抵抗器之搖柄以提升高壓至25, 00伏脫而始跳電者,即足爲合用之乾燥變壓器油.如不及此數,須再過濾,以上爲新油濾淨法.如變壓器正在應用亦可如第十一圖之裝置,連結油路.使油循環出入桶內,至其油能勝標準高壓試驗爲度.其時間視變壓器及濾油機之

第八圖

大小而定,甚爲簡捷,而油中水份及沉澱,可以排除盡量.

今列優等絕對乾燥變壓器之品質如下.

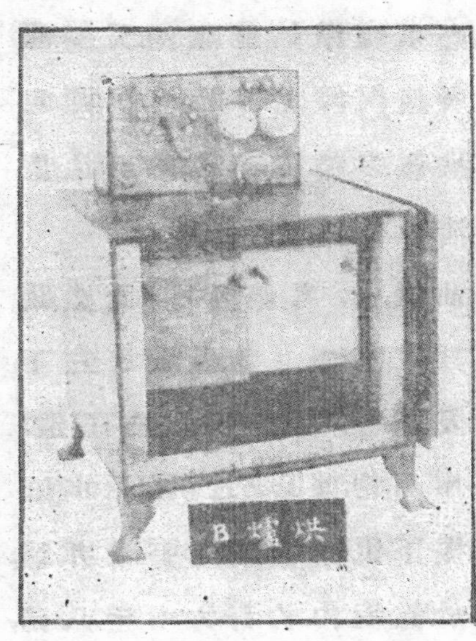

第　九　圖

第　十　圖

第　十　一　圖

沉　澱	於油溫度在150 C.且氣泡通過油中每小時 0.07 ½ 方呎時,其沉澱不得超過 0.049 % 於 120 C.時須毫無沉澱.
爆發點	(封固時)………………… 150°C.
〃	(開口時 ………………… 165°C.
液體阻力	Viscosity (15.5°C.時, … 180
〃	Viscosity (60 °C.時)…… 45
蒸發點	(10)°C.常溫於五小時之後)190
固化點	(凝固不流動)……………… -50°C.
沸　點	……………………………… 350°C.
比　重	(15.5°C.時) ……………… 0.870
〃	(60°C.時)………………… 0.848
比　熱	(1 .5°C.時) ……………… 0.460
〃	(60 °C.時) ……………… 0.175
膨脹係數	(15.5°C.時) ………… 692×10^{-6}
〃	(6 °C.時)………………… 725×10^{-6}
酸　質	……………………………… 無
硫　磺	……………………………… 無
絕緣量	在 1″圓板兩極相距0.1″時試驗…25,000伏脫或在 ½″ 球形兩極相距0.15時試驗……40,000伏脫

(三) 桶蓋

變壓器之鐵桶最要作用有三,即(1)傳播熱度;(2)盛放變壓器油;(3)保護內部.後二者不言可知前一項殊費研究.蓋設計時,務使滿載時變壓器油之温度,不得超過周圍大氣温攝氏55度,即桶壁傳熱或發散油中熱度於一定速率之下,始能使油不達此温度.然油之熱度乃自變壓器本身之鐵耗及銅耗 Core Loss and Copper Loss 而來,故桶壁傳熱之率,實與其每平方吋面積應傳之鐵耗及銅耗常數有關.今因增加桶壁面積起見,致油冷式變壓器桶壁形式甚多,有用波紋鐵皮者,有用數排鐵管者,有用附裝傳熱器 Attached Radiator 者,此常數隨各式而異,不能舉一反三.設計之當否,全視經驗之久暫,惟變壓器容量在 200—300 KVA 左右,當以波紋鐵皮桶壁,上下口均用生鐵澆出者爲最普通.(參看第一圖)

製造此種波紋鐵皮桶切須注意者至少有兩點如下:

桶壁原料　桶壁須求其傳熱之速,以限用薄鐵皮爲尙.但同時須兼顧本身之強固,以支持桶內之重量.此種鐵皮尤須特別製煉,使其不易生銹,而便於鑞銲及電焊.

桶壁油漆　尋常鑄鐵件表面恆帶毛點,油漆後塡平嵌緊不易剝落.鐵皮表面則不然,因其過於光滑,加漆後無處塡嵌,迨其漸燥,恆現小裂紋.久或整塊剝落.凡薄鐵皮桶壁漫不經意而加漆,不久即逼生銹斑,終至鐵皮洞穿,而全桶作廢.影響停電,損失不貲.故波紋鐵皮桶未漆之前,先須施以砂澆法 Sand Blast.使其表面發生如鑄鐵件之毛點.同時刷除一切雜質及膚殼,然後施以塡底之漆及烘乾,再施以最後之漆及烘乾,全桶始能歷久不壞.

(乙) 試驗情形

此225KVA變壓器曾經過全部工程試驗 Complete Engineering Test,今錄其試驗表於後.

Cont nuou ; Output, KVA	... ...	225
Normal Primary Line Voltage	... ...	6,400 Volts
Normal Secondary Line Voltage at no load	... ...	380/220 Volts
Number of Phases	... ...	3
Frequency	... ...	50 Cycles/Second
Primary/Secondary Connection	... ...	Delta to Star
Primary Tappings	... ...	One tap for 6,200 Volts
Secondary Tappings	... ...	None
Tappings Arranged for Constant Output	... ...	Yes
Tappings Arranged for Constant Current (See Separate Print No. .....for Connection Diagram)	... ...	———
Series-Parallel Connection on	... ...	———
Primary Neutral	... ...	None
Secondary Neutral	... ...	Yes, brought out
Mid. Wire to Carry Amperes	... ...	Full Load
Double or Auto Wound	... ...	Double Wound
Core Loss at Normal Primary Voltage.	... ... Watts	2,100
Copper Loss at Normal F. L. & U. P. F.	... Watts	2,350
Regulation at Normal F. L. & U P. F....	... %	1.06 %
Regu'ation at Normal F. L. & 0. 8. P. F.	... %	1.82
Impedance Voltage at Normal F. L & Ratio	... %	2.00
Reactance Voltage at Norm l F. L. & Rat o	... %	1.73 %
Efficiency at U. P. F. Full Load	... ... %	98.10 %
3/4	... ... %	98 01
1/2	... ... %	97.58
1/4	... ... %	96.16
Performance Reference Temperature °C.	... ...	55 °C.
*Normal Maximum F. L. Temperature Rise by Thermometer		45 °C.
Maximum Ambient Temperature	... ...	35 °C.
Maximum Overload Capacity with Ambient Temperature not exceeding 30°C.	Continuously	10 % Overload
	For 2 Hours	25
	For ½ Hour	50
(See Separate Performance Curve No... ...)	...	
Polarity ...	... ...	Subtractive

* Temperature Rise by Resistance 5°C. greater

REMARKS

Dielectric Test:	Between H. V. and L. V. at	14,000	Volts
(For 1 minute)	Between H. V. and Iron core,	14,000	Volts
	Between L. V. and Iron core,	2,000	Volts

木質房架接口之簡易計算法

著者：初毓梅

引言及總綱

作者自入天津基泰公司以來,經手所作之房架甚多,每感各教科書中所載之接口計算法,過於繁瑣,不甚合用.暇中研究,得一捷法,載之於斯,以備各工程同志之指正或採用.

(一) 普通設計法

設房架各部所受之外力,已用算法(Analytical Method)或畫法完全求出;各部大小尺寸,亦均已完全算定,即可從事計算接口矣(Joint).計算接口之理論(Principle),在求I-II及II-III(第一圖)之尺寸(Dimension),使BD中之壓力,壓於該二接觸面上(I-II及II-III面),皆甚安全.又I-II面與II-III面,應互相垂直;因垂直時,則二面均受純粹之壓力,而無滑力(Slipping)也.茲將普通計算法,綜合之如下:

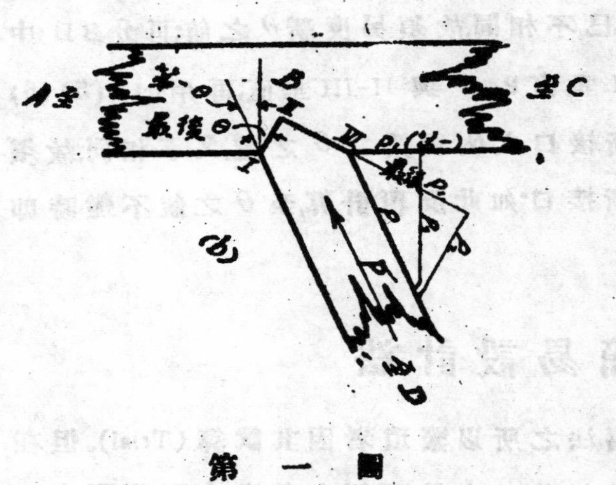

第一圖

(1) 將欲算之接口部分,按照預定之位置,及所算出之尺寸,畫出之,如圖一(b).

(2) 將BD所受之外力 P (已算出),分為二分力:一與AC平行(P_1)一與

AC垂直(P_2).

(3) 暫時假定接口之一邊,與AC垂直(如點線所示).用分度器(Protractor)將θ之角度量出之.

(4) 木質之受壓安全力S(Allowable Compressive Stress),可由

$S=S'\sin^2\theta+S''\cos^2\theta$ 之公式求出之.

S' = 壓力平行木紋時之受壓安全力(Allowable Compressive Stress with fiber).

S'' = 壓力垂直木紋時之受壓安全力(Allowable Compressive Stress cross fiber).

(5) 接口之深度 $I\text{-}II=\frac{P_1}{ts}$

$b=BD$ 之寬.

(6) 作I-II垂直II-III;並使I-II與所算出之深度相等.

(7) 現在θ之值,與以前θ之值,已不相同.故須另度新θ之值;再分BD中之壓力爲二分力:一與I-II垂直P_1,一與II-III垂直;再用(4)(5)(6)三條之法,求一新接口.此新接口之θ,與第二θ之值,又不相同.故須再用第三θ之值,再求一新接口.如此反覆計算,至θ之值不變時即得所求之接口.

(二) 簡易設計法

本法之所以簡易者,在用線圖;舊法之所以繁瑣者因其試算(Trial).但在線圖之先,須有公式.此公式各書中無載之者,故更須由基礎方面,先研究公式,以爲繪製線圖之用.

(甲) 基礎公式之研究 本公式可分兩部:一爲內力公式(Formula for internal Stresses),一爲外力公式(Formula for external Stresses).內力公式,即 $S=S'\sin^2\theta+S''\cos^2\theta$.各書中均有討論與記載,勿須再贅.玆研究者,乃外力公

式也.

計算接口時,須先知何部及何面規定設計.(Which member of the truss and which contact surface of the joint govern the design) 一接口由兩部構成:一爲AC之部(Chord),一爲BD之部(Strut).(第二圖)每部接口處有兩接觸面(Contact Surface):一爲I-II,一爲II-III.故共有四面:

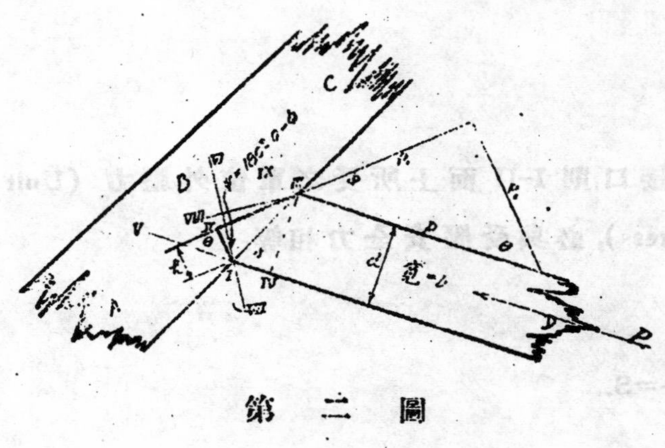

第 二 圖

(1) AC之I-II接觸面;

(2) AC之II-III接觸面;

(3) BD之I-II接觸面;

(4) BD之II-III接觸面.

設計時,須由內力公式 $S=S'\sin^2\theta+S''\cos^2\theta$,研究何面之受壓安全力爲最小,即何面規定設計.(同面異部,如AC之I-II面,與BD之I-II面相較;或AC之II-III面,與BD之II-III面相較.)有時更須比較I-II面與II-III面,必使各部各面均甚安全方可.爲便於解釋起見,先假定BD之I-II面規定設計,以便研究公式.其他各面,當以次分別研究之.

將BD所受之外力P,分爲二分力:一垂直I-II(P_1),一垂直一垂直II-III(P_2),

則 $P_1=P\cdot in\,\theta$,

作 III-IV ⊥BD (member BD),

則 $I\text{-}III=\dfrac{III\text{-}IV}{\sin\varphi}=\dfrac{d}{\sin\varphi}$;

故 $I\text{-}II=(I\text{-}III)\cos(280^0-\theta-\varphi)$

$$\cdot = \frac{d\cos(180^0-\theta-\varphi)}{\sin\varphi}$$

$$= -\frac{d\cos(\theta+\varphi)}{\sin\varphi}$$

令 I–II＝h,

$$\therefore \quad h = -\frac{d\cos(\theta+\varphi)}{\sin\varphi}$$

設 I–II–III 爲已算出之接口,則 I–II 面上所受之單位外壓力（Unit external Compressive Stress),必與受壓安全力相等.

故 $\frac{P_1}{h\,b} = S,$

即 $$\frac{P\sin\theta}{\left(-\frac{d\cos(\theta+\varphi)}{\sin\varphi}\right)b} = S.$$

簡之得

$$\frac{P}{S\,b\,d} = 1-\cot\theta+\cot\varphi.$$

又 $$h = -\frac{d\cos(\theta+\varphi)}{\sin\varphi}$$

簡之得

$$\frac{h}{d} = \sin\theta-\cos\theta\cot\varphi.$$

$S = S'\sin^2\theta + S''\cos^2\theta$.（內力公式見前）

以上數式,均根據BD部之 I–II 面規定設計之假定,而得出者.若 BD部不規定設計時,則上數式完全無用矣.故該部在何種情形之下,方能規定設計,不可不硏究之.

設 $S=S'\sin^2\theta+S''\cos^2\theta$. 式中之$\theta=0$ 時,則 $S=S''$; 設$\theta=90^0$ 時,則$S=S'$. 各種木料之S',均大於S''. 故S之最低限爲S'',最大限爲S'. 故θ遞大,S之值亦遞大,以S'爲極限.若θ遞小時,S之值亦遞小,以S''爲極限.

在圖二中,作 III–I–V 角之平分線VI–VII. 設I–II面與此平分線相合時,BD之I–II面之受壓安全力,與AC之I–II面之受壓安全力必相等;因θ與ψ之值相等,且AC及BD之木質相同也.若I–II在VI–VII之左時,$\theta<\psi$, BD之I–II 面

之受壓安全力,當小於AC之I-II面之受壓安全力;因θ愈小,其受壓安全力亦愈小也(前節),故BD之I-II面規定設計.反之若I-II在VI-VII之右時,$\theta>\psi$,照同理,AC之I-II面應規定設計.故此平分線,爲此兩種情形之限(Limit);亦卽$\theta=\psi$,爲此兩種情形之限也.

設　$\theta=\psi$.

但　$\psi=180^0-\varphi-\theta$ (圖二)

故　$\theta=180^0-\varphi-\theta$.

則　$2\theta=180^0-\varphi$,

∴　$\theta=90^0-\frac{\varphi}{2}$

代θ之值入$h/d=\sin\theta-\cos\theta\cot\varphi$式中,

則　$\frac{h}{d}=\sin(90^0-\frac{\varphi}{2})-\cos(90^0-\frac{\varphi}{2})\cot\varphi$.

簡之得

$$\frac{h}{d}=\frac{1}{2\cos\frac{\varphi}{2}}$$

φ之值最大爲90°,最小爲0°.

$\varphi=90^\circ$時,

$$\frac{h}{d}=\frac{1}{2\times.707}=.707.$$

$\varphi=0^\circ$時,

$$\frac{h}{d}=\frac{1}{2}=.5.$$

由上之結果觀之,設AC之I-II面規定設計時,則$\frac{h}{d}$之值,必須大於.707.或.5.但h=.5d(卽所刻AC之深度,等於BD之高之半.)時,所刻AC之深度已嫌過深,不適於用.若h>.5d時,更爲作法上所大忌.故AC之I-II面,規定設計實施方面,無用之者,可以不必深究矣.

今再研究接觸面II-III:作III-VIII垂直VI-VII(第二圖),則III-VIII必平分I-III-IX角故照前節之理,II-III面若與III-VIII相合時,則亦達於限.(卽AC之II-III面,與BD之II-III面,之受壓安全力相等).若I-II在VI-VII之左時,II-III亦在III-VIII之左;I-II在VI-VII之右時,II-III亦在III-VIII之右.換言之,卽BD之I-II面規定設計時,只須研究AC之II-III面;若AC之I-II面規定設計時,只須研究BD之II-III面.(因BD之I-II面規定設計時,則I-II必在VI-VII之左,II-III必在III-VIII之左,則AC之II-III面之受壓安全力,較BD之II-III面之受壓安全力爲小,故研究AC之II-III面已足.AC之I-II面與BD之II-III面同理.參看內力公式之解釋.)但前節已言,在普通情形之下,AC之I-II面,不可規定設計.故BD之II-III面因之亦不得規定設計,則是BD之II-III面亦無須深究矣.今只研究AC之II-III面已足.

爲便於研究起見,先假定BD之I-II面規定設計,然後細察在各種情形之下,AC之II-III面是否安全.

試察 $\frac{P}{Sbd}=1-\cot\theta\cot\varphi$ 式.此式可變爲 $P=Sbd(1-\cot\theta\cot\varphi)$ 之形.設 $\cot\varphi$ 不變時,則 θ 愈大, $\cot\theta\cot\varphi$ 當愈小,卽 $1-\cot\theta\cot\varphi$ 當愈大.又 θ 之值漸大時,S之值亦漸大.在一種情形之下,bd不變.故 φ 之值不變時, θ 愈大,P亦愈大.但 θ 之最大限 $=90^0-\frac{\varphi}{2}$ (見前),故 $\theta=90^0-\frac{\varphi}{2}$ 時,P爲最大.設 φ 等於任何值P等於其最大值時,則AC之II-III面,是否安全,今先研究之.

値,(一) $\varphi=90^0$,及P等於其最大值時:

P在其最大值時, $\theta=90^0-\frac{\varphi}{2}$.(前節)

$$\therefore\quad \theta=90^0-\frac{90}{2}=45^0.$$

此接口之形勢當如第三圖:

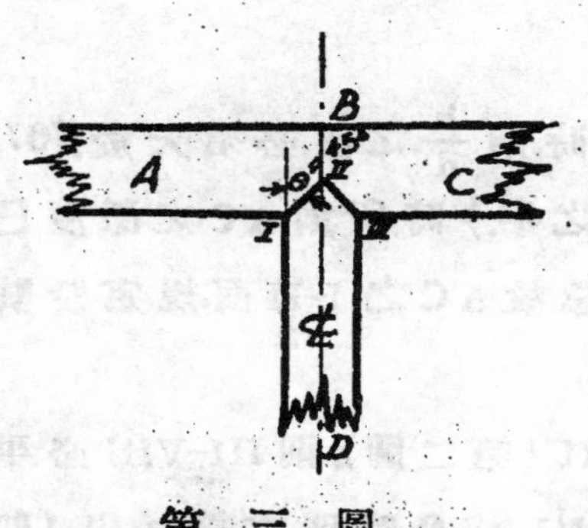

第三圖

試於BD之軸上(Axis),畫一中線.則此中線之左右相對稱.故I-II面旣安全,II-III面亦必安全.

(二) φ = 任何值(由 90^0 至 0^0),及P等於其最大值時:

$$\left.\begin{array}{l}\text{P在其最大值時}\ \theta=90^0-\frac{\varphi}{2},\\ \frac{h}{d}=\frac{1}{2\cos\frac{\varphi}{2}}.\end{array}\right\}\text{(見前)}$$

又 $\dfrac{P\sin\theta}{h\,b}=(S'-S'')\sin^2\theta+S''$,

故 $\dfrac{P\sin\left(90^0-\dfrac{\varphi}{2}\right)}{\dfrac{bd}{2\cos\dfrac{\varphi}{2}}}=(S'-S'')\sin^2\left(90^0-\dfrac{\varphi}{2}\right)+S''$,

簡之得 $\dfrac{P}{bd}=\frac{1}{2}(S'+S''\tan^2\varphi)$.

但 $P_2=P\cos\theta$, (第二圖)

故 $\dfrac{P_2}{bd\cos\theta}=\frac{1}{2}(S'+S''\tan^2\varphi)$,

亦卽 $\dfrac{P_2}{bd}=\frac{1}{2}(S'+S''\tan^2\varphi)\ \mathrm{cas}\left(90^0-\dfrac{\varphi}{2}\right)$,

$\therefore\quad P_2=\frac{1}{2}bd(S'+S''\tan^2\varphi)\sin\dfrac{\varphi}{2}$.

設在上式中φ之值由$90°$漸小,則$\tan\varphi$及$\sin\dfrac{\varphi}{2}$均漸小,故P_2亦漸小.

又 $\mathrm{II\text{-}III}=\sqrt{\left(\dfrac{d}{\sin\varphi}\right)^2-h^2}$.

但 $h=\dfrac{d}{2\cos\dfrac{\varphi}{2}}$,

$\therefore\quad \mathrm{II\text{-}III}=\sqrt{\dfrac{d^2}{\sin^2\varphi}-\dfrac{d^2}{\cos^2\dfrac{\varphi}{2}}}$.

設在上式中φ之值由$90°$漸小時,則$\sin^2\varphi$當漸小,故$\dfrac{d^2}{\sin^2\varphi}$必漸大;又$\cos^2\dfrac{\varphi}{2}$當漸大,故$\dfrac{d^2}{\cos^2\dfrac{\varphi}{2}}$必漸小,結果$\sqrt{\dfrac{d^2}{\sin^2\varphi}-\dfrac{d^2}{\cos^2\dfrac{\varphi}{2}}}$必漸大,故$\varphi$之值由$90°$漸小時,II-III必漸長.

綜上觀之,φ小於$90°$時,P_2必小於$\varphi=90°$時之P_2',而II-III必長於$\varphi=90°$時之II-III.今在$\varphi=90°$時,II-III面已甚安全(見前),故$\varphi<90°$時,II-III必更安全矣.

故P在其最大值時,φ等於任何值,II-III均甚安全.

P在其最大值時,II-III固甚安全矣.但P不爲最大時,II-III安全否?是不可不研究之.試用反證法:卽設P<最大P時,I-II至其安全限,II-III已超過其安全限(卽II-III面已不能支持外力).若是,則必引長II-III.但引長II-III,必縮短I-II(因直角三角形之弦不變).故I-II又超過其安全限,則此設計爲不可能.但P等於其最大值時,尙屬可能,今此P較小,當更爲可能,故假定

爲不合理.可知I-II面至其安全限時,II-III面必不超過其安全限也.

由上數節觀之,P及φ等於任何值,I-II安全時,II-III亦必安全,故研究I-II已足,II-III可以不必及之.

玆將所得之結果,綜合之如下:

(一) AC之I-II接觸面,不得規定設計,因AC所刻過深.

(二) BD之II-III面,不必顧及;因AC之I-II面,不得規定設計.

(三) BD之I-II接觸面規定設計.

(四) AC之II-III接觸面,可以不必顧及;因BD之I-II面規定設計時,AC之II-III面永遠甚安全也.

故繪製線圖時,只根據BD之I-II面規定設計時所得之各公式已足.

(乙) 設計線圖之作法 玆將應用之各公式,綜合之如下:

$$S=S'\sin^2\theta+S''\cos^2\theta \quad \cdots \quad \cdots \quad \cdots \quad \cdots \quad \cdots \quad (1)$$

$$\frac{P}{Sbd}=1-\cot\theta\cot\varphi \quad \cdots \quad \cdots \quad \cdots \quad \cdots \quad \cdots \quad (2)$$

$$\frac{h}{d}=\sin\theta-\cos\theta\cot\varphi \quad \cdots \quad \cdots \quad \cdots \quad \cdots \quad (3)$$

$$\theta=90^\circ-\frac{\varphi}{2} \quad \cdots \quad \cdots \quad \cdots \quad \cdots \quad \cdots \quad \cdots \quad (4)$$

爲繪製線圖較易計,將(2)(3)兩式中之φ消去之,得

$$\frac{P}{S\,bd}=\frac{h}{d\sin\theta} \quad \cdots \quad \cdots \quad \cdots \quad \cdots \quad \cdots \quad \cdots \quad (5)$$

著者所作之線圖,係以美國松爲標準;因天津方面,多用美國松,爲作房架之材料,著者爲便利個人設計起見,故以此爲標準.更附一矯正線圖,若房架非美國松,或雖爲美國松,而所定之安全受壓力不同時,可用之.(著者深知,線圖內之變數,不可過多.過多必亂,且使用時較慢也.如本線圖中之S′及S″之值規定後,至少可省去兩部以上之線,(Two series of lines)則線圖當然較清,使用時當然較簡.若讀者所作之房架甚多時,未嘗不可另作一圖,以爲已用.若不多時,用矯正圖已足.用矯正圖時,當然較不用矯正圖爲費時,但比之於普通法已快矣).

設令(3)式中之θ為定數，$\frac{h}{d}$及$\cot\varphi$為變數，則此式為一次式，故為直線。今令縱坐標(Ordinate)表示$\frac{h}{d}$，橫坐標表示$\cot\varphi$。若假定$\theta=5^\circ$，即得$\theta=5^\circ$之線；如假定$\theta=40^\circ$，即得$\theta=40^\circ$之線。看Diagram 1.

若(5)式中之θ一定時，則$\frac{P}{bd}$與$\frac{h}{d}$成比例。故在任一之θ線上，各$\frac{P}{bd}$線均相等。設在$\theta=40^\circ$之線上，求出$\frac{P}{bd}=600$之點。然後於此點及$\theta=40^\circ$之線與橫坐標之交點間，十二等分之，即得各$\frac{P}{bd}$線與$\theta=40^\circ$之線之各交點。同理可求各$\frac{P}{bd}$線與其他各θ線之交點。然後將$\frac{P}{bd}$之值相同之各點連結之，即得各$\frac{P}{bd}$線。

用(4)式可求出本線圖之限線(如長點線所示)。

因$\cot\varphi$，可由0變至無窮大故本線圖之$\cot\varphi$，係用兩種尺寸表示：即自90°至45°，係用1″=.4之尺寸；自45°至10°係用1″=1.5之尺寸。$\varphi<10^\circ$時，各$\frac{P}{bd}$線與橫標平行(相差甚小)，故未繪出。

矯正圖之原理甚簡，可解釋之如下。

設房架木料非美國松時，則其安全壓力，與美國松之安全壓力，自不相同。今令

S, S', S''，為美松之安全壓力(S, S', S''之區別見前)

S_1, S'_1, S''_1，為欲求之房架之木質之安全壓力。

$S_1=S_1' \sin^2\theta+S'_1\cos^2\theta$.

令　$S_1=ks$,

則　$ks=S_1'\sin^2\theta+S_1''\cos^2\theta$.

$$\frac{k}{S_1''}=\frac{\frac{S_1'}{S_1''}(\sin^2\theta)+\cos^2\theta}{S}$$

$$令\quad C=\frac{\frac{S_1'}{S_1''}\sin^2\theta+\cos^2\theta}{S}, \quad \dots \dots \dots \dots (6)$$

則　$k=S_1''C$,

$$故\quad S_1=CS_1''S. \quad \dots \dots \dots \dots (7)$$

$$但\quad \frac{P}{S\,bd}=1-\cot\theta\cot\varphi,\ (美松) \quad \dots \dots (2)$$

$$\frac{P}{S_1\,bd}=1-\cot\theta\cot\varphi.\text{（非美松）}\quad\ldots\ldots(8)$$

將（7）代入（8）

$$\frac{P}{CS_1'\,S\,bd}=1-\cot\theta\cot\varphi,$$

即 $$\frac{P}{SCS_1'\,bd}=1-\cot\theta\cot\varphi,$$

$\therefore\ \frac{P}{bd}$ 用 CS_1' 除之以後，Diagram I 即可應用矣，此爲矯正原理。

設（6）式中之 θ 規定之後，C 與 $\frac{S'}{S''}$ 爲兩變數，又因其爲一次式，故爲直線，故每一 θ 線中得兩點，即得所求之線。

矯正線圖之用法，另有舉例，參看自明。

（三）新舊法之比較

（甲）<u>誤差比較</u>

（1）新法　$\frac{h}{d}$ 之值，由圖中猜讀，其最大之誤差（Error）爲 .01.

設 $\frac{h}{d}=X$ 爲準確，

$h=dX.$

今讀作 $\frac{h}{d}=(X\pm.01)$，

$h=dX\pm.01d.$

故 h 之誤差 $=\pm.01d.$

在普通情形之下，d 最大 $=10''$，

故最大之誤差 $=\pm.01\times10=\pm.1''$.

在普通工作情形之下 $\pm.1''$ 之誤差，並不爲大，因工匠多或少下一鋸，即有 $\pm.1''$ 之誤差，故設計時過準亦無用也。

（2）舊法　爲半畫法（Semigraphical），其準確與否，在乎所用尺寸之大小。若在大樣（Detail Drawing）上求之，必不及新法之準確。因大樣之尺寸，最大不過 $1''=1'\text{-}0''$，差一線即有 $\frac{1''}{8}$，至於角度上之差，尙不論也。

（乙）<u>繁簡比較</u>　試看例題，即知新舊法孰簡孰繁矣。

(四) 例　題

(甲) 設房架之木料爲美松 $S'=1100$ lb./sq.in., $S''=250$ lb./sq.in., $P=8900$ lb., $\varphi=75^0$, $b=5''$, $d=4''$, 求接口:

(a) 舊法

1. 先將圖照所給之尺寸,及角度,繪出之.

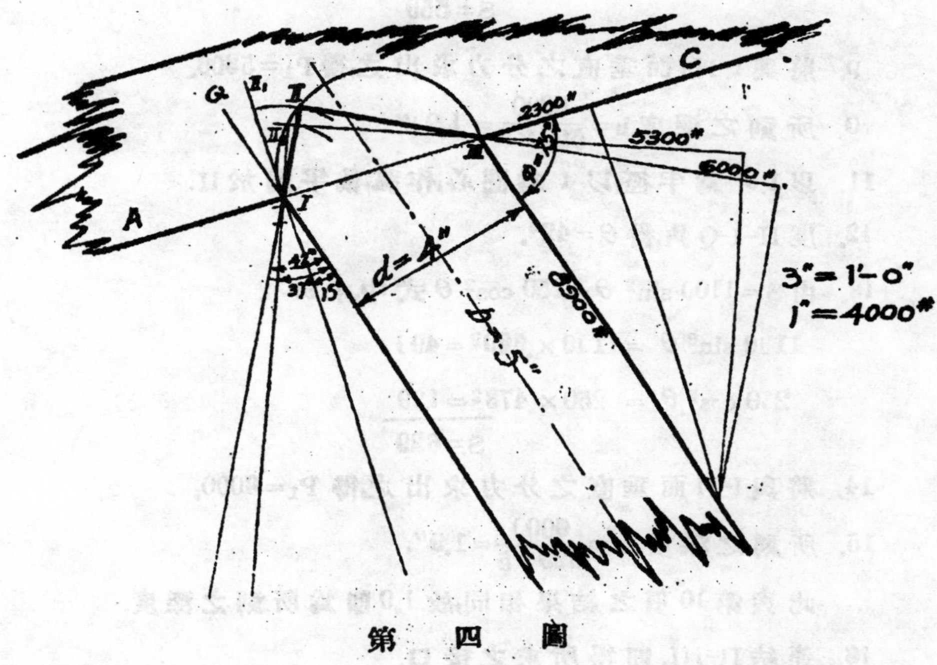

第　四　圖

2. 用畫法將與AC平行之分力求出之,得 $P_1=2300$ lb.

3. 度 II_1-I-Q 角,得 $\theta=15^o$. (II_1-I⊥AC)

4. 由 $S=1100\sin^2\theta+250\cos^2\theta$ 式中,求S:

$1100\sin^2 15^o=1100\times.259^2=1100\times.067=74$

$250\cos^2 15=250\times.966^2=250\times.930=232$

$S=306$ lb.

5. 所剜深度 $h=\dfrac{2300}{306\times5}=1.5''$.

4009

6. 以I-III為直徑,作二半圓,又以I為圓心,以1.5″為半徑,作弧,截半圓周於II_2.

7. 度II_2-I-Q角,得$\theta=37°$.

8. 由$S=1100\sin^2\theta+250\cos^2\theta$式中,求S:

$1100\sin^2\theta=1100\times.594^2=400$

$250\cos^2\theta=250\times.804^2=159$

$S=559$

9. 將與I-II_2面垂直之分力求出之,得$P_1=5300$.

10. 所剡之深度$h=\frac{5300}{559\times5}=1.9$″

11. 以1.9″為半徑,以I為圓心,作弧,截半圓於II.

12. 度II-I-Q角,得$\theta=42°$.

13. 由$S=1100\sin^2\theta+250\cos^2\theta$式中,求S:

$1100\sin^2\theta=1100\times.669^2=490$

$250\cos^2\theta=250\times.473^2=139$

$S=629$

14. 將與I-II面垂直之分力求出之,得$P_1=6000$.

15. 所剡之深度$h=\frac{6000}{629\times5}=1.9$″.

此與第10項之結果相同,故1.9即為所剡之深度.

16. 連結II-III,即得所求之接口.

(b) 新法

1. $\frac{P}{bd}=\frac{8900}{5\times4}=445$ lb./sq.in.

2. 以$\varphi=75°$,及$\frac{P}{bd}=445$,由Diagram 1中求

$\frac{h}{d}=.475$.

3. $h=.42d=.475\times4=1.9$″.

試觀舊法須16步方能得出結果,今只用甚簡單之3步,即可得出結果,新法之簡便可知.

(乙) 設房架之木料為 Hemlock, $S'=1800$ lb/sq.in, $S''=330$ lb./sq.in, $\varphi=71°30'$, $b=6''$, $d=5''$, $P=15300$, 求接口.

1. 先假定房架為美松:以 $\frac{P}{bd}=\frac{15300}{6\times5}=510$ lb/sq.in.及 $\varphi=71°30'$ 由 Diagram I 求得 $\theta=49°$.

2. 以 $\theta=49°$, 及 $\frac{S'}{S''}=\frac{1800}{330}=5.5$, 由 Diagram 2 求得 $c=.00485$.

3. 以 $\frac{P}{bd\,cs''}=\frac{510}{.00485\times330}=\frac{510}{1.6}=320$ lb./sq.in 及 $\varphi=71°30'$,

由 Diagram 1得求 $\theta=38°$, $\frac{h}{d}=.34$.

4. 以 $\theta=38°$, 及 $\frac{S'}{S''}=5.5$, 由 Diagram 2 求得 $c=.0048$.

5. 以 $\frac{P}{bd\,cs''}=\frac{510}{.0048\times330}=322$ lb/sq.in, 及 $\varphi=71°30'$, 由 Diagram I 求得 $\theta=38°$,

$\frac{h}{d}=.35$. 此與三項之結果相同,故此為答.

6. $h=.35\times d.=.35\times5=1.73''$.

本題較題(1)稍繁,但比之舊法簡單多矣.

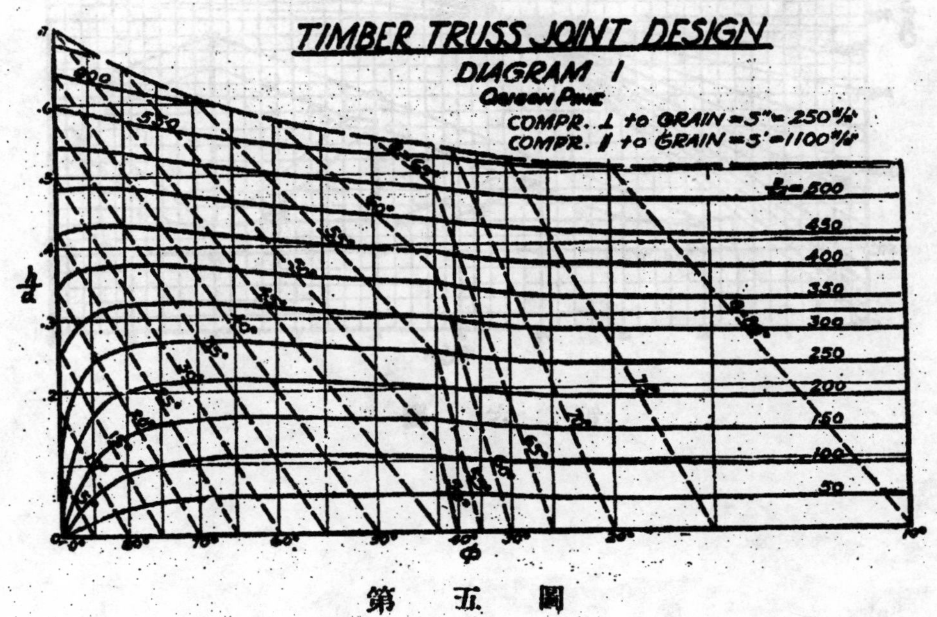

第五圖

(丙) 設第(1)題中之房架木料甚乾.且無甚多節子,則其受壓安全力,可增加30%.試求其接口.

以 $\frac{P}{bd\,(1+.3)}=\frac{445}{1.3}=342$, 及 $\varphi=75^{\circ}$;由 Diagram 1 中求得 $\frac{h}{d}=.37$.

故 $h=.37\,d=.37\times4=1.48''$.

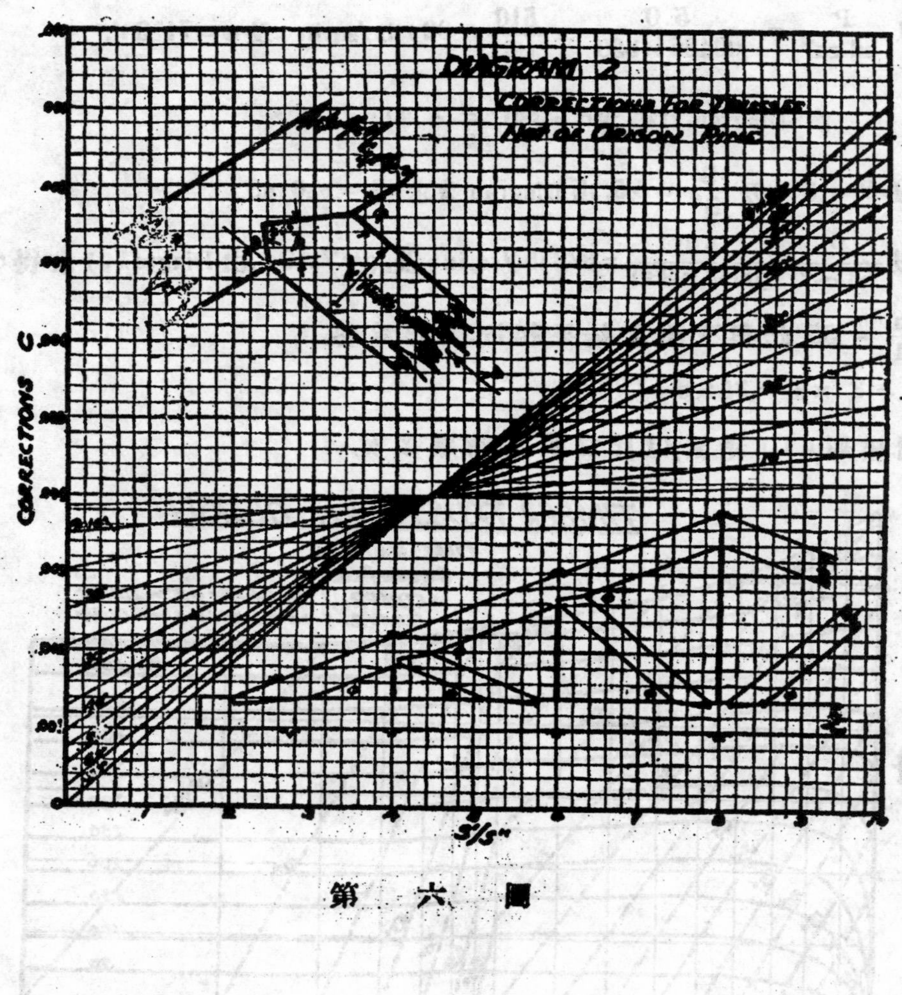

第 六 圖

隴海鐵路靈潼段新工紀要

(一) 緣起

自民十九內戰告終,舉國統一,於是西北開拓問題,乃爲國人所一致注意;良以濱海口岸之繁庶,與物質之文明,已臻極點;西北則地廣人稀,民智閉塞,工商業之待發展,礦藏之待開掘,與夫政治之待改進,實有急不容緩之勢,所謂西移計劃,殖邊問題,早成新時代唯一之研究矣.隴海路爲橫貫全國之一大幹線,東達海濱,西通關隴,凡有志於西北事業者,莫不切盼此路之早日完成,俾交通得以便利,一切工業原料及機械器具,俱可輸入,卽西北民衆,亦不至感關山之險阻,而視爲畏途.惟是隴海一線,興築幾念餘年,初厄於歐戰之發生,繼阻於內亂之不已,不能按照原定計畫,逐段完工.卽以最近靈寶至潼關一段而論,爲長祗七十餘公里,建築六七載,因軍事屢輟,迄未完成,良屬憾事.幸玆時局敉平,政府與國人對西北發展,又若是之狂熱,隴海本身,遂不得不應環境需要之督促,而積極趕修;雖銀團不能繼續投資,而鐵道部則竭力籌畫,必於最短期內,先通車至潼關,用利民行;然後再進行俄庚(發行公債)爲潼西路工建築自籌之資本.此路之重要,已受國人之注視如此,而工程之內容,尤爲國內一般工程家所急欲明瞭,惜平時從工人員,致力於工事之餘,鮮所貢獻於社會,以供衆覽,今特於工作之餘,將新工方面之一切概況,足資報告且能引起技術上之興趣者,輯爲斯編,或亦可爲全國關心路政與考察工程者之一助也歟!

(二) 靈潼段新工概要

靈潼全段新工,界於靈寶閿鄉,潼關三縣之轄境,傍黃河南岸,成東西平行線.沿線土山起伏,山麓又逼近河身,路線緣山而行,故所鑿之隧道獨多.大橋以函谷關前澗河爲較大,盤鄉與文底鎭次之.工程上所用砂石等料,大率於各山澗內就近取用,產量不多,運輸上亦至感困難.其於取料較遠之地各小涵洞,先用臨時木架便橋,敷軌通車者,卽職是之故.此段旣無石山,可以開採石料,故山洞橋墩多用混凝土建造.而沿河一帶之防護石工,混凝土耗費過鉅,遂在距靈寶六十五公里之硤石驛,開鑿塊石一萬五千立方公尺,爲築堤及護牆材料之用.鋼軌橋梁,悉由中比協定之退還比庚,向比國鋼廠購訂.

路工所購用洋灰,係與啓新於民國二年卽已訂定承辦合同,每包計重一百七十五公斤半,十二包爲一噸,每噸連蔴袋價洋二十元零八角,鐵桶灰每桶折合兩包,重爲三百七十五公斤,每六桶爲一噸,每噸計洋二十三元二角,並雙方訂定,隴海不能用他商洋灰,啓新亦不再增價.所訂灰價,悉爲唐山交貨,現時仍本此合同辦理.此次靈潼段於十九年冬月復工,預計工段,尙需洋灰一萬四千桶,向啓新續購.其一部分陸運,係由唐山轉北寧,平漢過鄭州,而達工次;其餘則由塘沽裝輪,運至大浦,轉靈寶應用.海輪運費每噸約計洋十元,較之車運昂甚.至於全線長度,建築年代,與建築經費等,分論於下:

(一)全線長度 計程七十二公里,由靈寶車站起,至潼關西關外車站止(Km 286 - Km 357).

(二)建築年代 建築始於民十三秋季,由陝州達潼關全線開工,原稱陝潼段工程.迨民十五冬,因工款告絀,未能積極舉辦.嗣又軍事發生,至民十六六月,卽完全停頓.其時以陝靈一段,業經敷軌行車,政府方面,收回隴海營業,自行設局管理,卽將靈寶以東,悉劃入營業段,而靈潼段工程,復於十八年一月,決行續修,預定爲一年完成;未及半載,又爲戰事所厄.秋間復工,豫戰又起;直至十九年底,全國始承統一,鐵道部乃限期將該段殘餘之工,先行完竣.綜計由民十三動工後,其間屢興屢輟,逵民二十冬季,可望粗成行車,歷時有七年之久,實際上工作期間,不足四年.

(三)建築經費 靈潼段建築原始之動機,係在民十三.觀陝段(觀音堂

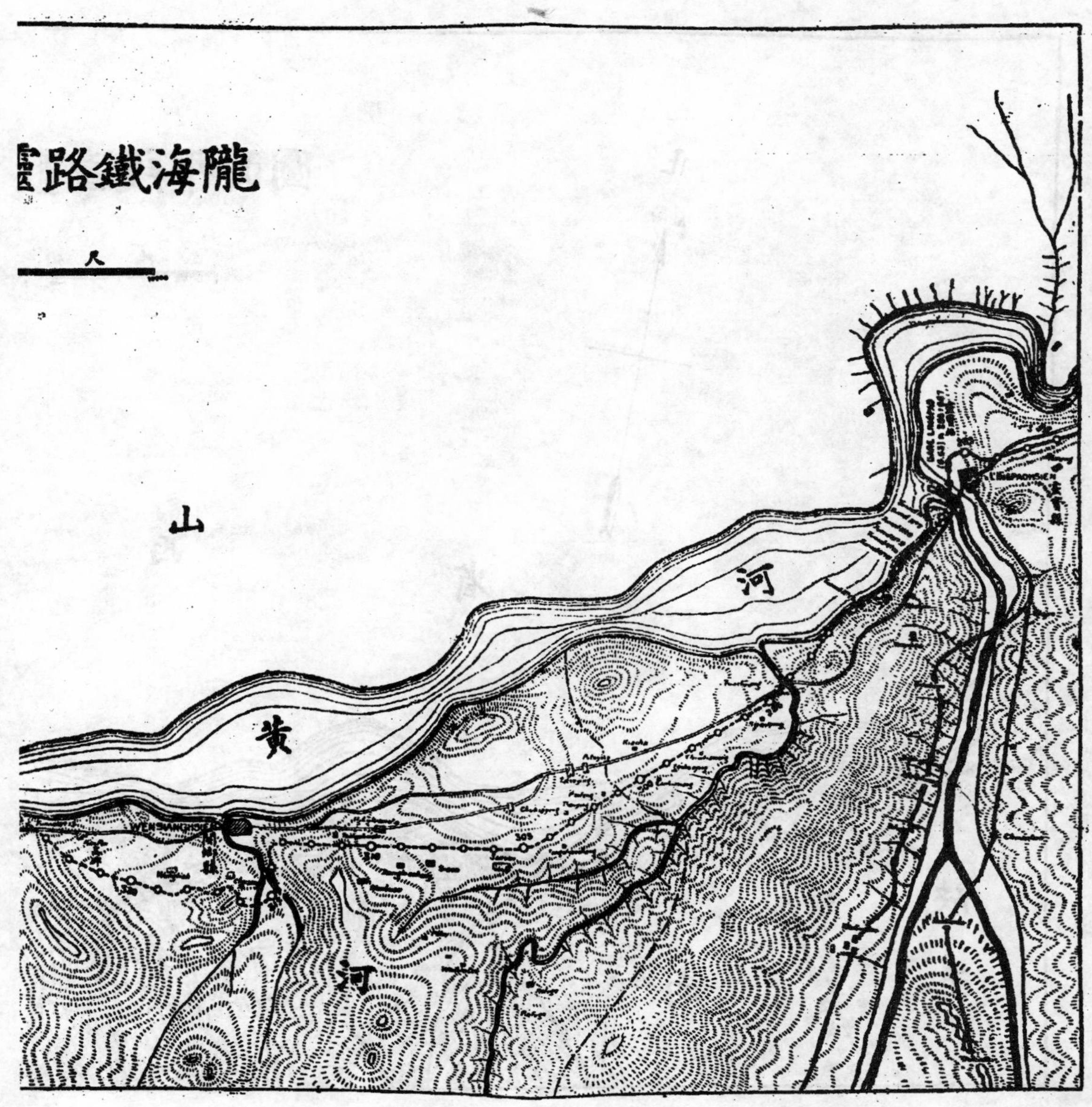

隴海鐵路圖
山
黃
河
河

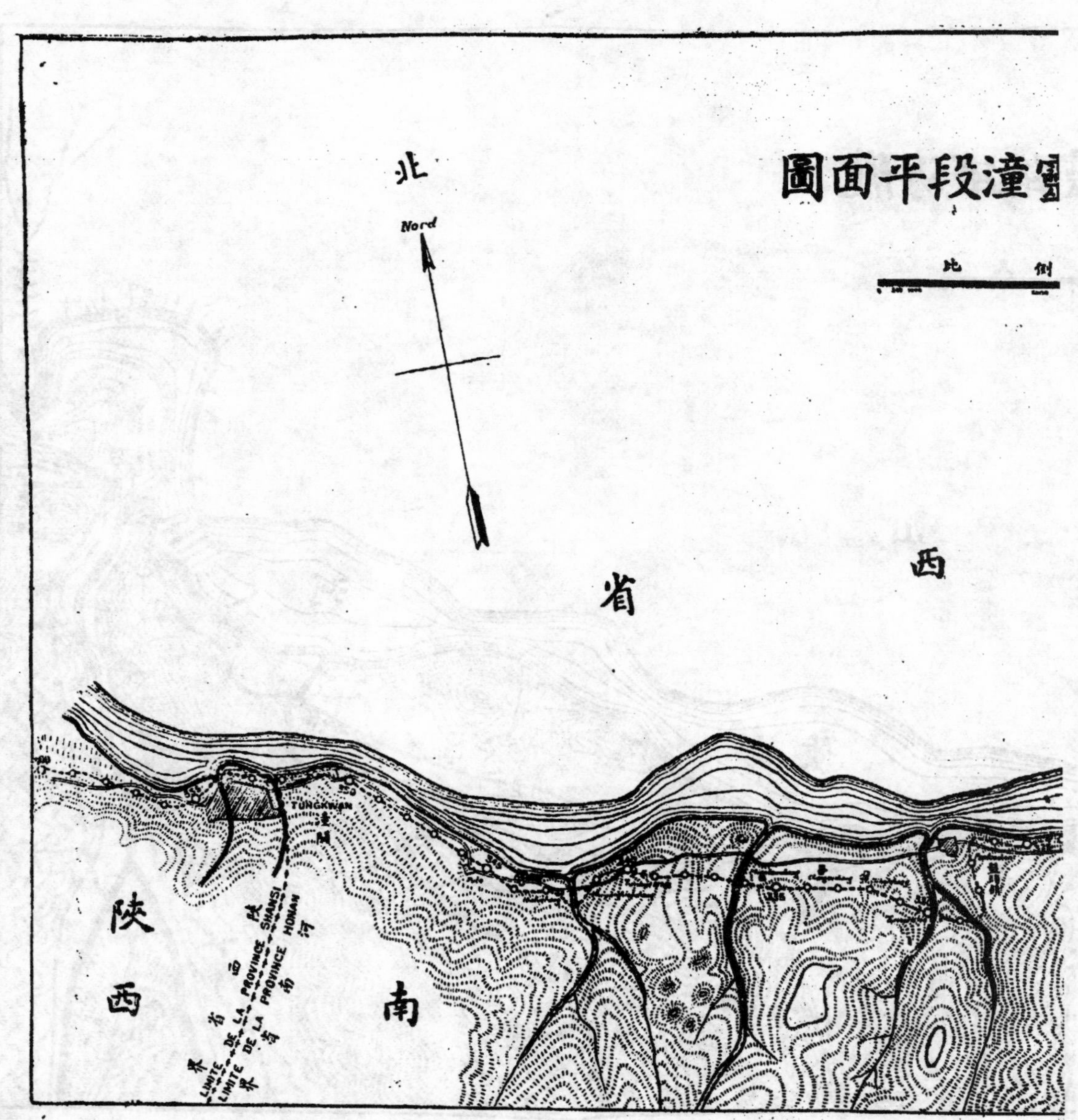

圖面平段潼
北
Nord
比 例
省
西
TUNGKWAN
潼關
陝西
南
LIMITE DE LA PROVINCE SHANSI
LIMITE DE LA PROVINCE HONAN

至陝州）竣工以後,中外方面均認陝安一段,(陝州至西安)於最短期間,有繼續完成之必要,庶隴海本身,一端通海,一端達陝西省會,營業上必日臻發達,建築上亦能告一段落.當時遂訂有八釐國幣之一千萬元,爲陝安全段建築經費,又比幣七千五百萬比伏郎,爲在比訂購材料之用.自協定以後,陝潼一段,即先次第興工.迄十五年底,其第一批所發行之五百萬元,已全數用罄,適值軍事方興,第二批之半數,乃停止發行,在比購料之比伏郎,亦形成泡影.其時陝靈段粗具規模,勉能通車,而所餘之靈潼工段,不得不中途停輟.自此以後,工程即完全以戰事爲轉移之中心,戰息則工興,戰起則工停.而在此屢興屢廢之時期內,工款來源,悉賴營業盈餘之撥助.總計靈潼段建築經費,除(一)前述華幣五百萬元內之一部外,(查銀行團墊款之五百萬元係包括陝潼全段所用之經費,現陝靈與靈潼之資本支出,並未劃分,故靈潼段單獨所用上項之銀行墊款,未能確計,其概數約爲四百萬元左右.)計(二)營業盈餘項下撥助約九十四萬元,(三)民十七中比協定之比庚退還,指定在歐購料而專用於靈潼新工方面者約一百八十萬元,(四)比庚項下各路攤還車輛費用之由部指定撥用者一百萬元,(五)原在歐用比庚已購機車讓售於京滬所得之車價約七十萬元,(六)比庚百分之三十五項下由部指撥者二批,計共約三十六萬元,(七)隴海管理局奉令協撥者六十萬元,綜上七項合計,則該段新工之建築經費總數,約在一千萬元左右.

(四)全段路線曲度　最大半徑2,000公尺.

最小半徑 350公尺.

(五)最大坡度　1.%

(六)土方總數　6,550,000立公方.

(七)較大橋工　三十公尺橋孔,計有三十孔.

二十公尺橋孔,計有二孔.

(八)山　洞　十一座計長8,042,940公尺(見第14頁附表)

(九) 臨時木架便橋　三座

(十) 車　站　計十一處(見附表)

車站號別	地　點	站　名
G. 63	285+807	靈　寶
G. 65	292+385	交車驛站
G. 67	300	常家灣
G. 69	308+225	驛　站
G. 71	314+014	閿　鄉
G. 73	322+238	高　碑
G. 75	331	盤頭鎮
G. 77	337+100	交車驛站
G. 79	343+700	文底鎮
G. 81	349+550	交車驛站
G. 83	356+550	潼　關

(十一) 黃河防護工程　約計長1,200公尺

所有全段各項工程,茲再分別敍述於次:

(一) 土工　土方工程計分挖土與塡土兩種.挖土最深者三十五公尺,其道基面寬九公尺,兩旁坡度,視土質而定,約爲¼.塡土最高者計二十五公尺,其道基面寬六公尺,兩旁坡度約爲½.至於普通路基,無論屬於塡土或挖土,均如下各圖.全段土質,屬於黃土性質居多,黏力不厚.挖土部分,兩壁坡度較陡,暴雨期內,每處崩落.惟閿鄉車站附近,遍地盡屬沙質,微風揚塵,能於數小時內,平地立成沙邱,殊足爲行車之害.

(二) 橋工　橋墩工事,因石料缺乏,皆用混凝土爲之.工作時,四周圍以木板,製爲模式,混凝土卽充實其間,層層夯結,每層以0.25 m至0.30 m爲限.混凝土組織成分,在本路以洋灰重量與沙石容量爲比例之配合.橋基橋墩所用

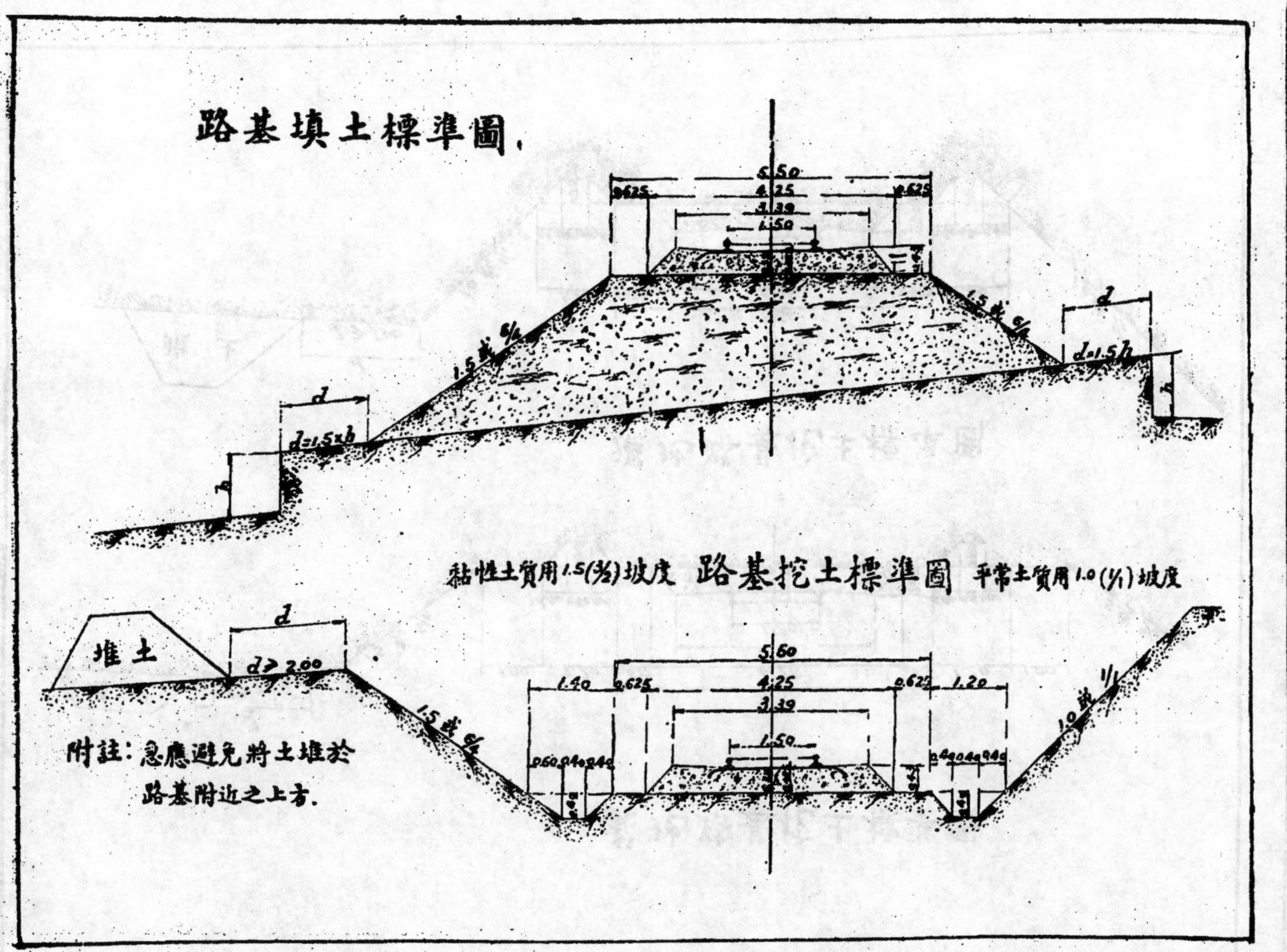
路基填土標準圖，
5.50
0.625
4.25
0.625
3.39
1.50
1.5 或 3/2
d
d=1.5×h
d=1.5h
h
粘性土質用1.5(3/2)坡度 路基挖土標準圖 平常土質用1.0(1/1)坡度
堆土
d
d≥2.00
5.60
1.40
0.625
4.25
0.625
1.20
3.39
1.50
1.5 或 3/2
1.0 或 1/1
附註：急應避免將土堆於
路基附近之上方.

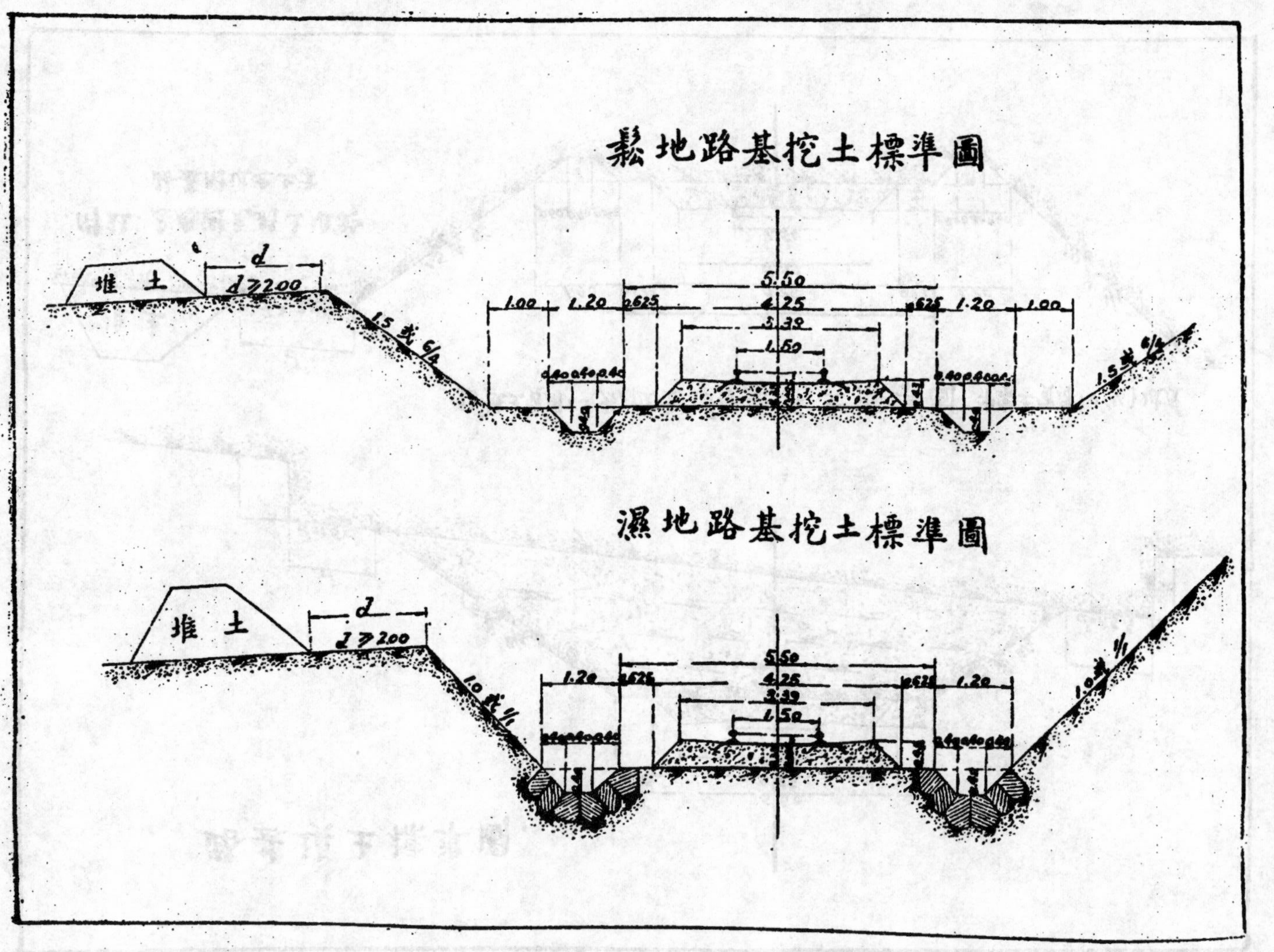

鬆地路基挖土標準圖
堆土
d
d ≥ 200
1.5 或 6/4
5.50
4.25
3.39
1.50
1.00
1.20
0.625
濕地路基挖土標準圖
10 或 1/1

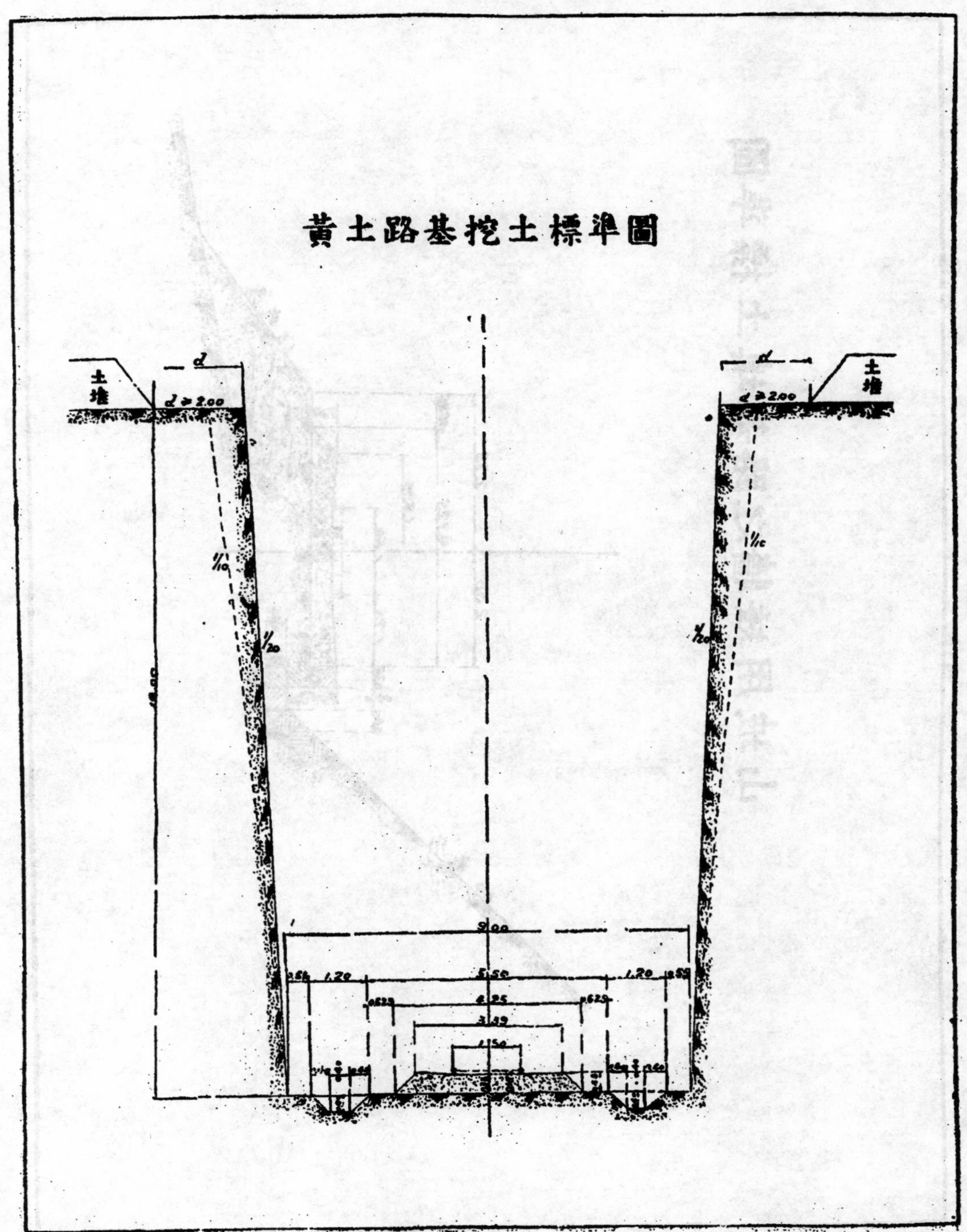
黃土路基挖土標準圖
土堆
d ≥ 2.00
1/10
1/20
9.00
5.50
1.20
1.20

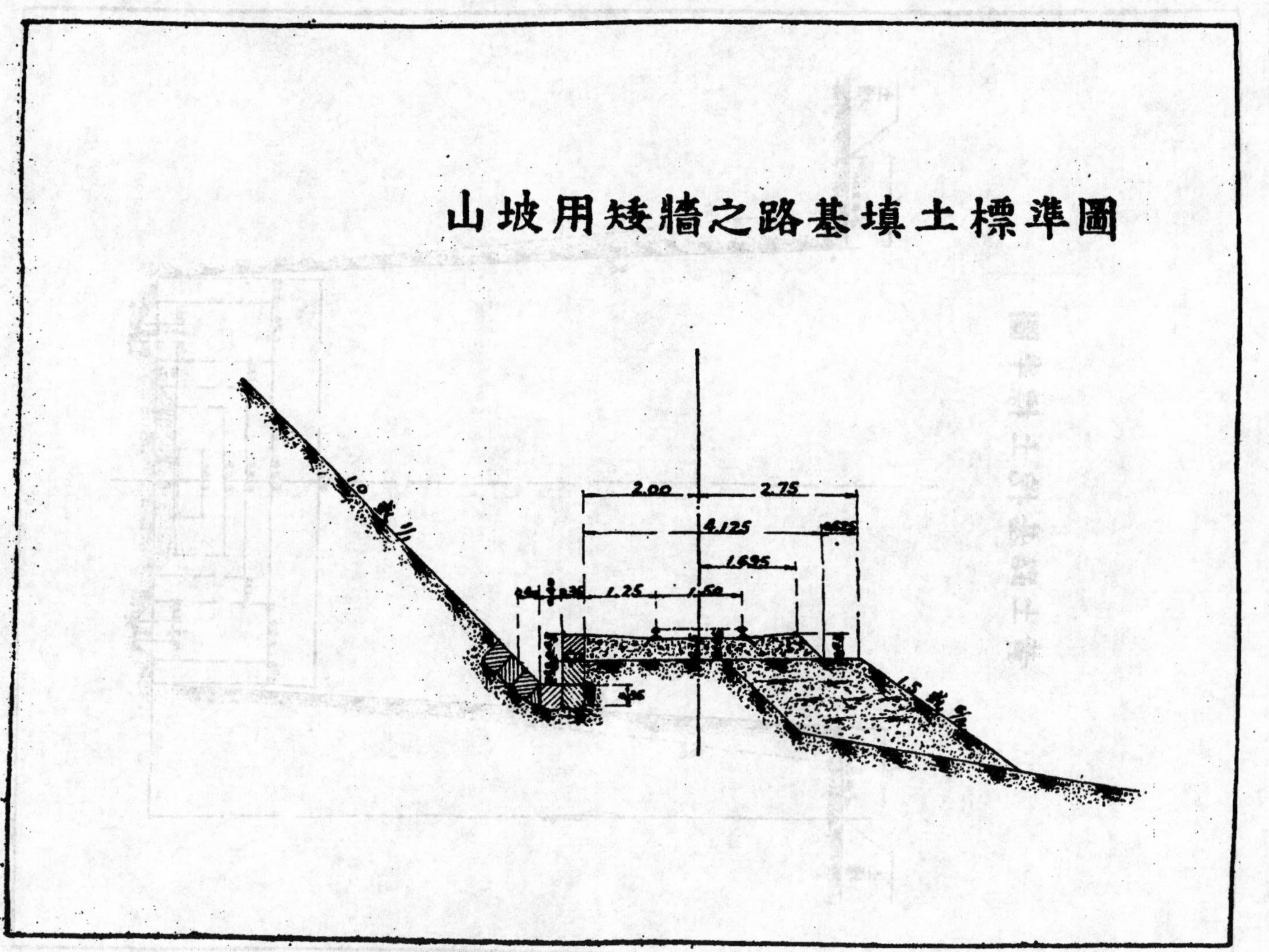

山坡用矮牆之路基填土標準圖

者,係混凝土配合表內之D號.(附洋灰漿及洋灰混凝土配合表)

洋灰漿成分配合表

灰漿類別	沙子成分 m³	石灰成分 m³	洋灰成分 重量	洋灰成分 桶數 (每桶170 Kgs)	備考
A	1	0.50			
B	1		170 Kgs	1	
C	1		255 „	1½	
D	1		340 „	2	
E	1		425 „	2½	
F	1		510 „	3	
G	1		595 „	3½	
H	1		765 „	4½	
I	1		850 „	5	
J	1		935 „	5½	
K	1		1,105 „	6½	

混凝土成分配合表

混凝土類別	石子量數	灰漿量數	灰漿種類	乾實之混凝土1.2 m³內所含洋灰重量	乾實混凝土1m³內所含洋灰重量	備考
b	1.00 m³	0.60 m³	B	102 Kg	85 Kg	
c	„	„	C	153 „	130 „	
d	„	„	D	204 „	170 „	
e	„	„	E	255 „	215 „	
f	„	„	F	306 „	255 „	
g	„	„	G	357 „	300 „	
h	„	„	H	459 „	385 „	
i	„	„	I	510 „	425 „	
j	„	„	J	560 „	465 „	

附註：攙和石子一方,灰漿0.60 m³,揉抖擊實以後,卽成混凝土1.20 m³.按D號組織成分,以容量計算,約爲1:4.4:8.8.工作時爲杜免工人攙伴不勻及夯擊不實起見,在實際上比例,約爲1:5:8.3.

4023

橋墩基礎之簽樁與否,率以地質情形而定.其堅實而無流沙者,可免用樁,否則須簽立六公尺至八公尺之木樁或洋灰樁.本路對試驗西段土質承受壓力之結果,每平方公分不得超過3 Kg (即每平方英尺不超過600磅壓力)爲設計之標準.至於簽樁時,或用人力來引重五十公斤至一百公斤之鐵錘,或用機器重力.本段凡三十公尺之橋工,大率皆用木樁以爲基礎.

橋基橋墩之混凝土,每人平均每日可作1立公方.竣工之後,其上部即爲鋼橋上梁工程.上梁時,先將橋梁在屯放之工場內安好,裝配齊全,然後於兩橋墩之空間,搭立木架或鐵架,與墩齊平,上置長梁兩根,復輔以軌道,然後將橋梁置於鐵輪架上,用機車拖引於橋面,徐徐落下,安放平正,而架橋工事可畢.

靈寶架橋之一

橋基橋墩工料,價以每立公方計.配裝橋梁時,工價以每噸重量計.安放時則以每孔計算.本段橋大率爲上軌式 (Deck Plate Girder) 約合E 45 Cooper's loading.其裝配

靈寶澗河架橋

估算工費如下:

裝配三十公尺鋼橋梁,計重74.215噸,每噸8.00元,共計593.72元.

安放三十公尺鋼橋梁於橋墩.

上包括架設鐵架等工費在內,　　每孔　　230.00元.

其每架重量如下表:

跨度	重量
30公尺	74.215噸
20公尺	41.600噸
10公尺	12.108噸
5公尺	6.577噸
3公尺	2.155噸

靈潼段較大橋墩工程價格表

橋工類別		包工姓名	橋工價格	備考
靈寶大橋	12×30 m	殷鈺臣	99,195元	
稠桑大橋	3×30 m	集成	110,220元	橋墩較高
閿鄉大橋	2×30 m	董方臣	43,583元	
盤頭大橋	3×30 m	董方臣	55,000元	
盤頭大橋	1×30 m+2×10 m	殷鈺臣	57,661元	
文底鎮大橋	4×30 m	陳源遠	70,000元	
文底鎮大橋	4×30 m	陳源遠	50,061元	
潼關大橋	1×30 m	協成	22,0[illegible]0元	
潼關大橋	2×20 m	未標出	35,000元	

(三) 山洞　靈潼段內山洞,均係土質,開鑿甚易.其挖土方價之大小,視其山洞本身洩土之遠近.有時由兩洞口向內開鑿,愈深則洩土愈遠,必擇一相當地點,開一或三數橫伸之地道,以便出土.求洞內唯一之安全,即爲多用木柱撑板,使其不易崩落.然後砌做石工.其次序爲先將弧穹用四十生的立方

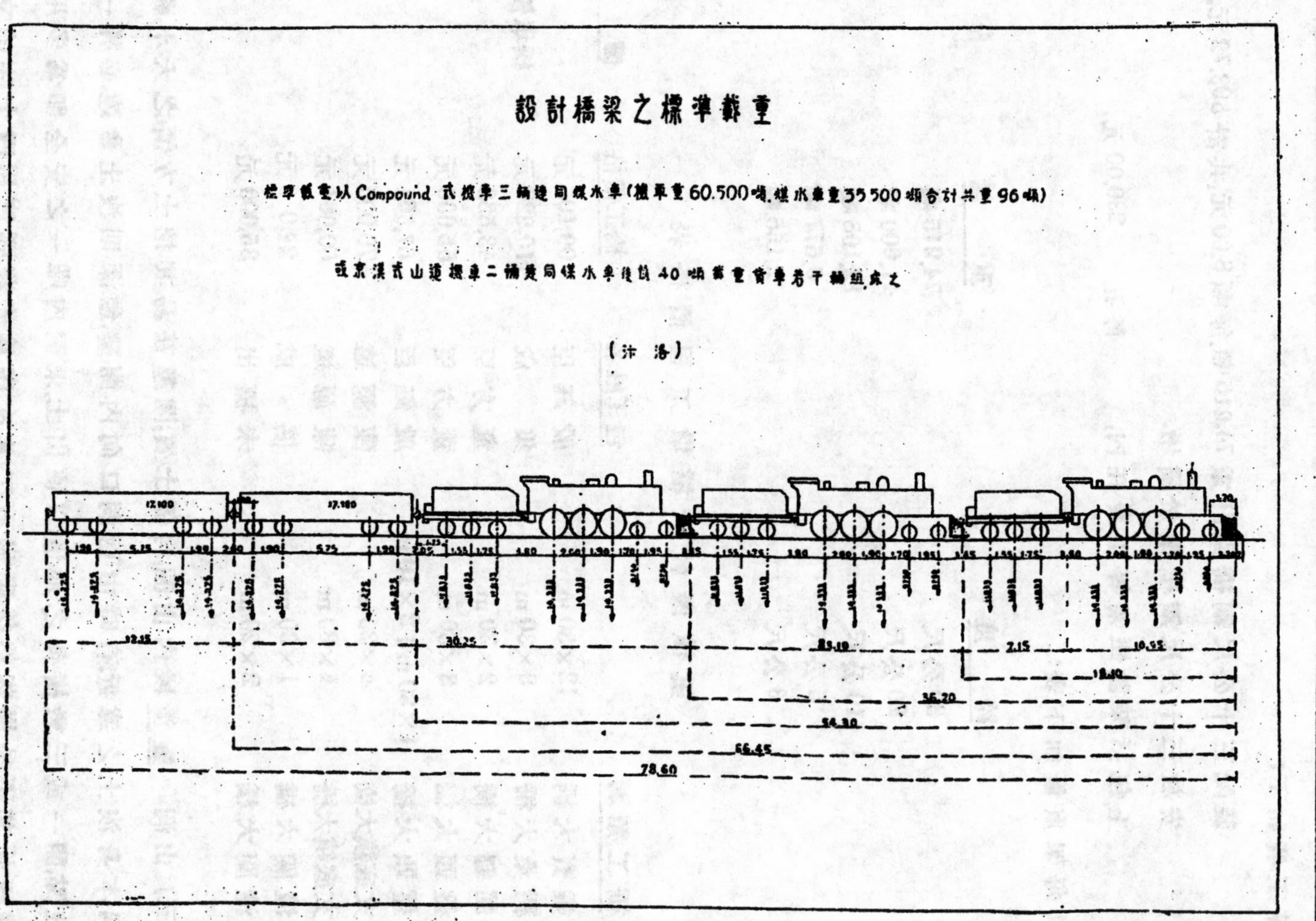
設計橋梁之標準載重
標準載重以Compound式機車三輛連同煤水車（機車重60.500噸，煤水車重35500噸合計共重96噸）
或京漢式山道機車二輛連同煤水車後隨40噸載重貨車若干輛組成之
（汴洛）

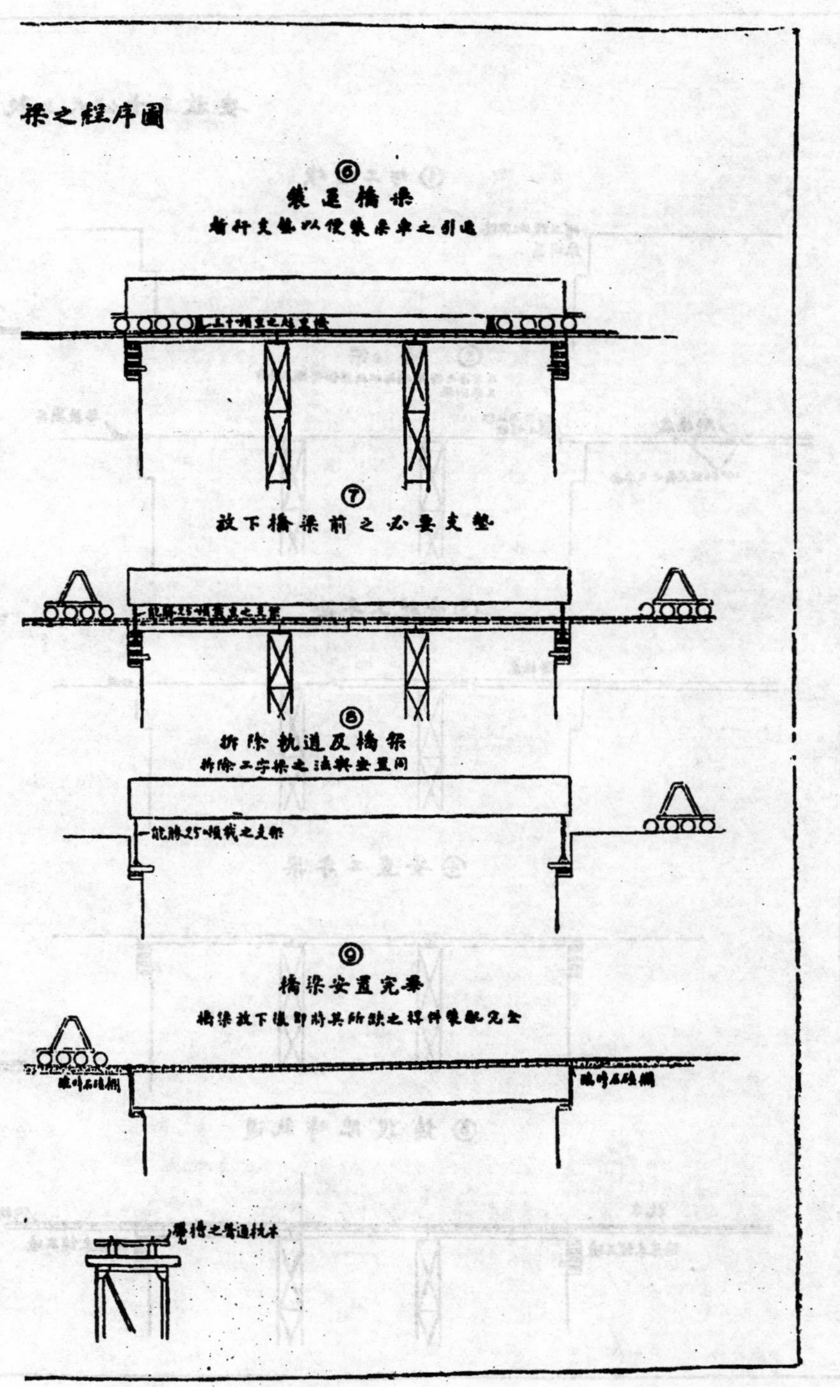
梁之程序圖
⑥
裝運橋梁
⑦
放下橋梁前之必要支墊
一能勝25噸載之支架
⑧
拆除軌道及橋架
拆除工字梁之法與安置同
一能勝25噸載之支架
⑨
橋梁安置完畢
臨時石碴欄
臨時石碴欄

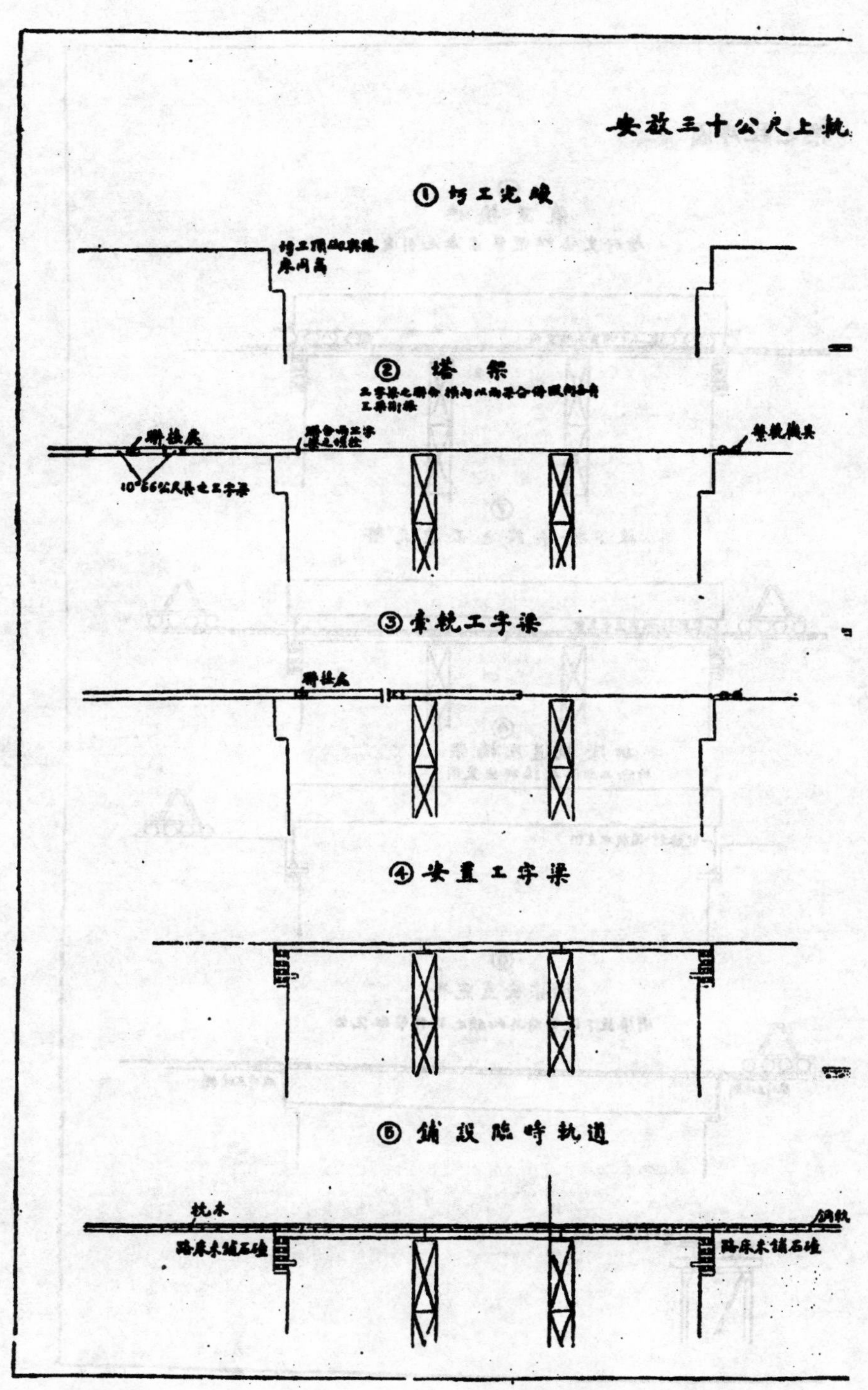

安放三十公尺上軌
① 圬工完竣
② 塔架
聯接處
10"56公尺長之工字梁
③ 牽軌工字梁
聯接處
④ 安置工字梁
⑤ 鋪設臨時軌道
枕木
鋼軌
路床未鋪石碴
路床未鋪石碴

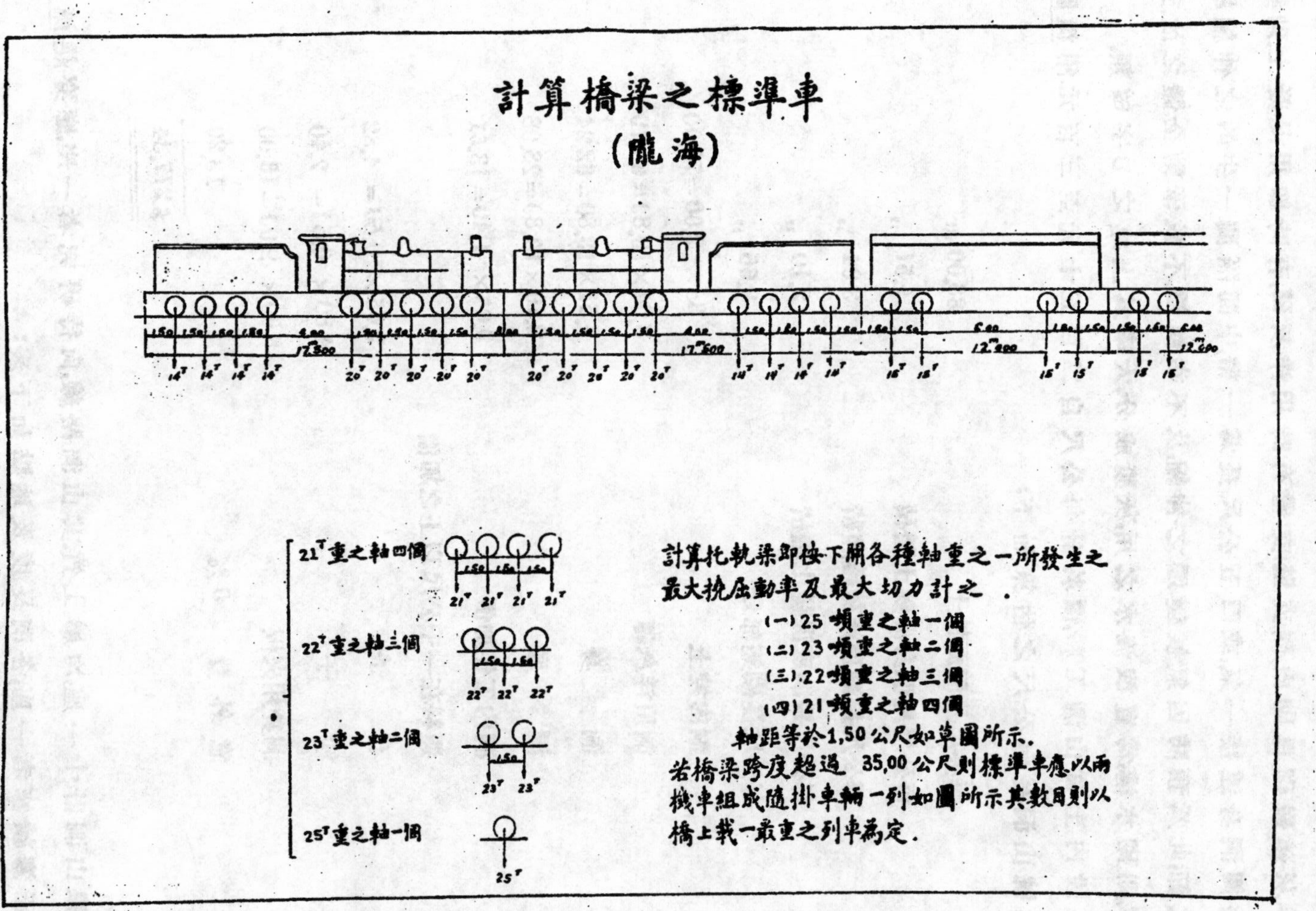
計算橋梁之標準車
(隴海)
計算托軌梁即按下開各種軸重之一所發生之最大撓屈動率及最大切力計之.
(一) 25 噸重之軸一個
(二) 23 噸重之軸二個
(三) 22 噸重之軸三個
(四) 21 噸重之軸四個
軸距等於1,50公尺如草圖所示.
若橋梁跨度超過 35,00 公尺則標準車應以兩機車組成隨掛車輛一列如圖所示其數目則以橋上載一最重之列車為定.
21T 重之軸四個
22T 重之軸三個
23T 重之軸二個
25T 重之軸一個

塊洋灰磚嵌砌.再砌兩側支壁;支壁有時用洋灰磚,有時卻用混凝土.砌築時,或整個洞身連接一氣,或四五公尺長爲一帶,中間隔開一生的之伸縮縫,再砌次節,此爲預防因氣候變遷之伸縮,及不幸傾壓,不致牽動全體之故.支壁之脚,則砌水溝,爲宣洩流水之用,灰漿成分,大率爲上述之D字種類.

該段內土質山洞工料,價約計每公尺自二百五十元至三百元.附潼關第十七號山洞,長每公尺之估算如下:

每公尺長洞面挖土	38.00 m^3
每公尺長洞身混凝土體積	4.51 „
每公尺長洞頂洋灰磚體積	3.04 „
每公尺長洞底混凝土體積	2.10 „
每公尺長砌築部分	9.65 „
洞內挖土	38× 2.00=76.00
洞頂洋灰磚	3.04×19.80=60.19
洞 身	4.51×13.80=62.24
洞 底	2.10×13.80=28.98
洞內工作加價	9.65× 1.40=13.51
運料在一千公尺以上之加價	
石 子	9.65× 0.50= 4.82
沙 子	4.80× 0.50= 2.40
洞身用夾板	8.00× 2.00=16.00
意外費 5%	13.20
	$277.34

全線山洞計十一座,均屬土質.其山洞路線,尙待研究者一座,應穿越潼關城牆,正籌續修者一座,各洞長度及地點如下表:

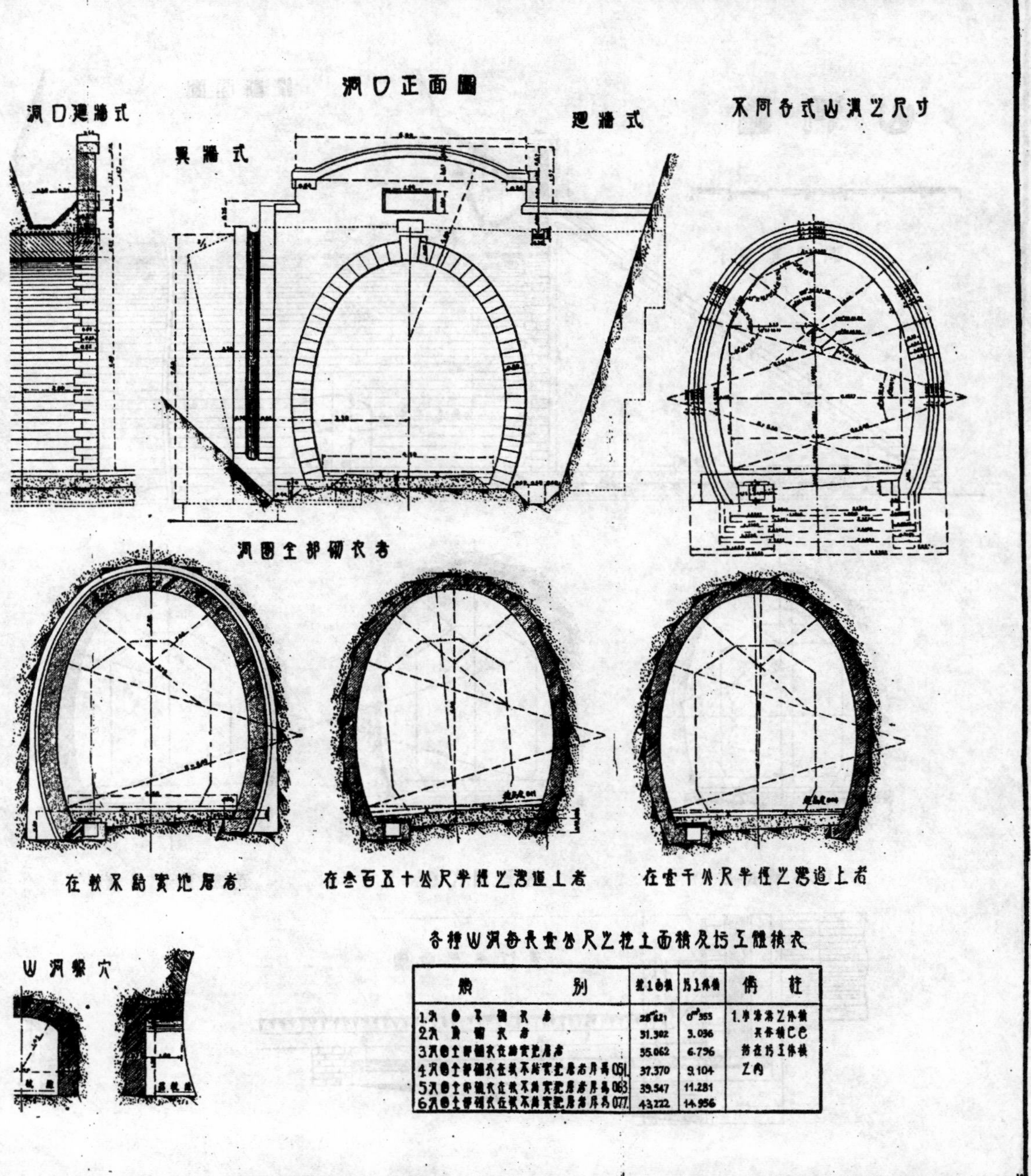

各種山洞每長壹公尺之挖土面積及圬工體積表

類別	挖土面積	圬工體積	備註
1.洞圈不砌衣者	28.621	0.353	1.水溝溝之体積其体積已包括在圬工体積之內
2.洞頂砌衣者	31.302	3.036	
3.洞圈全部砌衣在結實地層者	35.062	6.796	
4.洞圈全部砌衣在較不結實地層者厚爲0.51	37.370	9.104	
5.洞圈全部砌衣在較不結實地層者厚爲0.63	39.547	11.281	
6.洞圈全部砌衣在較不結實地層者厚爲0.77	43.222	14.956	

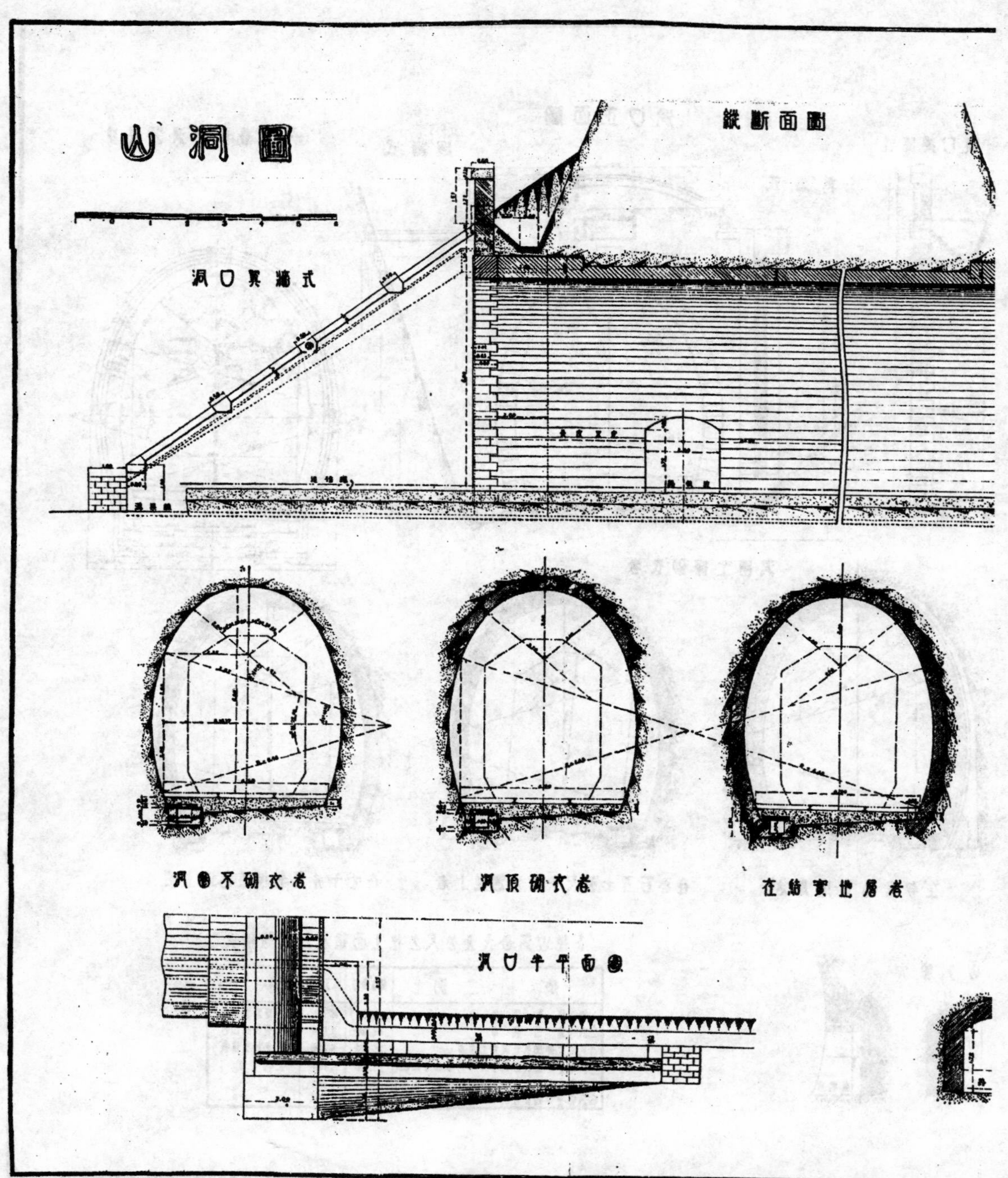
山洞圖
縱斷面圖
洞口翼牆式
洞圈不砌衣卷
洞頂砌衣卷
在結實地層卷
洞口半平面圖

靈潼段山洞表

山洞號別	地點	長度	包工姓名	承包價格	
No. 6	Km 386 bis +	90.50 m	集成	26,955 元	
„ 7	„ 287	621.20 „	集成	163,685 „	
„ 8	„ 287 +	90.30 „	集成	33,320 „	
„ 9	„ 287 +	107.40 „	集成	37,943 „	(1)
„ 10B	„ 289	2,800.00 „	未標出	988,000 „	(2)
„ 12	„ 291	622.60 „	集成	176,714 „	
„ 13	„ 324	695.00 „	董子紀	173,000 „	
„ 14	„ 330	631.84 „	殷鈺成	147,000 „	
„ 15	„ 339 +	395.00 „	陸廷熙	95,320 „	
„ 16	„ 351 +	909.10 „	協成	227,500 „	
„ 17	„ 353 +	1,080.00 „	未標出	324,000 „	(3)

備攷：(1) 第十第十一兩號山洞坍塌以後,曾有第十號A山洞之設計,未經適用即復測定第十號B山洞之路線.

(2) 山洞路線,正在研究,尙待解決.

(3) 正在籌劃續修.

(四)臨時便道　靈寶第十,十一兩號山洞,在民十五秋季,因洞身位於山坡之內,逼近黃河,業經坍塌,損失甚鉅.此次繼續興工,因該洞正在測量,尙待研究,故猶未定線.工程期限旣長,爲簡捷通車起見,勢難待其竣工,再行敷軌,乃於沿黃河山腰之上,測定臨時一線,繞越新建之山洞,即爲臨時便道,以應通車之用.該線長約二公里,雖不能耐久,暫時

靈寶車站西大橋橋墩山洞及函谷關全景

尙屬可用.不過在雨季期內,修養上應特別注意,甚或於必要時,須暫停車,以便修理路基.該線既位於山腰之側,線南傍山而行,已建有石工護牆,Retaining Wall 防山土為雨浸透而崩坍,線北濱臨河岸,竟有甚陡,現已擇要築以石塌,藉護堤基.該便道之估算如下:

靈寶西第八第九兩號
直線山洞

土工

1 塡土工程	3 ,0.,0 0方	0.14	4,200元
2. 挖淤及改水道	200,000方	0.17	40,800元
3. 坍塌加修挖土	40,000方		

涵洞

1. 一公尺明涵洞二座	3,000元
2. 二公尺明涵洞二座	4,000元

防護工程

1. 改水道及水溝砌石	3,000元
2. 護堤支牆砌石	80,000元
3. 牆脚乾砌石	15 000元
4. 防禦黃河工程	50,000元
共計洋	200,000元

(五)涵洞. 凡路基過高跨過深谷之處,其洩水量又不甚多,則建築石工橋墩,太不經濟,惟有造一涵洞,爲最適宜.現時金價暴漲,鋼鐵橋梁,咸從外洋運來料價尤貴.將來自籌自築,非但力避購用外洋材料,減輕受金漲之累,且宜就地取材,多築涵洞,及鐵筋混凝土橋,方可稍節漏巵,抑低成本.

門 涵洞洞身,應有各節厚度之不同,完全視上面路基之高度而定.在涵洞之縱斷面內,其中部所受路基土層之壓力最重,故亦以此部爲最厚,漸次向兩側伸展,則所受之土壓力亦漸輕.

涵洞底部基礎工程,須視該部分之土質如何及路基高度而定.其深淺普通估算時,約爲洞身方數之倍,用D字混凝土.每立方公尺工料,價約11.00元.如遇底脚見水,及挖有流沙,與四周應圍以木板樁等,當另行加價,俱在合同上載明.

(六) 敷軌工程　敷軌工程,應需材料之主要部分,厥爲鋼軌道枕.兩軌接縫之處,用二魚尾板連接,再以螺釘四枚或六枚固緊之.軌與枕之間,則有方形之鋼墊板墊之,藉免損及枕木.如用鋼道枕,則墊板可免,釘軌於枕上,或用螺釘,或用狗頭釘.

鋼軌有漢陽鐵廠製造,及外洋運來兩種,在靈寶以東各段,兩種均兼用之.靈潼段內敷設所用之鋼軌,係由比庚協定項下,在比國鋼廠購用,每根長十二公尺,在灣道內弦所用,間有每根十一公尺九二之長度,每公尺重42.164kg其每根軌之下,一律墊以木枕或鋼枕十七根.

軌枕在靈潼段內,計用鋼枕與木枕兩種.鋼枕及其他魚尾板,道釘,螺釘,墊板等附件,均由比庚項下與鋼軌一批釘購.計購五萬根,每根重量爲六十二.四五公斤.此外另在國內向上海標購美松木枕一萬根,又由部發交購料委員會代購木枕九萬八千根.以上鋼枕與木枕兩項總共數目,適敷全段軌線之用.在此段內凡屬灣道及山洞部分,爲欲求其耐久及便於營養起見概用鋼枕敷墊,其餘則用木枕.木枕每根重量約爲六十公斤,其尺寸爲6''×8''×9'-0''.又在各橋梁面部及車站轉轍器部分,則用美松特別枕木,其尺寸較普通枕木爲長,并塗以柏油,兩端用鐵皮箍緊,使其經久不朽.其敷軌應用材料之價值如下:

12m鋼軌	每根價	52.99元
鋼　枕	每根價	6.36元
螺絲釘	每噸	190.17元
魚尾板	每噸	142.71元

墊　板	每噸	146.34元
道　釘	每噸	171.13元
轉轍器	每付	649.69元*
橡枕木	每根	5.20元
美松木枕	每根	3.50－4.00元

* 以上各項價值,係十七年間之金價而在比國購訂者.按現時金價,則超過一倍以上.

關於敷軌工事,有時自行敷設,派一監工監督之.遇有長途敷設,則向來用包工制度,一切工具,由路借給,工畢交還.其工價約計如下:

用手推平車在四百公尺以內裝運料廠內料件並敷軌於鋼枕上	每公尺	0.15元
用貨車在四百公尺以內裝運料廠內料件並敷軌於鋼枕上	每公尺	0.15元
用手推平車在四百公尺以內裝運料廠內料件並敷軌於木枕上	每公尺	0.14元
用貨車在四百公尺以內裝運料廠內料件並敷軌於木枕上	每公尺	0.14元
裝配及安設轉轍器包括手推平車裝卸搬運及糾正轉轍器下之枕木與打孔鋸剖鋼軌鑽孔等在內	每付	12.00元
改鋸非轉轍器用之鋼軌	每次	0.20元
鋼軌鑽孔	每孔	0.10元
尋常枕木鑽孔每根四孔	每根	0.01元
尋常枕木敷設	每根	0.02元
手推平車運送鋼軌材料至四百公尺以外之補助費	每公里 噸	0.08元
路軌鋼料在貨車上裝費	每噸	0.15元
特別枕木刨捆及上油	每根	0.20元
橋面裝釘特別枕木	每根	0.25元

比庚協定項下在歐購訂靈潼新工材料表
(民十七年分)

材料類別	訂購數量	單位	單位價格	總價格
20m及30m橋梁	1,052,270	噸	187.32元	281,465.22元
3m 5m及10m橋梁	96,704	„	152.01	14,699.97
鋼軌	8,162,000	„	104.73	854,806.26
鋼枕	50,000,000	根	6.36	318,000.00
轉轍器	80	付	649.69	51,975.20
道釘	299,000	噸	173.13	51,765.87
魚尾板	504,000	„	142.71	71,925.84
螺絲釘	90,000	„	190.17	17,115.30
墊板	635,000	„	146.34	92,925.90
水櫃水鶴水管等				47,145.00
			總計	1,801,764.56元

備考：以上均十七年分金價折合當時每鎊合洋十元五角,現在金價暴漲,已超過上列各數一倍以上.

(七)臨時便橋　在路線之上,因緊急敷軌行車,不及待正式橋工築竣,於是在橋孔之位置中間,搭三數木架,其上敷軌,是爲臨時便橋.該項便橋上旣可不誤行車,橋下兩架所留之空間,仍可同時砌築石工,惟較爲危險而已.爲求安全計,有時在正線之左右兩側,另築一道基,兩端聯於正線上,中建便橋以洩水,此可與行車完全無礙.但極深之山溝內涵洞,與較高之土基,則不能適用此項便橋,以搭木架過高,不易穩定故也.在從前本路開徐段內各橋,値歐戰之際,工款告匱,利用地勢平坦,土基不高,曾採取此項便橋甚夥.此次靈潼段內,間有運料不便,行車緊急,即建有便橋三座,正式橋工,俟通車能將遠地材料,源源運來,再行建造.普通每一便橋,所需美松木料,約二十五立公方,每方按六十元計算,共費一千五百元,其他搭架截鋸工費所需甚微.

(八)車站　靈潼段內車站,除潼關車站較大,其餘各站均爲行車交軌及

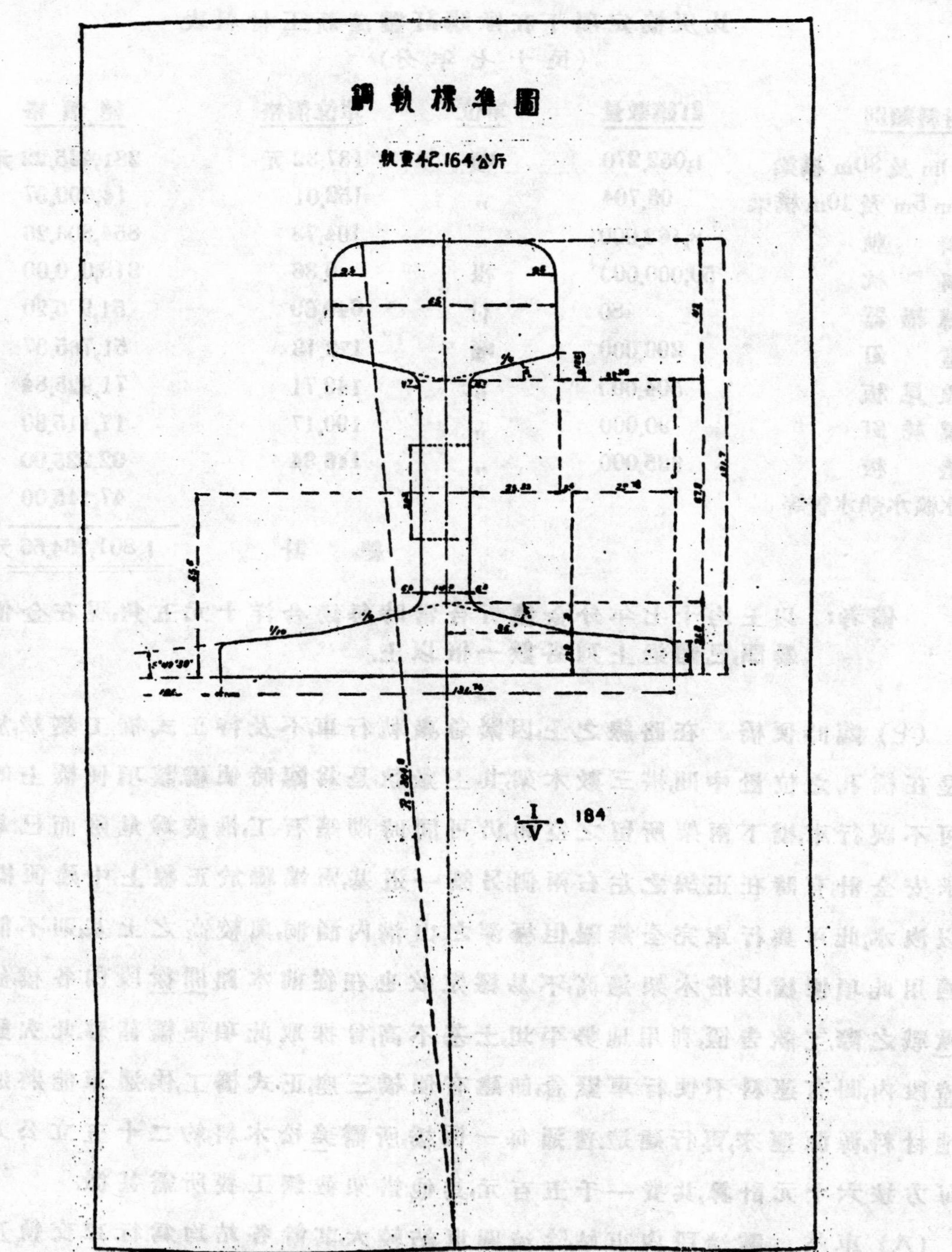
鋼軌標準圖
軌重42.164公斤
$\frac{I}{V}$ = 184

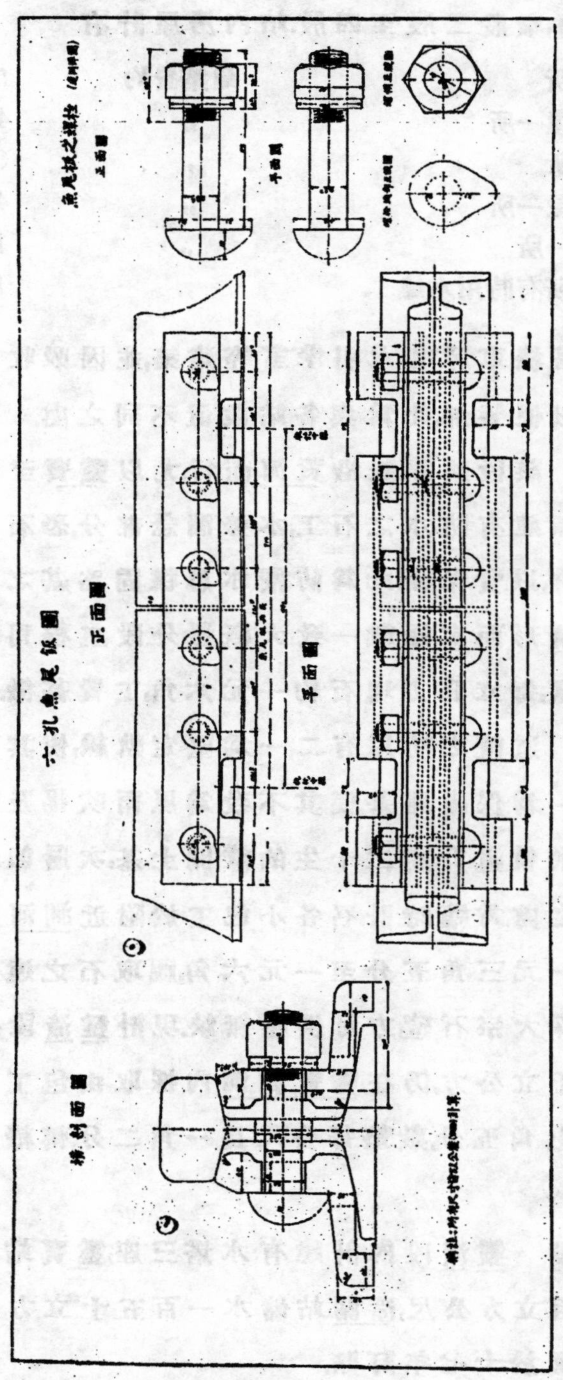
六孔魚尾板圖
正面圖
平面圖
側面圖

灌水之用.站內叉道,敷設二股至四股.站內房屋計有

票房一所	建築費約	5,000.00元
站長住屋一所	,,	4,000.00
警屋一所	,,	3,000.00
監工住屋一所	,,	4,000.00
道工房一所	,,	1,500.00
閘屋二所有時用木建	,,	1,000.00

潼關車站有水搭機車房唧水引擎室等建築,並因吸收河運之貨物,敷設枝線,引伸至河岸,以便裝卸.此為與各站設置不同之處.

(九) 防護工程　該段路線,旣沿黃河而行,尤以靈寶至潼關兩地,距河最近.求路基堅穩,必須建有防水之石工,水勢湍急部分,恐石工易為河流冲刷,應再簽立木樁一排.以資保衛,此為防禦水患穩固路基之設備.其石工高度與厚薄,視水流之情形而定,極難一致.大概憑分段工程司駐工視察力之判斷,計算時祇計料價,每立公方塊石約一元六角,工費甚微.

(十) 道碴鋪設　道碴之功用有二,一為堅定軌線,使其不致向兩側移動,減少行車之危險,一為保護路基,使其不致為風雨吹揚及侵陷.鋪墊時大率分為二層,初層為將軌道升高二十生的離開土基,次層即為將軌枕全部埋沒,僅露地面.取碴之處,為臨時計,召各小包工於附近澗河之內,拾取塊石,擊碎敷墊,每立方洋一元三角五分至一元六角,視取石之遠近而定.將來全線最後之敷碴,應開採大宗石碴,方可供應無缺.現計靈潼段全線,第一層敷墊,計需道碴五萬三千立公方,仍在靈寶澗河內採取,由包工承辦供給,每採一立公方碴料,價洋七角五分,敷墊每公尺長一角二分,轉轍器部分敷墊,每付工洋五元.

(十一) 供水設備　靈潼段內計建有水塔三座,靈寶站儲水五十立方公尺,閿鄉站儲水一百立方公尺,潼關站儲水一百五十立方公尺,此三水鶴連同唧水機,均在比國於十七年訂購.

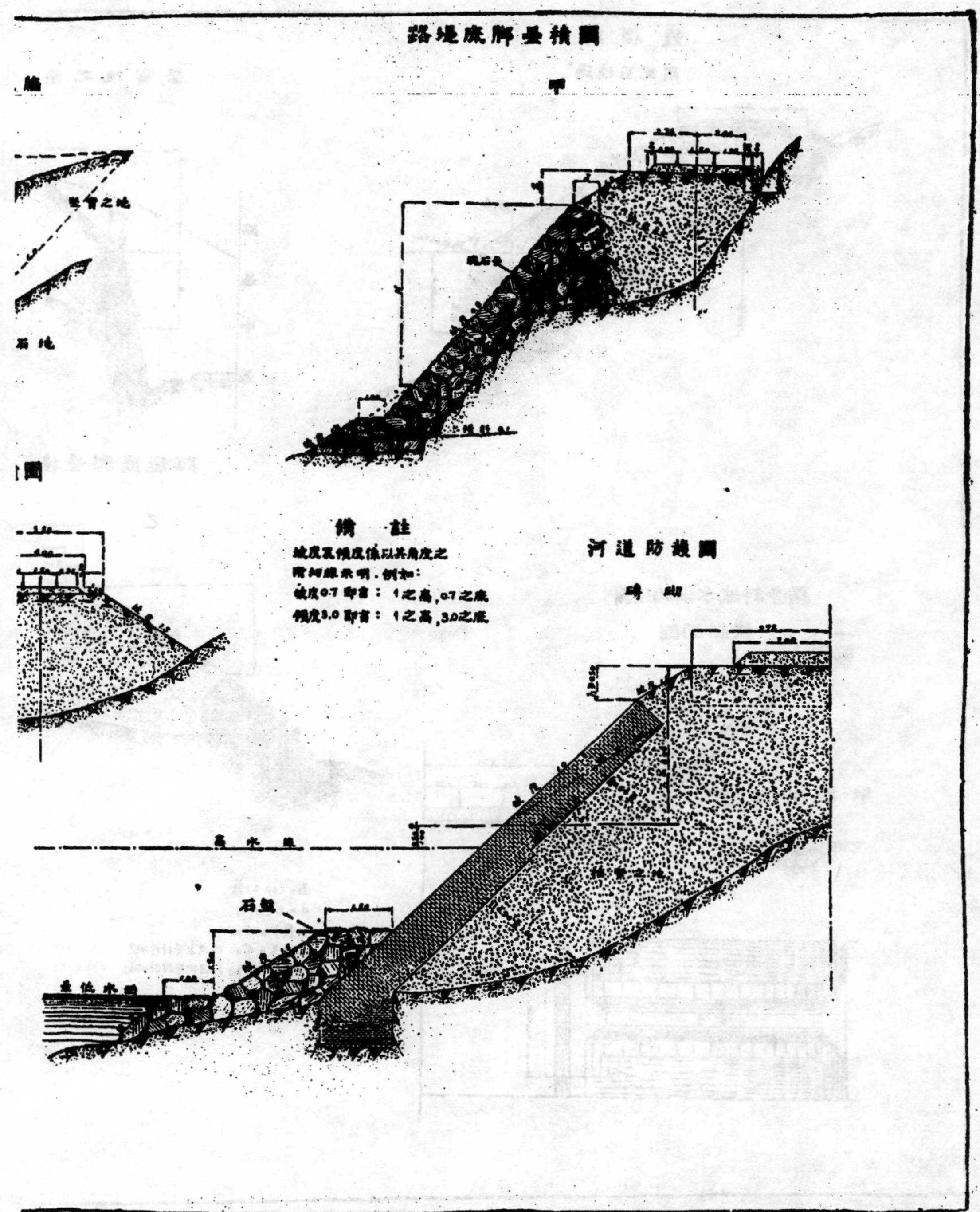
路堤底脚叠積圖
甲
備註
坡度及傾度係以其角度之
兩切線示明，例如：
坡度0.7 即言：1之高，0.7之底
傾度3.0 即言：1之高，3.0之底
河道防護圖
高水線
石盤

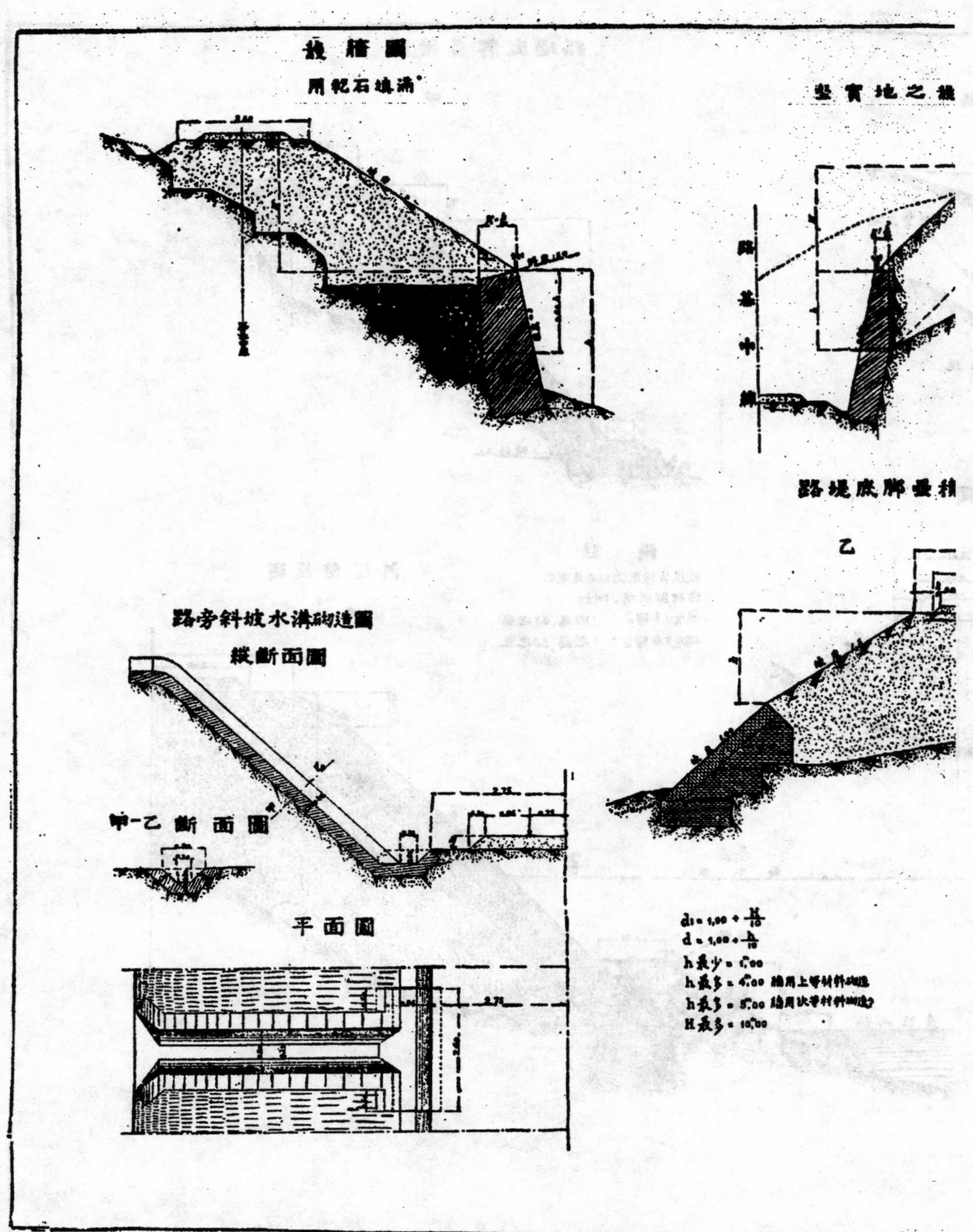

護牆圖
用乾石塡滿
堅實地之護
路基中線
路堤底脚叠積
乙
路旁斜坡水溝砌造圖
縱斷面圖
甲-乙斷面圖
平面圖
d₁ = 1,00 + H/10
d = 1,00 + h/10
h 最少 = 0,00
h 最多 = 4,00 須用上等材料砌造
h 最多 = 5,00 須用次等材料砌造
H 最多 = 10,00

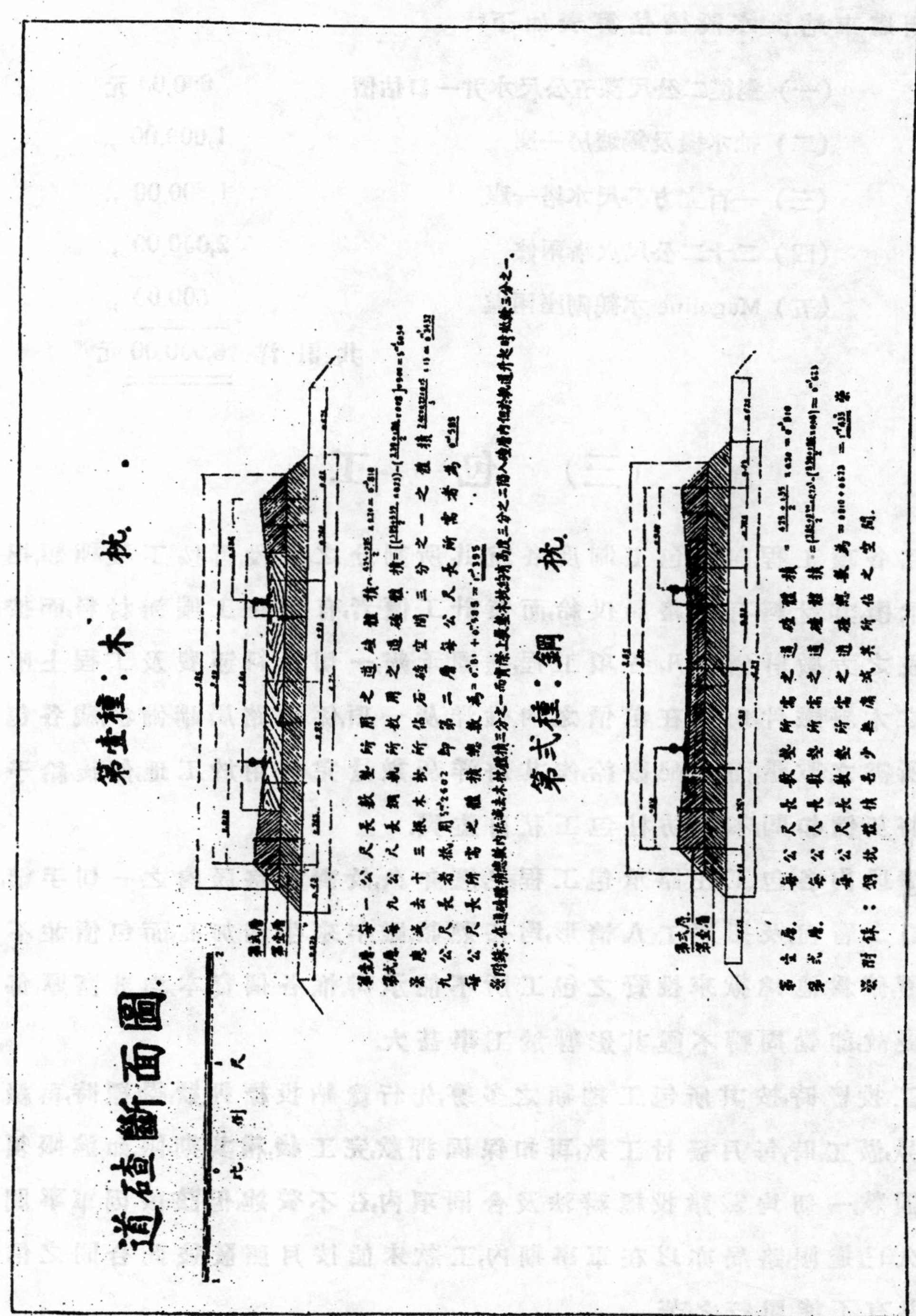
道碴斷面圖
第壹種：木　枕
第貳種：鋼　枕

附閿鄉車站供水設備估算表如下:

(一)	對徑二公尺深五公尺水井一口估價	600.00元
(二)	抽水機及鍋鑪房一座	1,000.00 ,,
(三)	一百立方公尺水堵一座	1,800.00 ,,
(四)	二十二公尺火溝兩條	2,000.00 ,,
(五)	Magoline 水鶴砌座兩處	500.00 ,,
	共計洋	5,900.00元

(三) 包 工

本路各項工程,向採包工制度,各就其所劃分之地段,再按工程種類,招工投標承攬.其材料有由路局供給,而僅計工價者,有由包工備辦材料,而按其所做成之方數計價者.凡各項工程,選標承辦,一切材料運費及工程上附帶所用之木料鐵件,均算在包價之內,惟洋灰一項,係由路局購備後,視各包工按期所需之數量,而分配發給,俟其將洋灰數量完全用於工地,然後給予工款,再將灰價扣回,不過防杜包工私自盜賣.

靈潼段內各包工在路承包工程,已歷年久,故對於路局內之一切手續,及工地風土習尚,及招雇工人情形,均甚熟諳.雖年來生活加高,而包價並不增昂其價,確爲他埠欲來投資之包工所不能承辦.惟各個資本,均非富厚,每遇時勢變故,即覺周轉不靈,其影響於工事甚大.

包工投標時,按其所包工總額之多寡,先行繳納投標押款,得標時,再繳承辦押款,做工時,每月發付工款,再扣保固押款,完工後,稽其期限,如逾限須再處以罰款,一切均載諸投標辦法及合同項內,茲不贅述.但該段因軍事關係,工期久已逾限,路局亦以在軍事期內,工款未能按月照發,彼此合同之信守,雙方俱有不能履行之苦.

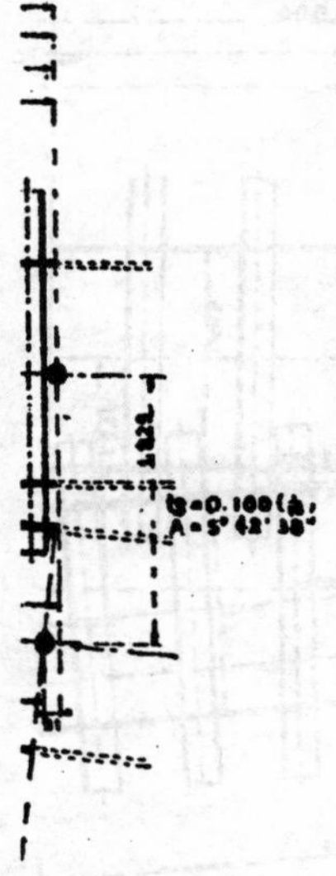

古式道岔所需之特别枕木

號數	類別	數目	尺寸			每塊枕木方數	總方數
			長	寬	厚		
1	特别枕木	1	2.70	0.30	0.15	0.122	0.122
6	〃	6	4.20	〃	〃	0.189	1.134
7	〃	2	4.50	〃	〃	0.203	0.406
	普通枕木	56					1.662

吉式道岔需之特别枕木

號數	類別	數目	尺寸			每塊枕木方數	總方數
			長	寬	厚		
1	特别枕木	1	2.70	0.30	0.15	0.122	0.122
6		6	4.20	〃	〃	0.189	1.134
7		2	4.50	〃	〃	0.203	0.406
	普通枕木	50					1.662

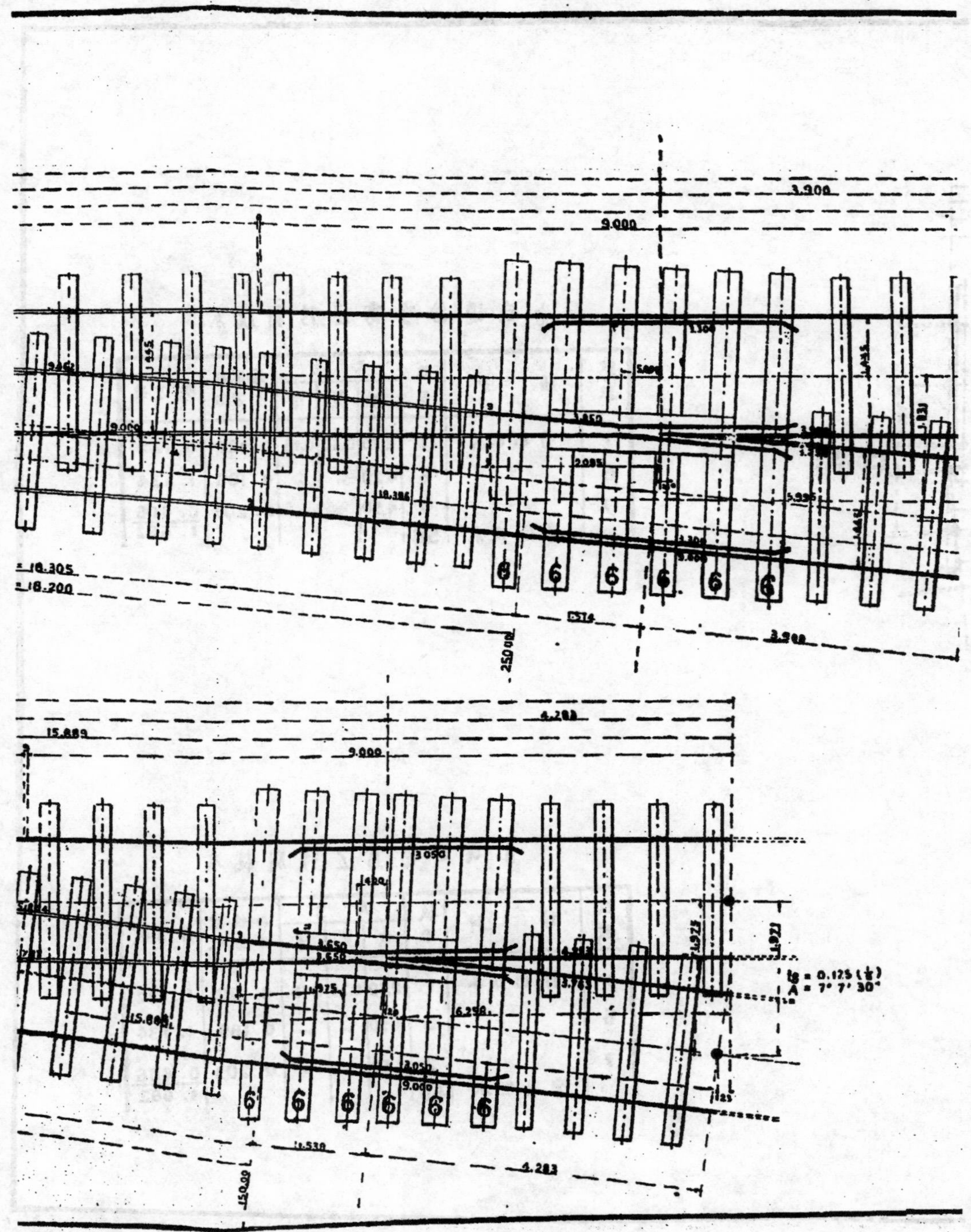

切綫等於 0.100(吋)之道岔鋪設圖

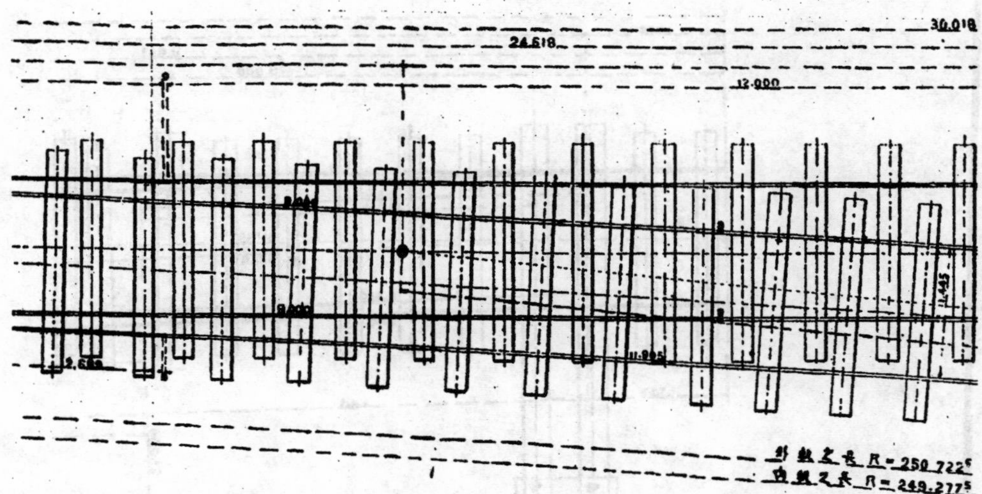

切綫等于 0.125(吋)之道岔鋪設圖

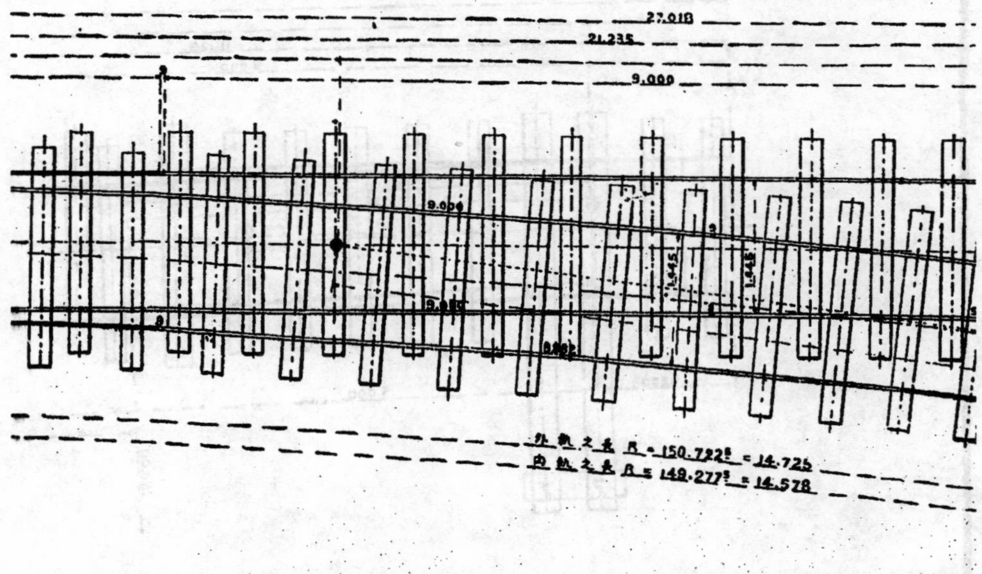

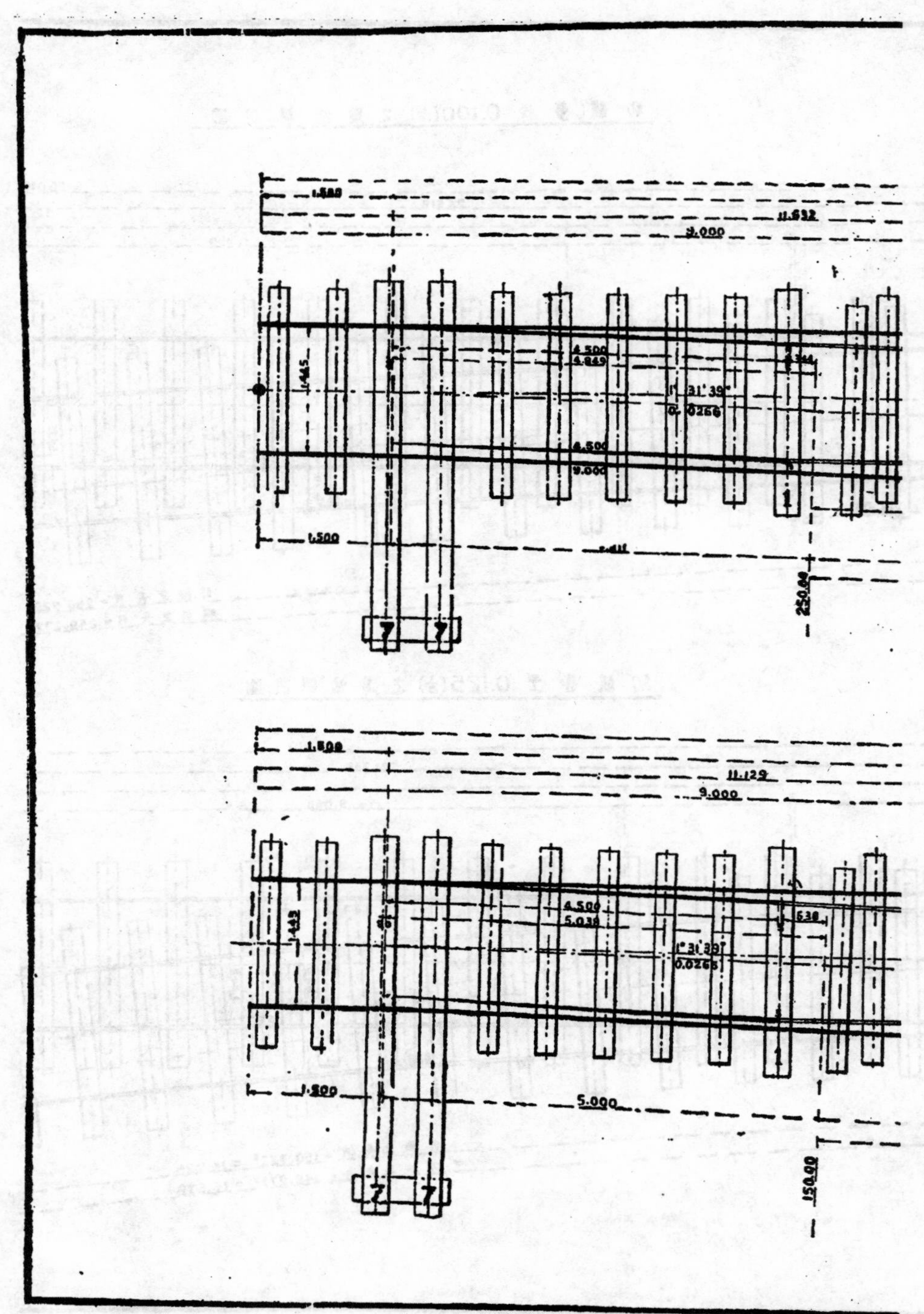

1.500
11.632
9.000
1.445
4.500
4.849
4.500
9.000
1°31'39"
0.0266
1.500
250.00
7
7
1
1.500
11.129
9.000
1.445
4.500
5.038
1°31'39"
0.0266
1.500
5.000
150.00
7
7
1

靈潼段各包工承辦工程一覽表

包工姓名	工程種類	標估總額	備攷
集　成	山洞,土工,涵洞,小橋及護堤工程等	1,150,000元	
協　成	仝　上	900,000元	
董方臣	橋工及[illegible]鄉車站供水設備	120,000元	
董子紀	山洞土工涵洞及小橋	360,000元	
陸廷熙	山洞,土工,涵洞,小橋及護堤工程等	550,000元	
朱楷亭	土　工	255,000元	
陳禹臣	涵洞及小橋	105,000元	
殷鈺臣	山洞,橋工,涵洞及土工	450,000元	
陳源遠	橋　工	120,000元	
劉振先	開採石料	18,000元	
黃紹彬	裝卸路料		

關於各包工投標及承辦時之各項手續,均有明文規定,遵守勿渝.但工程種類繁夥,故此項文件亦至多,玆爲便於閱者參考起見,摘錄投標須知,土工施工細則,及本段各包工工類單價表於下,庶覩其一可以概其餘,藉可比較與其他各路異同之點.

(一) 投標須知.

1. 包工應用華法文,將另表所有價目,按類塡齊,塡註不全之標,概作無效.

2. 凡願投標各包工,應在鄭州隴海工程局工務課,或上海九江路六號隴海鐵路辦公處,先繳投標保證金洋　　元,此款於開標後,若不得標,仍照數發還.

3. 收受保證金之機關,於收到保證金後,應塡發收據二紙,一紙應隨附標單之內.

4. 得標包工於簽訂合同之日,應按照合同第十六號之規定,將承辦工程保證金繳納鄭州工程局出納股,其投標時所繳之保證金洋　　元,即在此項承辦工程保證金內扣除之.

5. 投標文件及說明書,可向鄭州隴海工程局工務課,或上海九江路六號隴海鐵路辦公處取閱.

6. 標單極遲應於民國　　年　　月　　日上午十一時,送交或以掛號函由郵局寄交鄭州隴海工程局.

7. 包工得於前定開標日期及時間,前來列席開標.

8. 包工爲填造標單起見,應領取標册一份,惟須繳洋　　元.

9. 本路備有　　工程圖樣,以便包工研究.如包工欲購此項圖樣,本路亦可讓售,每份圖　　張計洋　　元.

10. 投標函件封面上,應書明(鄭州隴海鐵路工程局收啓),並須加注工程投標函件.　中華民　　年　月　日工程局局長　　總工程司

(二) 土工施工細則

1. 土工　鐵路之方向及其地位,皆以栽於中線上之樁橛,於地面上定明之.

包工在未動工之前,應將路線暨橫面圖,以及所發各種填土或填石工程圖樣,先行從事察驗,並須承認之.

如包工於未動工前,對於上項未提出何種請求,則路局所備之土地切面圖,無須採取其他手續,即成爲原訂包工合同之附件,並即用以實施工程及計算帳目.

樁橛一經驗明或添補,包工即應負責.

該包工對於樁橛手續辦清後,實施工程時,如有不合做錯及其他謬誤之處,並應負責.

2. 土工之形狀　土工之形狀,於開工時由總工程司定之,如中途須變更時,並得由總工程司決定.

關於此項包工,對於總工程司頒發之各種平面縱面圖,及發佈之命令,

靈寶潼關段包工合同主要單位工價表

Releve des prix unitaires principaux de Contrats entre Lingpao-Tungkwan

合同號數 N°s de Contrats	包工姓名 Noms des entrepreneurs	土方工程 TERRASSEMENTS		丙種混凝土 Béton C.	丁種混凝土 Béton D.	戊種混凝土 Béton E.	庚種混凝土 Béton G.	壬種混凝土 Béton I.	丁字灰漿人造片石 Moellon artificiel au mortier D.	平模板 Coffrage plat	彎模板 Coffrage courbe	山洞內土工 Terrassement du tunel	山洞內加價砌工 Supplément du tunel maconnerie	運送材料運價 Transport de matériau		
		土坡 Tranché	填土 Remblai											沙土 Sable	礫石 Gravier	片石 Moellon
130	集成公司 自 至 Tsi Tcheng (Km 286+523 au Km 287+600) (tunels, maconneries et terrassements) (山洞砌工及土工)	0.15	0.10 et 0.15	9.30	10.95	13.00	16.70	20.40	17.65	1.70	0.20	1.55	1.20	0.04	0.04	0.07
139	協成公司 自 至 Hsieh Cheng (Km 287+600 au Km 290+500) (tunels, maconneries, et terrassements) (山洞砌工及土工)	0.17	0.14 et 0.18	9.60	11.75	13.00	16.40	21.40	19.00	1.00	0.80	1.90	1.50	0.04	0.04	0.07
140	集成公司 自 至 Tsi Tcheng (Km 290+500 au Km 291+500) (tunels, maconneries et terrassements) (山洞砌工及土工)	0.17	0.14 et 0.17	10.10	11.80	13.70	16.10	21.40	18.50	1.40	0.40	1.90	1.50	0.05	0.05	0.07
附加合同第一號 140 av. 1	集成公司 自 至 Tsi Tcheng (Km 291+500 au Km 292+450) Terrassements 土工	0.145	0.145													
附加合同第一號 128 et av. 1	集成公司 自 至 Tsi Tcheng (Km 292+450 au Km 304+437) Terrassements 土工	0.165	0.165													
126	集成公司 自 至 Tsi Tcheng (Km 292+450 au Km 304+437) Maconneries 砌工			7.75	9.65	11.45	15.15	18.85		1.50	0.30			0.06	0.06	0.07
171	集成公司 Tsi Tcheng (Km 294+230) Pont metallique 3×3) m. 三孔三十公尺鐵橋			8.40	10.50	12.54	16.08	20.02		1.70	0.50			0.04	0.06	0.07
129	集成公司 自 至 Tsi Tcheng (Km 304+437 au Km 315+250) Terrassements 土工	0.115	0.138 et 0.16													
127	集成公司 自 至 Tsi Tcheng (Km 304+437 au Km 315+250) Maconneries 砌工			8.00	9.85	11.70	15.40	19.10		1.50	0.30			0.04	0.04	0.07
144	董方臣 Tung Fang Tcheng (Km 315+125) Pont metallique de 2×30 m. 二孔三十公尺鐵橋			9.00	10.80	12.38	15.55	20.50		1.70	0.50			0.06	0.06	0.07
133	朱楷庭 自 至 Tchu Kai Ting (Km 315+260 au Km 322+230) Terrassements 土工	0.13	0.12 et 0.155													

136	陳禹臣 自 至 Tcheng Yu Tcheng (Km 315+260 au Km 322+230) Maconneries 砌工			9 60	11.30	13.20	16.60	21.90		1.40	0 20			0.05	0.05	0.07
135	孟慶恆 自 至 Mong King Hung (Km 322+230 au Km 324+500) Tunel, Maconneries et terrassements 山洞砌工及土工	0.19	0.08	9.20	11.04	13.13	17.30	23.15	19.50	2.00	0.30	1.75	1.40	0.05	0.05	0.08
附加合同第二號 135 av. 2	董子紀 自 至 Tung Tse Ki (Km 322+230 au Km 324+500) Tunels, maconneries et terrassements 山洞砌工及土工	0.28	0.13	9.60	11.50	13.13	17.30	23.15	19.50	2.00	0.30	1.75	1.40	0.06	0.06	0.08
131	陸庭熙 自 至 Lu Ting Hi (Km 324+500 au Km 328+800) Maconneries et terrassements 砌工及土工	0.13	0.13 et 0.15	9.00	11.60	12.20	16.82	22.57		1.00	0.50			0.04	0.04	0.07
145	董方臣 Tung Fang Tcheng (Km 328+800 Pont metallique de 3×30 m. 三孔三十公尺鐵橋			9.10	11.00	12.60	15.76	20.90		1.75	0.50			0.06	0.06	0.07
132	殷鈺丞 自 至 Ying Yu Tcheng (Km 328+800 au Km 332+500) Tunnels. Maconneries et terrassements 山洞砌工及土工	0.13	0.115 et 0.185	9.30	11.10	13.70	17.80	23.50	19.50	1.65	0.50	1.65	1.40	0.06	9.06	0.07
135	朱楷庭 自 至 Tchu Kai Ting (Km 332+500 au Km 338+300) Terrassements 土工	0.10	0.13 et 0.16													
146	陳禹臣 Tcheng Yu Tcheng (Km 336+046) P. M. de 4×30 m. 四孔三十公尺鐵橋			9.70	12.40	14.20	18.10	23.70		1.40	0.70			0.05	0.06	0.07
137	陸庭熙 自 至 Lu Ting Hi (Km 338+300 au Km 341+550) Tunnels, Maconneries et terrassements 山洞砌工及土工	0.18	0.16	11.00	13.10	15.12	18.95	24.57	19.00	1.35	0.50	1.70	1.20	0.04	0.04	0.05
附加合同第二號 137 av. 2	陸庭熙 自 至 Lu Ting Hi (Km 341+636 au Km 343+600) Terrassements 土工	0.10	0.13 et 0.16													
147	陳源遠 Tcheng Yuan Yuan (Km 341+646) Pont metallique de 4×30 m. 四孔三十公尺鐵橋			9.70	12.40	14.20	18.10	23 70		1.40	0.70			0.05	0.06	0.07
138	協成公司 自 至 Hsieh Cheng (Km 350+500 au Km 352+500) Tunnels, maconneries et terrassements 山洞砌工及土工	0.17	0.12 et 0.18	12.00	13.80	15.80	19.65	25.27	19.80	1.20	0.80	2.00	1.40	0.05	0.05	0.07
附加合同第一號 138 av. 1	協成公司 自 至 Hsien Cheng (Km 343+600 au Km 350+500) Maconneries 砌工			12.00	13.80	15.80	19.65	25.27		1.20	0.80			0.05	0.05	0.07

應確切遵守.

3. 應行塡積之土地預備工作　應行塡積之土地,須小心打掃清潔,如有樹木暨棘籬根蒂,以及其他植物殘體,均須除去.

如在坡度甚大之土地上塡積,應將該項土地劃分階段,其地位及大小由工程段長定之.

4. 塡積及堤坡之作法　凡塡積之處,除或須命令添敷適宜土層外,不得留有植物根蒂及殘體,此項塡積橫面,須全部立卽做齊,以免須於兩傍堤坡上加土補足之.

堤坡暨塡積及開挖之兩傍平地,均須確切按照圖樣所載說明及形狀修整,其面部及頂部均須甚爲整齊.

當路堤粗成時,包工須將路基稍爲放寬加高,究竟放寬若干,加高若干,應由工程段長隨時指示之.如爲鞏固塡積起見,須將由壕溝內取出之材料,特加分配,或須將已經鋪好之塊積,重予整理,包工應遵照總工程司指示辦理,不得藉口要求任何津貼.

倘總工程司爲防備路工倒坍起見,認爲堤坡須隨工程進行程度,逐漸修整,以及開挖深度須加以修整,包工應卽遵照辦理,不給任何津貼.

遵照上開各項條件辦理必須之一切工價,均包括在單位價內.

如工程上需水,而泉水及雨水或致缺用,其備水之費,應包括在包臨時費內,由包工負擔.

5. 橋梁附近之塡積　橋梁附近之塡積,應格外小心辦理.此項塡積,至少須展長等於橋梁高度,並須按層鋪設,每層厚二十五公分,又須將前層詳愼打平搗緊,始可添鋪次層.至其物料,則僅用小石塊,或不含有沃土及鈣土之泥土,以沃土及鈣土,易使塡積縮壓或發生他項變動也.

穹洞頂部,必須俟兩邊塡滿至洞高三分之二之處,始可加土.總之每遇特別情形,由路局頒發包工之一切命令,該包工均應遵守.

(四) 工商情況

西北關山險阻,人民仍保守古時之農業生活,毫無工商業之可言.其工業原料,如棉花,獸皮,羊毛等,每年產量甚富,然皆悉數運出,而不知就地設廠製造,利用機械之原動力,改善人民之生活,故一入關中,祇見男耕女織,而巍然大觀之工廠,與高立雲表之煙筒,誠不稍見.而通商巨賈,更屬罕有.是以在隴西鐵路未通以前,猶能閉關固守,對內則農產穀實,足以自給,對外則無所需求,並不發生經濟上之恐慌.近來以天災人禍,交相逼迫,倚農爲生,已失生活上之屏障.於是強者流爲盜寇,老弱轉死溝壑,此即因無工商業可以調劑其環境之故也.

靈潼一段,岡巒起伏,無適宜之公路,以利民行.全省所恃爲貨物出入唯一之運輸,厥爲黃河之航道.在冬季結冰期間,則用牲畜,以代舟運.而商貨西運,因水流湍急之故,汽船又不適用,每遇逆風,舟運非十餘日不能達,殊不稱便.該段路工,實有發展三秦工商業之可能,及轉移陝民經濟狀況之力量.路線雖與河道平行,而車行時間之縮短,商貨運費之減省,確非黃河舟運,所得而拮抗之耳.

(一) 附靈潼潼西一帶工價表 (單位以元計)

工別	小工	木工	石工	泥瓦工
每日工資	0.20-0.3)	0.5)-0.80	0.70-1.00	0.50-0.80

(二) 附靈潼潼西一帶運價表 (單位以元計)

運具		每日運費
大車	(百里內)	6.00-10.00
推車	”	3.00
牛	”	0.80
驢	”	0.60
騾	”	1.00
馬	”	0.90

(三)附靈潼潼西一帶物價表(單位以元計)

品名	單位	價格
大米	每斗	2.50
小米	每斗	2.00
小麥	每斗	2.50
本地石油	每斤	0.30
外國石油	每斤	0.40
本地香油	每斤	0.30
鹽	每斤	0.14
烟煤	每百斤	3.00
無烟煤	每百斤	4.00
木炭	每百斤	6.00
鐵	每百斤	8. 0
石炭	每百斤	3.00
茶	每斤	0.40 1.00
磚	每百頁	3.00
頭等麵粉	每袋	5.60
二等麵粉	每袋	5.00
木柴	每百斤	2.30

(五) 物產

靈潼一段長祗七十餘公里,地瘠民貧,其間並無特殊之物產,即如靈寶附近之棗林,下鄉一帶之柿樹,以及潼關之胡桃,李,果等物,雖味尙甘美,然產量不豐,不足稱大宗物產.故該段行車之貨運,完全爲吸收陝省境內土產之輸出,惟年來西北災荒洊臻,人民易子而食,天然物產,受天災之影響不淺,今就民十五十七各年間所有陝省各縣羊毛棉花及煤炭,據調查之產額,其與本路商運有直接之關係者,分別列表於下,以饗閱者.至於甘肅之藥材皮毛,將來亦爲隴海貨運之源,其數量尙待調查.

(一)陝西各地羊毛產額一覽表

地名	物品	每年產額
靖邊縣	白羊毛	27,000
	羊絨	2,100
扶風縣	羊毛	16,048
耀縣	仝	3,000
富平縣	仝	2,000
白河縣	仝	20,000
乾縣	仝	2,000
邠縣	仝	3,000
隴縣	仝	10,000
米脂縣	仝	20,000
綏德縣	仝	30,000
清澗縣	仝	50,000
雒南縣	仝	1,000
栒邑縣	仝	10,000
咸陽縣	仝	5,000
蒲城縣	仝	3,000
涇陽縣	仝	5,000
	總額	209,148斤

(二)陝西各縣十五年十七年棉花產量表(單位以擔計)

縣名	十五年	十七年
長安縣	31,600	19,000
臨潼縣	50,100	7,125
朝邑縣	36,240	12,960
渭南縣	114,840	13,140
大荔縣	30,000	3,000
華縣	3,120	1,600
韓城縣	22,000	3,000
華陰縣	9,965	1,586

郃陽縣	20,000	3,000
潼關縣	4,000	144
澄城縣	1,200	
鄠　縣	550	130
蒲城縣	430	
郿　縣	2,600	180
富平縣	1,104	426
鳳翔縣	3,000	95
高陵縣	13,500	1,350
岐山縣	500	150
三原縣	13,000	6,440
扶風縣	700	630
涇陽縣	18,240	5,880
武功縣	600	460
盩厔縣	300	1,200
興平縣	970	740
醴泉縣	950	350
咸陽縣	1,300	720
藍田縣	3,900	980
總產量	381,109擔	81,268擔

(三) 同官煤鑛每日產額統計
(根據十七年調查)

所在地	紅土坡附近	黃保鎮附近西	陳爐鎮正東	黃保鎮正北	陳爐鎮正西三里同	每日總產額暨總值
每日產額	31,200斤	20,640斤	16,800斤	14,400斤	9,600斤	92, 40斤
總　值	78.00元	51.60元	42.00元	36.00元	24.00元	2 1.60元

(四) 同官煤價表

出產地		同官縣
產地煤價	(每百斤)	1.00元
西安煤價	(每百斤)	3.00元

(六) 名勝古跡

靈寶　處函谷關東約三里許,北臨黃河,與晉境相對,東望硤石,南鄰虢州,舊爲唐楊貴妃故里,以其地山清水秀,女子韶秀,有異附近他邑,故至今猶有小蘇州之目.城中有樹如蓋,下立一碣勒老子,著經處斕斑蘚跡,古色盎然,尤令人憑視摩挲不忍卽去.

函谷關　函谷倨兩峻嶺間,前阻澗河,更右爲黃河,峭壁崢嶸,下臨無地,更左爲秦嶺,復崇嶺屈曲.無徑可通.谷中尤偪隘,車至不能並駕.戰國時,屢嚮六國之兵,不能入關撻秦,天險之稱,蓋有由矣.谷口建城樓一座,下可通人.樓前勒函關二字,兩側有嘉慶間某邑侯題聯,爲「未許田文輕策馬,願逢老子再騎牛」二句,丸泥封固,豪壯可知.至本局計畫路線,係以多數山洞穿嶺而過.山洞最長者,達二千八百公尺.他日客商來往,越險如夷,貨運萬鈞,指顧已達,又豈公路所可同日語者.關頭書紫氣東來,指老聃過秦事.關前有關龍逢塚,商諍臣遺址也.

函谷關

閿鄉　縣城在靈寶西約六十里,地臨黃河在城西二十餘里之隔岸爲永樂村,有呂祖故里屬晉境.閿鄉城中有唐武后祠,楹聯「六宮粉黛無顏色,萬國衣冠拜冕旒」工整無似.

盤豆鎮　處閿鄉西二十里,亦近黃河,舊傳唐明皇幸蜀,曾過是鎮,野人以盤豆獻,故名,未知確否.鎮有楡樹數株,大可合抱,風景甚佳.

潼關　潼關負山爲城,氣勢奇壯,黃河縈帶其下,俯瞰如練.隔岸爲風陵渡屬晉境,其他有軒轅氏風后之陵.北對太行,橫亙天際頂平如鏟,嵐光掩映,蒼紫如繪.西望爲河,渭合流之口.其處河廣流急,聆之貌然神往.城東五里爲第一關,行旅往來,孔道亦屬險要,豫秦交界處碑,即在其處.此城經明代重修,至今雉堞井然,光整如新.城東西各橫一額,曰「屏藩兩陝,控制三秦」,當時此關重要,可概見矣.

潼關風景

華山　在潼關西南約三十里,巉巖突出,高插入雲者,五嶽中之西嶽華山也.山分東西南北中五峯,東峯最高,高出附近地面約一千八百公尺.峯上有絕壁呈巨靈掌石月等跡,駢指六出,銀蒜一彎,晴日用遠鏡自下眺之,彌覺清晰.山上險勝如迴心石,英雄進步千步幢,百尺梯,猢猻愁,擦耳巖,鷂子翻身等處,傍旋則有玉泉院爲宋陳摶逃辟處.院中有無憂亭,無憂樹,希夷塚及石雕陳摶臥像等古蹟.附近復有胡林翼塚,亦偉壯可觀.山下即華陰縣城,城中有文廟,已就頹敗.城東二里爲華嶽廟,已改作修械廠址,廟以城環,成形長方,城垣絕新.廟內有一三級樓,稱萬歲樓.其旁更有前清那拉太后與德宗同鑾之行宮舊址.

(七) 工程人員組織

(一) 駐鄭本局辦事者

工程局長	淩鴻勛	
代理總工程司	格來士(法籍)	
工務課課長	洪觀濤	由平漢調用
副工程司	潘保申	
幫工程司	劉澄厚	兼工務課工事股主任
幫工程司	王江陵	兼工務課港務股主任
繪圖主任	陳克雄	

(二) 駐工段辦事者

總段工程司	弗樂利(法籍)	駐靈寶
總段副工程司兼第一分段工程司	李儼	駐靈寶
幫工程司	曾昭桓	"
第二分段工程司	陸廷瑞	駐盤豆鎮
幫工程司	章臣梓	"
第三分段工程司	江博沅	駐潼關
幫工程司	王超鎬	"

黃河泥沙免除之管見

著者：張含英

(一) 緒論

引黃之最大問題,無論其目的爲開溝渠以利灌溉,爲濟他河以便交通,爲分黃流以減水患,厥爲泥沙之處理.是故昔日運河航運未停之時,每年借黃濟運,自張秋鎮至臨清凡二百餘里,每借黃一次,河身淤塞而平.故於次年借黃之時.必重新開挖.靳輔論賈讓策有言曰:「就河善淤,自古記之.通河引灌,雖極卑窪之地,一過而平,再過而高,不出數年,且深谷爲陸.若歲歲開挑,出自國,則爲費不貲;出自民,則民力勢有所不能.…」可見黃河之不能利用乃因泥沙處理問題之不得解决也.

據華北水利委員會在民國七年至十年間,測驗黃河含沙量,其盛漲時,以重量計,在陝州爲百分之十一,在山東十里鋪爲百分之七,在洛口爲百分之三.似此下游含沙量之遞減,或爲下游淤塞日甚之現象(若然,殊爲危險).又據該會十七年七月至次年一月在洛口測驗之結果,八月間可至百分之七.平均計之,似亦在百分之三至五.既欲引黃,則不能避免洪漲時期,而停止工作,則對於最大含沙量,不能視爲管理設計之重要因素.今姑不以最大量作標準,以百分之五計,(以體積計約純沙爲百分之二•五,若以飽和沙之體積計,約爲百分之十)若按所設計之引黃計畫,引渠之流量爲每秒一百八十立方英尺,每日即可攜純沙近四十萬立方英尺.然因引渠之流速小,沙必沉,而引渠因之淤,若用人夫挑之,每日需四千工,每日以五角計,每日需款二千元,若以飽和泥沙計,每日約需款八千元,其費用之巨殊可驚人!若此問題,而不能解决,則引黃之計畫永不能實現也.

影響於流量攜沙之條件甚多:

(1) 河道之深淺　愈淺則攜沙量愈多.

(2) 泥沙之大小　泥沙之直徑愈大,攜帶之量愈少,愈小則愈多.

(3) 泥沙之重量　比重愈大,其攜沙量愈小,比重愈小,則攜沙量愈大.

(4) 流速之大小　流速愈大,則攜沙量愈大,流速小,則攜沙量小.

(5) 流量之大小　流量愈大,則攜沙量愈大,愈小則愈少.

(6) 其他　其他之條件若岸坡之大小,及其他種種關係.

含沙量之因素,既若是之多,而欲用一公式表示含沙量之多寡,勢所不能.是故以下之討論,不得不根據前述華北水利委員會之測驗,而加之推測也.

黃水糞田.典籍屢見,然既不能開黃河之堤使正洪漫流,則引渠內之速率,又不能盡如黃河之大,即能之,支渠之速率,或亦不能若是其大,即或能之,而小渠之速率,必不能若是之大也.是故攜帶之泥不沉澱於幹渠,亦必沉澱於支渠或小渠,而不能分之於田地也.若欲引黃以淤鹹鹵之田,則乃另一問題.非本篇所論.

泥沙處理之方法,就討論所得,約有七種,然其原則不外乎三:(1)濾水法,(2)取河面之水法,(3)沈澱法.玆分別論其利弊:

(二) 柳篦濾水

於虹吸管口周圍,安置三角形式柳編篦壩,兩面近水,一面卽係堤岸.柳壩分五層,每層用木樁鑲柳條編製之篦子.其最外週之篦空較大.約二英尺之方,次空爲一英尺五寸之方,最小者爲四英寸之方.柳篦用柳條扎把,外細以鉄絲,每結扣處.均用鉄條細實.週圍聯於鉄製夾板,每扇寬六英尺,(卽木樁間之距離)高四英尺.鉄製夾板則聯於木樁內之鉄片凹槽.木樁之高約十五英尺,必三扇相連而成一整扇.其上部之兩角,備有鉄環繫以鉄絲繩,聯於起重機.起重機設置於樁頂相連之木板上,以手搖之,可使柳篦任意上下移

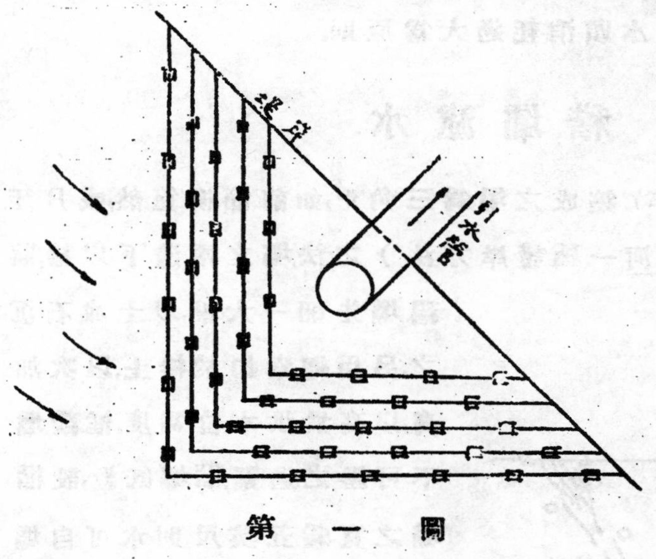

第一圖

動.第五道籠之邊長爲一百英尺.籠與籠之間之距離爲五十英尺.

於此籠修成以後,柳籠完全放下,則水經過第一道柳籠時,必因生阻礙,而流速較減,流速減則有少數沙泥沉澱.俟通過第二道至第三道時,當能減去大部（究竟幾分之幾,雖不敢一定,但能減去一大部可斷言也）.此種泥沙當沉澱於兩道籠間之空地.若干時後,必將該籠底部淤高進水之量必見減小,而水溜則順第一道籠向河心斜流,水管吸收之量,定感不足,若欲免除此種現象,則必於籠基沙淤稍高時,先將第一道籠提高,阻力旣減,則水溜必大.自可利用水溜急刷,向河心斜行,於是第一及第二籠間之淤可順流而下,再將第二籠提起,以次而作.總之,籠之起閉,其作用在於停淤放淤,其效用之大小,全視管理安裝之情形而定.

設第一道籠至第五道籠間之面積,約言之,爲八萬平方英尺（由前假定之大小約計之）.經過柳籠之流量爲每秒二百立方英尺再按含沙量平均百分之十(以飽和體積計),設可濾去百分之八,則每點鐘沉澱於籠間之沙泥,爲五萬七千六百立方英尺,平均計之,籠間每點鐘只淤深約八英寸之沙泥.然或不勻等,深處亦不過約十二英寸.黃河最大時之平均速率爲每秒十一英尺,低水位時,亦有每秒五英尺之速率.若利用每秒五英尺以上之流速,冲刷數英寸之積淤泥沙,數分鐘內,當能盡之,故於工作絕無影響.

此種辦法,雖屬濾水,似亦利用沉澱之理,其費用甚省.且用自然淤刷之法,亦稱便利.惟管理較煩,籠壩之位置必得適當之選擇.第五道籠壩每邊之長短因流量之大小而定,以不至水頭消耗過大爲原則.

(三) 稭壩濾水

稭壩者,卽以稭壩(卽高梁竿)築成之壩爲三角形,如前節柳笆然,或月牙形.其法如築圈堤,按修埽(黃河一種護岸方法)之法,壩之兩邊下以樁,將稭壩先細一大細,以土或石沉之,另用繩索扣於樁上.以次加高,以高於洪水位爲度.惟稭壩不可壓之過緊,稭壩旣鬆,設稭壩之寬爲五英尺,則水可自壩濾過,而流入於引水管之進口處.

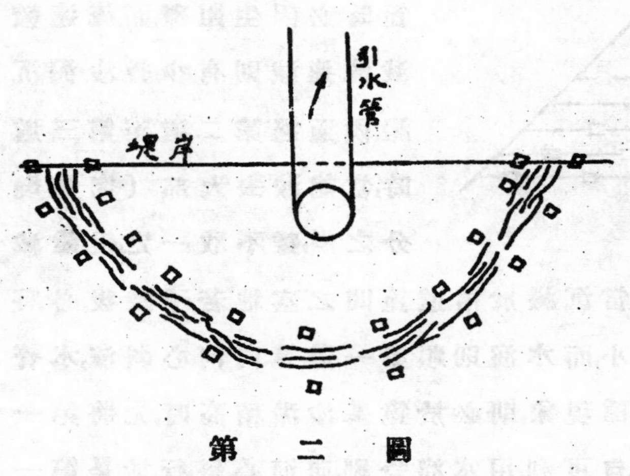

第 二 圖

此法之作用亦極簡易,而濾水亦甚清.故曾有人主張用之者,惟其困難之點亦多:(1)阻力消耗過巨;(2)流量過小;(3)淤沙冲刷頗難.

欲水之濾清,則必經過稭壩之阻力,水之濾愈清,其阻力必愈大,換言之,卽稭壩以內之水面,必較河中水面愈低也.此等損失,對於利用虹引之影響甚巨,蓋引水管進水口之水面愈高,愈有利於虹引作用,而流量亦愈大也.此水面之差不有實驗,雖不敢一定,然其差必甚大,敢斷言也.

稭與稭之間雖可流水,然必較密,然後始可濾清.因其密則流緩.然欲有每秒二百立方英尺之滲濾,則壩之周圍必甚大,若壩長,則費用固大,而管理亦難.

水旣濾淸,則沙淤壩之周,欲免除此沙,一則用挖泥機器,或將進水口安置洪流之處,以便洪流冲刷,則沙不至淤停.若用挖泥機,則費用較大,若置之洪溜則管理頗難.再則若於細沙填滿稭埽之空隙時,恐有絕流之患.

然若用之於引水流量較小之處,而水之需要必甚清時,似可用之.

(四) 管之進水口可上下移動以取河面之水

河面之水其速率必較平均速率爲小,更較最大流量爲小.緩溜處河面之速率,亦必較急溜處之水面速率爲小.然水之含沙量,約與流速之三次方(?)成比例.則在緩溜處河面之含沙量自必爲數甚少.例如河面流速當平均流速百分之八十,以三次方計,則河面之含沙量,約平均五十一,而緩溜處之河面速率,設爲平均河面速率爲百分之八十,則緩溜處河面之含沙量必當全河平均含量,約爲百分之二十五.若平均含沙量以重量計爲百分之五,則緩溜水面之含沙量必約爲百分之一又二五.此等假設,因尚不知黄河有否此項統計,究竟此數之大小,尚待考求.是故若能引取河面之水,亦爲免除泥沙之一法.

管之進水口上下移動之法,亦有二種:(1)水漂法;(2)用起重機升降法.

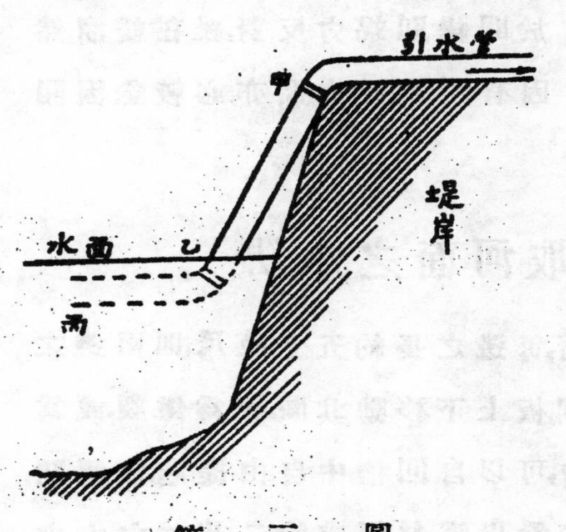

第三圖

如第三圖示設于引水管之甲處及乙處各置有活節,於丙處置有浮漂,如中空之鉄球,浮船等可隨水面之高下而浮沉.進水口永遠在水面下相當之深度.爲防備引水管前後搖動起見,可以用木架裹之,或用鉄練繫之.

若用起重機升降法,如前圖,乙丙間一段可以省去.只用一活節於甲處.爲防備進水口處前後搖動起見,必用木架裹之,然後以起重機繫於進水口,按水面之高以升降之.

活節之構造法,亦有數種,用節扣,用牛皮,用橡皮皆可,要以不通氣,耐久爲要.

此法之管理似較濾水法爲便,且可免除水頭之損失亦無須冲刷等手續.若水管較小,似可應用此法.故另一計劃,水管爲廿四英寸,擬採用水漂之法.惟以每秒引水二百立方英尺,直徑五十八英寸之水管,用之頗有不適.蓋以該管活節以下之長約二十五英尺,卽以直徑五十八英寸,長二十五英尺,所盛之水之重,卽爲十四噸,鐵管尙不計也.以如此重量之管,而欲用起重機使之升降,則活節處,必易受損傷,再則必有極堅固之支架.然在黃河堤岸,實爲不易再則以如此重之管,其前後移動及處理似亦不易.

若進水口用數隻水管,如第四圖示,按水面之高低以啓閉之,則可省去升降之勞,且無活節,管之生命亦必長久,實較有活節者爲佳.然在黃河上,於現在之環境下,頗不適用.蓋以此項辦法,非利用虹吸,乃必開堤,然後水管埋於地下.人民對於開堤則竭力反對.雖在緩溜,然因有引水,卽水溜亦必較急因開此涵洞,堤岸有無影響,亦須詳爲考查.

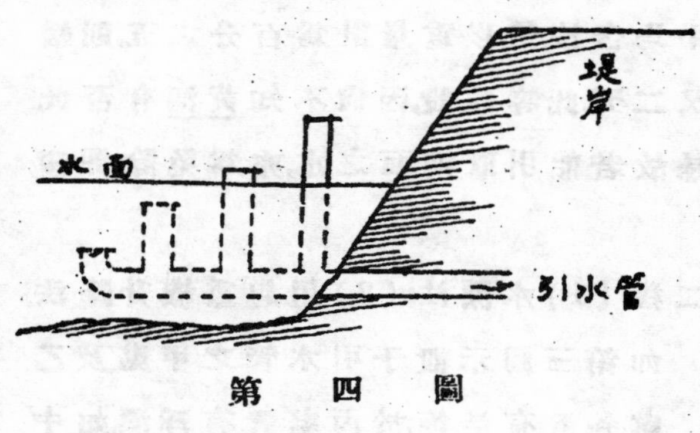

第四圖

(五) 利用板閘以取河面之水法

於進水口之周圍,按三角形之兩邊打樁,每邊之長約五十英尺,則兩邊之長約一百英尺.於每樁上,安有凹槽,以便閘板上下移動其間.板爲鐵製,或爲木製.按水面之漲落,閘板用起重機或人力,可以自凹槽中自由提起,以便閘板較低水面低於相當之深度.閘板之內,安置以鐵絲編成約二英寸之方之篦,以防水面浮物,流入管中.

若按本問題之設置,可用寬四英寸,深一英尺,長九英尺三英寸之木閘板.

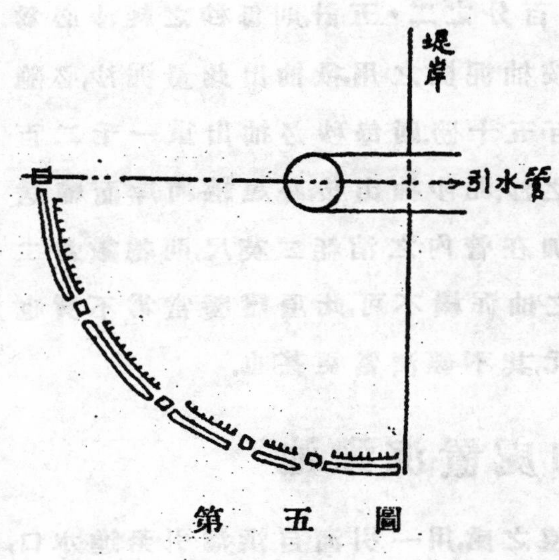

第　五　圖

樁與樁之間相距九英尺半.若用鐵製板,可用深三英尺者五層,即足用.若爲五層三英尺者於水深十三英尺時,可將第一層之鐵板提起,則可有一尺之水頭,深入於閘中.若水深於十一英尺時,可再第二層提出則可有二英尺之水頭,深入於閘中,以此類推.鐵閘板提高之法,用人力或用升降機起落之.若用木閘板,其起落法亦如之.

此法似甚適宜,頗可採用之.惟當設計之時,應極注意各邊之長短,不可過小.例如本問題,吸水量爲每秒二百立方英尺,則漫閘而流入之水,亦必爲此量,而閘內外之水面差,又不可過大,(因閘內之水面愈低,愈不利於虹吸,前曾言之)今設閘壩兩邊之長爲一百英尺,欲其流量爲每秒二百立方英尺,則按凹口公式($Q=3.33 L H^{3/2}$)水頭必約爲八英寸半.換言之,即經過此閘,水耗頭之消耗爲八英寸半也.故閘壩之邊長,不可過小,過小則水頭之降落愈大.

(六) 於出水口處置沉澱池

於進水口處不加特別之工程,令水經過引水管,至出水口處,注入於廣大之沉澱池,因水流緩緩,其沙必沉,然後再將清水引於幹渠之中.

沉澱池之平面及大小設計,自當按問題之不同而定.惟此法實不適用於本問題.於泥沙沉澱於池後,其取出之法可有兩種:

(1)用抽泥機,(2)用人力,玆分論之.

若用抽泥機,則每立方英尺之抽出物體,其中不過百分之二十爲泥沙,其他八十爲水(?).若沙含量以體積計爲百分之二·五計,則每秒之純沙必爲五立方英尺,其重約爲六百二十五磅.按抽泥機之用,欲抽出此量泥沙,必隨二十立方英尺之水,其重約爲一千二百五十磅.則每秒必抽出重一千二百五十磅之物,方能不至令池中有存餘之沙.泥沙抽出後,必超越河岸而輸送於黃河內.黃河之堤約高二十英尺,外加在管內之消耗二英尺,則總數爲二十二英尺.欲作工作非有七十三馬力之抽泥機不可,此項護養當爲不貲也.

若用人力,則曾言之,每日需款八千元,其不經濟當更甚也.

(七) 於進水口處置沉澱池

將引水管進水口安置於距洪溜較遠之處,用一引河自洪溜引至進水口,於進水口處建一較大之沉澱池.水自洪溜入引河,則必逐漸沉澱,至池則沉澱更甚,水自澄清.

欲用此法必用挖泥機,逐日工作.然其最大問題,則爲無適當之進水口位置.黃河河床不定,魯省沿黃多屬平原.洪溜今年靠近左岸,遇有變遷,則將左岸淤澱,而滾於右岸,如此變化,乃屬於常事.且引河更不易挖,地皆流沙,今日成河形者,明日恐湮沒矣.若無適宜之地點,此法亦不適用.

(八) 進水口處之水管豎立用自然沉澱法

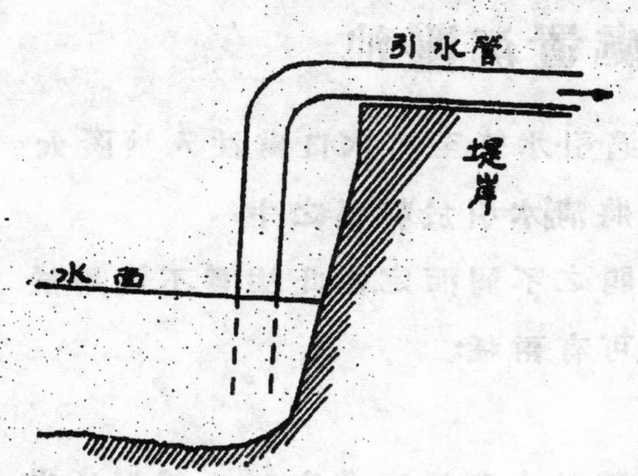

如第六圖示,進水口處引水管之部分,爲豎直者,於引水時,水雖上升,但若速率較小時,沙泥或可因其重量落下.如此則流入於管中之水必較清.

此法實不適用,蓋以欲令沙沉澱,則上升之速率必極小.

(九)　結　論

一切方法決非徒憑空言所敢斷定者,蓋以治河之法,雖皆本諸學理,然必以實驗爲根據.對於黄河既乏過去之張本,又無試驗之機會,其水勢浩大,變遷無恆,居然討論其泥沙處理之法,或爲識者所笑.然於無辦法之中,必思有以治理,於難解決之間,必詳加以討論.且虹吸計劃,一面既可溉田,再則以試驗各法之應用結果,則其應詳加討論,至顯然也.

以愚見所及,對於本問題之應用,似以柳笆濾水及利用閘板,以取水面之水法,較爲適宜.然前者尤爲適宜故擬先試用之.

海外道路摭談

美國公路之闊大　世界各國汽車之產額及消量,首推美國,故其道路亦特別寬大.林肯公路闊及百五十尺,兩旁除植樹木並佈置花園外,其沿途又有停車號誌,及休息亭等.米憩根省亦有一公路闊一百二十尺.內分急行車道與慢車道.紐約市復有橫過大河之水底公路管,闊二十餘尺,二旁有人行道,中可通車,實爲世界唯一之水底公路管也.

意國公路之專道　意大利爲道路建築之首創者,羅馬比亞道建築於三千年前,至今尚有遺跡,路面用大石塊鋪造,故當時有道路所及,即羅馬勢所及之說也.及至現代之意大利,仍以世界上最先進最完美之公路自誇,於普通公路外,又有所謂汽車專道.其運輸法以路線橫貫數省,全用混合土及土瀝青建築,兩旁有物圍護,並自備車輛,依時行駛,在相當地點,設立車站,有電話及旗號等之設置,迨與鐵道無異,亦可爲別開生面之公路也.

津浦鐵路藍鋼車之氣壓給水統系

The Pueumatic Plumbing System of Blue Express. T. P. R.

著者：李金沂 周勱

我國各鐵路之客車,多無冷熱水之設備,三等車無論矣,卽頭二等亦不過洗盥間內備有冷水管而已,甚爲不便.惟津浦鐵路藍鋼車之頭等臥車餐車及客廳車等,皆有極完備之給水統系,每房間均備有面盆及冷熱水管,總水箱則裝於車架下側,上水旣易,修理亦便.水則利用空氣輪掣系(Air Brake System) 壓縮空氣 (Compressed Air) 之壓力,輸送至車上.冬日復藉暖汽之熱力,將一部分水加溫,設備甚簡,而功用頗大.各路客車不妨仿造裝置,所費無幾,而旅客當益可舒適便利也.茲將其構造略述如下:

(一) 水箱: 水箱爲圓柱形,長 8'.0",外徑 22." 係用 1/8" 鐵鈑鉚成,兩端鐵鈑則用電銲銲住.全箱用U形鐵條(1/4"×2")及螺絲縱吊於車架 (Underframe) 下旁.箱外另加鐵皮套,以保護之.套之內面加厚毡一層,以防冬日受凍.

(二) 副壓縮空氣箱 (Auxiliary Reservoir): 車輛之裝有空氣輪掣設備者,皆有比箱.惟另須加一鐵管通至水箱.

(三) 備用空氣箱 (Air Pressuere Tank). 平時車輛與車頭接連時,車頭壓氣機 (Air Compressor)輸出之壓縮空氣,可灌滿副箱及此備用箱.給水統系則利用副箱之壓縮空氣.當車輛與車頭分離後,副箱之壓縮空氣已用完時,則備用箱所存之壓縮空氣,亦可注入水箱,供給水壓.

(四) 壓縮空氣統系: 此統系中之主要機關,爲一六道閥, (Six way Cock) 如圖中所示.第一圖爲閥塞,第二圖爲閥體.壓縮空氣由副箱(Auxiliary Reservoir) 來,經過空氣閥 (Air Valve), (參閱第四圖) 濾氣器. (Air Strainer 此器內有細絲網一層,能將灰塵濾清), 及調節器. (Governor) 再分爲兩股,一股注入備

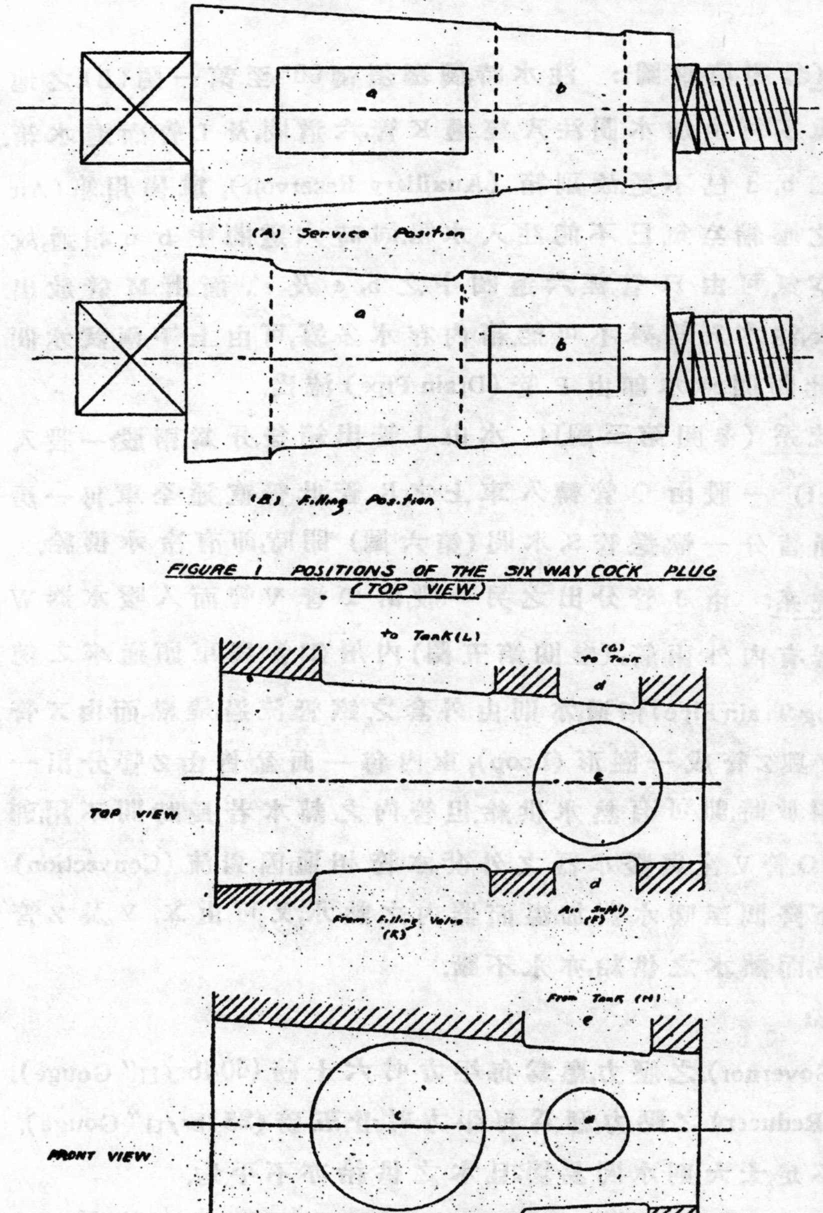

FIGURE 1　POSITIONS OF THE SIX WAY COCK PLUG (TOP VIEW)

FIGURE 2　SIX WAY COCK BODY

用箱, (Air Pressure Tank) 一股經過一減壓器 (Reducer); 及一逆流攔阻閥 (Check Valve), 而至六道閥.當閥塞在第一圖 (A) 之地位時,閥塞之a孔與閥體(第二圖)中之c道不通,故水箱已與加水閥 (Filling Valve) 隔絕.但閥塞之b孔則與閥體之d孔通,故壓縮空氣經過逆流攔阻器後,可由F管入六道閥.經d b孔,而出六道閥,再經G管及H管直進水箱.水面受此壓力後,水即可由J管輸出.

（五）注水統系（參閱第四圖）：注水時,閥塞須轉90°至第一圖（B）之地位.是時a與c已通水可由加水閥注入,經過K管,六道閥,及L管而進水箱.但同時六道閥中之b, d已不通,故副箱（Auxiliary Reservoir),或備用箱（Air Pressure Tank）中之壓縮空氣已不能注入水箱.同時六道閥中b e相通,故箱內餘剩之壓縮空氣可由H管經六道閥中之b, e及e',而由M管放出.否則箱內壓力甚大,注水入箱,殊不可能.箱內存水多寡,可由上下兩試水閥（Gauge Cock）試出.此閥開時,水卽由P管（Drain Pipe）流出.

（六）冷水供給統系（參閱第三圖）：水由J管出箱後,分爲兩股:一股入暖水器（Steam Jacket）一股由Q管轉入車上之R管.此管直通全車每一房間,廁所面盆,及便桶皆分一輸送管S,水閥（第六圖）開時,卽有冷水供給.

（七）熱水供給統系：由J管分出之另一股,由T管V管而入暖水器W（Steam Jacket）此器有內外兩管,（參閱第五圖）內層銅管與車頭通來之總暖汽管（Steam Heating Train Pipe）相通,水則由外套之鐵管流過,變熱,而由X管轉入車內之Y管.Y與Z管成一圈形（Loop),車內每一面盆皆由Z管分出一輸送管A,熱水閥開放時,卽可有熱水供給.但管內之熱水若長時間不用,卽易冷却,故Z管仍由O管V管與暖水器之外套水管相通,因對流（Convection）作用,水冷後,卽可下降回至暖水器加溫.而器內之熱水,又可由X, Y及Z管上升備用,循環不易,而熱水之供給,亦永不斷.

（八）應注意之點:

（1）調節器（Governor）之壓力,應爲每平方吋六十磅（60 lbs/□" Gouge).

（2）減壓器（Reducer）之壓力,應爲每平方吋廿五磅（25 lbs/□" Gouge).過小則水之壓力不足,太大則水閥易壞,且水之供給亦不平均.

（3）暖水器（第五圖）之兩端,皆備有石綿塞（Asbetos Packing),以防漏水

（4）水箱須常洗滌,以免淤塞.箱之後端有視察孔（Inspection Hole),及水閥各一.洗滌時,須先將存水放淨,然後開啓視察孔,用水管伸進冲洗.

（5）冬日停車後,水箱及水管內之存水,必須放淨,以免結冰,致管箱破裂.

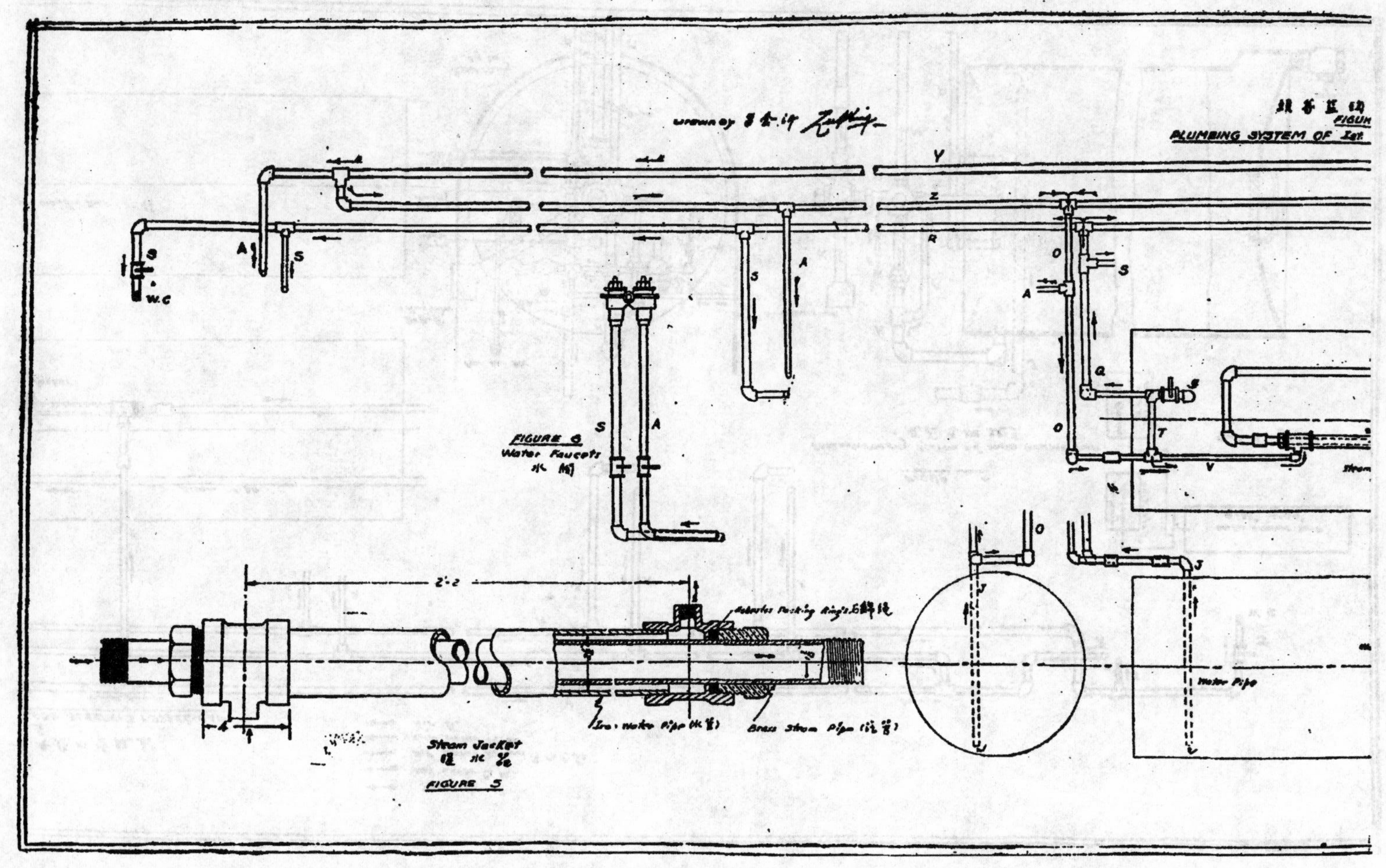
PLUMBING SYSTEM OF
FIGURE 6
Water Faucets
FIGURE 5
Steam Jacket
Water Pipe

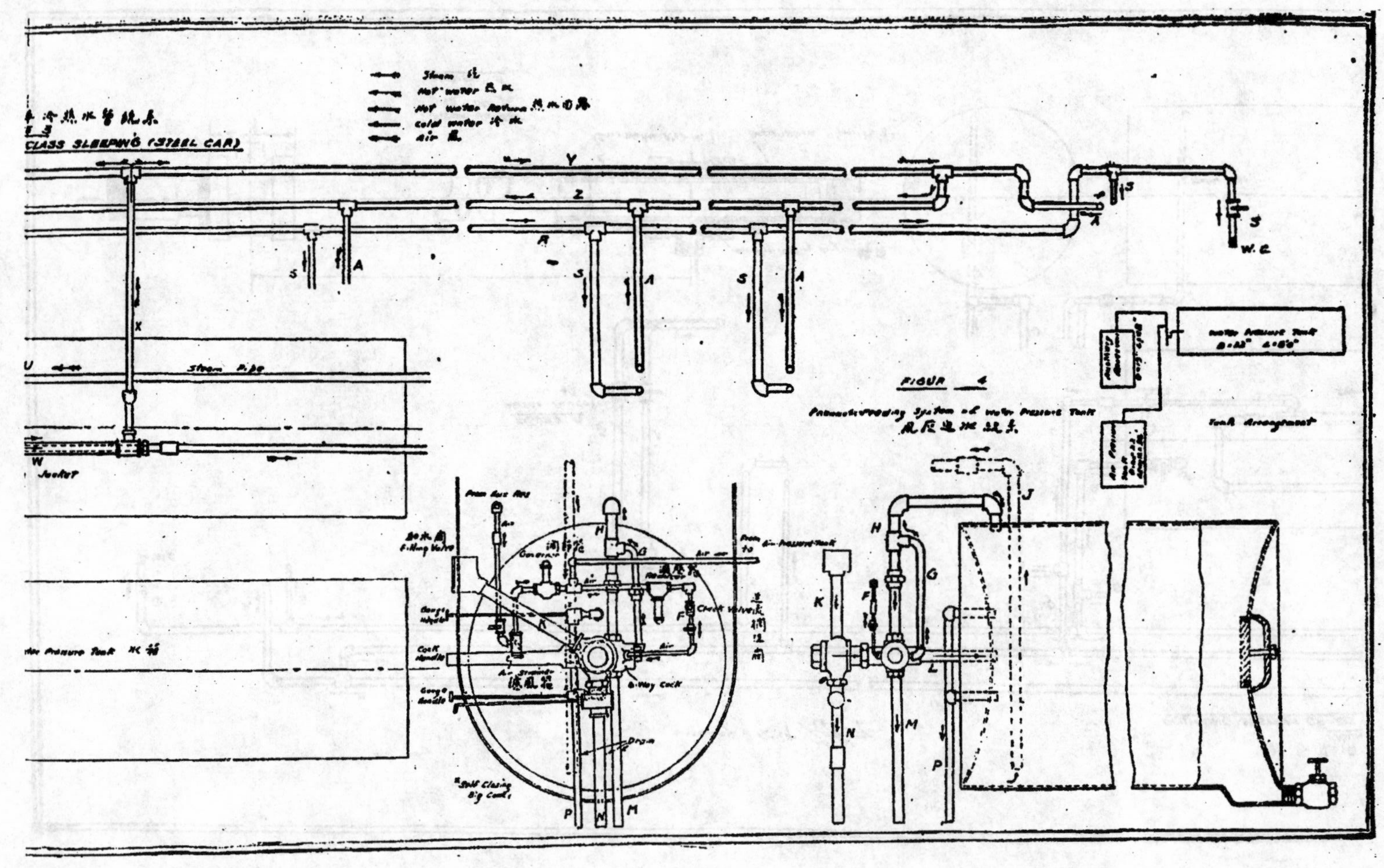
CLASS SLEEPING (STEEL CAR)
Steam
Hot water
Hot water Return
cold water
Air
Steam Pipe
Jacket
W.C.
FIGUR 4
Tank Arrangement
Filling Valve
Check Valve
3-Way Cock
Drain
Self Closing Big Cock

A Mew Method for the Precise Measurement of a Capacity in Series with a High Rasistance

BY K. F. SUN (孫國封)

Let an alternating electromotive force E be impressed upon two parallel circuits, as shown in Fig. 1.

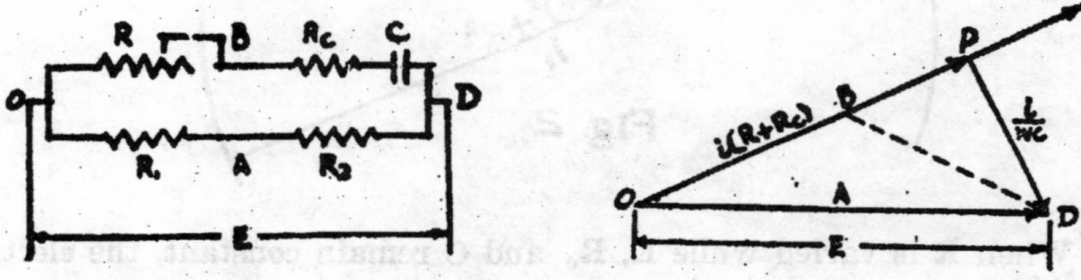

Fig. 1

Let one of these circuits consists of a variable non-inductive resistance R in series with another non-inductive resistance Rc and a capacity C; let the other consists of two equal similar resistances R_1 and R_2 in series. The potential drop in the first circuit OBD, due to the resistances R and Rc, may be represented by a vector OP having a value $(R+R_c)$ i, where i is the current flowing in the circuit OBD. The drop, $i/\omega C$ due to the capacity C is in quadrature with OP and is represented by the vector PD. Here ω is $2n$ times the frequency. The total drop is, accordingly, represented by the vector OD, which is equal to the impressed electromotive force E. As the current flowing in the resistance circuit OAD is in phase with the impressed electromotive force E, the potential drop through the resistances R_1 and R_2 must coincide with and be equal to the electromotive force E. Since $R_1=R_2$, the point A must be at the middle point of OD, as is shown in Fig. 1.

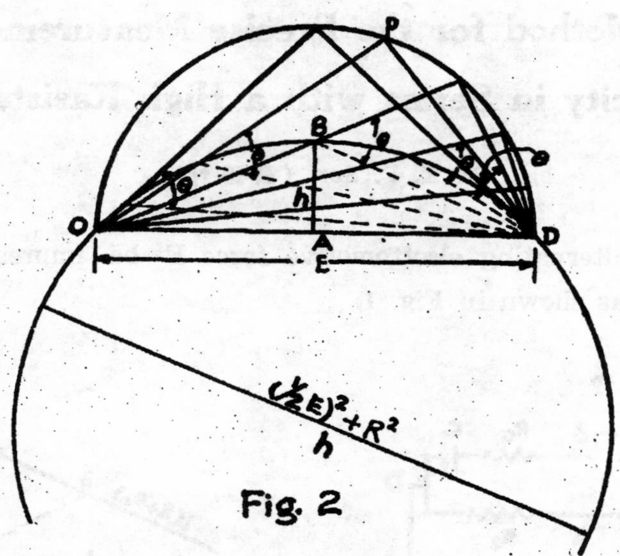

Fig. 2

When R is varied while E, R_0 and C remain constant, the electromotive force diagram may be represented[1] as in Fig. 2. The phase angle Θ in the impedance circuit BD is constant as long as the values of R , C and ω are constant. Hence, Θ is not affected by the variation of R, and the angles supplementary to Θ are all equal. The locus of the point B will, accordingly, lie on an arc of the circle passing through O and D. The diameter of this circle is $(\frac{1}{4}E^2+h^2)/h$ where h is the length of the perpendicular drawn fiom the middle point A of the line OD to the middle point B of the arc OBD. By the geometry of the figure, the perpendicular h is the shortest distance from the point A to the arc OBD.

If there were no capacity in the circuit BD (see Fig. 1), a point B could be found so that no current would flow through a detector connected

[1]Bedell and Crehore, Alternating Currents, p. 275;

LaCoour, Theory and Calculation of Electric Cureits, p. 49; and other texts.

between this point and the middle point A of the other parallel circuit. With capacity, however the electormotive force at the terminals of the detector can only be reduced to a minimum value h in quadrature with the electromotive force E, the magnitude of h depending upon the value of E and the phase angle Θ If, then, we had an idiostatic voltmeter that would read to very small fractions of a volt with high accuracy, we should have only to adjust for a minimum and read the value of h. E and R being known, the values of Θ, Re and C could then be obtained. As such an instrument does not seem to be available, the use of a balancing or zero method with a suitable detector seems most desirable.

In precise direct current measurements, zero methods have been used almost to the exclusion of others. The necessity of such methods is, however, even greater in alternating current measurements: first, on account of the limitations of alternating current indicating instruments; and second, on account of the practical imposibility of obtaining alternating currents as steady as the direct currents furnished by the storage battery. An electrodynamometer lacks sensitiveness as the current approaches zero. By separately exciting one of its coils, the sensitiveness may be greatly increased but means must then be provided for bringing the excitation in phase with the current to be measured. Used as a zero detector, a telephone is very sensitive under proper conditions, and has the advantage of simplicity and cheapness; but the capacity effects of the various parts of the circuit reduce the accuracy at low frequencies on accunt of the harmonics present. The vibration galvanometer is also very sensitive as a zero detector, when properly turned to the fundamental or other harmonic to be balanced; but it is not satisfactory as regards ease of manipulation, and with a change of frequency there is a great decease in sensibility.

Direct current galvanometer, with synchronous commutator, as detector. A direct current galvanometer, in combination with a reversing commutator driven by the generator or by a synchronous motor, makes possible the same accuracy in alternating as in direct current measurement. In the present investigation, a 1000-ohm galvanometer with a sensitiveness of 1.7×10^{-6} volt millimeter of the scale at a distance of one meter, was used. The rectifying commutator used was fully described in the paper by F. Bedell.[2] It hat four segments, alternately connected, and is directly driven by a four-pole synchronous motor. The alternating current to be measured is connected to the commutator segments through slip rings. Direct current is taken by two brushes bearing upon the cummutator and accurately set so as to be separated from each other by a distance of one commutator segment. These brushes are so mounted that, without changing the distance between them, they can be shifted by a tangent screw and their position read upon a graduated scale. By such a device the brushes can be adjusted to any position with respect to the poles of the motor so that the reversal occurs at any desired phase of the current. By shifting the brushes to a proper position, the glavanometer may be made to respond to any given component of the current while it is insensitive to the component in quadrature with it. It will be noted that each degree of mechical shifting of the brushes corresponds to two degrees of electrical phase shifting when, as in this case, the driving motor has four poles.

It was suggested by Professor F. Bedell[2] that the resistance of brushes and contacts should be as low and uniform as possible. The fact was pointed out by C. H. Sharp and W. W. Crawford that the apparent resistance of the sliding brush contact under working conditions tends to become very high as the current falls to a low value. To reduce this resistance, the brushes were made of seven strips of soft copper soldered, at the end remote from the surface, which is a bevelled edge.

The commutator was conected to the motor by a fibre shaft, 20 cm long, which gave high insulation. The direct current for the field of the motor was brought by a circuit separated as far as possible from the testing circuits. The bearings were carefully fitted to eliminate axial motion and a small fly-wheel was used to eliminate hunting.

Details of method. The method of introducing a variable alternating current electromotive force between the points A and B in quadrature with the impressed electromotive force E, and balancing the voltage h by means of a zero detector connected between A and B was tried with various arrangements of the apparatus.

[2]Devised by Ayrton and Perry, Phil. Mag., 12, 297, 1906; see also; L. T. Robinson, Trans. A. I. E. E. 28, 1024, June, 1909; C. H. Sharp and W. W. Crawford, A. I. E. E. 29, 1517, July, 1910; F. Bedell, Jour. Franklin Inst., p. 385, Oct. 1913.

The arrangement finally used is shown in Fig. 3, and more completely in Fig. 4. As in Fig. 1, the variable resistance R is in series with resistance Rc and capacity C; in parallel therewith are two equal resistances R_1, R_2 in series. The electromotive force E applied to these two circuits in parallel, is brought through a reversing switch SR (see Fig 4) from adjustable taps on an anto-transformer 3′, 4′, which is supplied with current from taps 3, 4, on an another auto-transformer 1, 2, 3, 4, 5, having four equal coils. Taps 2, 4, are connected to the supply, in this case the A. C. end of an inverted rotary convector the speed of which, and hence the frequency of the alternating current as shown by a frequency indicator, is controlled by field adjust ment.

In order to measure h, a known alternating electromotive force was introduced between A and B which was adjsted until a zero reading was obtained on the zero detector, consisting in this case of the direct current galvanometer and commutator already described. Various arrangements were tried, whereby the known electromotive thus introduced should be in quadrature with the terminal electromotive force E. This quadrature electromotive force was most satisfactorily obtained from the secondary of a Brookes inductometer M, conected as shown, having a range of 1 to 10 milli-henries, the primary and secondary each having resistance of 4.5 ohms. In taking readings, the brushes of the synchronoushes communtator are properly set and M is adjusted until the galvanometer reads zero. The

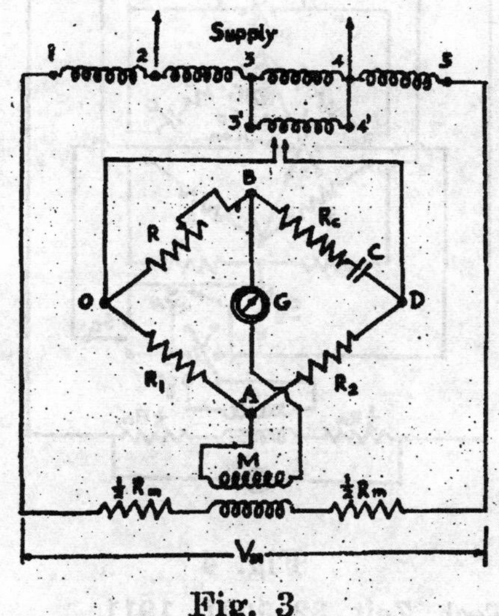

Fig. 3

secondary inductometer voltage (and hence *h*) when the galvanometer reads zero, is $M\omega I_1$. The value of M is read from the inductometer, the value of I_1 is calculated from primary resistance and voltmeter reading.

A high non-inductance resistance was placed in series with the primary so that the primary current would be in phase with E. Hence, the secondary electromotive force, in quadrature with the primary current was in quadrature with E, as desired. It was found that accurate readings were obtained only when the prinary resistance was divided into two equal parts symmetrically placed as shown.

Elimination of error. The complete connections are shown in Fig. 4. With the galvanometer connected to A (instead of B) and inductometer to B (instead of A) a slight difference in readings was found. This was made small by reducing as much as possible leakage and induction between different parts of the system. Errors due to thermal and contact electromotive forces in the commutator were eliminated by averaging the readings taken in the positions. A Wagner[3] earth connection, through equal resistances r_1, r_2, as shown, was introduced to eliminate a small zero deflection of the galvanometer.

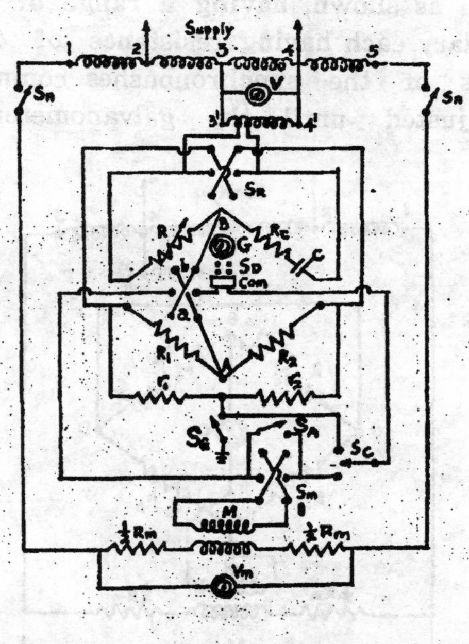

Fig. 4

[3]K. W. Wagner, Elect. Tech. Zeit. 32, 1001, 1911.

It is important that R_1' and R_2 be precisely alike. Curtis resistances were used for this purpose and were interchanged by the switch S_R.

Operation. As a preliminary, r_1' and r_2 were balanced against R_1' and R_2 by adjusting r_1' and r_2 until the galvanometer read zero. The commutator brushes were set so that the galvanometer was most sensitive for change of resistance. Switches S_G, S_A, S_M, are open; ab is closed at a; S_c is up. The main measurement is then taken. Switches S_G, S_M are closed; S_c is closed down. A resistance balance, R, R_c against R_1', R_2 is first obtained by adjusting R. The commutator brushes are in the position of maximum sensitiveness for resistance adjustment; no change of M then affects the galvanometer. The brushes are then shifted ninety electrical degrees, giving maximum sensitiveness for M and zero sensitiveness for zero galvanometer deflection was thus determined. By using the various reversing switches, eight readings for M and four for R were taken and averaged. Two or three minutes were sufficient for a set of readings. By this method, made possible by the synchronous commutator, the adjustments of M and R are separately made and are entirely independent of each other. With most zero detectors there is not this impedance, and a tedious adjustment of M and R simultaneously is necessary in order to bring the galvanometer to zero.

A careful study of the conditions for sensitiveness shows that for maximum sensitiveness OB=BD[4], the resistance in the adjusting arm being equal to the impedance. Of the arm to be measmed the sensitiveness increases with decrease in frequency and increase in electromotive force so long as Θ is less than 45°. The method, therefore, is well adapted for use at commercial frequencies, whereas a higher frequency is commonly required by other methods.

The accuracy of the method was tested under the conditions which

4. Kuo-Feng Sun, A Theoretical Study for the Sensitive Conditions of this Method.

apply when the capacity in series with a high resistance is being measured.[5] This was done by putting a known capacity of 1 microfarad in series with various knwon resistances and then measuring the values. The error in the measured value of C was found to be less than 1 per cent for values of Rc up to and including 20,000 ohms. For Rc=40,000 ohms, a 3 per cent error in the value of C was observed. The error in the measured value of Rc was found to be less than 1 per cent for values of Rc from 1,000 up to 40,000 ohms.

SUMMARY.

1. The method might be used as a direct instrument for measuring the phase angle of any electrical vector.

2. It is applicable for simultaneous measurement of capacity and resistance or of inductance and resistance.

3. It gives the precise results for the measurement of capacity in series with a high resistance up to 100,000 ohms at the low frequency of 60 cycles.

4. The most sensitive condition in using this method is the equalling the resistance of the adjusting arm to the impedance of the arm to be measured.

5. In the case of the measurement of capacity or inductance, if a large permissible impressed E. M. F. has to be applied, the sensitiveness will be accordingly be increased.

6. By omitting the grounding circuit, this method can be used as a simple and precise instrument for the phase angle and ratio tests of any auto-transformer which takes the place of the adjusting and measuring arms.

7. The precise measurement of capacity and resistance of electrolytic cells, has satisfactory determined by this method.[6]

[5] Kuo-Feng Sun, The Experimental Verification of this Method.
[6] Kuo-Feng Sun Physical Review, Vol. 23, No. 5.

廣州市馬路改艮經過及保養情形

著者：余季智

I 馬路之類別

廣州市爲南中國最大通商口岸,闢路於民國初元.其時市政尙未發展,馬路係屬草創,加以軍事影響,經費束縛,故以花砂路爲多.隨以花砂路面易損壞,且難保養,是以逐漸改善,其新闢者亦甚少花砂路面,現計全市馬路以材料方面言之,約分爲五種:(一)鋪腊青路面,(二)塗掃臘青路面,(三)花砂路面,(四)三合土路面,(五)坭路面.市內馬路俱是第(一)至第(四)種,郊外馬路則有第(二)第(三)及第(五)三種,而以第五種爲最多.

II 路面之建築

廣州市建築各種馬路所用材料,除臘青外,均本地所出或國貨.

臘青 所用臘靑,有亞細亞,美孚及德士古三行之貨.普通鋪臘青路面多用Penetration 40—50,而塗掃臘青則用Penetration 60—70.且規定數則,如與相符,方爲合格.例如:

(A) Specific Gravity (25°/25°C) 或 (77°/77°F) 1,050 to 1,070;

(B) Flash Point 不得少過 175°C 或 347°F;

(C) Penetration 60 -70 (因氣候及用途隨時更改);

(D) Ductility 不得少過 40;

(E) Loss at 163°C (325°F) 五個鐘頭內不得過 30 %;

(F) Total Bitumen Soluble in Carbon Disulphide 不得少更 9.45 %.

士敏土 廣東士敏土廠及啟新泰山等洋灰公司出品爲多,須經工務局試驗及格,給有執照,方能採用.如必要時,曾經工務局試驗優良之外國士敏土,亦得採用.

黄砂　共分三種:(一)粗砂,(二)中砂,(三)幼砂;均要尖利起峯,無雜質混合.

黑石　黑石以英德所出之灰石,不染坭質爲限.

白石　普通所用之白石碎,由一分至一寸半,以堅硬之蔴石不染坭質者爲限.

人工　工人工金視其所造工作而異:

工　類	每日工金
坭　工	$ 0.40—$ 0.50
花砂路工	„ 0.65
三合土工	„ 0.75
臘青工	„ 0.85—„ 1.50
執路面工	„ 0.75

造價　廣州市馬路,俱開標由商承建.普通以每百平方尺計算,工料費若干,及加工承商利益百份之一十;或以全段工程總價若干.另加承商利益百份之一十.

路　面	厚　度	每百平方尺工料價
鋪臘青	1 in.	$ 12.21
鋪臘青	2 in.	„ 16.03
鋪臘青	2 in.	„ 21.52
三合土	5 in. (1:3:5)	„ 42.00
塗搽臘青		„ 3.75—$ 4.15
花　砂	6 in.	„ 17.25
坭　路		„ 5.33

右表所列各價,俱以廣東毫銀爲本位.

尺度 廣州市工務局各種工程,均以英尺爲本位;量度材料,亦以英尺計算.

路面斜坡 各種路面,其斜坡各有不同.如路面平滑者,則其斜坡較少;反之,則其斜坡較大.蓋平滑路面,流水自易,粗糙路面,雨水流動不靈,故其斜度必須增加.

路面	路面斜坡每斜低(由中線起計)
鋪臘青	由1/8″至1/4″
塗掃臘青	由1/4″至1/2″
三合土	由1/4″至3/8″
花砂	由1/2″至3/4″
坭路	由1/2″至1″

(一)鋪臘青路面 鋪臘青路面工程,其造價雖昂,倘做法妥善,完成路面,雖行車至五六年之久,仍不變動,夏季炎日晒之,亦不溶化.普通其厚度,由一寸至三寸:一寸厚者,則用臘青砂,而一寸以上者,則分兩層.玆將兩層構造法,略述如下:

須先將路底用坭鋪平,汽轆轆至十分平實,及無低下之處,方可用黑英石角約六寸方四寸厚鋪墊,路基四寸厚,再用二寸大英石碎鋪回二寸厚,然後用轆轆至平實,以便鋪士敏三合土路面.

路基上三合土造法 鋪路基上之三合土,厚五寸,兩邊四呎,內厚六寸,其水平後道路中部高三寸,以備道路中部鋪臘青之用.並須分段鋪造,每段長五十尺,每段內之三合土,必須一次鋪妥,否則,須鋤起從新鋪造.又每段所落之三合土,必須跳格,並須隔日後,始可接續落鄰段之三合土.士敏三合土份量,用一份士敏土,三份粗淨黃砂,五份六分至一寸大白石碎.石料須用水洗淨,及用篩篩妥,方可採用.其混合法先將士敏土及黃砂和勻,然後落石碎,再撈透徹,方得用花洒灌洒淨水,繼續撈勻透開,至粘質充足爲度,切不可落水量過多.隨即將三合土鋪蓋路面,依照規定厚度一次落足,隨落隨用鏟背打實然後用灰匙過於滑面上,至將凝結時,須常用水洒濕,或用蔴包漏水蓋面,護三日後,始可脫去.

鋪臘青路面工作及次序　鋪臘青路面,共分兩層:底層粗石,臘青厚二寸半,其份量每用半寸白石一立方呎,六分白石碎一立方尺,須用粗砂一立方呎,幼砂一立方呎,及臘青一十六磅混合.面層臘青砂,砂厚半寸,其份量每用幼砂三立方尺,須用白石粉或士敏土八份,三立方尺,及臘青二十五磅混合.

炙臘青混合物　臘青混合物及臘青砂,須炙至華氏表三百五十度,至四百度熱,始可鋪上路面.既鋪造并壓實後,仍不得少過華氏表二百七十五度.惟所有臘青,如遇炙至高過華氏表四百度熱時,無論曾否與他物料混合,切不可用.因熱度太高,臘青則減少其彈性 (Ductility) 成爲硬脆而無膠力也.落粗石臘青時,須用鐵鏟先將鑊底之臘青混合士抽起取用,并要將鏟面之臘青混合士,反鋪在路基上.

鋪築臘青路面　如天氣低過華氏表五十度,或下雨水時,或路基潮濕不潔,均不宜鋪造.未鋪臘青之先,必須將路基掃洗乾淨後,用炙熱之淨臘青油塗在路基上,及各渠邊石上,進人井蓋及其他自來水蓋旁,然後將炙至適熱度之粗石臘青混合物,用手車運至.鋪臘青之三合土,上用二寸半高,照路面斜坡之木枋作厚度標準,然後將臘青混合物鋪好,用木板將面刷平,即用熱鐵轆乘熱壓實,壓實後厚度二寸半.乘熱再鋪面層臘青砂,用三寸高,照路面斜坡之木枋作厚度標準,然後將臘青砂鋪上,用木板將面刷平.即用熱鐵轆乘熱轆實,轆實後厚度半寸,再用淨臘青塗過路面一次,上鋪以幼砂,用輕汽轆(六噸重或八噸重機) 先由兩邊起直轆,漸向路中.每次須循向路中最少半個轆位,直轆後,在路面上交角轆兩次,至不見轆痕止.轆時須遲緩,其凹入部份,須臘青混合物填妥再轆.每一千八百平方尺路面,須轆一小時之久.用汽轆時,將適合之水或油,塗在轆上,以免損壞臘青.如近渠邊石電桿,不能用汽轆之處,須用壓實器壓實.

(二) 塗掃瀝青路面　此種路面,多由(1)花砂路面及(2)舊三合土路面改掃而成.蓋花砂路面之壽命,長不過六閱月,短則三閱月,遂開始破壞.保養此種路面,頗覺煩難,且不經濟.至於三合土路面,其兩段接口處,每易破裂.久之則成小坑,車行不便.故此二種路面,改塗掃以瀝青,雖三四年之久,不用大修,即使再行塗掃,其價亦較翻造花砂及三合土爲平也.

塗掃瀝青方法及次序　先將沿路路基,如有低陷地方,須用瀝青石碎混合,或用瀝青混合修補,用汽轆轆實,使其一律適合平水.如天氣低過華氏表五十度,或下雨,或路基不潔,及潮濕時,均不宜興工.所有路面浮坭,先用竹掃把掃至乾淨,然後用椰衣掃把擦掃路面,至不能擦出坭塵爲止.再用蔴包打淨路面上一切坭塵,及必須乾潔全無水氣,將煮至華氏表三百五十度至四百度熱,瀝青貯在有嘴罐內,斟溉路面,每百平方尺,須瀝青四十磅,每次斟溉,先從路中心,順向路邊而下.即用麥桿掃,將瀝青掃勻後,鋪以粗砂(中砂均可)一片,用木扒將粗砂掃勻,再鋪幼砂蓋面.然後用輕汽轆轆過,使至平妥爲止.

(三) 花砂路面　花砂路面工程所用石料,較爲易得.然石質不可不研究.如石質堅硬,則路面壽命較長,否則如重車行過,輾成石粉,日久則凹凸不平,常須修理.但做法單簡,價又甚廉,故以此種路面爲數最多.

作工方法及次序　先將路底平至適合平水,如要填時,須用淨坭,(即無拉扱混合者)或瓦礫磚碎填足,用汽轆轆至平實,始可鋪路.底石用黑石角四寸方三寸厚,再用二寸黑石碎一寸厚鋪上,用汽轆轆至平實,然後用一寸至一寸半白石碎鋪面三寸厚,并用四分白石碎塞入石罅內.用汽轆從傍邊漸次轉向路中,轆至石碎不能移動爲止.然後加鋪以幼砂及石粉或坭粉半寸厚,用竹掃擦勻,洒水再轆至路面堅實,及適合水平爲止.

(四) 三合土路面　各種路面造價,以此種爲最昂,其壽命雖至三四年而不變.惟於兩段接口之處,最易損壞.因此全路受其影響,新築馬路,多不採擇此種路面.

作工方法及次序　路底已妥,鋪上黑石角六寸方四寸厚再鋪上黑石碎一寸厚,用汽轆轆至平實,然後鋪路面三合土,其成份爲一份士敏土,三份粗淨黃砂,五份半寸黑石碎(或白石碎)五寸厚,并須分段鋪造,每段長度五十尺,每段內之三合土,必須一次鋪妥.又每段所落之三合土須跳格,并須隔日後.始准接續落鄰段之三合土.三合土之混合法,先將士敏土及黃砂和勻,然後落石碎,再乾撈透澈,方得用花洒罐洒淨水繼續撈透,至粘質充足爲限,切不可落水量過多,隨卽將三合土鋪蓋路面,照規定厚度一次落足,隨落隨用鏟背打寔.然後用灰匙熨滑面上及將凝結時,須用水淋濕面上,每日兩次如是者五六日.或用蔴包濕水蓋護三日後,如可除去,蓋使面上勿乾太速,須與下層三合土同時乾結,則不易爆裂.鋪妥路面,於四星期後,方可通車.

(五)坭路面　建築坭路,最爲簡單,其壽命則視坭質而定.如山坭含有小石,約百份之三十至四十者,及坭含有粗幼砂混合者爲最佳,數年不變;田坭次之.此種坭路,太陽晒之而爆裂,經車輪輾過而成坭塵,大雨後路面泥濘不堪,每經大雨後,須修理一次,保養誠不易也.

作工方法　路基所填之坭,均由附近採用,或由別處路基之餘坭填補.路邊斜坡,一律高一尺,平開一尺半.倘因地位狹窄,以致斜度尺寸不足,則須用山石結砌斜坡,以免爲水冲崩.填坭路基高度,應按照規定水平外,另多加高百份之一十至二十五,以備填妥後,坭身收縮.路基所掘出之坭土,均用作填路基之用.路基之外,掘坑一條,以便接路面斜坡,及兩傍斜坡之水.如路之一邊近山,則基斜坡之上,另掘水坑一條,以免山水流入斜坡內.兩傍所掘之斜坡,均須每高一尺,平開一尺.路面造妥後,用石轆或用六噸汽轆,轆至平寔,路面斜坡,須使於下雨時路面雨水,卽可向兩傍坑渠流去,始爲妥當.

III 養路之情形

（一）養路股工作　廣州市自開闢馬路以來,每年續闢新路,漸次增多,尤以近數年進步最速.而養路工作,亦成爲正比例.近年各項工作,亦因此而增多.廣州市已成馬路,其路線之長度,略述如下:

塗掃臘青路面,長九萬四千二百八十四尺;鋪臘青路面,長三萬五千四百七十六尺;花砂路面長六萬二千六百四十一尺;三合土路面,長五千二百六十八尺;坭路面長共二十英里.而坭路多在郊外,路途遙遠,每次修理,殊非易事.每月養路費規定三萬元,現欲速將市內花砂路面改良起見,每月路費內,除修理花砂等路費萬餘元外,餘款作塗掃臘青之用.

（二）工人組織　養路工人,分隊作工,每隊工人二十名,派監工或工目管理之.監工或工目之職責,爲監督工人作工,及將每日該隊工作成績,報告一切.工隊則分爲數種:如花砂隊,則修理花砂面;臘青隊,則修理臘青路面,渠務工隊;則修理各馬路渠道.各種工隊,均各司其專責.

（三）作工時間及工金　每日工作八小時:由早八時起,至下午五時收工.工人工金,每日由四毫起至八毫半.視其所造工作而異,每星期發給工資一次.

（四）器具　修理各種路面所用器具不同.所有應用器具,均須購儲於工廠,以備各隊領用.每隊泥車一架,運料汽車三輛,爲運料供給各隊之用.另碎石機一架,汽轆四架.

（五）工作及報告　每晨各監工等按照每日預定工程表,領隊前往施工地點作工;完工後,須將做妥面積,已用材料若干,及工數,須詳述於每日修路報告中.工作大都係修理各種損壞路面,及翻造花砂路面,及建築或改修坭路面.如改鋪或塗掃全路臘青工程浩大者,則開標由商承建.工程較小者,及修補臘青路面,悉由養路處辦理.現養路工人約三百餘,僅足支配.每年春夏

二季,雨水甚多,路面因此多被損壞,故修補工作,以秋冬二季爲最忙.此項工作較少,而築路成績略高,玆將每月成績,另表列明:

民國十九年廣州市工務局修繕股
每月修路成績表

類別	月份	面積(平方尺計)
花砂路面	一月	80,000
	二月	102,000
	三月	132,000
	四月	119,000
	五月	139,000
	六月	108,000
坭路面	一月	111,000
	二月	80,000
	三月	198,000
	四月	280,000
	五月	280,000
	六月	132,000
塗掃臘青路面	二月	17,000
	三月	66,000
	四月	148,000
	五月	49,000
	六月	86,000
三合土路面	三月	2,000
	六月	1,000

潤滑油黏度試驗

著者：張延祥

潤滑油（Lubricating Oil）爲原動機所必須用者,無論爲蒸汽機,煤氣機,黑油機,汽車,飛機,以及一切機械相磨擦面,均用之以減少磨力及損耗.潤滑油多爲礦物質油,由石油中提煉而得,種類極繁雜,其潤滑鐵質之功用,全賴其有黏性（Viscoity）.油厚者黏度大,薄者黏度小,何種機器須用何種黏度適當之潤滑油,一部分與他部分之需要不同,夏季與冬季又不同,故選用潤滑油須十分留意.

潤滑油之試驗,分黏度,燃燒點,色澤,炭燼,等數種,以黏度爲最重要.美國最通用者,爲西波氏黏度試驗儀器(Saybolt Standard Viscosimeter),曾經美國商部及全國石油會（N. P. A.）美國石油協會(A. P. I.)等機關承認,作爲標準儀器及試驗法.其油之黏度,即以西波氏秒數（Saybolt Seconds)稱之. New & Revised Tag Manual for Inspectors of Petroieum 一書內,詳載試驗方法,因亟譯之,以爲國內機械工程師之參考.至于此項儀器,國內有否,尚待調查.

西波氏黏度試驗儀器之構造

西波氏黏度儀器有二種,一種爲試驗普通潤滑油者,稱之爲世界式（Soy-bolt Standard Universal Viscosimeter),又一種係試驗厚油之用,（Saybolt Standard Furol Viscosimeter）構造除油空一小一大外,完全相同如圖:

此項儀器全用金屬製成,J爲油管,上有溢量盤E,油管四圍浸在水中或他種之油內.油管下底有一小管,即出油管,試驗之油,經此小管之出油空,滴入下面之玻璃瓶R內,玻璃瓶瓶頸上有一刻度,瓶內所盛容量,至刻度線止爲六十立方公分,其容量不許相差0.15立方公分上下.小管外面,有一大管

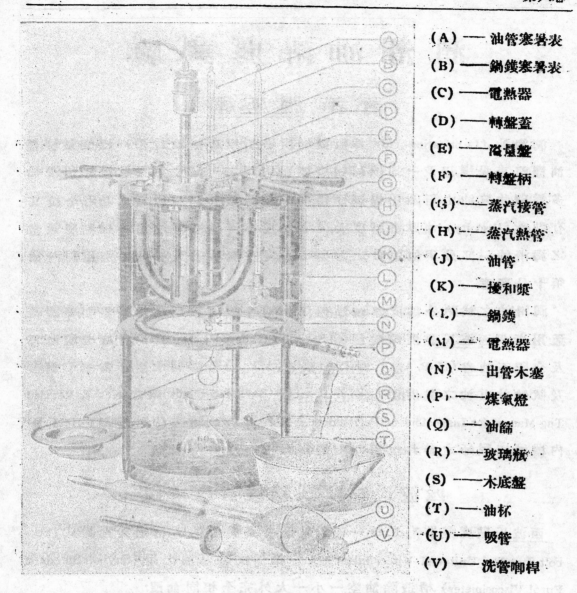

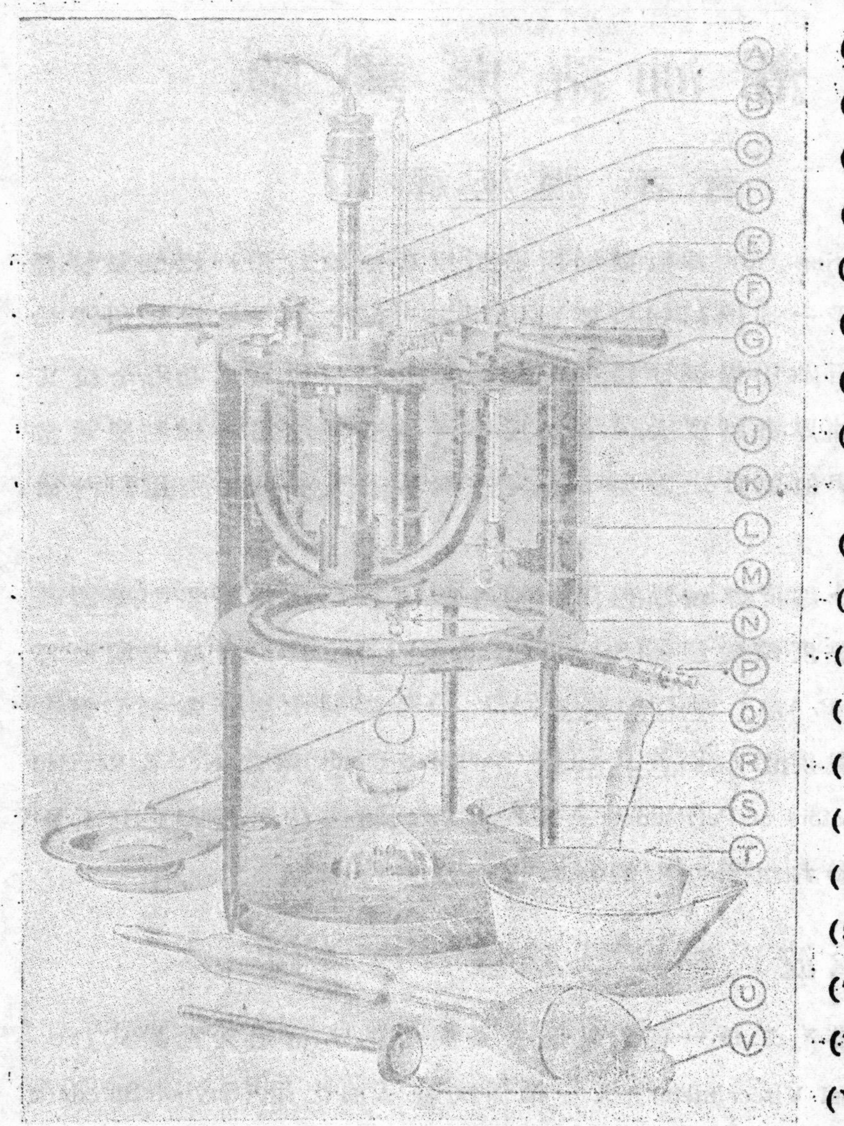

(A)——油管寒暑表
(B)——鍋鑊寒暑表
(C)——電熱器
(D)——轉盤蓋
(E)——溢量盤
(F)——轉盤柄
(G)——蒸汽接管
(H)——蒸汽熱管
(J)——油管
(K)——攪和漿
(L)——鍋鑊
(M)——電熱器
(N)——出管木塞
(P)——煤氣燈
(Q)——油篩
(R)——玻璃瓶
(S)——木底盤
(T)——油杯
(U)——吸管
(V)——洗管唧桿

及軟木塞栓 N 塞住,試驗時拔去木塞,油卽滴下.油管內試驗之油之溫度,以及油管外鍋鑊盛水或油之溫度,各有寒暑表一支以示之.鍋內之水或油,用電力,或蒸汽,或煤氣,以增加其溫度.試驗滴油之時間,用一跑馬表計時.油管上面溢量盤內之油,則用一吸管 U 吸取之.

潤滑油黏性於溫度高底頗有關係,普通試驗,定在華氏表一百度,或一百三十度,二百十度,三種,視油質厚薄而定.

試驗手續

1. 試驗時室中忌通風,室內溫度亦不可有劇烈變動.

2. 試驗儀器須平放在試驗桌上.鐵架之三脚適撑開在底下木盤之外週.

3. 鍋鑊內盛油(約華氏 350° 至 400° 之燃燒點)以加蓋後滿口爲度.

4. 以旋轉蓋 D 安上.

5. 置有空小銅片於油管內,凹面向上,如⌣形.鍋鑊之寒暑表插入三叉管內,寒暑表之水銀球,須靠在三叉管底.

6. 若有電燈處.可將電熱器之電線,插入電燈燈頭內,以通電流.若電熱器取出油鍋必須將電健關斷.因此種熱器若不浸在油內或水內,而通電流,不數分鐘,卽毀壞也.

7. 鍋鑊內之油漸漸增加熱度,以至所規定之試驗熱度爲止.

8. 鍋鑊內溫度須均勻,可時時用轉盤之手柄 F 轉動四分之一轉.

9. 鍋鑊內溫度,不可高過於各個寒暑表之剩度.

10. 試驗之油先經油篩 Q,傾入油杯 T 中,略加溫度.

11. 用洗管唧桿 V,及將試驗之同種油少許,洗淸油管 J.上 5 節內所云之有空小銅片,可暫取出.

12. 出管木塞 N 塞緊.

13. 欲試驗之油,從油杯 T 中,經油篩 Q 傾入油管 J 內,滿至溢量盤 E 爲止.

14. 油管內之溫度須保持規定試驗度數.

15. 油管寒暑表,A 可用作攪和桿,使其溫度均勻.

16. 試驗時先拔去寒暑表 A.

17. 用吸管 U 將溢量盤 E 內之油吸出,使油管 J 適滿一管.

18. 以玻璃瓶R置在木底盤S之中央空洞內.

19. 拔去木塞N,同時揿動跑馬表,注視油管內之油滴入玻璃瓶.

20. 瓶內之油,逐漸增加,至瓶頸60立方公分處,油彎面底到刻度處,乃停止跑馬表,表上時間秒數,卽西波氏黏度爲該油在試驗溫度時,滴下六十立方公分之時間.

21. 試驗完後,將油管蓋住.

22. 若無電氣而用蒸汽以加熱鍋鑊內之油,則蒸汽可由蒸汽接管G通入.若須鍋鑊減低熱度,則可通入汽水.

T3. 若用煤氣以熱鍋鑊,則可燃點煤氣燈P.

他種試驗儀器

西波氏世界式黏度儀器外,另有試驗厚油用之一種,所不同者祇在油管及出管小空擴大,以易油之滴下.厚油試驗,大概定在華氏122°之溫度,卽攝氏50°,其度數稱爲 Saybolt Furol,卽該油在試驗之溫度時,滴下六十立方公分之時間秒數也,亦有在佛氏77°試驗者.

潤滑油在25秒 Saybold Furol 以下者,應在世界式黏度儀器試驗之.

潤滑油在32秒 Saybolt Universal 以下者,不能算爲燃燒油.

英國所通用之黏度計,爲 Redwood 賴特胡氏之黏度計.歐陸用 Engler 恩格勤之儀器,稱爲 Engler degree.玆將各家黏度比對表列后,以備參考.表係 Dr. T. G. Delbridge 所編者

潤滑油黏性試驗各家度數比對表

(Approximate Viscosity Conversion Table)

西波氏世界牌秒數 Saybolt Universal Time, Seconds.	西波氏厚油秒數 Saybolt Furol Time, Seconds.	拜貝氏每小時立方公方數 Barbey Degrees C. C. Per hour.	賴特胡氏第一號秒數 Redwood No. 1 Time Seconds.	賴特胡氏海軍儀秒數 Redwood Admiralty Time, Seconds,	恩格勒氏度數 Engler Drgrees
32		5320			1.05
34		3060			1.11
36		2170	33		1.16
38		1700	34		1.21
40		1105	36		1.26
45		1000	40		1.40
50		788	45		1.53
55		656	49		1.66
60		564	53		1.79
65		498	57		1.92
70		446	62		2.05
80		371	70		2.32
90		319	78		2.58
100		282	87		2.85
120		228	104		3.38
140		192	121		3.92
160		166	138		4.46
180		146	155		5.00
200		131	172		5.55
225		116	194		6.22
250	28	104	214	21	6.91
275	30	94	236	23	7.59
300	33	86	258	26	8.27
325	35	80	279	27	8.96
350	38	74	300	29	9.64
375	40	69	322	32	10.32
400	43	64	344	34	11.01
450	47	57	386	38	12.38
500	52	52	429	42	13.75
550	57	47	472	46	15.11
600	62	43	515	50	16.49
700	72	36	601	59	19.23
800	82	32	688	67	21.97
900	92	28	773	75	24.71
1000	102	25	860	84	27.46

水電資產估價之原則

著者：錢慕寗

估價之主旨

水電事業之服務機能，基於物質設備在其壽命過程中，全盤資產之價值，常因種種關係而有估量之必要，其最顯著者，不外下列四項目的：

（一）更易所有權　爲適應環境之要求，商公司與商公司，或商公司與政府之間，換替所有權時，其財產之代價，遂成交易之關鍵。就普通情況而論，自以實值價格爲要點。然有時亦視其對於買主所收之實益爲準繩。例如因設備之一部，勢難繼續適用，買主購得後，尙須改造或竟毀棄，則買主願給之價，當較實值爲低。或因市場日趨發達，事業將更進展，則賣主要索之價，當較實值爲高。

（二）決定出品之公平價格　水電事業，製備飲料電力，銷售於用戶，取值自以公允爲第一要義。蓋其貨品爲日常生活之需，倘索價過昂，則用戶之負擔，未免繁重；過低則營業坐以虧損，勢難美滿供應。欲求公允之售價，須先知精確之成本，而資產費用，實爲成本之重要部份，故宜切實估計，以期正確。

（三）計算損益及稅額　營業之損益，係收入與全部轉運費用及資產費用之差，誠與資產價值有密切關係。而政府捐稅如所得稅等，常按淨利之多寡爲標準。故徵收適當之捐稅，亦以資產實價爲先決要件。

（四）擔保債務　水電事業常須發行債券，藉作創辦或擴充經費，或以塡補營業虧損之用。政府及債權人，對於擔保品之資產，每欲知其確實價值，以杜流弊，而維利權。

II 價值之憑證

通常貨物之價值,係賴金錢表示,其市價恆受供求定律之支配.當供求達平衡狀態時,價值之高低,應依成本之多寡爲轉移.惟水電事業,交換不常,非如普通貨物,隨時均有交易所之記錄,可供考據,且有時受法律之限制,問題較爲複雜.

水電資產價值之計算,向無一定之方式或成規,可以隨處適用.故所謂價值者至多不過代表一種根據觀察及經驗之意見,而非絕對之數量.然在歐美各國,因估價涉訟之結果,法庭對於資產價值 Value 之憑證,曾認定若干原則,視各地環境情形之不同,而分別採用.略如下述各項:(一)成本 Cost,(二)實益 Worth,(三)事業之盛衰,(四)設計之優劣,(五)管理之良窳.

水電事業性質繁雜,範圍宏廣.其資產估價,非如米一斗或煤一噸之簡易,故估評任務,宜委託專家辦理.誠以估價者除備充實經驗外,尚須具有下列三項學識:

(一)經濟　關於投資習慣,物價標準,以及各項商情之統計,費用之核算,等等.

(二)工程　關於工廠之設計,建造,及轉運情形,並施工之方法,工作之效率,成本之高低等等.

(三)法律　關於產業管理之原則,公用事業在法律上之地位,與成法之運用,以及各項管理,監督之法令,規章,等等.

III 估價之式方

(一)再造價值法　產業之過去或將來價值,雖可爲價值之佐證,然其現時之價值,實爲估價問題之主要關鍵.故在現時工作方法工料價格之下,再造同樣設備所需之費用,按年代之久暫,分別折舊後,所得之現值,應與事理

最相吻合.玆將估價程序中之各重要事項,申述如下:

(甲) 再造價值

再造價值之計算,根據資產數量又單價兩項.關於資產之彙集,尤當分別眉目,詳愼審核,倘有圖表足資考證者,亦當附入.單價之採擇,多以市價爲標準惟遇價格漲落不常時,每有採用最近五年之平均數者,意在求與最近將來之實情相符.此項計算,與承包工程人之預算標價,原理相同,務求其能代表眞實情況,過低過高,皆足爲失敗之由.

(乙) 折舊

(1) 折舊之類別: 凡設備因陳舊所減低之價值,謂之折舊,大別爲下列兩種.

物質折舊 卽設備因運用年久,消磨損壞所失之價值.

效用折舊 卽設備因時勢變遷,效用減少,所失之價值,如舊式機器,本身雖新,然因新式機器發明後成績更爲優勝,舊式機器遂致無形失其一部或全部之價值.

(2) 折舊之計算: 各項水電資產之有用壽命,依經驗所得,由十年至百二十年不等.折舊費之實際計算,係依各項資產已歷之有用壽命,並根據下列之折舊辦法,以推求之:

直線法 Straight-Line Method: 此法係假定每年折舊之數,與有用壽命成正比例.如某項設備之壽命,預定爲二十年,則每年折舊爲原價二十分之一.直線法用於短壽命之資產,較爲適宜.若施諸長壽命之資產,則最初數年之折舊,實屬過高,致有負擔不稱之弊.

年金加利法 Sinking Fund Method: 根據多數工程師之觀察,一切設備在初用時,損壞之速率最微.其後漸次增加,至老年爲最大.故此法假定每年之折舊費,適與同期之年金加利基金逐年增益之數相等.因其與實在情形極爲符合,故結果平允,而採用亦廣.玆爲表示資產價值,隨運用年載而跌落之

程度,並上述兩法差異之處,特製定 (A) (B) 兩表及 (a) (b) 兩圖如下,以資比較

(A)表: 資產之原價,折舊,及其現值;壽命十年,直線法

經歷年載	原價或再造價值	折舊年金	折舊累積	現　值
0	$ 100,000			$ 100,000
		$ 10,000		
1	100,000		$ 10,000	90,000
		10,000		
2	100,000		20,000	80,000
		10,000		
3	100,000		30,000	70,000
		10,000		
4	100,000		40,000	60,000
		10,000		
5	100,000		50,000	50,000
		10,000		
6	100,000		60,000	40,000
		10,000		
7	100,000		70,000	30,000
		10,000		
8	100,000		80,000	20,000
		10,000		
9	100,000		90,000	10,000
		10,000		
10	100,000	$ 200,000	100,000	0

(B)表: 資產之原價,折舊,及其現值;壽命十年

年金加利法(複利四厘)

經歷年載	原價或再造價值	折舊年金	折舊累積之年利	按年折舊總額	折舊累積	現值
0	$100,000					$100,000
		$8,329		$8,329		
1	100,000				$8,329	91,671
		8,329	$333	8,662		
2	100,000				16,991	83,009
		8,329	680	9,009		
3	100,000				26,000	74,000
		8,329	1,040	9,369		
4	100,000				35,369	64,631
		8,329	1,415	9,774		
5	100,000				45,113	54,887
		8,329	1,805	10,134		
6	100,000				55,247	44,753
		8,329	2,210	10,539		
7	100,100				65,786	34,214
		8,329	2,631	10,960		
8	100,000				76,746	23,254
		8,329	3,070	11,399		
9	100,000				88,145	11,855
		8,329	3,626	11,855		
10	100,000	$83,290	$16,710		100,000	0

(a)圖：折舊年金及累積之百分率直線法

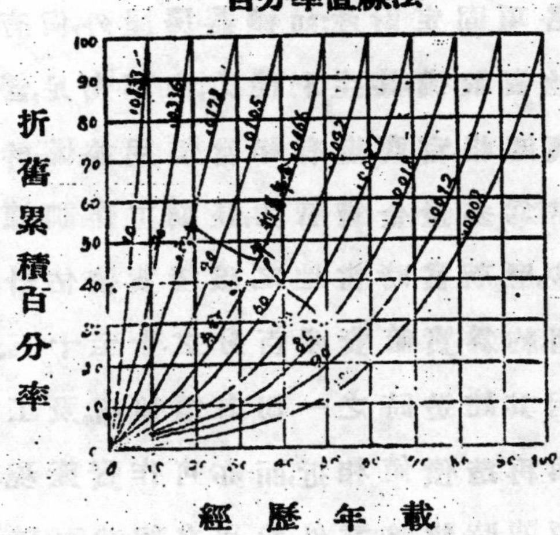

(d)圖：折舊年金及累積之百分率年金加利法(複利四厘)

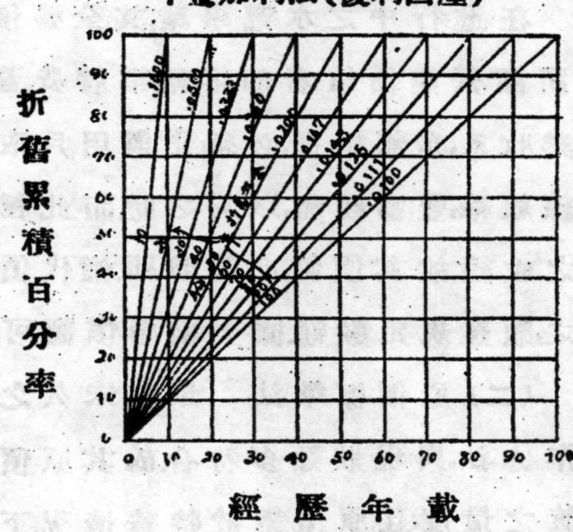

實地察驗法：評定折舊程度之法,尚有全憑觀察者.倘全部設備均顯露於外時,則按此法所得之結果,或且較上述兩法爲正確,惟資產情況,常有非耳目所能普及者;如埋藏地下之水管及各項基脚等,且非具有充富經驗,評斷亦難適當,故實施頗爲不易.

(丙)配佈用費

凡事業未創立之前,必經過各種組織,設計,調査等籌備手續;建造期中又須管理技術,爲之指導一切.此種費用,與廠內之一磚一木同爲資產之一部分,未容忽略,因其性質配佈一切,故在計算全盤事業之再造價値時,必須酌量加入.計算之法,多按實有資產之百分率,估計如下:

籌備費	百分之一至二
工程技術	五至七
意外費用	四至六
管理	二至三
建造時期利息	三至五
總計	百分之十五至二十三

(丁) 營業價値

在進行中之水電事業,其全盤價値,除各項固定財產:如機器房屋外,尙有所謂營業價値者卽俗稱招牌費.蓋僅有物質設備,其生利能力,尙不充足,營業收入,亦無把握,必須先與用戶取得切實連絡,官商獲有相當信用.並備幹練組織,豐富經驗,以爲之助.而此種情況,常爲多量金錢智力,長期工作訓練之成績,故其價値,亦甚明顯.惟代價之多少,應就當時當地之環境實情估計之,頗難規定數額.據美國習慣認可之範圍,約爲實値資產百分之十至十五.

(二) 原價標準法　創立未久之事業,因其建造時之一切市價情况,及工作方法,與現狀類多符合.故其原價每易與再造價値相近,而亦可作資產現値之量度.且原價對於特殊情况下之各項臨時建築,意外費用之紀載,更較他法爲詳實,是其優勝之點.惟採用原價之最大滯礙,在簿記方法之常欠精確,如資產與維持費用之錯亂混列,或僅有機器成本.而乏裝置費用等類.苟根據資產之過去歷史,先求得原價之概數,再就今昔市價之差異,而酌量增減其値,並除去相當折舊,亦可得與現値近似之數量.

(三) 憑利估本法　通常事業交替產權時,每有視其贏餘之厚薄,而定價値之標準者;水電事業之買賣,自亦難避此範圍.惟公用事業出品之售價,未許由私家任意增加,而其獲利之程度,亦易受法律拘束.故營業之贏餘,用爲估價準繩,頗受限制,以作參考資料.則效用甚廣.

IV 實 例

玆列舉資產估價實例三種,以明上述原則及方法之運用.

(C)表……推算折舊及現値.

(D)表……更換所有權.

(E)表……評定出品之公平價格.

(C)表：美國某水廠估價細賬，1924年

項數	項目	再造價值	壽命	折舊年金		折舊累積			現值
				百分率	數量	經歷年載	百分率	數量	
(1)	(2)	(3)	(4)	(5)	(6)=(3)X(5)	(7)	(8)	(9)=(3)X(8)	(3)-(9)=(10)
320	製水設備								
	1.進水管	$2,350	18	3.9	$ 90	13	65	$ 1,530	$ 820
	2.新進水管	2,440	40	1.1	30	1	1.1	30	2,410
	3.渾水管	1,270	40	1.1	10	1	1.1	10	1,260
	4.砂池水管	1,300	29	1.9	20	21	73	950	350
	5.洗池清水管	830	29	1.9	20	24	73	610	220
	6.定水池	11,460	40	1.1	130	1	1.1	130	11,330
	7.加綠氣器	500	10	8.3	40	3	23	130	370
	8.洗池打水機	1,300	18	3.9	50	13	65	840	460
	9.電線	300	18	3.9	10	13	65	200	100
	10.救急設備	290	7	13.0	40	2	27	80	210
	11.木質砂濾器	36,000	20	3.8	1,350	15	65	23,340	12,660
	共計	58,040	21	3.1	1,790	12	48	27,850	36,190
321	送水管路	52,330	53	0.56	2,940	19	15	77,130	446,200
322	白鐵小管	55,830	20	3.4	1,900	10	40.3	22,500	33,330
323	用戶水表	31,330	30	1.8	560	12	27	8,460	22,870
324	救火海淀	21,360	40	1.1	230	26	46	9,830	11,530
各項總數		$689,890	40	1.07	7,420	15	21.1	145,770	514,120
折舊累積之本年利息					5,830			.04	
本年資產折舊之總額					13,250			5,830	

說明一，(5)(8)兩項百分率，均係參照壽命年限，由前章之(b)圖量得。

說明二，折舊年金下之各項數量，僅計年金，未合利息，至表底始將折舊累積之利息，一次加入，以求該年應負資產折舊之總數。

4103

(D)表: 英國某水廠估價總結之比較, 1922年

項　目	原價		再造價值		買主實益	
	新造	現值	新造	現值	新造	現值
製水設備	$7,425	$6,562	$10,321	$9,250		
送水設備	36,440	31,232	58,833	50,582	$51,645	$43,383
地產改良	180	142	280	220		
	44,045	37,936	69,434	60,652	51,645	43,383
意外及工程費用(10%)	4,405	3,794	6,943	6,005	5,165	4,338
管理費用 (2%)	881	559	1,389	1,201		
建造總計	49,331	42,289	77,766	67,258	56,810	47,721
組織費用 (2%)	881	881	1,389	1,389		
地　產	1,200	1,200	2,500	2,500		
建造時期利息損失						
建造費項下 (1%)	493	423	778	673	568	477
組織及地產費項下(2%)	42	42	78	78		
全體總計(存料在外)	51,947	44,835	82,511	71,898	57,378	48,198

評價工程師提議採用之價值 = $57,000

說明一, 原有製水設備簡陋過甚,買主於取得所有權後,須全部拋棄,另建新廠,故所列買主實益之總價,較再造總值為少.

說明二, 本廠創設在<u>歐</u>戰前,而出售在<u>歐</u>戰後,因工料成本增加之故,再造價值較原價超出甚多.

說明三, 評價工程師提議採用之數,係折衷三項估價,並酌加營業價值所得,亦即雙方最後同意之標準.

(E) 表: 中國某水廠收支情況及水價概算, 1929 年

資產現值	$ 4,000,000
轉運費用	1,240,000
捐　　稅	100,000
資產折舊 (2.5 %)	100,000
事業純利 (20 %)	800,000
總　　計	2,240,000
全部收入估算	2,240,000
總出水量 (千加倫)	4,000,000
不取費水量 (千加倫)	800,000
取費水量 (千加倫)	3,200,000
平均每單位應收水價	0.70

說明一, 依美國法律之規定,水電事業所定售價應足使其收入總額除供給全盤開支外,尙能令公平估價之資產,獲得相當之利息,並能保持其全部資產,使在事業終了時,其價值仍得與開辦時相等.

說明二, 國民政府十八年十二月二十二日公布之民營公用事業監督條例,第九條,規定營業純利許每年達到實收資本總額百分之二十五,逾此則次年應減少收費,或擴張設備.

說明三. 本表計算水價,係假定除市政消防外,其他用水一律憑表計算,所有應收水費,又均能切實收取,不受絲毫損失.

中國工程學會職業介紹委員會
介紹職業簡章

(一)宗旨　本委員會以介紹相當技術人材發展工程事業爲宗旨

(二)範圍　本委員會介紹人材以曾經專門技術訓練或具有相當經驗者爲限

(三)手續　凡招聘或待聘者均須先向上海寧波路四十七號領取委託書或志願書填明寄交本委員會指定之審查委員詳細審查分別登記後介紹相當人材或位置

(四)用費　凡委託本委員會代聘技術人員者概不取費惟經委託者之同意登載廣告或發送電報等費用須由委託人負擔之其有經本委員會介紹而得有位置者永久會員得自由捐助普通會員應捐第一月薪十分之一非會員十分之三以資彌補

(五)責任　本委員會關於招聘者與應聘担保事項概不負責應由雙方自行辦理之

(六)證書　應聘者如有證書或像片等物經本委員索取或自行寄交本委員會者請附帶寄回所需郵費否則本委員會不負寄回之責

(七)附則　本簡章得隨時修改之